Louise Zapala
H.F.C.C.

# Biology:
# The Science of
# Life

The Macmillan Core Series in Biology
General Editor: Richard A. Goldsby

# BIOLOGY:
# The Science of Life

Joan E. Rahn

**Macmillan Publishing Co., Inc.**

NEW YORK

**Collier Macmillan Publishers**

LONDON

Macmillan Publishing Co., Inc.
866 Third Avenue, New York, New York 10022

Collier-Macmillan Canada, Ltd.

Library of Congress Cataloging in Publication Data

Rahn, Joan Elma
  Biology: the science of life.
  (The Macmillan core series in biology)
  1. Biology.   I. Title.
QH308.2.R34        574        72-14071
ISBN 0-02-397670-X

Printing:   2 3 4 5 6 7 8      Year: 4 5 6 7 8 9 0

This book is dedicated to
Ginny and Munroe Winter,
for their patience with me
during the years of reclusion
that produced it.

# Preface

It is an old cliché that the world becomes smaller every day, but it is nonetheless true. As the numbers of human beings living on the earth increase, and as their technological capabilities expand, their acts have increasingly serious consequences that affect more and more persons. This is true on any level—family, local, or national. The decision of a family to have one more child can affect the neighboring families as well as the one producing that child; the decision of a city on how to dispose of garbage or sewage can affect neighboring cities; the decision of a country to build a dam may affect a neighboring country; and all these decisions affect future generations. Nearly all the habitable regions of the earth now are called home by some people, and there are few places one can go to escape the pollutants produced by his fellow man. The only escape today is to search for the solutions to these problems rather than to avoid them.

This book does not solve these problems. However, the students of today, who will be the voters and leaders of tomorrow, will have a hand in solving them. To do so intelligently they must have some basic knowledge about the biological nature of man and the animals and plants with which he shares this planet and on which he depends for his continued existence. Most of the voters of tomorrow, however, will not be biology majors. Most of them will take perhaps only one biology course during their college years. This book is aimed at such students.

Wherever possible, the introduction of technical terms has been avoided. The student may not believe it, but his instructor will. Rather, the emphasis is on principles—how things work and how they are interrelated in a functional way. For this reason the book contains no detailed phylogenetic material. An outline of the living world in the appendix gives some idea of the diversity of living things but gives little indication of their evolutionary relationships.

Each chapter begins with a list of statements that give the reader a preview of the material to follow. The textual material expands on the material summarized in the statements. This approach is used because it has been the experience of the author that many students, especially the nonscience majors, too often fail to see the forest for the trees. Perhaps it is because we have pointed out too many facts to be memorized and the student loses interest in the generalities and theories based on those facts. For this reason each chapter of the book begins with generalities and proceeds to details.

Many of the chapters can stand independently of each other. This allows the instructor flexibility in assigning reading material to students and should permit him to follow his own preferred schedule.

I would like to thank the following persons who read part or all of the manuscript during its preparation and to whom I am grateful for guidance and suggestions:

M. MICHELLE BAKER, Colorado Mountain College

RICHARD K. BOOHAR, University of Nebraska

KENNETH DERIFIELD, Northern Arizona University

DAVID G. DIXON, Los Angeles Valley College

GERALD G. FARR, Southwest Texas State University

PAUL H. GEBHARD, Indiana University

RICHARD A. GOLDSBY, University of Maryland

ALBERT J. GRENNAN, San Diego Mesa College

WERNER G. HEIM, The Colorado College

JAMES J. HUBER, Lorain County Community College

NEIL Y. KANEMOTO, The Loop College—City Colleges of Chicago

GEORGE H. KIEFFER, University of Illinois

RHODA LOVE, Lane Community College

LINDA MCGINNIS, Yuba College

ROBERT P. OUELLETT, Massasoit Community College

ELEANOR L. POTORSKI, Suffolk Community College

S. ARTHUR REED, University of Hawaii

JOSEPH H. RUBINSTEIN, New York University

ALLEN B. SCHLESINGER, Creighton University

ALLAN A. SCHOENHERR, Fullerton Junior College

JOHN H. STANDING, Delaware Valley College

W. G. WEAVER, Miami-Dade Junior College

Finally, I wish to thank the biology editor, Charles E. Stewart, Jr., whose energy and enthusiasm helped to bring the work to a conclusion.

JOAN E. RAHN
*Lake Forest, Illinois*

viii

# Contents

ix

# The Living World:
# An Introduction

# Part I

The many variations that life takes are almost staggering to the imagination. Biologists have discovered over a million species of plants and animals, and certainly more remain to be discovered. Plate I illustrates only a few of these.

Although the average person can expect to see only a small fraction of all the types of **organisms** (living things) on earth, in his ordinary life he is likely to see a moderate assortment even if he lives in a city: potted geraniums and philodendrons; pigeons and starlings; the seeds in a bag of bird food; pet dogs and cats and their fleas; moss growing in cracks in the sidewalk; the mosquito that annoys on a summer night; mold covering an old melon or strawberries; the unwelcome cockroach in the plumbing; green algae on shady, damp sidewalks or tree trunks; trout, clams, and shrimp on display in the food market; the bacteria that cloud the unchanged water in a several-day-old bouquet; spiders that spin webs in undisturbed corners; the ferns that seem always to grace weddings and funerals; earthworms that emerge from the lawn with any prolonged rain; weeds; ants; grass; flies; Christmas trees; and, of course, people.

Variations exist within species also. Males and females commonly are different enough to be easily recognized. In some species two forms of one sex may occur, as in bees, which have fertile queens and nonfertile female workers. A single living thing may take several forms within its life span. A caterpillar spins a cocoon around itself and enters a dormant period as a pupa; later it emerges as a moth.

If the variety of life is amazing, even more amazing is the basic uniformity that all these forms share. There is no single feature that separates living things from nonliving things, but certain features are typical of all or nearly all organisms.

All organisms have a certain degree of **organization.** Except for viruses, they all are composed of units called **cells,** each of which is a small mass of living material (called **protoplasm**) surrounded by a membrane. Within the protoplasm are some

# 1

# Characteristics of Living Things

1 Living things take a great variety of forms, but all share certain fundamental characteristics.

2 Characteristics of living things include:
   a a high degree of organization
   b a certain chemical composition
   c metabolism
   d growth
   e movement
   f irritability
   g adaptability
   h homeostasis
   i rhythmicity
   j constant expenditure of energy
   k reproduction

3 Living things are all interdependent, one species requiring the waste products of another.

3

subcellular units common to all cells and others that are unique to certain groups (Chapter 17). The simplest single-celled organisms and even the noncellular viruses are more highly organized than nonliving things in the natural world. In more complex organisms the cells are organized into tissues (groups of similar cells), and the tissues are organized into organs. In many animals the organs form organ systems.

The parts of viruses and cells are composed of **chemical compounds,** such as carbohydrates, fats, and proteins (Chapter 8) and additional compounds derived from them. The variations within these groups—that is, which carbohydrates, which fats, which proteins, and so on, are present— account for differences among organisms and among the cells of a multicellular organism.

Reactions between the chemical compounds in a cell are necessary for the maintenance of life. The total of these chemical reactions is called **metabolism.** In young, rapidly growing cells the metabolic rate typically is high. In dormant cells the metabolic rate is very low, sometimes being barely perceptible. The complete cessation of metabolism results in death. Metabolic activities are classified into two groups: anabolism and catabolism. In anabolic processes, complex chemical compounds are synthesized from simple compounds; anabolism requires energy. Photosynthesis (Chapter 11) is an example of anabolism. In catabolism, complex chemical compounds are converted into simpler compounds; catabolic processes release energy. Respiration (Chapter 12) is an example of catabolism. Anabolism and catabolism frequently occur simultaneously in a cell or in an organism, for the energy required for anabolic processes often comes from the energy released in catabolic processes.

**Growth** is a characteristic of living things. When anabolic processes occur more rapidly than catabolic processes, new protoplasm is formed and growth occurs, and the organism ordinarily increases in weight and size. Growth may continue throughout the life of some organisms, such as redwood trees; or it may be confined largely to the early part of the individual's life, as in the case of humans; but most adult organisms retain at least limited ability to grow new tissues when some part of the body is damaged. If catabolic processes prevail, the organism usually loses weight and shows a decrease in size.

Most organisms show **movement** of some type. Many animals are capable of locomotion—movement of the entire organism from place to place. Some plants are capable of growth movements, such as the bending of stems toward a light source. The protoplasm of many cells moves within the cell.

Irritability and adaptation are related features of living things. **Irritability** is sensitivity to at least some changes in the environment. Plants and animals are known to be sensitive to changes in such factors as light, temperature, moisture content, pressure, many chemicals, and gravity; sensitivity to each of these factors varies with the species. **Adaptability** is the ability of the organism to adjust to environmental changes; for example, an animal may adjust to the cold weather of winter by hibernating, by migrating to a warmer climate, or by growing a thicker coat of fur. Irritability and adaptability are essential for the survival of organisms. In the past many species became extinct because they failed to adjust to environmental changes.

In spite of the drastic changes that may occur in the environment, most organisms change relatively little during most of their adult lives. The maintenance of a nearly constant condition is called **homeostasis.** Homeostasis within an organism depends on the integration of the many internal chemical and physical processes; this is usually regulated by hormones and, if it is present, a nervous system. If homeostatic control breaks down, the organism may become so poorly adapted to the environment that illness or death results.

Organisms exhibit **rhythmicity.** Cyclical changes contribute directly or indirectly to homeostasis. For instance, an organism actively seeks food and

4

eats at one time or certain times of day and sleeps at another time; both of these events contribute to maintaining its well-being. Rhythmical changes occur on the cellular level, the organismal level, and the community level.

All organisms are maintained by a **constant expenditure of energy.** Many of the chemical reactions essential for life require energy. For most plants the energy source is sunlight. For animals the energy source is the food they eat; bacteria and fungi also use food that they themselves have not made.

Energy is utilized by means of chemical reactions that change the environment in two general ways. First, the organisms all require some substances which they remove from their environment. Most plants require the gas carbon dioxide, water, and minerals (they also require the gas oxygen, but healthy plants produce more oxygen than they use). Most animals require oxygen, water, and food (in the form of plants or other animals). Second, they all produce waste products which they add to their environment. Fortunately the waste products of one organism are the materials required by another. Most living plants produce oxygen as a waste product, and living animals produce carbon dioxide. Both release water and minerals to the environment during life, and their dead bodies disintegrate into carbon dioxide, water, minerals, and a few other substances. Thus there also is a homeostasis on the community level. This constant recycling of chemical compounds not only prevents the accumulation of poisonous waste products, but also creates a dependence of organisms on other types of organisms. For instance, animals are totally dependent on plants for their food and oxygen and would become extinct if all plants were to die.

Finally, all organisms have the power of **reproduction:** they produce more organisms of the same species. Reproduction is of two main types, sexual and asexual. Sexual reproduction usually (though not necessarily) involves two parents, which usually are differentiated as male and female. Each parent contributes one cell, the sperm of the male and the egg of the female (both cells are identical in some algae, fungi, and single-celled animals). The two cells fuse, forming a single cell (zygote) which is the first cell of the offspring. The sperm and the egg contain the parents' genetic contributions to their offspring, which at maturity resembles them enough so that it is readily recognized as a member of the same species; but it usually differs from the parents in minor ways (such as height or color), because its heredity is not identical with that of either parent. With the exception of some unicellular organisms, most species are capable of sexual reproduction.

In asexual reproduction, only one parent is involved, the offspring developing from one or a few cells of the parent. In some cases special asexual structures, such as spores, are formed, but in other cases it is difficult to distinguish asexual reproduction from ordinary growth; an ivy plant that under one set of circumstances might remain intact, in other circumstances may become broken into several plants that take root and survive. An asexually produced offspring has the same hereditary make-up as the parent, and any differences between parent and offspring usually may be attributed to different environmental effects on the two individuals. Many species of plants and lower animals are also capable of asexual reproduction, but asexual reproduction is lacking among the most highly evolved animals such as mammals and birds.

In all successful species, adults not only have the ability to reproduce themselves, but they have the ability to reproduce themselves many times over. A pair of human beings can, in their lifetimes, produce a dozen or more offspring; a pair of dogs may produce several dozen; a corn plant may produce several hundred; an orchid plant or a pair of fish may produce thousands or millions. It is obvious that if all offspring were to survive and to reproduce like their parents, the world soon would be smothered in layers of organisms. That this does not happen is due to the early death of

5

nearly all offspring. They die because they cannot find enough food, because they are eaten by other organisms, or because parasites kill them; some succumb to accidents; others reach maturity but fail to find mates and thus fail to reproduce. Although the numbers of surviving offspring vary from time to time and from place to place, on the average one pair of parents produces in their lifetimes only two organisms that survive *and* reproduce. In this way the population of any species remains relatively constant over long periods of time. The balance is maintained by interaction with other species which are their food or their predators and parasites, for a given species tends to increase in numbers if food is plentiful and to decrease in numbers if plagued by many predators or parasites.

If, for some reason, this balance should be upset and one species experience a population explosion, disaster may result. For example, the increasing numbers of one species of animal may consume so much food that in a few years there is insufficient food; mass starvation results, and the population declines suddenly and often tragically. If the environment has been irreparably damaged, not only this species but others also may become extinct in the area. If the damage is less severe, the community may in time recover.

We the members of the human species currently are experiencing a population explosion that threatens through the polluting effects of agriculture and industry to do serious harm to our environment. We like to think of ourselves as superior to other forms of life, and perhaps we are; but we still are bound to this planet with its soil, its atmosphere, and its seas. Man has learned to culti-

vate plants and to herd animals, he has learned to increase crop yields and to preserve otherwise perishable food, but he has not yet learned to *make* food. For this he is as dependent on plants as he was a million years ago, and assuming he would like a good deal of variety in his diet, he is also dependent on animals. These plants and animals (and also man himself) are dependent on clean soil, clean atmosphere, and clean seas. It remains for our superiority, which has given us agriculture and industry, with their polluting side effects, to devise ways of reducing the pollutants while permitting us still to enjoy the products that only the human mind ever envisioned and only the human hand has produced.

The remaining chapters of this book discuss in brief form the characteristics of ourselves, the living things with which we share this planet, and many of the interrelationships among organisms and between them and their environment.

## Questions

1. Make a list of all the living things you saw in the last 24 hours. In what way did they influence your life?
2. What are the characteristics that distinguish living things from nonliving things?
3. How many of these characteristics do nonliving things exhibit?
4. Do any nonliving things exhibit all these characteristics?
5. Can you see all the characteristics of life in your pet dog or cat? Can you see them in a potted plant?
6. Why is it so difficult to define life?

*6) There is no single feature that distinguishes living things from non-living things, but certain features are typical of nearly all organisms. The many variations that life takes. There are over a million species of plants + animals that have been discovered.*

In the interest of orderly communication among biologists, we name organisms and classify them. The basic unit in biological classification is the **species.** A species (both singular and plural forms are spelled *species*) is difficult to define but usually is considered to be a group of organisms with a common ancestry and with the potentiality of interbreeding with each other and producing fertile offspring. Members of different species ordinarily do not mate with each other; but if they do, usually either no offspring are produced or the offspring are sterile. (Whether sterile or fertile, the offspring of a cross between different species is called a **hybrid.** One of the hybrids most useful to man is the mule, the rarely fertile offspring of a female horse and a male donkey.) Fertile hybrids sometimes occur among plants, but they are rare among animals.

## Naming Living Things

Closely related species are classified in the same **genus** (plural *genera*). The **scientific name** of any organism includes the names of both the genus and the species to which it belongs. These names are in Latin (or Latinized forms of words from other languages) and are supposed to be descriptive of the species. In practice, the latter requirement is not always fulfilled. The horse and the donkey both belong to the genus *Equus,* but the horse is a member of the species *Equus caballus* and the donkey belongs to *Equus asinus.* Dogs and their relatives belong to the genus *Canis.* Included in this genus are the domestic dog, *Canis familiaris;* the wolf, *Canis lupus;* and the coyote, *Canis latrans.* Maple trees belong to the genus *Acer.* Among them are the sugar maple, *Acer saccharum,* and the red maple, *Acer rubrum.* Human beings belong to the species *Homo sapiens.*

Generic and specific names always are italicized. The generic names always are capitalized, but the

# 2

# Naming and Classification of Living Things

**1** Biologists name and classify all known living things. The classification is so constructed as to reflect evolutionary relationships.

7

specific name usually begins with a lower-case letter.

Because generic and specific names are strange to the average person, this book uses common names wherever it is convenient to do so, but not all organisms have common names. This is especially true of microscopic organisms and other organisms not seen by most persons in their daily lives.

It should be pointed out that the use of common names has a disadvantage. Common names vary from place to place and from language to language. Several different common names may be applied to one organism in different parts of the same country, or the same common name may be applied to different organisms. For this reason common names are not accepted internationally among biologists, whereas scientific names are.

## Classifying Living Things

At least 1 million species of living things inhabit the earth. Of diverse types, they run, walk, and crawl on land, swim in the sea, fly in the air, or live their entire lives attached to one spot of land or sea bottom.

A list of 1 million items is imposing; it also is unwieldy. Breaking a large group into several smaller ones, the members of each group having some characteristics in common, makes the larger group somewhat easier to comprehend. If the smaller groups are still inconveniently large, they can be divided into still smaller groups, and so on. The criterion for separating the members of a large group into smaller ones depends on the interests of the person doing the job. An anthropologist may think of people as caucasoids, negroids, and mongoloids; a clergyman classifies them as Catholic, Protestant, Jewish, Muslim, and so on; a politician considers whether they are Democrat, Republican, or of some other political party.

Biologists look on classification of organisms as more than a convenience, which, of course, it is; but if it were to be only a convenience, we might group organisms by some very easily observed traits such as their size and color. Most persons would agree that this could lead to absurdities, for a brown standard poodle then would be in the same group as a brown Shetland pony rather than with dogs of other sizes and colors. Biologists prefer to classify organisms by a system that reflects their evolutionary relationships with each other, closely related organisms being classified in a common group; that is, organisms believed to have a common ancestor are placed in the same group. Although the idea is a simple one, universal agreement never has been reached on how closely or how distantly related some groups are.

Our classifications are based on data gathered from several sources: anatomy of both living and fossil organisms, cellular structure, biochemistry, and metabolism. The data from these separate fields usually support each other and so strengthen our ideas about our scheme of classification, but occasionally they lend themselves to different interpretations, and for this reason different biologists embrace somewhat different classifications.

The two-kingdom classification is very old and recognizes only the plant and animal kingdoms. This works well enough for relatively large, familiar organisms such as dogs, butterflies, potatoes, and dandelions; and indeed most kinds of organisms that we encounter in our everyday lives are either plants or animals. Some of the differences between these two groups were readily observed even in ancient times; others have been discovered in relatively recent years only after the microscope and modern chemistry became tools of biology. The differences between plants and animals are correlated with their two different modes of existence. Plants generally are stationary, light-absorbing organisms that manufacture food by a chemical process called photosynthesis. Most animals are organisms that move about seeking food and eating it.

The cells of plants are surrounded by cell walls that contribute somewhat to the rigidity and support of plant tissues. The locomotion of animals would be hampered by complete rigidity of the body, and their cells have no walls. Plant growth movements are relatively slow, requiring at least a few minutes, and often hours, to be noticeable. Movements of animals can be very swift.

The green color of most plants is due to the presence of the pigment chlorophyll, which is lacking in animals (the green color of some animals, such as birds with green feathers, is not due to chlorophyll). Chlorophyll is a light-absorbing pigment that is essential for photosynthesis, a process that cannot proceed without light energy; plants are thus capable of manufacturing their own food, whereas animals are not. Animals are dependent on plants for their food; animals eat either plants or plant-eating animals.

Plants generally have unlimited growth; they may continue to grow through their entire lives. Many of them have no definite life span but continue to grow indefinitely; death is often due to some external cause, such as destruction by grazing animals, storms, disease, lumbermen, and so on. Animals generally have limited growth; they usually grow to an adult size characteristic of the species and then stop growing; some reptiles and fish which continue to grow as they age are exceptions. The life span of animals is fairly definite also; if death does not occur because of external causes, failure of internal organs usually brings on death. Many plants grow only in their meristems; these are localized areas in which frequent cell divisions produce new cells. Animals, while they are growing, grow throughout most of the body.

Many exceptions to the preceding statements exist. Most of the exceptions are the so-called lower plants and animals. The fungi, for example, are plants that lack chlorophyll and thus do not manufacture food. Some protozoa (single-celled animals) possess chlorophyll and do manufacture food. Though many of the algae (sometimes popularly called seaweeds) obviously are plants, some

of them are similar to protozoa. Sponges and sea anemones are aquatic animals that have no power of locomotion and remain attached to rocks or other solid objects as plants do. These organisms possess such a mixture of plant and animal characteristics that it often is difficult to decide whether they are plants or animals; but for the most part they can be included, if somewhat arbitrarily, in either the plant or animal kingdom. The fungi and most of the algae usually are considered to be plants, and the protozoa are thought to be animals. There seems to be little doubt that the sponges and anemones are animals.

Three groups are somewhat more troublesome: bacteria, blue-green algae, and viruses. The cellular organization of the bacteria and blue-green algae differs significantly from that of all other cellular organisms (Chapter 17), and so they are classified here as a separate kingdom, the Monera. These are single-celled organisms, but their minuteness does not make them unimportant to us. The diversity of chemical capabilities (or incapabilities), especially among the bacteria, makes some of them essential for the recycling of materials in the natural world; others modify our foods and beverages, thus producing desirable products like cheese and wine; still others cause illnesses ranging from minor stomach upsets to bubonic plague and syphilis.

The viruses constitute a third group of organisms not readily classifiable as plants or animals; indeed, some biologists do not consider viruses to be living. Unlike all other organisms, they possess no cellular structure and can be crystallized and stored in a bottle in an apparently lifeless condition, characteristics that cause us to place them in a fourth kingdom by themselves. All viruses are parasites, and when the crystallized, "lifeless" viruses are introduced into the proper hosts they exhibit some of the characteristics of living things, including reproduction. Viruses have been useful objects of research, especially in the field of genetics, but otherwise they have little to recommend themselves to humanity. Many serious diseases

9

(for example, poliomyelitis, rabies, and almost certainly some forms of cancer) are caused by viruses; some viruses seem not to harm their hosts, but none seems to confer any benefit.

This book, then, recognizes four kingdoms of living things:

Viruses
Monera—bacteria and blue-green algae
Plants—algae (other than blue-green algae), fungi, mosses, vascular plants (ferns, cone-bearing plants, and flowering plants)
Animals—protozoa (single-celled animals) and multicellular animals

Other books may present somewhat different classifications, for their authors have different opinions about how to classify living things. The student should not be confused or overly concerned about this. Rather let him be impressed with the diversity of living things which does not lend itself well to being forced into the pigeonholes of our classification schemes.

By itself, the grouping of all organisms into four kingdoms does little to relieve the confusion of more than a million species. Therefore, kingdoms are divided into smaller groups called phyla (singular, phylum). Phyla are divided into classes, classes are divided into orders, orders into families, families into genera, and genera into species. The several orders of rank thus are

kingdom
    phylum
        class
            order
                family
                    genus
                        species

Subgroups may be incorporated wherever convenient, a phylum being divided into subphyla, a class into subclasses, and so on.

Two individuals of a species are considered to be more closely related to each other than are two individuals from different species within the same genus. Two species within the same genus are considered to be more closely related to each other than two species in different genera within the same family, and so on.

## Questions

**1.** How do plants and animals generally differ from each other? How many exceptions can you find to your answer?
**2.** How do the Monera differ from other living things?
**3.** How do viruses differ from other living things?
**4.** Of the four major divisions of living things—viruses, Monera, plants, and animals—which two do you think resemble each other the most?
**5.** Which of these four groups do you think is most different from the other three?
**6.** Are there any objections to the use of common names in biology?
**7.** What are the advantages of giving each species a universally accepted name?
**8.** What is a species? A genus?
**9.** What is the difference between the natural system of classification and an artificial one?
**10.** Is our present system of classification a completely natural one? Why or why not?

## Suggested Readings

Bailey, L. H. *How Plants Get Their Names.* 1933. Reprint. New York: Dover Publications, Inc., 1963.

Hanson, E. D. *Animal Diversity,* 2nd ed. Englewood Cliffs, N.J.: Prentice-Hall, Inc., 1964.

Simpson, G. G. *Principles of Animal Taxonomy.* New York: Columbia University Press, 1961.

Whittaker, R. H. "New Concepts of Kingdoms of Organisms," *Science* 163: 150–160 (1969).

10

# The Environment

# Part II

Organisms do not live in a vacuum but rather in an environment that influences them and that they, in turn, act upon. Included in the environment of any organism are other living things and the physical environment. **Ecology** is the study of the interactions among organisms and between them and their physical surroundings.

At any moment innumerable interactions involving living things are occurring. A cow grazes on grasses and clover; a lion and a lioness mate; the green leaves of a corn plant manufacture food from water and carbon dioxide; a little girl inhales oxygen and exhales carbon dioxide; a mosquito bites a man; dead leaves fall from a maple tree in autumn; bees pollinate the flowers on an apple tree; an epidemic of influenza strikes the people of New York City; a dog urinates at the base of a tree; a mare nurses her foal; fishermen haul in nets laden with salmon; a corn crop is ruined by corn blight; sewage is processed and sold as fertilizer; a mother moose protects her calf from hungry wolves; an eagle soars on updrafts; a fox sleeps in his den.

The total of all these activities and others like them tends to maintain a homeostatic condition within the individual, within the species, and within the entire community. By eating food, the wolves, the cow, the mosquito, and the men obtain the nutrients necessary to support their activities and to maintain their bodies in reasonably good health. Regardless of its good health, however, no organism lives forever; and by mating, the lion and the lioness produce a new generation that will replace them. The lioness, like the mare and the moose, will tend her young, thus increasing their chances of survival. The wastes of any species, whether maple, dog, or human, replace the minerals that it or another species removed from the soil; so, too, does its death. This enrichment of the soil provides nutrients used by plants, which directly or indirectly feed all animals. Thus each species maintains not only itself, but other species also.

These many interactions are so numerous and

# 3

# Interactions Among Organisms of the Same Species

1 Interactions among members of the same species tend to ensure the survival of the species.

2 These interactions permit the survival of only a few members of each species, and thus they tend to regulate population size.

13

a    b

**3.1** Animals that lead a solitary existence. (a) The weasel is a carnivorous animal usually hunting alone at night. (San Diego Zoo Photo.) (b) The common argiope spider also lives alone. It obtains food by spinning webs and devouring insects and other small animals that become trapped in the webs.

**3.2** A pair of black-necked swans and one of their offspring. A mated pair remains faithful for life. (San Diego Zoo Photo.)

so complex that biologists today understand only a few of them well, and they are aware that many undiscovered relationships also exist as well. For convenience in discussion the topic of interactions has been divided here into four sections: interactions with organisms of the same species (this chapter), interactions with members of different species (Chapter 4), successions and communities (Chapter 5), and interactions between organisms and their physical environment (Chapter 6). None of these chapters pretends to do more than touch on a few of the myriads of interrelationships that exist; later chapters touch on more.

## Social Structure

Some examples of animal behavior that involve primarily interactions between the members of the same species are herding, mating and courtship, the raising of young, establishment of rank and of territory, competition, and communication.

Some animals, such as woodchucks, hamsters, and many insects and spiders (Fig. 3.1), lead solitary lives, adult males and females coming together only for brief periods to copulate; the offspring are reared largely by the mother, or, as in the case of many insects, the offspring are independent from the time of hatching. In other species the members come together in large groups only at migration time. Ducks and geese are only a few of the many kinds of birds that flock together in large groups and travel together for long distances. Then they disperse in pairs or small groups until it is time for the next migration (Fig. 3.2).

Still other animals, such as caribou (Fig. 3.3), zebra, antelope, and elephants, are more gregarious and stay in herds where they have mutual protection from predatory animals. Another advantage of group living is that food or water found by one member of the group can be shared by all;

**3.3** Caribou are gregarious animals living in herds. These animals are in migration. (Charlie Ott from National Audubon Society.)

**3.4** Fur seals. A bull lives with a harem of several females and young.

this reduces somewhat the time and effort spent by each animal in search of these commodities. While the rest of the group is feeding, a few members may act as sentries that sound an alarm when danger threatens. Because a predator is likely to take the easiest prey, the stampeding of a frightened herd ensures that only the weakest or the slowest to react will be caught, whereas the strongest and fastest escape. Elephants, which are less susceptible to attack by predators, show great concern for other members of the herd and will stop to assist the young and injured.

The **family life** of animals runs the gamut of all possible domestic arrangements. In a few species, such as wolves, foxes, swans, and Canada geese, monogamy is practiced, with the two partners remaining faithful to each other for life. In many common songbirds, such as sparrows, the pair splits up at the end of the breeding season, and the following year the two birds find new mates. House wrens, who often raise two broods of young each year, may even find new mates for the second brood. Seals, walruses, and elk are polygamous (Fig. 3.4), a male keeping a harem of many females, and he may have a different harem every year. Total or nearly total promiscuity may be the rule in still other animals, such as rodents. Although promiscuity is frowned upon in most human societies, it is nonetheless practiced to some extent.

In some species the males and females are noticeably different from each other in either size or coloring or both (Fig. 3.5). This **sexual dimorphism** enables animals to recognize the sex of other individuals. A male bird of a sexually dimorphic species will attack a stuffed male bird of the same species or anything resembling it. A male robin was observed doing his best to drive off a tuft of orange feathers about the size of a male robin's breast, and male cardinals have fought futilely with their own images in window panes. On the other hand, a male bird may attempt to mate with a stuffed specimen that resembles a female. The sight of sexual markings thus elicits appropriate responses by animals when they meet each other; it eliminates wasted effort that might otherwise be spent in chasing away prospective mates or attempting to copulate with a member of the same sex. In species without marked sexual dimorphism males and females may be recognized by different scents or different behavior patterns.

Males frequently arouse females through **courtship** rituals that may involve more or less elaborate dances, or, in the case of some birds and bees, aerial acrobatics. In some sexually dimorphic animals the males display their colors to great advantage. In some species both males and females engage in these courtship rituals in which both individuals become sexually aroused; fireflies of both sexes signal to each other by flashing their lights.

a

b

**3.5** Examples of sexual dimorphism. (a) A bull elk bears antlers, but cow elk have none. (Photograph by Richard J. Naskali.) (b) The luxuriant mane of a male lion makes him easily distinguishable from a lioness. (San Diego Zoo Photo.)

17

Sexual dimorphism sometimes provides protective coloration to females. Female cardinals and indigo buntings have none of the brilliance of their mates, and their drab colors make them inconspicuous as they sit on their nests. This protection is passed on to the eggs or young that are covered by their mothers.

The **care of young** is as variable as are family arrangements. Many insects, such as tent caterpillars the larvae of which make their own homes, and many fish, such as salmon, receive no parental care, for their parents die shortly after the eggs are laid. The eggs contain sufficient food to support the young embryos until they reach a stage of maturity in which they can feed themselves. White-tailed deer, bears, and many cats receive care only from their mothers, the fathers having

**3.6** Adult coyote brings a jackrabbit to a pup waiting at the den. As the pup grows older it will be taught by its parents to hunt its own food. (Photo by E. R. Kalmbach. Courtesy of Bureau of Sport Fisheries and Wildlife.)

departed after mating. Polygamous males with many mates rarely give attention to their offspring other than the general protection they provide for the entire herd. Many waterfowl are capable of feeding themselves soon after hatching and require only protection and guidance from their parents. The chicks of doves and many other birds are tended by both parents who bring food to their offspring. Wolf and fox pups and many other young carnivores not only are fed by one or both parents (Fig. 3.6), but are trained by them in the art of hunting. Primates, including gorillas, chimpanzees, and monkeys, are raised in large, loose family groups that include aunts, uncles, and cousins; although the mother takes the most interest in her child, adults are solicitous of infants in general and unite to protect young ones from danger (Fig. 3.7). Humans have a long period of infancy relative to the life span; during this time they usually are tended by both parents.

Within a gregarious species there is nearly always some kind of social organization. The young usually are subordinate to the adults. Females often are subordinate to males, although this is not always so; a herd of red deer, for instance, sometimes is led by a mature female.

**Dominance** often is established by mock battles in which little serious injury is inflicted, although blood sometimes does flow, and death occasionally results. In domestic chickens dominance is established by pecking; in cattle and bighorn sheep it is accomplished by butting contests. Physical contact is not always necessary between the combatants; aggressive postures, display of feathers, snarling, or other signals of intent to do battle if pressed may be sufficient bluff to make one of the animals acquiesce and assume a social position inferior to that of the victor.

Through many meetings of the several members of a group a social hierarchy (sometimes called pecking order) is established. The animal at the top, called the alpha animal, can intimidate all other members of the group with impunity. The animal of second rank, or beta animal, can domi-

**3.7** Family of chimpanzees with several adults and offspring.

nate all but the alpha animal. The animal lowest in the hierarchy is subordinate to all. In an established group all members recognize each other and repel invaders, but a persistent stranger eventually may be accepted into the group—usually at a very low rank.

Rank changes when illness, injury, or old age causes a previously dominant animal to lose in a confrontation with one of its subordinates. In some species, such as wolves, a female assumes the same rank as her mate; and if his rank should rise or fall, hers does accordingly. In general, dominant animals produce more and healthier offspring, and the young of dominant animals are more likely to become dominant when they mature.

In a social hierarchy rank has its privileges. The alpha animal generally has the first choice of mate (or mates, in a polygamous species) and the privilege of dining first and eating his fill; beta animals have second choice; and so on. This undemocratic custom has advantages for the species. When food is plentiful, no one is deprived. When there is insufficient food to go around, at least some animals are assured of having enough food to survive even if lesser members of the group die. In this way, some members, usually the more fit, live long enough to produce progeny, and the species survives. If a meager food supply were shared equally among the members of the group, all might die,

and the population would die out. Were this to happen to all populations within a species, the species would become extinct.

## Territory

Just as there is social order within a group, so also is there order in the distribution of each species throughout its range. Each breeding group occupies a particular **territory.** A breeding group may be a mated pair (as a pair of Canada geese), a male and his harem (as a male elk and his several cows), or other family arrangement. Often the territory initially is established by a male, and it is later defended either by the male alone, by the pair, or by the entire group. Individuals establish and defend territories in much the same way that status is established—by confrontations with much threat and bluff but with little or no bloodshed. Some of the beautiful singing of songbirds which sounds to humans like joyful exuberance celebrating the coming of spring in reality has the function of a "No Trespassing" sign that indicates a territory is occupied and no other males are welcome. Many mammals mark the boundaries of their territories by urinating or defecating at its

19

**3.8** A colony of gannets. These birds nest very close together, each mated pair defending only a few inches of territory in any direction. These birds obtain their food from the ocean waters and not from the immediate vicinity of the nest site. (Courtesy of Dr. J. B. Nelson.)

borders or by depositing scented substances produced in special glands.

If an intruder enters an established territory, the owner defends his territory vigorously. Almost invariably in these confrontations the owner is the victor, and the intruder retreats. In the rare event that the outcome is otherwise, and the owner loses, he relinquishes the territory to the invader.

The size of a territory varies with the species. A pair of small songbirds may defend an acre of woodland, and a mountain lion or a pack of wolves may range over 100 square miles. The exact size of the territory tends to vary somewhat with population density and food supply, but it does not go below a certain minimum for the species.

Some birds, such as gannets or murres, that live on rocky cliffs, have nests crowded within a few feet or even inches of each other (Fig. 3.8); and it is only the immediate vicinity of the nest that is defended by the owners. All the adults of the colony share a common feeding ground some distance away; this latter area may be defended by the group as a whole.

The advantages of maintaining territories are several. It provides some degree of protection for the young, and competition for food is reduced. At the same time, it prevents the overutilization of resources; for example, only a fraction of the available grass is eaten by grazing animals, and this permits the reproduction of the grass and ensures a future supply for the animals. Like the phenomenon of rank, territoriality ensures that at least some individuals are fed well enough to survive and reproduce. It also acts as a method of population control, for only those males with established territories are successful in obtaining mates. Those without territories lead a bachelor existence and do not reproduce. These excess males are "spares" ready to challenge any aging or sick males or to replace one that dies. A female who has not found a mate with territory becomes part of this "stockpile" too.

Any effort to eradicate animals from a small portion of their range by killing them usually is useless, for the territory left undefended soon is occupied by animals from adjacent territories. Setting mousetraps in one's backyard is at best a stopgap measure if the neighbors have mice.

## Competition

**Competition** exists among members of the same species, especially if the population density of that species is high. Competition among animals for food, rank, territory, and mates has been mentioned earlier. Plants, too, compete with each other. If several seeds fall near each other and all germinate, there will be competition for light, water, and minerals. If all the seedlings grow equally well, they are likely all to be thin and spindly with little chance to survive long enough

to reproduce. Rarely, however, do all have exactly the same environment. If one is more favorably placed than the others, and receives a little more sunshine, it grows faster than the others and soon shades them. It survives, and they die. So, too, if the root system of one of the plants lies in soil with a somewhat more favorable concentration of water or minerals, that plant is more likely to compete successfully with its fellows. Competition ensures the survival of a few well-favored individuals rather than the death of all.

An interesting mechanism that avoids the necessity for competition for water exists among desert plants. On the desert, where water is extremely scarce, roots secrete compounds that inhibit the germination of seeds or the growth of seedlings within a certain distance of each established plant. This leads to fairly regular spacing among the plants and conserves water, again ensuring the survival of a few members of the species.

## Communication

**Communication** among animals has survival value, and nearly every species has developed some means of exchanging information by sight, sound, or chemicals.

Visual signals are so many that only a few will be mentioned. Patterns that indicate the individual's sex have already been discussed. The white rumps of antelope and elk serve as signals by which animals can follow each other; this keeps a frightened herd from scattering.

The sight of a gaping mouth invites a parent bird to deposit food in it. The wider the gape the more likely is the mouth to receive some food. Because the hungriest chick is somewhat more likely to have the widest gape, this mechanism ensures an equitable distribution of food in the family. A chick that is ill, however, and fails to demand his share is unlikely to be served; he dies without

21

wasting much food that could nourish his brothers and sisters.

An exceedingly complex visual signal is the dance used by bees to communicate to members of their hive the location of a newly discovered food supply; the bees perform dances which, by their orientation, indicate the direction in which the food lies and, by their speed, indicate the distance to the food.

Sound signals are also numerous, with different species communicating by roars, grunts, snorts, squeaks, chirps, and other sounds. Some species have as many as two dozen different sound signals indicating, among other things, hunger, fear, threat, annoyance, sex, courtship, and parental concern. Nocturnal and marine animals make frequent use of sound signals for many of them live in darkness where vision is of little value. Bats and whales both use sonar in determining the locations of objects, including other animals.

Scented compounds released by animals transmit information to others; such compounds are called **pheromones.** They influence the activities or development of other animals of the same species. Compounds in urine, feces, and musk deposited at territory boundaries fall in this category. Ants lay chemical trails containing pheromones that other members of the colony follow. So sensitive are these small insects that they are able to determine the direction in which the previous ants moved. If a section of a trail is removed, turned around, and replaced, the ants become confused at the reversed section.

Many animal species have breeding seasons at only certain times of the year. These are so timed that the young are produced at seasons most favorable for their survival. Females are receptive to males for only a few days in the year, and males are attracted to them sexually only during this time. Some sex attractants are pheromones that signal receptivity of many female animals; these compounds prove irresistible to males of the same species. Although these substances become very much diluted as they diffuse in many directions

22

through the atmosphere, males often are able to detect receptive females from distances of several miles. The survival value of this for the species is obvious, for it increases the chances of finding mates and reproducing.

The production of such sex attractants by female insects opens the possibility of controlling insect pests by baiting traps with females or with a small quantity of their scents. Unlike the use of insecticides, most of which kill many beneficial as well as harmful insects, this method affects only the target species and does not threaten to exterminate others.

The many interactions among members of a species have but one function: the survival of the species. Though many of these interactions seem to many persons to be cruel, nature wastes no sympathy where sympathy will not help. In the long run the production of more offspring than can possibly survive and reproduce, coupled with the elimination of the weak and injured, is a benefit to the species as a whole.

# Questions

1. Define the following terms.
   ecology
   territory
   dimorphism
   dominance
   social hierarchy
   alpha animal
   pheromones
2. In what ways do the following contribute to the survival of the species?
   herd living
   solitary life
   sexual dimorphism
   parental care of young
   dominance
   establishment of a territory
   competition within a species
   communication

**3.** For several days observe a pet dog or cat that runs free in your neighborhood. Can you determine the extent of his territory?

## Suggested Readings

Davis, D. E. *Integral Animal Behavior.* New York: Macmillan Publishing Co., Inc., 1966.

Eibl-Eibesfeldt, I. "The Fighting Behavior of Animals," *Scientific American* 205(6): 112–122 (December 1961).

Estes, R. D. "Trials of a Zebra Herd Stallion," *Natural History* 76(9): 58–65 (November 1967).

Guhl, A. M. "The Social Order of Chickens," *Scientific American* 194(2): 42–46 (February 1956).

Johnsgard, P. A. *Animal Behavior.* Dubuque, Iowa: William C. Brown Company Publishers, 1967.

Lorenz, K. *King Solomon's Ring.* New York: Thomas Y. Crowell Company, 1952.

Todd, J. H. "The Chemical Languages of Fishes," *Scientific American* 224(5): 98–108 (May 1971).

Wallace, B., and A. M. Srb. *Adaptation,* 2nd ed. Englewood Cliffs, N.J.: Prentice-Hall, Inc., 1964.

Watts, C. R., and A. W. Stokes. "The Social Order of Turkeys," *Scientific American* 224(6): 112–118 (June 1971).

Wilson, E. O. "Pheromones," *Scientific American* 208(5): 100–114 (May 1963).

_____. "Animal Communication," *Scientific American* 227(3): 52–60 (September 1972).

# 4

# Interactions Among Species

1 **Interactions among species tend to maintain fairly steady population levels over long periods of time, though there may be fluctuations in population sizes over short periods.**

2 **Competition regulates the number of species that can live in an area.**

3 **Predation helps to regulate the population size of the species preyed upon, and the population size of the prey species helps to regulate the population of predators.**

4 **Symbioses are intimate relationships between members of two species in which one member benefits and the other member benefits (mutualism), is neither benefited nor harmed (commensalism), or is harmed (parasitism).**

The preceding chapter dealt with interactions between members of single species, but in practice it is not likely that any given act involves only one species, for every event has far-reaching consequences that go beyond the boundaries of the immediate situation.

If a lioness has cubs to feed, she may kill a zebra. If the zebra has a nursing colt, the colt is likely to die, too, thus providing food for hyenas and vultures who may compete with each other for the prize. The grasses that the two dead animals might have eaten become available to others. The deaths of the zebras means that any parasites they harbored will have to find new hosts or will die. Any remains of the zebras that are not devoured quickly by large animals are consumed by a variety of insects and microorganisms. The minerals that were present in the bodies ultimately return to the soil where they nourish the grasses and other plants of the plains. The plants will provide food for more zebras and other grazing animals. Thus, in the act of feeding her offspring, the lioness sets in motion a series of events that has no foreseeable end (Fig. 4.1). All the organisms that profited by the death of the zebras—lions, hyenas, vultures, ants, microorganisms, and grasses—will reproduce, bringing to the world new generations of individuals that will continue to play their roles in a complex web of interactions.

The interactions of individuals of one species with those of another range all the way from the most casual of contacts to competition and predation and on to intimate relationships in which members of two species live together in close physical contact.

## Competition

**Competition** exists between species just as it does within a species. The competition may be for food, nesting places, or other requirements. Cattle

**4.1** The hunting of prey has more consequences than just the obtaining of a meal or two. Directly or indirectly, all the organisms in the community are affected, if only slightly. See text.

and sheep are grass eaters, and if placed together they compete with each other for food, a fact that led to many a bitter dispute between cattlemen and sheepherders who moved into the newly opened plains of the United States in the last century. Kangaroos, which cattle would not meet under natural conditions, eat grass too, and if placed with cattle will compete with them. None of these animals would compete with moose, who enjoy water lilies and the tender twigs and buds of willow trees, or with the meat-eating wolves and mountain lions. Several species of small songbirds may compete with each other for nesting sites in the shelter of tree branches, but they do not compete with the cliff-dwelling gannets and murres or with the red-winged blackbirds that nest in cattail marshes.

It is a general rule that two competitive species cannot occupy the same area if their population densities are high. The amount of food, the number of nesting sites, and the supply of other commodities available are always finite and limited, and one of the species is likely to be more efficient than the other in utilizing these resources. The less favored species dies out, and the other survives. If their population densities are low, the two species may be able to coexist.

## Predation

**Predation** is the way of life of meat-eating animals. Wolves and mountain lions eat deer, moose, rabbit, squirrel, and other animals. Eagles, owls, and hawks consume mice, rats, rabbits, and even larger prey. Insect pests are kept in check partly

25

by many of the smaller birds of field and woodland, and some insects dine on other insects, as do many spiders.

Predation contributes to the control of population size of the species preyed upon. A wolf may take an occasional sheep from a farmyard and an eagle may fly off with a stolen chicken, but their contributions to the maintenance of population sizes of other species more than compensates for this. Without predators the prey species soon might overrun the land.

The fecundity of rabbits, though surpassed by that of many other animals, is great enough to have earned them the reputation of reproducing at a rapid rate. The mental image of a pair of rabbits rarely fails to produce hundreds of imaginary rabbit grandchildren and great-grandchildren. Yet, in nature, two rabbits almost never have so many descendants in so few generations. Insufficient food supplies may limit their numbers part of the time, but hungry wolves and other predators enjoy a meal of rabbit. As they prey on the rabbit population, that population decreases in numbers.

If a predator population is relatively high, the prey population soon decreases in size. Extinction of the prey usually does not happen, for as its numbers fall, the predators have difficulty finding sufficient food. Their numbers decrease as many of them starve to death. This relieves some of the pressure on the prey, whose numbers begin to rise again. As they do, the predators have a better chance of finding food, their numbers increase also, and the cycle begins again. So we see that predators and prey tend to control each other's populations.

In nature, cycles like this rarely are so simple, for it is not just a matter of two species existing together and isolated from all others. Each is influenced by other factors. Wolves eat more than rabbits, and more than wolves prey on rabbits. The food supply available to the rabbits is another factor that influences their population density.

One of the relatives of the weasel is the fisher. Its luxurious fur and its habit of stealing fisher-

men's catches led to widespread trapping of this animal and to its near extinction. Its disappearance from many woodlands was followed by an alarming increase in porcupine damage to trees. The sharp-quilled porcupine is avoided by all but the hungriest of most predators. The fisher, however, has the rare ability to attack and eat porcupines in safety. Without fishers to keep their numbers in check, porcupines multiplied rapidly. Gnawing on tree trunks to obtain the inner bark for food, the porcupines killed the trees and destroyed the forests. Fishers have been reintroduced into some areas in efforts to re-establish the ecological balance.

## Symbiosis

Food—and sometimes shelter and other benefits—may be obtained in more intimate relationships called symbioses. *Symbiosis* means living together, and it involves two different species. Usually the two organisms live in physical contact, one living on or in the other. Three main types of symbiosis exist: **mutualism** (once called symbiosis), **commensalism,** and **parasitism.**

### Mutualism

Mutualism is a relationship in which both partners derive some benefit from the association. One well-known mutualistic relationship and one of far-reaching significance to man and other organisms is that which exists between any of several leguminous plants and nitrogen-fixing bacteria of the genus *Rhizobium*. Leguminous plants include several important crop plants—among them are beans, peas, and alfalfa—as well as some forest trees such as black locust and honey locust. *Rhizobium* is a bacterium that can live in the soil, but when it encounters the root of a legume, it enters the root. It causes the production of nodules,

**4.2** Roots of a leguminous plant that grew in soil containing mutualistic nitrogen-fixing bacteria. After the bacteria enter the root of a legume, the root responds by forming tumorlike nodules. In the absence of the bacteria, no nodules form. (Courtesy of O. N. and Ethel K. Allen, University of Wisconsin.)

which are enlargements on the root (Fig. 4.2). In these nodules the bacteria grow and multiply. Here they also fix nitrogen, something they do not do when they live alone in the soil. **Nitrogen fixation** is a process by which the nitrogen gas ($N_2$) of the air is converted into ammonia ($NH_3$), which then is converted into some organic form such as amino acids (Chapter 13). Nitrogen gas is of no direct use to most of the plants and animals of the world, for they do not have the ability to metabolize it; yet they all need nitrogen. The nitrogen compounds that *Rhizobium* produces are useful not only to the bacterium itself, but also to the leguminous plant which absorbs some of them and so flourishes (Fig. 4.3). *Rhizobium*, like most other bacteria, is not photosynthetic and cannot manufacture food; this it absorbs from its host. In this way both organisms benefit from their association.

Farmers have learned to utilize leguminous plants when they rotate crops. For two or three years nonleguminous plants are grown in a field. The next year the crop is a leguminous plant. After the crop is harvested, the roots and other unused parts are plowed back into the soil, thus enriching it with useful nitrogen compounds. Then the cycle is repeated.

On the roots of many flowering plants grow the filamentous bodies of several species of fungi. The symbiotic association between the root and the fungus is called a **mycorrhiza.** The nonphoto-

**4.3** Mutualism between plants and bacteria. These two groups of red clover plants are the same age. Seeds were planted in sterile sand and watered with a nutrient solution lacking nitrogen but containing other minerals required by the plants. To the pot at the right mutualistic nitrogen-fixing bacteria were added. Note the difference in the amount of growth. (Courtesy of O. N. and Ethel K. Allen, University of Wisconsin.)

27

synthetic fungus receives food from the roots of the host plant. The fungus absorbs water and minerals from the soil and transfers them to the host; the fungus seems to be more efficient in this than is the root of the host, which usually grows much better with its mycorrhizal fungus than without it. Pines, blueberries, and orchids are a few plants that commonly have mycorrhizal associations. Some orchids cannot be grown from seed without their mycorrhizal fungi.

Another well-known example of mutualism is the lichen. It is customary to speak of a lichen as a single plant, but in reality it consists of two species: one a fungus and the other an alga. The fungus harbors within itself many single-celled algae. The algae are photosynthetic and produce the food that the fungus requires but cannot make, and the algae receive some shelter and a moist environment. Lichens are very hardy plants and grow where few others could survive. They grow on bare rock surfaces that are extremely dry except during rains, extremely hot under the summer sun, and cold at night or during winter. Lichens invade such territories, and when they die, the disintegration of their bodies leaves a thin layer of soil in which other plants are able to grow. One lichen, the so-called reindeer moss, comprises the dominant vegetation of extreme northern North America and Eurasia; through much of the year it is the major food of the caribou and the reindeer, a domesticated form of caribou. These two animals are, in turn, a major food source of Eskimos, Laplanders, and some Siberians.

Mutualistic relationships also exist between plants and animals. Insect-pollinated plants are dependent on visits by insects for cross-pollination. Although some of these plants can set seed if self-pollinated, cross-pollination is essential for others, and these plants are absolutely dependent on pollinating insects. The insects, in turn, obtain some food from the plant, often nectar which is secreted by the flower when the pollen is ripe. Many fruit trees, such as apples and pears, as well as some other crops, are pollinated by bees. In a year when the bee population is scanty in an area, the crop produced there is correspondingly low. Applications of insecticides in agricultural areas often result in the death of many bees. To ensure adequate pollination farmers sometimes rent hives of bees during the critical pollination periods.

Another example of mutualism is that existing between cattle and the protozoa (single-celled animals) and bacteria living in their stomachs. Although cattle feed exclusively on plant materials that are rich in cellulose, they are unable to digest this material by themselves. The protozoa and bacteria, however, digest the cellulose to compounds that both they and the cattle can use. In a similar relationship termites, which are notorious for their ability to consume wood, are completely dependent on their intestinal protozoa for the digestion of the cellulose in the wood.

African tick birds or oxpeckers accompany game animals such as rhinoceros (Fig. 4.4), giraffe, and zebra as well as domestic cattle. The tick birds eat external parasites on the skin of these animals. In this way the birds receive food, and the cattle and game are freed of parasites. The Egyptian plover seems a little more daring as it braves the open mouths of crocodiles; here it feeds by picking clean the crocodiles' teeth. The crocodiles, of course, benefit by the removal of scraps of meat that might otherwise increase the rate of decay of their teeth.

Many fish are cleaned of their parasites by other fish or by some species of shrimp that feed on the parasites. Some cleaners seem to maintain "cleaning stations" to which the parasitized fish come. The latter have been observed to open their mouths and gill covers and to allow the small cleaners to enter with safety.

## Commensalism

Commensalism is a relationship in which one member benefits while the other is neither helped nor harmed. The remora fish is peculiar in having near the top of its head a fin modified into a suction disc. With this disc the remora attaches itself

**4.4** Mutualism between two species of animal. African tick birds associate with rhinoceros and obtain some of their food by eating ticks, flies, and other external parasites of the rhinoceros, which thus are relieved of their pests.

to a whale, shark, or large turtle. The remora benefits by being transported from meal to meal, feeding on scraps left by its host. Sharks are particularly untidy feeders, and they scatter ample food which their passengers eat. The remora is relatively small and does not harm its host, nor does it seem to interfere with the host's movements.

Other commensal hitchhikers are the barbed and spined seeds and fruits that become attached to the fur of animals. They can be transported long distances before they fall off. In this way the plants become widely disseminated, and the animals that carry them are not affected by their temporary presence. Seeds and small fruits often are transported in the mud that clings to the feet of birds. During migration times these seeds can be transported across oceans or continents this way.

## Parasitism

Parsitism is a relationship in which one of the members benefits and the other is harmed. The parasite nearly always absorbs food from its host, and it may receive water, minerals, and shelter as well. Many, but not all, parasites are pathogens; that is, they cause disease.

Nearly all species of plants and animals are parasitized by at least one species of parasite and usually by many. Diseases of humans that are caused by parasitic bacteria include typhoid fever, cholera, tuberculosis, syphilis, and gonorrhea. Viruses are responsible for smallpox, polio, influenza, and the common cold. Amoebic dysentery, African sleeping sickness, and malaria are caused by protozoa. Malaria is still one of the major communicable diseases of the world. Parasitic worms include Chinese liver fluke, blood flukes, and tapeworms. Ringworm, in spite of its name, is a skin disease caused by a fungus, as is athlete's foot.

Plants are not immune to disease; some fungus-caused diseases of crop plants and ornamentals are wheat rust, southern corn leaf blight, corn smut, and Dutch elm disease (Fig. 4.5).

Like predation, parasitism tends to regulate populations. A dense host population provides the opportunity for epidemics that could cause many deaths. In the reduced population the parasites are not so likely to find new hosts, and the chances of a new epidemic occurring are correspondingly low. If the host population increases to its former density, epidemics become more likely again. The two species thus tend to regulate each other's population densities.

The amount of damage done to the host varies a great deal and depends on the resistance of the host and the virulence of the pathogen. If the host species and the pathogen species evolve together for thousands of generations, they may develop a

29

**4.5** A row of dead American elm trees killed by Dutch elm disease. Because the trees were planted close together, the disease could spread quickly from tree to tree. Dutch elm disease is caused by a fungus the spores of which are carried by a beetle that lives under the bark of American elm trees.

mutual tolerance for each other, the hosts managing to survive even when harboring the parasites. It is actually to the benefit of the parasite not to kill its host, for the demise of the host deprives the parasite of food. Unless a new host is found, the parasite is likely to perish also.

When European settlers came to America they brought with them their diseases, among them the so-called children's diseases such as measles and chicken pox. The viruses causing these diseases had lived for many generations in the ancestors of the settlers. For these people these diseases caused discomfort for a few days, but complete recovery usually followed. Smallpox, a disease that was, and still is, greatly dreaded, was also introduced to America by European settlers, some of whom had become immune to it by having had the similar, but milder, cowpox. The Indians and

Eskimos, who had had no previous contact with these diseases, suffered severely from them, and many died. Although there is no doubt that violence and treachery contributed to the defeat of the American Indians, the white man's diseases took their toll also. Sometimes treachery and disease were combined, as when the settlers gave the Indians the blankets of patients ill with smallpox or other diseases.

Plants, too, are likely to be most susceptible to pathogens with which there has been no previous contact. The American elm is suffering today from just such a problem. The fungus that causes Dutch elm disease was accidentally introduced into the United States from Europe about 1930. Since then it has spread throughout the northeastern United States, killing American elms slowly but inevitably. No cure and no effective methods of prevention are known.

Many parasitic animals have much-reduced digestive systems or none; this is especially true of parasites that live within their hosts. They have little need for any, for they usually absorb food that the host has already digested. Organs of locomotion also are reduced or absent, for parasites often remain on or within their hosts and do not move from place to place. (A few parasites, like fleas, are exceptions.) There is the problem, however, of the transfer of a new generation of parasites to a new host. Many leave the host through natural body openings. Parasites of the respiratory tract are sneezed or coughed into the air where they may be inhaled by another animal. Parasites of the digestive system are passed in the feces; they contaminate irrigation or drinking water and may reach new hosts in their food or water. Parasites of the blood can be transferred from host to host by insect bites. Venereal diseases are spread by sexual contact.

The transfer of parasites from host to host is an unsure thing. Any single bacterium that has been sneezed into the air or any single tapeworm egg that has been passed into a lake has relatively little chance of reaching a new host. Compensation for this disadvantage lies in the extraordinary powers of reproduction possessed by most parasites. Some of them seem to be little more than reproductive systems producing hundreds, thousands, or even millions of offspring.

The likelihood of reaching a suitable host is increased through the transport of some parasites by alternate hosts. The malaria organism, for instance, is transported from one human being to another by the *Anopheles* mosquito. During its

**4.6** Life cycle of the malaria organism (*Plasmodium vivax*). This obligate parasite has no means of locomotion but is transferred between its two hosts, man and mosquito, by bites of the *Anopheles* mosquito. Like many internal parasites, it is prolific. The asexual cycle (upper right) produces ten to twenty new merozoites from each dividing merozoite once in each cycle. The cycle requires 48 hours for completion, and toxins are released when the host's red blood cells burst. This accounts for the 48-hour intervals between attacks of fever and chills typical of tertian malaria.

Some merozoites mature into sexual cells, but these do not produce eggs and sperms unless they have been sucked into the digestive system of a mosquito. The resulting zygote burrows into the stomach wall and after encysting there divides into hundreds of sporozoites. These migrate to the salivary glands and are injected into the human bloodstream by another bite by the mosquito. A sporozoite enters a red blood cell, becomes ameboid for a while, and then divides into merozoites.

This parasite thus produces numerous offspring in both of its hosts. This compensates for the fact that many individual cells of the malaria organism fail to be transferred to a suitable host.

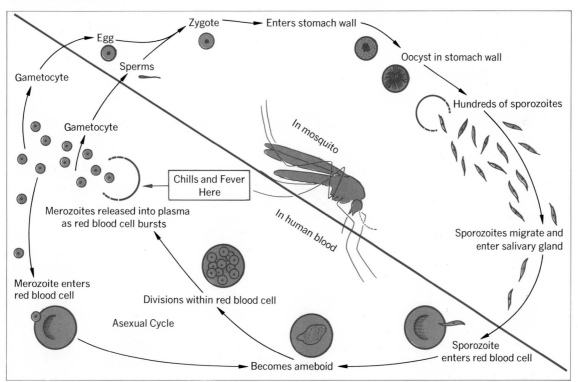

31

stay in the mosquito, the malaria organism goes through several stages in its life cycle, only one of which is capable of infecting a new human being; but this stage develops in the salivary glands of the mosquito, from which it is transferred at the next bite (Fig. 4.6).

The urinary blood fluke, which is common in the Middle East and much of Africa, also has two hosts: men and snails. The adult fluke lives in the veins of the urinary bladder of humans, and the eggs are passed in the urine (Fig. 4.7). In the ab-

sence of modern sewage treatment the eggs survive and hatch into a larval stage called a miracidium. In quiet waters it is likely to find a new host—any of several species of fresh-water snails. In the body of the snail the miracidium undergoes several transformations, eventually emerging as a cercaria, another larval stage. After the cercaria escapes into the water, it enters the human body merely by burrowing through bare skin. Persons wading barefoot in the water are especially likely to become infected; so, too, are village women who do

**4.7** Life cycle of the urinary blood fluke (*Schistosoma haematobium*), which causes bilharzia (schistosomiasis). The adults live in the veins of the urinary bladder of human beings. The female fluke holds the slender male in a special groove, an adaptation that almost ensures he will be available to fertilize her eggs. Fertilized eggs are passed in human urine. In ponds or streams they mature into swimming miracidia that can enter fresh-water snails. In the snail each miracidium matures into a single sporocyst, but each of these produces several second-generation sporocysts. Each of these produces several cercaria. The cercaria leave the snail, swim through water, and burrow into the skin or mucous membranes of humans swimming or wading in the water or drinking it. Reproduction in both hosts produces many offspring, thus ensuring that at least some will find a suitable host. Bilharzia and malaria are the most prevalent infectious diseases in the world today.

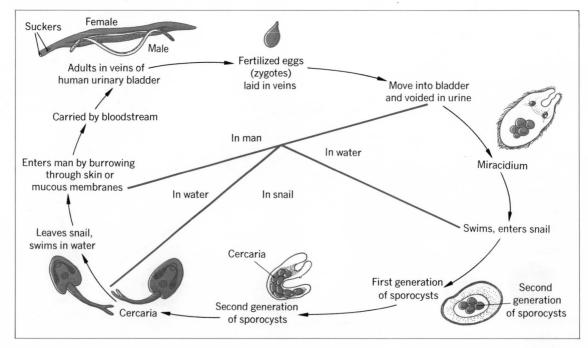

their laundry in quiet pools. Drinking of untreated water is another source of infection, for the cercariae penetrate the lining of the mouth. The cercariae mature into adults that eventually settle in the veins of the urinary bladder. The disease that they cause is known as schistosomiasis or bilharzia; the wall of the bladder is irritated and bleeding results.

Before the advent of modern methods of irrigation, about 5 per cent of the people of Egypt had this disease. The snails, which are necessary for the perpetuation of the fluke's life cycle, grow best in quiet water, and the building of dams across the Nile River has greatly increased their numbers. About 75 per cent of the farmers in the Nile Delta are infected, and the life expectancy has dropped severely. Within a few years of the building of the first Aswan Dam, the incidence of schistosomiasis rose to about 75 per cent in the area it served, and it is assumed that the Aswan High Dam will have similar consequences.

The Chinese liver fluke, which is a serious problem in China, Korea, Japan, and Southeast Asia, has a similar life cycle, but it requires three hosts: man, snails, and fish. The adults live in the human liver, and their eggs are passed in human feces. In many parts of Asia it is customary to use human fecal material as fertilizer in rice paddies, for the farmers are too poor to purchase chemical fertilizers. In the quiet waters of the flooded paddies live the snails that become infected by eating the eggs. Several stages in the life cycle of the liver fluke develop within the body of the snail, and the cercariae escape into the water. A cercaria cannot infect man, but it can infect fish; it bores directly into the body of a fish and becomes encysted in the muscles. Asians often stock their rice paddies with fish, for they provide some protein in their ordinarily meatless diets. It is the custom to eat fish raw; and once inside the human digestive system the still-living, encysted cercaria matures into an adult liver fluke that lodges in the bile ducts of the liver. The symptoms of the disease may be only mild inflammations of the liver, but if many flukes are present, the bile ducts become blocked. The disease is a chronic, slowly progressing one with death often the final outcome.

# Interference with Natural Relationships

This chapter has given only the briefest of surveys of the numerous ways in which plants and animals affect each other. The total of their relationships is an extremely complex web of interactions. Every organism is dependent on others, and any interference with this web is certain to produce some consequences. Sometimes the effects are slight, and it is difficult to notice them. This usually happens with a competing species is available to move in and replace a missing one. In other cases great tragedies come to pass. The case of the fishers and porcupines has been cited already; no other porcupine-eating animal was capable of replacing the slaughtered fishers.

The introduction of an exotic plant or animal into a new area may also have drastic consequences. The introduction of a few prickly pear cactuses into an Australian garden resulted in their spreading over millions of acres of pastureland that could no longer be used for domestic animals. It was only after the introduction of an insect enemy of the cactus that this plant was brought under control (Fig. 4.8). Australia suffered similarly from the introduction of a few rabbits that became the ancestors of millions that were shot, poisoned, trapped, and finally infected with a deadly virus in efforts to keep them from ruining the land. (In this case the proverbial fertility of rabbits had a chance to express itself, for in Australia the rabbits had plenty of food and no natural enemies until the introduction of the virus.)

These are only a few of the examples of tragedies that could have been prevented if man had not interfered with the system of intricate rela-

33

a

b

c

**4.8** The effect of introduction of a few cactus plants into Australia. (a) October 1926: A forest of trees has been invaded by a dense stand of prickly pear cactus. (b) October 1929: The introduction of the parasitic insect *Cactoblastis cactorum* has killed nearly all the cactus. (c) December 1931: After the cutting and burning of the timber, the same area now supports a dense growth of Rhodes grass. Man's activities have changed the area completely. (Courtesy Queensland Department of Primary Industries.)

tionships that required a few billion years to evolve to its present highly developed state. The addition of pollutants to our lakes and streams has had similar results. Where fishermen once cast lines for pleasure or let down their nets for profit, nothing but the so-called trash fish can be found. The fact that the Cuyahoga River in Cleveland is so replete with flammable compounds that it catches fire is of concern only to the humans who still navigate its murky waters, for no fish can live there.

Another matter of great concern is whether we are interfering with interactions among the soil microorganisms on which the fertility of our fields and forests depends (see also Chapter 13). At the present there is little information about what the additions of herbicides and other pesticides may do to soil fungi and bacteria. If mycorrhizal fungi were lost, many of our forests would be lost as well. Even more frightening is the possibility that nitrogen-fixing organisms could be affected adversely. It is conceivable that if nitrogen fixation were to cease, life on the earth would gradually come to a halt.

It is only with caution that we should add new chemicals to the environment. Before they are released in any quantity to water, soil, or air, their effects should be known not to be harmful. Like the wild rabbits of Australia, once released they cannot be recalled easily—if at all. Recent evidence indicates that soil microorganisms react with sulfur dioxide and some other gaseous pollutants and remove them from the air. If this proves to be true, these microorganisms provide yet another valuable service for us.

# Questions

**1.** Define the following terms.
predator
prey
symbiosis
mutualism
commensalism
parasitism
host

**2.** The benefit of predation to a predator is obvious. Of what benefit is it to the species preyed upon? To other species?

**3.** The next time you observe an interaction between two species in your neighborhood, consider what consequences this has on other species.

**4.** Describe some examples of mutualism, and tell how each partner helps the other.

**5.** Describe the unique role of nitrogen fixers in the environment. How do they benefit other organisms?

**6.** What is crop rotation? How does it benefit farmers?

**7.** How is a high population density of a species likely to influence the population density of the species preying on it? The species on which it preys? The species that parasitizes it?

**8.** Answer the previous question after you substitute *low* for *high.*

**9.** Describe the ways that parasites travel from one host to another.

**10.** What is schistosomiasis? Why has its presence in Egypt increased rapidly recently?

**11.** Explain why the eradication of one species in any one area could have serious ecological consequences.

**12.** What is the difference between predation, parasitism, and competition among species? In what ways do they resemble each other?

**13.** What causes species in nature to tend to maintain fairly constant population levels over long periods of time?

**14.** Give some examples of man's activities that resulted in very drastic changes in population levels of certain species.

# Suggested Readings

Ahmadjian, V. "The Fungi of Lichens," *Scientific American* 208(2): 122–132 (February 1963).

_____. *The Lichen Symbiosis.* Waltham, Mass.: Ginn/Blaisdell, 1967.

Alvarado, C. A., and L. J. Bruce-Chwatt, "Malaria," *Scientific American* 206(5): 86–98 (May 1962).

Bell, R. H. V. "A Grazing Ecosystem in the Serengeti," *Scientific American* 225(1): 86–93 (July 1971).

Cheng, T. C. *Symbiosis.* New York: Pegasus, 1970.

Davis, D. E. *Integral Animal Behavior.* New York: Macmillan Publishing Co., Inc., 1966.

Estes, R. "Predators and Scavengers," *Natural History* 76(2): 20–29 (February 1967).

_____. "Predators and Scavengers: Part II," *Natural History* 76(3): 38–47 (March 1967).

Johnsgard, P. A. *Animal Behavior.* Dubuque, Iowa: William C. Brown Company, Publishers, 1967.

Lapage, G. *Animals Parasitic in Man,* rev. ed. New York: Dover Publications, Inc., 1963.

Limbaugh, C. "Cleaning Symbiosis," *Scientific American* 205(2): 42–49 (August 1961).

Maddox, D. M., L. A. Andres, R. D. Hennessey, R. D. Blackburn, and N. R. Spencer. "Insects to Control Alligatorweed," *BioScience* 21: 985–991 (1971).

Odum, E. P. *Ecology.* New York: Holt, Rinehart and Winston, Inc., 1963.

Read, C. P. *Animal Parasitism.* Englewood Cliffs, N.J.: Prentice-Hall, Inc., 1972.

Yoeli, M. "Animal Infections and Human Disease," *Scientific American* 202(5): 161–170 (May 1960).

Zinsser, H. *Rats, Lice, and History.* Boston: Little, Brown and Company, 1935.

35

# 5
# Successions and Communities

1 **When an area previously devoid of life is invaded by living things, a succession of communities occupies the area.**

2 **Each community except the final one changes the environmental conditions sufficiently that it no longer can reproduce itself; another community then succeeds it.**

3 **The final or climax community of a succession reproduces itself indefinitely and is affected only by very great environmental changes.**

A volcano emerges from mid-Atlantic, its hot lava and steam sterilizing every bit of land on the new island. An Alaskan glacier, slowly retreating as a warming climate melts its forward edge, reveals the smooth surface of rock that its advance many years before had scoured clean of all life. The shoreline of Lake Michigan changes its shape as the action of wind and wave moves its tiny particles of sand from place to place. Lava, bare rock, and shifting sands may seem unlikely places to support abundant life; yet, given sufficient time, each of these areas will have a flora and fauna appropriate for its climate. The island Surtsey, born of volcanic eruptions from 1963 to 1965 in the Atlantic Ocean, already is colonized by a few plants and animals brought to it by winds or ocean currents. A thick Sitka spruce forest now grows where a glacier relinquished its icy grip about 300 years ago. Like the other Great Lakes, Lake Michigan once was larger than it is now; and as it shrank, its old shorelines became covered by evergreen forests at its northern end and by deciduous forests (forests with trees that lose their leaves at one time of year, such as autumn) and prairie at its southern end.

## Successions

The arrival in a new area of seeds, spores, and small animals like insects and spiders depends somewhat on winds or ocean currents, but change in the community over the years is anything but fortuitous. A definite, orderly, generally predictable sequence of communities occupies the area. This sequence is called a **succession** or **sere.**

A primary succession is one that begins in an area devoid of life; successions on bare rock and sand dunes are examples. Secondary successions occur where an earlier community has been destroyed or damaged, but some organisms and a considerable amount of organic matter remain;

successions on abandoned farm fields and lumbered forests are examples.

The first community to occupy a new area in a primary succession is a **pioneer community.** It changes the environment enough that a different set of organisms can move in; this changes the environment again, and a third community occupies the area, and so on (Fig. 5.1). These temporary communities are called **seral communities.** Eventually a final community called the **climax community** appears. The climax community reproduces itself and continues to occupy the area indefinitely unless destroyed by events of catastrophic proportions such as violent storms, landslides, farming, or lumbering. If left to themselves thereafter, such disturbed areas undergo secondary succession and eventually reach the climax stage once again.

The constitution of all these communities is affected greatly by the climate and geology of the region. Trees cannot survive the extremely cold winters of northern Canada or the tops of high mountains. There no forests exist; but within these areas, where the summers are mild enough for snow and ice to melt away, a type of low vegetation called tundra exists (see p. 42). A rocky coastline is a firm substrate for large algae (seaweeds) which in turn provide food and shelter for many aquatic animals. A sandy beach, on the other hand, supports relatively little life, for young algal plants can find no support on the sand grains shifted about by waves. Earthquakes or other geological forces can raise or lower land sufficiently to change communities. Excavations under present-day forests or grasslands often expose fossils of ancient marine organisms, indicating that the land once was under water that supported a completely different kind of community.

Because it is the plants that determine to a large extent the types of animals that live in a given area, a community usually is described in terms of its dominant plants, but this should not be construed to mean that animals are not there or are not important to the community.

**5.1** Succession of communities. This pine forest (tall trees) is being replaced by an oak forest (small trees). In time the old pine trees will die, and the oaks will have grown and taken their place.

On the Illinois shores of Lake Michigan, wherever civilization has not disrupted the environment very much, the natural succession leads from open beach to deciduous forest (Plates II and III). On the beach immediately adjacent to the lake no plants grow, for the shifting of sand by waves prevents seedlings from taking root. Beyond the upper limit of wave action several species of beach grasses act as pioneers. Their extensive root systems stabilize the sand somewhat against the action of wind. A sandy beach has little humus, but the death and decay of the grasses contribute a small amount of humus to the sand. Seedlings of cottonwood trees take root among the grasses, and their root systems serve further to stabilize the sand. The decomposition of their fallen leaves and branches and eventually the dead trees themselves add more humus. In time, junipers replace the cottonwoods, and still later a black oak forest replaces the junipers. Eventually the area is covered by a climax forest consisting mostly of oak

37

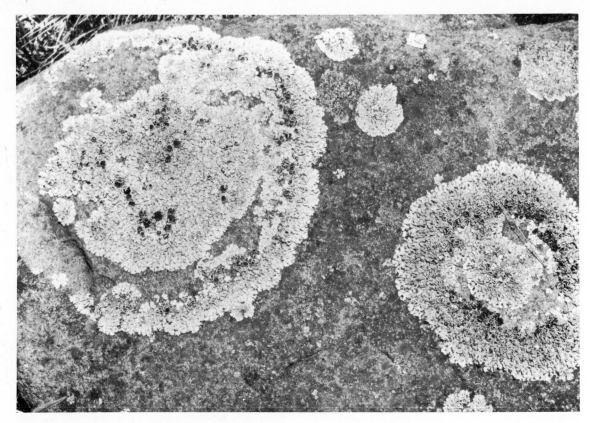

**5.2** Bare rock often is colonized first by lichens, which usually are the only plants hardy enough to live in such an inhospitable environment.

and hickory trees. In the absence of catastrophic damage, the oak–hickory forest perpetuates itself indefinitely. The climax represents a homeostatic condition on the community level.

In much of northern Illinois and southern Wisconsin lichens are the usual pioneers on bare rock outcroppings (Fig. 5.2). These hardy plants are fungi harboring single-celled algae within them; together the two can withstand extremes of heat, cold, and drought that would kill other plants. Dust and soil begin to accumulate among the lichens, and this forms enough of a substrate that mosses can grow there. The mosses continue the disintegration of the rock, and they form a somewhat thicker mat than the lichens. They hold more

moisture and retain soil particles in place as they grow, and later ferns and small shrubs take root (Fig. 5.3). The growth and death of these plants increase the depth and water-holding capacity of the soil and permit the rooting of tree seedlings. A succession of trees then may occupy the area, but finally an oak–hickory climax develops. By this time the soil is relatively deep. Depending on the size and arrangement of the original rocks, they may be completely hidden under the soil and a covering layer of decaying leaves and other organic material.

A pond succession in the same area begins with open water occupied by algae and other submerged plants. Gradually these are replaced by

38

**I.A.**

**I.B.**

**I.C.**

**I.D.**

**I.E.**

**I.F.**

**Plate I.** Each of these photographs shows at least two species of organisms. Can you recognize them even though you may not know them by name? Some are not obvious at a casual glance. (For answers see p. 447.)

**Plates II and III.** Sand dunes succession in northeastern Illinois.

**II.A.** Lake Michigan and storm beach. Wave action on the storm beach prevents plants from taking root here, but very small animals may burrow in the sand.

**II.B.** Pioneer community consists of beach grasses. Their roots help to stabilize the sand.

**II.C.** Cottonwood trees follow the beach grasses.

**III.A.** Junipers follow the cottonwoods. Junipers usually are erect trees, but the variety here is a prostrate form.

**III.B.** Black oak woods follows the junipers.

**III.C.** The climax community is an oak-hickory forest; the oak trees are mostly white oak and northern red oak.

**Plates IV and V.** Some climax communities of North America.

**IV.A.** Alpine tundra in the Rocky Mountains. The vegetation is low, consisting of lichens with an admixture of low herbs. (Courtesy of Dr. Elizabeth Lunn.)

**IV.B.** Coniferous forest in Ontario, Canada. Nearly all the trees are spruce.

**IV.C.** Deciduous forest in northern Illinois. This oak–hickory forest has mostly broad-leaved trees. In addition to oak and hickory, this forest has sugar maple, ash, and other trees and shrubs.

**V.A.** Deciduous forest in south-central Wisconsin. This is primarily a sugar maple–basswood forest with some oaks. Deciduous forest of eastern United States usually display bright colors in autumn before the leaves fall.

**V.B.** Grassland in Illinois. Prairie is the climax community in much of the area between the Mississippi River and the Rocky Mountains, but isolated patches of prairie exist as far east as Ohio. Prairies do not receive enough rainfall to support forests, but along the edges of streams and ponds trees grow.

**V.C.** Desert in Nevada. The spacing between the plants is an adaptation to a dry habit. (Courtesy of Dr. Shirley Tucker, Louisiana State University.)

**A.** The effect of water on the distribution of life. Except where water is locally available, Egypt is a desert. The Nile River flows to the right of the picture. Notice the sharp line of demarcation between the relatively moist soil, which supports vegetation, and the drier soil. which supports few or no plants.

**B.** The ruins of Jerash. This ancient town once was part of the civilization that thrived in the Fertile Crescent. Today, because of poor use of the land, this area has become a dry desert with only the sparsest of vegetation.

**PLATE VII.** Pollination.

**A.** Portion of a corn tassel. Corn is wind-pollinated, and its anthers dangle freely in the breeze.

**B.** Young ear of corn. The silks are the styles and are covered by stigmatic surfaces. Exposed to the breeze, they receive wind-borne pollen.

**C.** Orchid flower. This large, showy flower is insect-pollinated.

**D.** Butterfly weed. Although the individual flowers are small, they are massed into brightly colored groups that attract insects.

**PLATE VIII.** Dissemination of seeds and fruits.

**A.** Winged fruits of maple.

**B.** Milkweed seeds with hairs.

**C.** Cattail fruits with hairs.

Wind dissemination.

Animal dissemination.

**D.** Beggar-ticks fruits with barbs.

**E.** Burdock fruits with hooks.

**F.** Red raspberry, a juicy, fleshy fruit.

plants such as water lilies with their roots anchored in the mud bottom and their leaves floating on the water. As these plants die, their remains drop to the bottom of the pond, thereby raising the level of the bottom. As the water becomes shallower, cattails replace the water lilies. Gradually, the accumulated debris rises above water level, and grasses and sedges replace the cattails. These are followed by low shrubs and then one or two seral stages of trees. Finally an oak–hickory climax on dry land occupies the site of the original pond.

Successions with different origins follow different seral stages but in a given area the climax stage is the same. In Massachusetts bare rock, open ponds, and sandy beaches develop toward beech–maple forests which are the climax community there, whereas forests of coniferous (cone-bearing) trees are the climax of succession in much of Maine, and grasslands are the final stage between the Mississippi River and the Rocky Mountains.

A pioneer community consists of relatively few species. It is not very stable and is susceptible to disruption by such agents as storms, drought, and overgrazing by animals. With each succeeding seral stage, the number of species usually increases, and the total volume of living material increases. Because of greater possibilities of interactions among their more numerous members, the communities become more and more stable as the succession advances (see also Chapter 14). The total amount of humus (partially decomposed organic material) in the soil also increases with each seral stage, and this contributes to stability by holding much of the water that the community receives during rains and retaining it as a reserve that can be used during periods of meager rainfall. The humus also holds a store of minerals readily available to the roots of the plants. The stability of the climax prevents it from being very much disturbed by a few weeks or even a few months of unusually hot or cold weather, drought, or flood. Unlike a corn field or rose garden (which in their simplicity resemble pioneer communities), which

**5.3** A mat of moss covering a rock provides a suitable environment for the growth of young shrubs or trees.

can be severely damaged by the several weeks of heat and drought that are not uncommon in Temperate Zone summers, a climax community survives year after year without benefit of artificial irrigation. Its seeds need not be taken indoors to survive the winter, nor do its plants require any but the natural mulching provided by the dropping of their own leaves. It requires no weeding to prevent its members from being choked out by undesirable plants.

39

**5.4** Where the terrain is flat or nearly so, the succession has reached the forest stage. Where the slope is steep, bare rock still is visible.

However, because the nature of the climax community depends on the climate and the geology of the region, any change in these factors brings about corresponding alterations in the community, and a new climax community composed of different species becomes characteristic of the area. Climatic changes and many geological changes ordinarily are so slow that they are not noticeable during a human lifetime, but occasionally violent events like earthquakes can alter the terrain drastically in only a few minutes.

Primary successions usually require at least several decades and often centuries to run to completion. Generally more time is required in cold climates than in warm climates. Succession usually proceeds faster on flat ground than on steep hillsides; where the slope is very steep there is almost no opportunity for accumulation of soil, and so succession is correspondingly slow (Fig. 5.4).

In some places it is possible to see at one time all the stages of a given succession from pioneer stage to climax. For example, pond succession begins near the edges of the pond, and so the edges fill in first, the center last. At any one time, differ-

**5.5** Several stages in succession. This pond has some open water. Water lilies are adjacent to the open water, and cattails grow beyond them. Trees can be seen in the distance.

ent seral stages are likely to occupy different distances from the center of the pond, the most advanced stage being the most distant from the center. If there is still open water, the observer can see concentric rings of vegetation corresponding to the seral stages. As he proceeds from open water to the shore and then to the climax community, he sees within a few hundred feet all the stages that occurred over a period of a few hundred years where the climax forest now stands (Fig. 5.5). Were his life long enough, he could watch the center of the pond fill up and become a climax community.

Secondary successions (Fig. 5.6) ordinarily require less time for completion than do primary successions in the same general area, for soil already has been formed. An abandoned orchard or burned prairie, for instance, begins succession at a stage comparable to one of the intermediate seral stages of a primary succession rather than starting with a pioneer community.

Successions can be hastened artificially. Addition of nutrients to a lake, called **eutrophication,** speeds the succession, or, in common parlance, "ages" the lake. It is the natural fate of lakes to become filled in as a result of succession and so "die." In the case of a large lake the process is a long, slow one requiring under natural conditions thousands of years for completion. In areas of high population density, artificial eutrophication of several lakes by the addition of sewage, garbage, and runoff from chemically fertilized fields has advanced succession at an alarming rate. Lake Erie has aged so much in recent years that it is considered to be dying.

It is ironic that by reason of its beautiful clear waters, a lovely lake nestled in woods or mountains attracts to its shores vacationers whose sewage hastens eutrophication of the lake and decreases its desirability. Lake Como in Italy is only one example. The threat to Lake Tahoe on the California–Nevada border seems to have diminished with the construction of a modern sewage disposal plant.

a

**5.6** Two examples of secondary succession. (a) This area once was a lawn but then was abandoned; it is in its fourth year of unhampered growth. (b) A tennis court abandoned for several years.

b

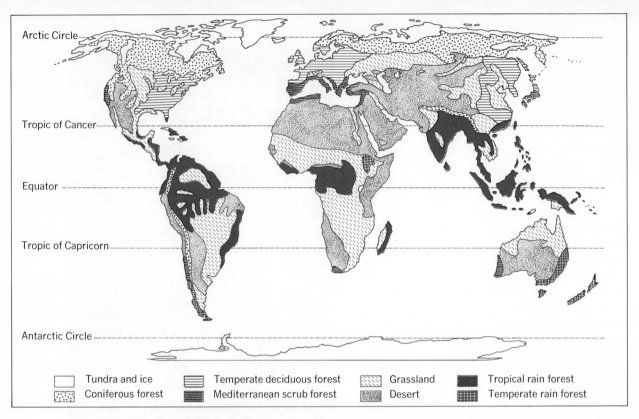

**5.7** Map of the natural climax communities of the world. In some areas the natural vegetation has been destroyed by man. This is especially true of temperate deciduous forests and grasslands.

Legend:
- Tundra and ice
- Coniferous forest
- Temperate deciduous forest
- Mediterranean scrub forest
- Grassland
- Desert
- Tropical rain forest
- Temperate rain forest

## Climax Communities

Climax communities on land are of four different major vegetation types: tundra, forest, grassland, and desert (Plates IV and V). Each is further subdivided (Fig. 5.7).

**Tundra** occurs in extremely cold climates where the subsoil is frozen the year around in a condition called permafrost. The upper layer of soil freezes in the winter, but it thaws during the short summer. Partly because of the extremely cold weather and partly because of the shallowness of the thawed soil, which often does not exceed a foot in depth, the vegetation never grows very tall. In the driest areas it consists primarily of lichens, grasses, and sedges; stunted willow trees may reach heights of only 2 or 3 inches. In wet tundra, shrubs get to be 2 or 3 feet high. Some tundra animals, like many species of birds, escape the rigors of winter by migrating to warmer areas. Polar bears, foxes, and other mammals that remain during the winter develop thick fur as a protection against the cold.

Arctic tundra extends across the northernmost parts of Alaska, Canada, Scandinavia, and Siberia. Alpine tundra occurs at high altitudes on mountains such as the Rocky Mountains of North America, the Andes Mountains of South America, the Alps in Europe, and the Himalayas in Asia. The altitude for tundra varies with latitude; at the equator it occurs only near the tops of the highest mountains, where the weather is cold throughout most of the year, but in Arctic regions it is found at sea level as well as in the mountains.

**Forests** grow in areas of the Temperate Zone and the tropics where rainfall exceeds about 30 inches a year. Several types of forest occur in different parts of the world, but three types—coniferous, deciduous, and tropical rain forest—account for most of the acreage in timber.

Most coniferous trees bear their seeds in more or less woody cones; they produce no flowers or fruits. They include pine, spruce, fir, and hemlock, all of which are popularly called pines by persons not familiar with them. Most conifers are evergreen, retaining their needle-shaped leaves throughout the winter—a characteristic that makes some of them favorites as Christmas trees. Coniferous forests occupy a broad band extending across Alaska, Canada, and Eurasia just south of the tundra. They also grow on mountainsides below the altitude occupied by tundra. Frequently for many square miles there are only one or two dominant species of tree.

Coniferous forests contain valuable timber, but because of their distance from population centers, much of the coniferous forests of Canada and Siberia remains untouched. However, in the western United States and in other places in which they are more easily reached, coniferous forests are falling. In their haste to harvest tree crops, lumbermen have taken to clear cutting; all trees regardless of size are removed from an area. On hillsides the inevitable result is erosion; previously clear streams become polluted with soil washing into them. As the hillsides lose their soil, so they lose their ability to retain water, and thereafter stream flow becomes more irregular, flooding more severely after heavy rains and drying up after a drought. Persons living downstream then become subject to the vagaries of the stream. We are learning that the value of a forest lies not only in its timber.

Deciduous forests consist of broad-leaved trees that drop their leaves in autumn and produce new ones the next spring. Usually a given woods contains two or three dominant tree species such as beech and maple in the northeastern United States; oak and hickory in western Ohio, Indiana, Illinois, and Missouri; and oak, chestnut (now nearly extinct), and tulip tree in Tennessee, Kentucky, and West Virginia. In addition to the dominant trees, several other species of trees are present in most deciduous forests.

Deciduous forests once covered much of the eastern United States, Europe, Japan, and China, but a great deal of these have been cut down with the advance of civilization. One huge forest is said to have covered the entire eastern United States from the Atlantic coast to the Mississippi River before the coming of Europeans; except in mountainous areas, this forest now is almost completely replaced by farms and cities. Similar damage was experienced earlier by the deciduous forests of Europe and Asia. In the most densely populated areas of Asia, even the hillsides were denuded of their trees and then terraced to grow crops.

Tropical rain forests, popularly called jungles, consist of broad-leaved, evergreen trees. They lie on either side of the equator: in the Amazon River Valley, Central America, central Africa, southeastern Asia, and the East Indies. Compared with temperate forests, their variety is amazing; no species or group of species dominates an area. In some cases as many as a hundred species have been found in a single square mile. Rubber, mahogany, Brazil nuts, and cacao (from which chocolate is made) are only a few of the commercially important products of these forests.

High temperatures and an annual rainfall that often exceeds 100 inches give tropical rain forests a hothouse quality and permit rapid growth of plants. They also allow rapid decomposition of dead organisms. Despite their lush appearance, tropical rain forests of South America and Africa do not hold realistic promise of the same intensive agriculture that now is carried out in areas for-

43

merly occupied by temperate deciduous forests. Many of these rain forests grow on lateritic soils, which have the unhappy property of hardening to a permanent bricklike consistency if the forests are cleared.

**Grasslands** are the climax vegetation where the annual rainfall is from about 10 to 30 inches, the tallest grasses growing where the rain is most abundant. The world's two largest temperate grasslands lie in the North Temperate Zone. One stretches from eastern Europe across southern Russia, Iran, Afghanistan, and into China. The other occupies the great plains of the United States and southern Canada. Smaller areas of grassland occur in South America, Australia, and Africa.

Grasslands are highly fertile; and the tall-grass prairies, if properly tended, are excellent for raising corn, wheat, and other cereals, which are themselves tall grasses. The short-grass prairies are best adapted to grazing; attempts to farm them nearly always result in disasters such as the famous Dust Bowl of the 1930's, when tons of topsoil deprived of its natural cover dried and blew away in tremendous dust storms. Soil from Kansas and Oklahoma is said to have been carried to the Atlantic seaboard.

A savannah is a type of grassland with occasional trees that give it a parklike aspect. Savannahs are primarily tropical; they occupy much of Africa south of the Sahara Desert and the southern part of Brazil.

**Deserts** exist in those parts of the world in which the annual rainfall is less than about 10 inches. Most of the world's large deserts lie near 30° north latitude and 30° south latitude. The Sahara Desert, the Arabian Desert, and the Gobi Desert form a large discontinuous strip of desert across Africa and Asia. In the United States much of Arizona, Nevada, and Utah and parts of New Mexico, California, Idaho, Oregon, and Montana are desert. Central Australia, the western coast of South America, and southern Africa have smaller deserts.

Many, but not all, deserts are hot by day, and

44

so what little water may be present soon evaporates. Characteristically desert vegetation is sparse, and the plants have water-conserving properties: thick cuticles (waxy layer secreted by the outermost cells), small leaves, or thick fleshy stems that store water. Desert animals frequently retire to burrows or rest in the shade during the hottest hours of the day and become active only after the sun goes down. The kangaroo rat conserves water by secreting highly concentrated urine; it does not even require liquid water to drink, but can survive on the water produced within its tissues by the metabolism of its food. The camel, on the other hand, requires water, but within a few minutes it can swallow enough water for its needs for several days.

Not all communities occupy dry land. Lakes and streams have their own inhabitants, as do the oceans. Some of the most fertile natural communities are **estuaries;** these are bays or wide river mouths leading into oceans. Because of their proximity to the ocean they are subject to tides which keep in constant circulation the minerals and other substances brought down to them by their rivers. Many estuaries support large stands of cattails and other emergent plants that provide shelter for a wide variety of fish, oysters, shrimp, and other seafood as well as waterfowl. Some animals merely begin their lives in the relative calm of the estuaries and venture out to the open sea after reaching maturity. Despite their great fertility, estuaries usually have not been harvested directly by man and so have not been valued greatly. Rather, many of them have become convenient dumping grounds for sewage, garbage, and rubbish. Even if not so mistreated, many estuaries are the recipients of similar materials dumped into the streams that empty into them, and marine life quickly smothers under the offensive burden. Sewage-contaminated seafood also presents the danger of epidemics, especially of typhoid fever and other water-borne diseases.

On the other hand, hindering the natural flow of nutrients in streams can be just as disastrous to aquatic communities downstream. Each stream carries with it minerals and organic compounds in solution, and larger particles of clay, sand, and gravel are moved along by the force of the water's flow. The rate of movement of solid particles depends on their size and on the rate of flow of the water. Heavier particles require faster flow to be moved. When a river is dammed, a lake forms behind the dam, and because the lake is wider and deeper than the original stream, movement of water within the lake becomes slower. Gravel, sand, and clay particles settle to the bottom and accumulate behind the dam. So do any minerals adhering to the surface of the clay. The water passing through the dam thus is poorer in nutrients than the water that previously had flowed freely. The Aswan High Dam recently built in Egypt is one of the most outstanding examples.

From the beginning of history, Egypt has depended on the yearly floods of the Nile River which brought silt-laden waters rich in nutrients to the fields along its shores. When the floods subsided, the fields had been fertilized without much effort required by the farmer. As the Nile emptied into the Mediterranean Sea, its waters also fertilized those of the sea, and through the food chain there supported an abundance of fish.

A glance at a map of Egypt shows it to be a country squarish in shape, but if we consider only the populated areas, Egypt becomes an extremely long, thin country following the length of the Nile River and extending only a mile or so in either direction from its banks and then following the shores of the Mediterranean Sea and the Red Sea. The width of the inhabited region along the Nile depends on how far from the river its waters soak into the soil and provide enough moisture and minerals for agriculture.

Throughout much of history, Egyptians have attempted to increase their agricultural acreage by irrigation and to control by dams the irregularities of the floods, which failed in some years and were too heavy in others. Until recent years, efforts to control the Nile had relatively little effect on the Mediterranean Sea, but in the 1960's the Aswan High Dam was built. Behind it is forming Lake Nasser, which accumulates silt and minerals behind the dam. The Mediterranean Sea, deprived of the minerals, now is unable to support its former abundance of life, and the fishing industry has suffered correspondingly. The 18,000-ton annual sardine catch in the eastern Mediterranean dwindled to virtually nothing after completion of the dam.

The damage done by this project might be justified if its gains exceeded its losses. But the increased acreage in agriculture has not been able to keep pace with Egypt's population growth, and the amount of food raised per person is less than it was before. The money and effort might have been better spent in a campaign to reduce the country's rapid population growth.

Obviously the stability of even climax communities is not immune from the effects of the hand of man. His activities have been damaging to natural communities since he changed from hunter and fisher to farmer and manufacturer. At one time his imprint on the face of the earth made itself felt slowly, with hundreds of years often being required to change drastically the nature of communities over large areas.

Today through his technology man has become a veritable geological force, changing the course of rivers and removing even the mountains. Some effects have been intentional; others, like the destruction of life in the Mediterranean Sea, are the by-products of projects with other goals. Regardless of intention or lack of it, these activities disrupt successions and destroy climaxes.

We would do well to examine our values to determine if we really want all these things to happen. In a society in which doctors, nurses, and all manner of life-saving equipment and medicines are employed for weeks, months, or even years to maintain the life of a single sick human being, it seems inconsistent to write off merely as the cost of doing business the destruction of precious natu-

45

ral resources that someday may be needed desperately for the survival of many.

## Questions

1. Define the following terms.

| | |
|---|---|
| succession | climax community |
| pioneer community | eutrophication |
| seral community | |

2. How does a primary succession differ from a secondary succession?

3. Why are climax communities stable whereas pioneer and seral communities are not?

4. What stage in a succession does a wheat field most resemble? Explain.

5. Why do the plants in a vegetable garden require tending whereas those in a vacant lot do not?

6. Why are two successions beginning on bare rock and in an open pond in Iowa expected to reach the same climax community, whereas two successions beginning on bare rock in Iowa and bare rock in northern Alaska are expected to reach different climax communities?

7. Lakes are said to grow old and die. How does the death of a lake fit into a natural succession? Why then does the impending death of Lake Erie cause so much concern?

8. Describe briefly tundra, forest, grassland, desert. On a world map locate the major expanses of each of these vegetation types. What factors determine which of these occupies a region?

9. What is the dominant vegetation in your area? In what ways has it been altered by man?

10. Why is raising corn likely to be much more successful in Iowa and Illinois than in western Kansas and Oklahoma?

## Suggested Readings

Amos, W. H. "The Life of a Sand Dune," *Scientific American* 201(1): 91–100 (July 1959).

Cloudsley-Thompson, J. L. *Desert Life*. Oxford: Pergamon Press, 1965.

Costello, D. F. *The Prairie World*. New York: Thomas Y. Crowell Company, 1969.

Deevey, E. S., Jr., "Bogs," *Scientific American* 199(4): 115–122 (October 1958).

Gómez-Pompa, A., C. Vázquez-Yanes, and S. Guevara. "The Tropical Rain Forest: A Nonrenewable Resource," *Science* 177: 762–765 (1972).

Neill, W. T. *The Geography of Life*. New York: Columbia University Press, 1969.

Odum, E. P. *Ecology*. New York: Holt, Rinehart and Winston, Inc., 1963.

Powers, C. F., and A. Robertson. "The Aging Great Lakes," *Scientific American* 215(5): 94–104 (November 1966).

Storer, J. H. *The Web of Life*. New York: The Devin-Adair Co., 1953.

Swan, L. W. "The Ecology of the High Himalayas," *Scientific American* 205(4): 68–78 (October 1961).

Went, F. W., and N. Stark. "Mycorrhiza," *BioScience* 18: 1035–1039 (1968).

Whittaker, R. H. *Communities and Ecosystems*. New York: Macmillan Publishing Co., Inc., 1970.

All organisms live in a physical, nonliving environment that affects them and that they affect. They exist in water, soil, or air, each of which contains a variety of chemical compounds. Light and temperature vary with the seasons and times of day. Winds, water currents, rain, snow, and fire have their effects on organisms. The distribution of organisms depends in part on the distribution of the materials that they require and the conditions that they can or cannot tolerate.

## Water

Water is universally required by organisms. Most species of organisms live in water, and all carry a supply within their bodies. Water comprises up to 99 per cent of living things, most active organisms having a water content of about 60 to 90 per cent. Dormant seeds and spores may have a water content as low as 10 per cent. The amount of water available in any given area has a great influence on the number of kinds of organisms as well as the number of individuals that can live there (Plate VI). The unusual properties of water may have made life possible.

At temperatures ordinarily encountered at or near the surface of the earth, nearly all substances exist as solids or gases. Water is the only naturally occurring compound found in the liquid state at temperatures compatible with life, and it is the only compound that is likely to be found in any appreciable quantity in all the three physical states: solid (ice), liquid (water), and gas (water vapor and steam). Water freezes at 0°C (32°F), a temperature that kills some tropical organisms but which many living things can survive. If kept in a closed container, water is converted to steam at its boiling point of 100°C (212°F). At this temperature and even at somewhat lower temperatures, many organisms are killed, but a significant num-

# 6

# The Physical Environment

1 **Organisms are affected by the physical factors in their environment.**

2 **Organisms change their physical environment.**

47

ber of bacteria can survive these high temperatures.

Water is the only substance available to serve as a solvent in which minerals, oxygen, and some foods can move through soil, oceans, lakes, and streams. Water is also the solvent in which many waste products are transported away from organisms. Within the bodies of plants and animals, water is the major solvent: In the human body food and oxygen are transported in blood; wastes are carried by blood and urine; and in the uterus, a baby develops within its own small private ocean, the amniotic fluid.

Chemical reactions occur most rapidly in solution. Not only do most of the chemical reactions in living things occur in aqueous solutions, but water is one of the reactants in many chemical reactions (photosynthesis, Chapter 11, and digestion, Chapter 8) and the product of others (aerobic respiration, Chapter 12). As a solvent, water is important in osmosis and in the maintenance of the turgidity of living things (Chapter 18).

It is a general rule that as compounds become warmer, they expand, and their densities decrease. As they become colder, they contract, and their densities increase. For water this generality is only partially true. Above the temperature of 4°C (39.2°F), water follows the general rule, but as it is cooled below this point it expands and its density decreases. Thus water has its greatest density in its liquid state (at 4°C) rather than in its solid state as do most other substances. This characteristic causes a body of water to freeze at the top first, and it keeps ice floating on liquid water rather than sinking to the bottom. A lake freezes from top to bottom rather than from bottom to top. Small bodies of water commonly freeze solidly from top to bottom in extremely cold weather, but in large lakes and in oceans a thick layer of ice on the surface acts as an insulating blanket that retards further cooling of the waters below. The temperature at the bottoms of large, ice-covered bodies of water frequently is about 4°C. At this temperature many aquatic organisms can survive.

Some, like fish, pursue active lives in the cold water; others, like frogs, go into a dormant state but are not killed by the cold temperature. Some aquatic plants survive as spores, seeds, or other dormant structures that spend the winter under water.

If bodies of water froze from bottom to top, many more of them would freeze solidly each winter. The temperature of the ice would drop below 0°C in areas with severe winters, and the number of aquatic species surviving a winter would certainly be much lower than it is now.

When heat is applied to a substance, its temperature rises, but the amount of heat necessary to raise the temperature a certain amount is not the same for all substances. The amount of heat necessary to raise 1 gram (about 0.035 ounces) of a substance 1°C is its **specific heat.** The specific heat of water is much higher than the specific heat of many common substances, and this has great advantages for living things. One calorie[1] is required to raise the temperature of 1 gram of water 1°C, while the specific heat of sand is 0.2 calorie per degree per gram. To state it another way, a given amount of heat will raise the temperature of a given quantity of sand five times as much as it will the same quantity of water. One need only walk barefoot on a beach some sunny day to become aware of this. The dry sand exposed to direct sunshine is uncomfortably hot, but the wet sand and the water are pleasantly cool even when they are directly illuminated. Its high specific heat protects water and living things from excessive heating in sunshine.

Similar effects are seen in the amount of heat required to convert ice to liquid water. When ice has been brought from a lower temperature just up to 0°C, 80 calories of heat must be applied just to melt 1 gram of ice to water. The temperature

[1]A calorie is a unit of energy. It is defined as the quantity of heat required to raise the temperature of 1 gram of water from 14.5°C to 15.5°C.

of the water has not been changed, the 80 calories of heat per gram are needed just for the change of state. The quantity of heat required to melt a substance is called its **latent heat of fusion.**

Winters of heavy snow are often followed by springtime floods as the snow and ice melt into large quantities of water that many river beds cannot accommodate. Persons living in the valleys of the Mississippi River and some other large rivers have this annual worry, but the problem would be even greater if water had a low latent heat of fusion like ethyl alcohol (25 calories per gram) or mercury (3 calories per gram), and the snow and ice melted several times faster than they actually do.

One takes advantage of the high latent heat of fusion of water when he uses ice cubes to chill a drink. The beverage cools not only because it is in contact with cold ice, but because the melting of the ice withdraws heat from the liquid.

The latent heat of fusion absorbed by water is released when it freezes again. The freezing of lakes in early winter moderates the temperature in the immediate vicinity.

To convert liquid water to steam requires an even greater amount of heat. The **latent heat of vaporization** is the amount of heat required to change 1 gram of a substance from the liquid to the gaseous (or vapor) state. For water at 100°C, this is 540 calories per gram; again, these calories do not warm the water but merely change its physical state. The evaporation of water to water vapor at lower temperatures requires slightly more calories. By comparison, ethyl alcohol has a latent heat of vaporization of only 204 calories per gram and mercury requires only 71 calories per gram to vaporize. This means that as water evaporates, it absorbs tremendous quantities of heat from its environment, thus cooling it. This is why the summer air is cooler near an ocean or a large lake than it is in the center of a continent (at low altitudes). This also is why a desert is so hot in the daytime, for there is not enough water to bring about appreciable cooling by evaporation. The evaporation of perspiration on the skin has a cooling effect in the dry desert air, although excessive evaporation of this water can dehydrate a person and cause prostration or even death. If water had a lower latent heat of vaporization than it does, dehydration and resulting death would occur sooner and more frequently than they do. Perspiration has little cooling effect in humid weather, for then the rate of evaporation is greatly reduced.

The heat absorbed by water when it evaporates is released when the water vapor or steam condenses to liquid water again. This phenomenon keeps the regions near oceans and large lakes warmer in the winter than it is farther inland. A desert usually is very cold at night, for in the low humidity there is little water vapor that might condense and release heat to the environment. Oceanic islands have particularly constant temperatures; their yearly temperature extremes do not approach the daily extremes in the continental deserts where the daytime temperature may go over 40°C (104°F) and the night temperatures approach freezing.

The **capillarity** of water causes it to adhere to many substances, including rock, sand, and clay. This property causes water to move upward in small spaces as it does in a fine soda straw. The smaller the diameter of the space, the higher will the water move. In soil, where the spaces between the particles are very fine, the water rises above the water table and comes into regions where it can be absorbed by the roots of plants growing on the surface. Soils with a high clay content permit a higher rise of water by capillarity than does sand, for clay has exceedingly fine spaces between its particles.

Water is not an unmixed blessing. Most floods cause erosion and leach valuable minerals from the soil. A heavy snow breaks branches of trees, and a hailstorm in summer damages leaves and young twigs. If water freezes within living tissues, it forms crystals which dehydrate protoplasm by reducing the amount of liquid water available to it.

49

# Light

**Light** from the sun, moon, and stars shines on nearly all organisms. Exceptions are those organisms that live their entire lives in caves or in the depths of the oceans, where light from the celestial bodies does not penetrate. But even here some organisms produce their own light.

With light nearly universally present it is not surprising that organisms have evolved many different responses to it. Because these are so varied and because some of them deserve detailed discussion, some are considered in greater detail in other chapters of this book, but they are briefly mentioned here.

Photosynthesis (Chapter 11) is the manufacture of food by plants containing chlorophyll. The energy necessary for this process comes from visible light reaching the earth from the sun. Photosynthesis is the main process that "runs" the biological world. It has the unique function of converting radiant energy of sunlight into the energy that is stored in foods and that all living things require.

Many animals have developed eyes, organs of vision (Chapter 37) that are sensitive to light and that are capable of focusing an image. Animals vary in the acuteness of their vision. Humans have better vision than many animals; dogs and bears, for example, rely more on their sense of smell than their sense of sight. Vision is exceptionally acute in hunting birds such as eagles and hawks, which can detect mice from high altitudes, and owls, which hunt at night and can see in light of only 1/100 the intensity required by humans. Most animals do not have color vision as humans do but see the world much like a black and white movie. Some insects have color vision that can be used in identification of flowers from which they collect nectar or pollen. Insect vision differs in another way from that of most animals. The insect eye is a compound eye that registers many images of the object seen rather than a single one. The possible confusion that might result from this arrangement is offset by the increased likelihood that even a small movement in the environment will be detected; this increases the chances of escape from danger.

Some organisms, including some flatworms and oysters, have eyes that do not form images but are sensitive only to the intensity of light. Several single-celled algae and protozoa have light-sensitive organelles called eyespots; these, too, are sensitive only to light intensity.

Phototropisms (Chapter 29), the growth movements of plant parts toward or away from a light source, are adaptations that place leaves where they are most likely to receive the light they require for photosynthesis. Roots, which do not require light, either grow away from it or do not respond to it.

Phototaxes, the movements of entire organisms toward or away from a light source, have similar functions and place the organisms at their optimum light intensities. Phototaxes are common among the algae and protozoa that have eyespots.

Photoperiodism (Chapters 44 and 45) is the response of plants and animals to day length. Daily cycles of sleeping, wakefulness, exercise, eating, and discharge of waste products are adjusted to the alternate periods of daily light and darkness. Reproduction of many species coincides with a particular time of year because reproductive processes are initiated only by certain day lengths critical for the species. Migrations of some animals also is initiated by appropriate day lengths.

A few organisms—fireflies and a few bacteria and fungi—have the ability to produce their own light. In many cases the function of this light is not known. In fireflies the lights are recognition signals between the sexes. Some deep-sea fish live in a symbiotic relationship with some luminescent bacteria that grow in special pockets under the skin of the fish. Some of the fish have flaps of tissue that they can raise or lower over these pockets, thus turning their lights "on" and "off." These lights may serve as searchlights in the darkness of the ocean depths, or, like firefly lights, they may be recognition signals.

## Radiations Other than Light

Visible light is only one of several radiations that reach us from the sun and other natural sources. **Invisible radiations** include gamma rays, X-rays, ultraviolet light, infrared (heat) radiations, and radio waves. All of these (including visible light) are called electromagnetic radiations, and each has its own characteristic wavelengths (Fig. 6.1). The wavelengths of visible light range from about 400 to 700 nanometers (a nanometer is one-billionth of a meter). Gamma rays, X-rays, and ultraviolet light have wavelengths shorter than those of visible light; and infrared radiations and radio waves have longer wavelengths.

The shorter its wavelength, the more energy is carried by the radiation. In general, the more energy carried by the radiation, the more damage it can do to living cells. For this reason unnecessary

exposure to ultraviolet light, X-rays, and gamma rays should be avoided. These radiations can cause mutations (Chapter 46), changes in the genetic material that may be inherited from generation to generation. Occasionally a mutation is beneficial, but most of them by far are harmful. Prolonged exposure to these radiations also causes the development of both benign and malignant tumors. Radiations from the sun consist, in part, of these invisible radiations, but because most of them are filtered out by the gases in the earth's atmosphere, life proceeds on the surface of the earth and in the oceans with relative safety. Farmers and others who spend most of their days in sunshine are more likely to develop cancer of the skin than are other people.

The use of X-rays in medical diagnosis or treatment is justified because the benefits outweigh the dangers. Care must be taken, of course, not to expose the patient to any more X-radiation than is necessary.

Radiations from the decay of radioactive compounds consist of several types of short-wave radiations that are discussed in Chapter 14, but at this point it should be pointed out that many of them cause mutations and cancer, too. There is

**6.1** Electromagnetic spectrum. Note that the scale used here is a logarithmic one, each segment having a value ten times larger than the one to the left of it. Visible light accounts for only a small portion of the total electromagnetic spectrum.

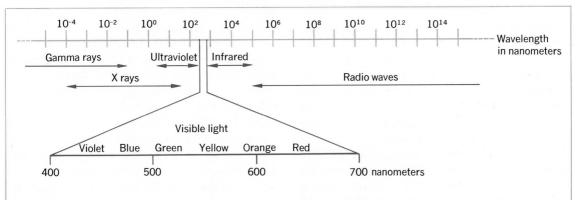

51

always a low level of natural radiation from such radioactive compounds as uranium and radium which exist in the rocks of the earth's crust. Undoubtedly this radiation is responsible for some small amount of the spontaneous mutations and malignant growths to which living things always have been subject. The explosions of atomic bombs and hydrogen bombs and the peaceful use of radioactive compounds in medicine and nuclear power plants have increased the amount of radiation in the environment.

Long-wave radiations, like infrared and radio waves, have relatively little energy. They are not harmful to organisms at intensities ordinarily encountered, but at very high intensities they can be harmful.

## Temperature

Within certain limits, an increase in temperature increases the rate at which biological processes proceed, and therefore it can increase the rate at which some organisms grow. For each species there is a temperature range within which it can grow. Many organisms can survive temperatures below freezing, although they may not grow under such cold conditions.

Thick fur or wool serves as insulation against external temperature extremes. The luxuriant furs of mink, sable, and fox are protection against the bitter cold of arctic and subarctic winters. On the other hand, the camel's woolly coat provides insulation against the heat of the desert.

Sweat glands are abundant in many animals that live in hot climates, for perspiration has a cooling effect when it evaporates. Man, who is a nearly hairless animal of tropical ancestry, is covered with sweat glands. Dogs, which are covered with thick fur, perspire through their mouths, with the evaporation of water from their nasal passages, lungs, and tongues removing heat from their bod-

ies; for this reason they pant in hot weather. Some small rodents cool themselves by licking their fur and wetting it with saliva that then evaporates.

Behavior patterns of many animals adjust the loss or conservation of heat as is appropriate at the moment. The reader might observe a pet dog or cat sleeping on a hot summer afternoon. The animal lies on its side with head, legs, and tail stretched out, exposing to the air a maximum amount of surface area through which heat can be lost. In cold weather the same animal curls up with its legs folded and drawn close to the body; the tail curves forward and covers the nose. The body forms a compact ball with little surface area exposed; this helps to conserve heat. The reader might compare his own sleeping positions in summer and winter.

Some desert animals confine their activity to the cooler hours of night or twilight, and they rest in burrows or in shade during the hot hours of midday. Huddling together in nests or burrows helps small animals to conserve heat in cold weather. Shivering is a response to cold; by increasing muscular activity it produces heat that helps to warm the body.

Temperature has an indirect effect on growth of aquatic organisms through its effect on the oxygen concentration of water. Gases, unlike minerals, are more soluble in cold water than in warm water. The oxygen-rich waters of arctic and antarctic regions support greater quantities of living things than do many warmer tropical oceans. The higher carbon dioxide content of the colder waters also permits a high rate of photosynthesis that produces the food on which all the aquatic organisms depend. Along the west coast of South America, where the Humbolt Current flows northward carrying with it the icy antarctic waters, mineral-rich waters rise from the ocean depths. Here the ocean teems with marine life of all kinds. Anchovies and tuna fish are only two fish that support fishing fleets from the United States, Japan, Peru, and other nations. Also feeding on this bounty are innumerable cormorants, pelicans, gannets, and

52

boobies that roost on the rocky coasts of Peru and Chile. In their crowded rookeries their excrement, called guano, accumulates in deposits that are as much as 150 feet deep in some places. This nitrogen-rich material is collected to be sold for use as fertilizer in agricultural areas.

## Wind

Winds have great effects on the weather of the lands over which they blow. In the North Temperate Zone, the prevailing winds blow from the west. At 44° north latitude, the westerly winds that reach the Oregon coast are relatively mild in winter and cool in summer, for these winds are tempered by their journey over the waters of the Pacific Ocean. The weather along the coast here is pleasant throughout the year. In the middle of the continent South Dakota lies at the same latitude, but its westerly winds, having blown many miles over land, have no such tempering effect. South Dakota, like many of its neighboring states, has uncomfortably cold winters and scorching summers.

Winds can bring damage. Dry winds dehydrate organisms, and moisture-laden winds can bring excessive rains or heavy snows. In the form of hurricanes or tornados, winds wreak damage on plants and animals alike.

## Fire

Fires set by lightning in dry woodlands are a natural part of the environment, and surprisingly enough to many persons these fires can be beneficial. When dead limbs and underbrush are burned away by small forest fires every few years, little damage is done. The fires remain near the

ground, and the bases of the trees are protected by their bark. Minerals from the burned material return to the soil. If there is no fire for many years, the accumulation of dead, dry brush provides such large quantities of tinder that bigger, hotter forest fires are likely to occur. These fires do a great deal of damage; they burn the crowns of trees, killing them and completely devastating the area, which then is subject to erosion and concomitant loss of minerals. Fires seem to be a normal part of the coniferous forests in the southeastern United States and of the chaparral (drought-resistant, shrubby vegetation) in California. Ironically, by suppressing the smaller fires that clear away underbrush, man actually encourages the more destructive fires.

The many physical factors of the environment occur in many combinations. Between the hot, humid, tropical rain forests, where the temperature varies only a few degrees throughout the year and the "dry" season is only somewhat less rainy than other seasons, and the frigid polar regions with eternal ice and snow are many other habitats (Chapter 5) with other combinations of physical factors. In the 3 billion or more years that life has existed on the earth, organisms that walk, swim, and fly have successfully invaded land, sea, and air. Understandably, some inhospitable environments such as the hot sands of the Sahara and the icy land of Antarctica do not exactly teem with life, but even here sturdy, resistant animals like camels and penguins survive and reproduce and seem in no immediate danger of extinction unless it be by the hand of man.

## Effects of Organisms on Their Environment

53

Every organism has an effect on its physical environment. Even in a seemingly inactive state like sleep, an animal breathes. In so doing it absorbs oxygen from the air (or from water, in the case of aquatic organisms) and releases carbon dioxide. The concentrations of these two gases in the environment thus are changed by the animal. During periods of intense activity, similar changes occur, but the volume of gases exchanged is greater.

The accumulation of carbon dioxide and the exhaustion of oxygen in the air is prevented by the photosynthesis of green plants. In sunshine they absorb carbon dioxide and release oxygen.

The urine and feces of animals, the leaves that drop in autumn, and the dead bodies of both plants and animals are decomposed by bacteria, fungi, and other small soil organisms. The decomposed material forms humus, a mixture of organic compounds in varying stages of decomposition.

Humus is an important constituent of soils. Unlike sand, it can hold a great deal of water against gravity. Unlike clay, it does not form heavy clods when wet, nor does it dry to a bricklike consistency. Soil with a high humus content usually is fertile soil. Under natural conditions humus accumulates slowly over the centuries in grasslands and forests. Forests are particularly valuable for their water-holding capacities. Rather than letting the water run off immediately after a rain, the humus releases it to streams slowly, giving a fairly steady flow to streams that can be relied upon to have at least a certain minimum flow throughout the year.

Not only do generations of plants and animals contribute to the formation of topsoil with its humus, but burrowing activities of small animals aerate the soil, and the intricate entanglement of roots of many plants holds it in place. When forests are cut down, the roots die and the unprotected topsoil is washed away by rain. Consequently floods are to be expected after rains, and extremely low flows of streams accompany the dry seasons.

When grasslands are plowed and planted to crops, much the same thing can happen, especially if the crops are planted in rows with considerable space between them. The bare soil is exposed to rain, which erodes it. Erosion is greatest where farmers plow in straight lines that run up and down hillsides; each furrow is a small gully that conducts rainwater downward. Contour farming, in which the lines of plowing follow the contours of the land, reduces the rate of erosion, for each furrow tends to act like a small dam that retards the flow of water downhill.

Trees and other plants cast shadows and thus create shade where there would otherwise be none. In the shade the temperature is lower and the humidity is higher than they are in direct sunlight. The plants and animals that live under the shelter of trees are different from those that can withstand the heat and intense light of sunshine.

Some organisms so change their environment that they or their offspring can no longer survive in it. White pine seedlings, for instance, require sunshine; because the seedlings do not do well in the shade of the parent trees, a white pine forest is not likely to replace itself.

Man has often been considered to be an organism that cannot live in conditions of his own making. The ancient civilizations that thrived in the Middle East and the Mediterranean area—Babylon, Assyria, Egypt, Greece, Rome—depended on the rich forests and agricultural lands that are now deserts and dry, cutover hillsides. The consequences of poor use of the land by the ancients still plague their descendants today. In the once lush Fertile Crescent of the Middle East poor sheepherders and goatherds tend their flocks, which barely manage to find sustenance from the sparse vegetation (Plate VI). The impoverished villages of Greece perch on bare mountainsides. In Italy the Arno River arises in long-denuded mountains above Florence. Every few years, when the rainfall is somewhat higher than usual, the Arno floods the low-lying portions of Florence. The flood of November 1966 was responsible for irreparable damage to many great art treasures.

From ancient times to the present, irrigated

54

lands sooner or later have become too salty to support agriculture. The water diverted for irrigation always contains some minerals. When brought into a dry area, much of the water evaporates, leaving the minerals in the soil. The deposit of mineral salts each year is small, but gradually it accumulates. Eventually the soil becomes too salty to support crops. The Imperial Valley of California, which owes its fruitfulness to irrigation and which supplies the rest of the United States with much of its fruits and vegetables, is beginning to show signs of salting.

Until recently, the destruction of forests and plains seemed of little consequence, for there were always more forests and plains into which to move. The history of the United States is a good example of this. It received the excess population of Europe that moved in on the country from both east and west. Now nearly all the available and usable land is occupied both in this country and others. Much of the little that remains should be preserved as wilderness. Man soon must learn to affect his environment in such a way that he can survive in it, or he will become extinct.

Before pursuing the matter of man and his environment further, the next few chapters digress to give some background in the basic chemical and physical processes characteristic of living things.

# Questions

1. Define the following terms.
   specific heat
   latent heat of fusion
   latent heat of vaporization
   relative humidity
2. Why is perspiration cooling on a dry day but uncomfortable on a humid day?
3. List as many reasons as you can why life without water probably is impossible.
4. List several ways in which light affects living things.
5. What kinds of radiation are harmful to organisms? What kinds are harmless or relatively harmless?
6. How do winds benefit and harm organisms?
7. How do fires benefit and harm organisms? Why is the effort to suppress all forest fires harmful to forests?
8. In what ways do organisms alter their environment? List some ways that man specifically has changed his environment.
9. Why does a mixture of sand and clay make a better soil for growing plants than does either sand or clay alone?
10. What affect does temperature have on the oxygen concentration of water, and how does this affect the populations of various seas? Does this help you explain why penguins and Eskimos can survive in the polar regions?

# Suggested Readings

Buswell, A. M., and W. H. Rodebush. "Water," *Scientific American* 194(4): 76–89 (April 1956).

Carson, R. *The Sea Around Us,* rev. ed. New York: The New American Library, Inc., Signet Science Library Book, 1961.

Cooper, C. F. "The Ecology of Fire," *Scientific American* 204(4): 150–160 (April 1961).

Davis, K. S., and J. A. Day. *Water, The Mirror of Science.* Garden City, N.Y.: Doubleday & Company, Inc., Anchor Books, 1961.

Ellison, W. D. "Erosion by Raindrop," *Scientific American* 179(5): 40–45 (November 1948).

Irving, L. "Adaptations to Cold," *Scientific American* 214(1): 93–101 (January 1966).

Jacobsen, T., and R. M. Adams. "Salt and Silt in Ancient Mesopotamian Agriculture," *Science* 128: 1251–1258 (1958).

Johnston, V. R. "The Ecology of Fire," *Audubon* 72(5): 76–119 (September 1970).

Wald, G. "Life and Light," *Scientific American* 201(4): 92–108 (October 1959).

Wolf, A. V. "Body Water," *Scientific American* 199(5): 125–132 (November 1958).

# Some Chemistry and Physics: A Background for Life

# Part III

Fifty or one hundred or more years ago the study of biology was easily distinguished from the study of chemistry. To be sure, it was long recognized that living things are composed of chemical compounds and that many chemical reactions must be occurring within them, but relatively little of the chemistry of living things had been studied. Courses in biology consisted mostly of classifying and identifying plants and animals, studying their external anatomy, and then as microscopes and their use improved, studying internal, microscopic anatomy. Today one of the main fields of interest in biology is metabolism—the chemical reactions within living things. Another interesting field is concerned with the effects of one organism on another; some of these effects are caused by the chemical compounds they produce. For this reason we must digress and brush up a little on some basic chemistry.

## Elements and Atoms

All material things are composed of chemical substances. A substance that cannot be broken down into simpler chemical substances is called an **element.** There are 92 elements known to occur naturally in the universe; a few of them are hydrogen, oxygen, carbon, gold, and silver. The smallest unit of an element is an **atom.** It cannot be broken down into smaller units without changing its properties. An **atomic symbol** is the abbreviation for an element (or one of its atoms). The atomic symbol is usually the first letter or the first two letters of the name (in English or some other language) of the element. The atomic symbols of most of the elements known to be present in and required by living things are recorded in Table 7.1.

Every atom is composed of several **elementary particles.** In recent years a large number of elementary particles have been discovered, but we will be concerned with only three of them: **pro-**

# 7

# A Little Chemistry

**1** All substances are composed of atoms.

**2** In a chemical reaction, atoms react with each other by sharing, gaining, or losing electrons.

**3** In an ordinary chemical reaction, matter is neither created nor destroyed.

59

**Table 7.1**

**Some Elements Present in Living Things**

| Element | Atomic Symbol | Atomic Weight | Most Important Valences |
|---|---|---|---|
| Carbon | C | 12 | 4 |
| Hydrogen | H | 1 | 1 |
| Oxygen | O | 16 | 2 |
| Nitrogen | N | 14 | 3, 5 |
| Sulfur | S | 32 | 2 |
| Phosphorus | P | 31 | 3, 5 |
| Potassium | K | 39 | 1 |
| Calcium | Ca | 40 | 2 |
| Iron | Fe | 56 | 2, 3 |
| Magnesium | Mg | 24 | 2 |
| Copper | Cu | 64 | 1, 2 |
| Boron | B | 11 | 3 |
| Zinc | Zn | 65 | 2 |
| Chlorine | Cl | 35 | 1 |
| Sodium | Na | 23 | 1 |
| Manganese | Mn | 55 | 2, 4, 7 |
| Cobalt | Co | 59 | 2, 3 |
| Iodine | I | 127 | 1 |
| Molybdenum | Mo | 96 | 3, 6 |

*( wt for 1 atom ) Divide by 2 for = Protons & neutrons ( Electrons same )*

**tons, neutrons,** and **electrons.** The protons and neutrons are located in the center, or **nucleus,** of the atom, and the electrons orbit around the nucleus. The protons of all elements are the same; they all have a positive electrical charge and a weight[1] of 1. Neutrons also have a weight of 1, but they have no electrical charge; that is, they are electrically neutral. The electrons of all elements are the same; they all have a weight of 1/1,836, and they all have a negative electrical charge.

The **atomic weight** of an atom is equal to the total of the weights of all its protons and neutrons. (Electrons may be ignored in calculating atomic weights; because they are so light, they have little influence on the weight of the atom.) Atomic weights of elements in living things are recorded in Table 7.1.

In an atom the number of protons equals the

[1]The units of weight of elementary particles have no names such as pounds or grams. Protons and neutrons have been arbitrarily assigned the weight of 1.

number of electrons; therefore the charges balance and an atom is electrically neutral.

The structure of a few atoms will be studied here. Hydrogen is the simplest element. It has an atomic weight of 1. One atom consists of one proton around which one electron moves:

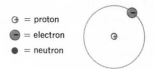

⊕ = proton
⊖ = electron
● = neutron

Hydrogen atom

The path of an electron around a nucleus is its **orbit.** A carbon atom has six protons and six neutrons in its nucleus; its atomic weight is 12. Its six electrons move around the nucleus in two orbits; the inner orbit has two electrons, the outer orbit has four electrons:

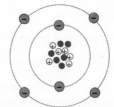

Carbon atom

A nitrogen atom has seven protons and seven neutrons in its nucleus; its atomic weight is 14. Its seven electrons move around the nucleus in two orbits; the inner orbit has two electrons, the outer has five electrons:

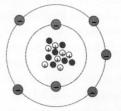

Nitrogen atom

Oxygen, with an atomic weight of 16, has eight protons and eight neutrons in its nucleus. Its eight electrons move around the nucleus in two orbits; the inner orbit has two electrons, the outer has six electrons:

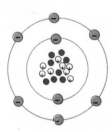

Oxygen atom

## Compounds and Molecules

A **compound** is a substance usually composed of two or more elements bound together in definite proportions by chemical bonds. The smallest unit of a compound is called a **molecule;** it is composed of two or more atoms. A **chemical formula** is an abbreviation for a molecule (or a compound). The subscript number in a chemical formula indicates how many atoms of the preceding element are present in the molecule. Some chemical formulas are given below:

$H_2$   Hydrogen. One molecule of hydrogen is composed of two atoms of hydrogen.

$O_2$   Oxygen. One molecule of oxygen is composed of two atoms of oxygen.

$H_2O$   Water.[2] One molecule of water is composed of two atoms of hydrogen and one atom of oxygen.

[2] We do not use a subscript if only one atom of an element is present in a molecule. Thus we write $H_2O$, not $H_2O_1$.

$C_6H_{12}O_6$   Glucose. One molecule of glucose is composed of six atoms of carbon, twelve of hydrogen, and six of oxygen.

If there is a variation in the structure of a molecule, then the nature of the compound is changed. For example, $H_2O_2$ is not water, but hydrogen peroxide, which has properties different from those of water.

The **molecular weight** of a molecule is the total of the atomic weights of all its atoms. Molecular hydrogen has a weight of 2, molecular oxygen 32, water 18, and glucose 180.

## Chemical Reactions

Compounds are formed by **chemical reactions** between two or more elements or compounds. In chemical reactions we are concerned primarily with the number of electrons in the outer orbit of an atom. For the most part, the other electrons and neutrons will not concern us here. For this reason, hereafter we will indicate only the outer orbit of electrons and indicate the nucleus by the atomic symbol of the element.

Each orbit is "complete" with a certain number of electrons. The innermost orbit is complete with two electrons. The outermost orbit is complete with eight electrons. In large atoms that have four or more orbits, some of the orbits may have more than eight electrons, but the outer orbit is always complete with eight.

Atoms with complete outer orbits tend to be chemically inert; that is, they do not react readily with other atoms. But atoms with incomplete outer orbits tend to share, gain, or lose electrons. This sharing or transferring of electrons results in the formation of chemical bonds between the atoms, and we say that a chemical reaction has occurred.

61

The simplest compound is molecular hydrogen. Each atom of hydrogen has only one electron in its only orbit. Because this orbit is incomplete, two hydrogen atoms may share their electrons. Each atom then shares in two electrons, thus completing its orbit:

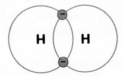

Hydrogen molecule

The hydrogen that is present in the air (in minute amounts) is molecular hydrogen; it is much more stable than atomic hydrogen.

Because drawing the structure of each atom is tedious and time-consuming, a bond between two atoms may be indicated in a chemical formula by a pair of dots (for the pair of electrons in the bond) or by a single dash. H:H and H—H both indicate the same thing: two hydrogen atoms with a single chemical bond between them.

Each atom of oxygen has six electrons in its outer orbit. It tends to share two of its electrons with another oxygen atom (which also shares two of its electrons):

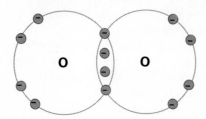

Oxygen molecule

Each atom now has a complete outer orbit with eight electrons. The two chemical bonds (two pairs of electrons) in a molecule of oxygen may be indicated by O::O or O=O.

Unlike atoms may also share electrons. In a molecule of water, $H_2O$, an oxygen atom shares one pair of electrons with one hydrogen atom and another pair of electrons with the other hydrogen atom:

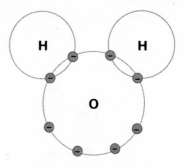

or H:O:H or H—O—H

Water molecule

The oxygen atom then has a complete outer orbit with eight electrons, and each hydrogen atom has a complete outer orbit with two electrons.

A carbon atom has four electrons in its outer orbit. A molecule of the gas carbon dioxide, $CO_2$, has one carbon atom that shares two pairs of electrons with each of two oxygen atoms:

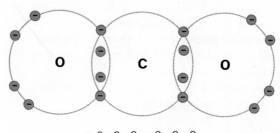

or O::C::O or O=C=O

Carbon dioxide molecule

Atoms that have only one or two electrons in the outer orbits often lose them rather than share them; and atoms that have six or seven electrons in the outer orbit frequently gain electrons. For example, common table salt, sodium chloride, NaCl, is composed of sodium (Na) and chlorine (Cl). Sodium has one electron in its outer orbit and chlorine has seven; the sodium loses its electron to the chlorine:

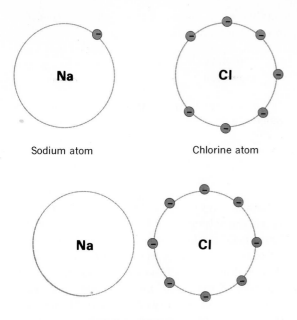

Sodium atom          Chlorine atom

Sodium chloride molecule

arrow that points to the chemical formula(s) of the product(s) that are formed:

$$CO_2 + H_2O \longrightarrow H_2CO_3$$

One of the basic principles of physics is that matter is neither created nor destroyed in ordinary chemical reactions. This means that for every atom that appears on the one side of an equation, one atom of the same element must appear on the other side. (This is not true in equations involving nuclear reactions, but we are not concerned with these here.) If we check the preceding equation, we find that on each side there are one carbon atom, three oxygen atoms, and two hydrogen atoms.

Let us consider another equation. If the gas methane, $CH_4$, is burned completely in the presence of oxygen, carbon dioxide and water are formed:

$$CH_4 + O_2 \longrightarrow CO_2 + H_2O$$

But on checking our equation, we see that something is wrong. There is one atom of carbon on each side of the equation, but the left side has four atoms of hydrogen and the right side has only two, and the left side has two atoms of oxygen and the right side has three. Some hydrogen seems to have disappeared, and some oxygen seems to have been formed from nothing. Obviously this is an impossible situation, and the equation is not balanced.

We can balance the hydrogen atoms by indicating two molecules of water on the right side:

$$CH_4 + O_2 \longrightarrow CO_2 + 2H_2O$$

This gives us four hydrogen atoms on each side, but the situation with regard to oxygen seems even worse than before—two atoms on the left side and four on the right. This can be remedied by indicating two molecules of oxygen on the left side:

$$CH_4 + 2O_2 \longrightarrow CO_2 + 2H_2O$$

Because they do not share electrons, the sodium has a positive charge (having lost one negatively charged electron but having retained all of its positively charged protons) and the chlorine is negatively charged (having gained the electron). Sometimes sodium chloride is written $Na^+Cl^-$ to indicate these charges. Atoms or molecules that have gained or lost electrons and thus bear electrical charges are called **ions.** Sodium chloride thus consists of sodium ions and chloride ions.

The number of its electrons that an atom tends to share, gain, or lose is called its **valence.** Thus hydrogen, sodium, and chlorine each have a valence of 1; oxygen has a valence of 2, and carbon has a valence of 4. The valences of some other elements are listed in Table 7.1.

A **chemical equation** is a method of indicating what occurs in a chemical reaction. If, for instance, we bubble carbon dioxide through water, some carbonic acid, $H_2CO_3$, will be formed. Rather than writing all this out in words, we find it convenient to write an equation by placing the chemical formula(s) of the reactant(s) on the left side of the

The equation is now balanced, with each side having one carbon atom, four hydrogen atoms, and four oxygen atoms.

Balancing an equation is not merely a mathematical exercise. When methane is completely burned, two molecules of oxygen actually are required for each molecule of methane, and one molecule of carbon dioxide and two molecules of water are produced. (As with subscripts, the figure 1 is not used before a chemical formula indicating only one molecule. Thus we write $CH_4$, not $1\ CH_4$, in the preceding chemical equation.)

The speed with which a chemical reaction proceeds depends in part on the temperature at which it occurs. In general, the higher the temperature, the faster the reaction occurs. An often-used rule of thumb is that the speed of a chemical reaction is doubled or tripled by a 10°C rise in temperature. Conversely, the lowering of the temperature by 10°C slows the rate of a chemical reaction to one half or one third of what it was.

## pH

**Ionic compounds** are those compounds that tend to ionize, or dissociate, in water. Sodium chloride, for instance, ionizes into its sodium and chloride ions:

$$NaCl \xrightarrow{H_2O} Na^+ + Cl^-$$

($H_2O$ is written above the arrow to indicate that water is present but that it is not a reactant or a product in the equation.)

Water also ionizes to a slight extent. In any sample of water a few molecules ionize into hydrogen ions and hydroxyl ions:

$$H_2O \longrightarrow \underset{\substack{\text{hydrogen} \\ \text{ion}}}{H^+} + \underset{\substack{\text{hydroxyl} \\ \text{ion}}}{OH^-}$$

When acids ionize, they release hydrogen ions into the solution:

$$\underset{\text{acetic acid}}{CH_3COOH} \xrightarrow{H_2O} \underset{\text{acetate ion}}{CH_3COO^-} + \underset{\text{hydrogen ion}}{H^+}$$

When bases (alkaline substances) ionize, they release hydroxyl ions into the solution:

$$\underset{\substack{\text{ammonium} \\ \text{hydroxide}}}{NH_4OH} \xrightarrow{H_2O} \underset{\substack{\text{ammonium} \\ \text{ion}}}{NH_4^+} + \underset{\substack{\text{hydroxyl} \\ \text{ion}}}{OH^-}$$

Both hydrogen and hydroxyl ions are generally present in solutions. Acid solutions are those that have more hydrogen ions than hydroxyl ions. Alkaline (basic) solutions have more hydroxyl ions than hydrogen ions. Neutral solutions have equal numbers of hydrogen and hydroxyl ions. Some acid substances are citrus fruit juices, vinegar, and the contents of a normal stomach. Some basic substances are sodium bicarbonate (often used by people who complain of "acid stomach"), milk of magnesia, and the contents of the small intestine. Pure water is neutral, for it has equal numbers of hydrogen ions and hydroxyl ions.

The strength of a solution in terms of its acidity or alkalinity is its pH. The pH scale runs from 0 to 14. Neutral solutions have a pH of 7. Acid solutions have a pH of less than 7. The more acid the solution, the lower the pH; pH 5 is more acid than pH 6, pH 4 is more acid than pH 5, and so on. Alkaline solutions have a pH higher than 7. The more alkaline the solution, the higher the pH; pH 9 is more alkaline than pH 8, pH 10 is more alkaline than pH 9, and so on.

The pH of a solution can determine what reactions may occur in that solution. Some chemical reactions require acid solutions, others neutral solutions, and others alkaline solutions. This is especially true in living cells where many reactions are controlled by enzymes (Chapter 10). Enzymes are very sensitive to pH and can be inactivated by a relatively small change in acidity or alkalinity.

# Questions

1. Define the following terms.
   element
   compound
   atom
   molecule
   ion
   acid
   base

2. Match the following elementary particles with their charges.
   \_\_\_\_ Proton       a. Positive charge.
   \_\_\_\_ Neutron      b. Negative charge.
   \_\_\_\_ Electron     c. No charge.

3. An atom is electrically neutral because it has equal numbers of
   a. Protons and neutrons.
   b. Neutrons and electrons.
   c. Electrons and protons.

4. What is a chemical reaction? *sharing o gaining or losing electrons*

5. By indicating protons, neutrons, and electrons, diagram a molecule of ammonia ($NH_3$). Nitrogen has five electrons in its outer orbit.

6. Are the following equations balanced or not? If not, balance them.

$$CO_2 + H_2O \longrightarrow H_2CO_3 \quad ok$$
$$H_2 + O_2 \longrightarrow H_2O \quad H_2 + 2O$$
$$CO_2 + H_2O \longrightarrow C_6H_{12}O_6 + O_2 \quad CH_2O_3$$

7. Indicate whether the following are acid, basic, or neutral.
   a. pH 8.  *alkaline*
   b. pH 3.  *acid*
   c. pH 7.  *neutral (water)*
   d. pH 13. *alkaline or basic*

# Suggested Readings

Baker, J. J. W., and G. E. Allen. *Matter, Energy and Life.* Reading, Mass.: Addison-Wesley Publishing Co., Inc., 1965.

Friedan, E. "The Chemical Elements of Life," *Scientific American* 227(1): 52–60 (July 1972).

White, E. H. *Chemical Background for the Biological Sciences.* Englewood Cliffs, N.J.: Prentice-Hall, Inc., 1964.

$$H \equiv N \equiv H$$

$$H : : : N : : : H$$

# 8

# A Little More Chemistry: Foods

**1 Three important groups of organic compounds that serve living things as foods are**
  **a  carbohydrates**
  **b  fats**
  **c  proteins**

Carbon, as we have seen in the previous chapter, has a valence of 4. Not only does carbon form bonds readily with oxygen, hydrogen, nitrogen, and other elements, but carbon atoms also tend to share electrons with other carbon atoms, which in turn share electrons with still other carbon atoms. This permits the formation of long chains of carbon atoms that form the backbones of more

complex molecules than are formed by other elements. Some of these long-chain molecules are branched; others occur in the form of rings.

When it was learned that complex carbon compounds were formed by organisms, these substances were called organic compounds. Today chemists usually define an **organic compound** as any compound containing the element carbon. Nearly all organic compounds also contain hydrogen, a great many contain oxygen, and a considerable number contain nitrogen, phosphorus, sulfur, and other elements as well.

**Inorganic compounds** are those that lack carbon. Some examples are water ($H_2O$), hydrogen ($H_2$), oxygen ($O_2$), sodium chloride (NaCl), ammonia ($NH_3$), and hydrogen sulfide ($H_2S$). A few very simple carbon compounds, such as carbon dioxide ($CO_2$), usually are considered to be inorganic.

Living things are composed almost entirely of water and organic compounds. Many of the inorganic compounds that they absorb from their environment are incorporated into organic compounds.

Organisms synthesize and utilize a vast variety of organic compounds. In this chapter we will consider only three groups of organic compounds: carbohydrates, fats, and proteins. These are the main groups of foods on which all organisms depend. Some other organic compounds will be considered in later chapters.

# Carbohydrates

Carbohydrates were named for the fact that they contain carbon, hydrogen, and water, the last two being in the same ratio that they are in water. Carbohydrates include monosaccharides, disaccharides, and polysaccharides.

The simplest of the carbohydrates are the **monosaccharides** or simple sugars. One of the most common monosaccharides is glucose (also called dextrose); its chemical formula is $C_6H_{12}O_6$. Its structural formula

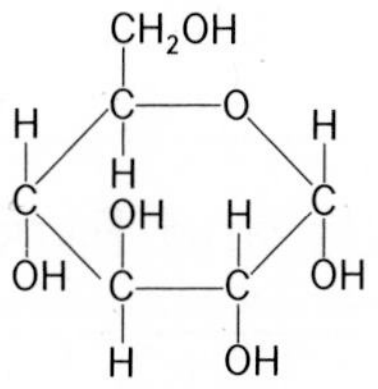

Glucose

shows that it is somewhat more complex than the inorganic compounds mentioned in the preceding chapter. The carbon atoms are bound to each other; and five of them, together with an oxygen atom, form a ring. Other atoms are attached to members of the ring. Another monosaccharide is fructose; it has the same chemical formula as glucose, $C_6H_{12}O_6$, but its structural formula is slightly different:

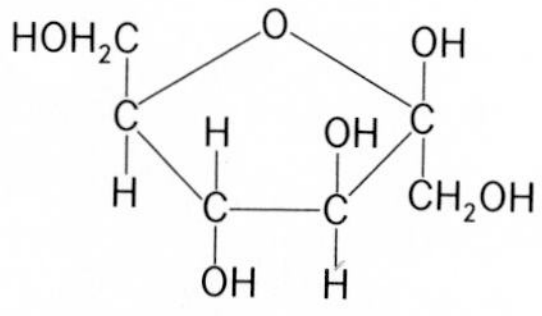

Fructose

Ribose is a monosaccharide with five carbon atoms, $C_5H_{10}O_5$:

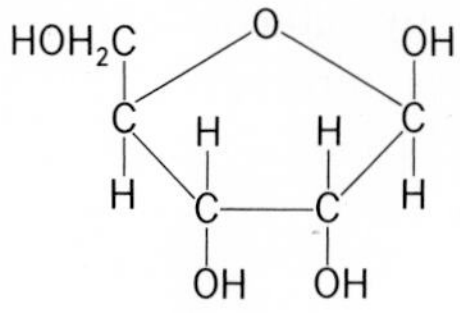

Ribose

Because it is tedious and inconvenient to indicate every atom in a structural formula, the following abbreviations often are used:

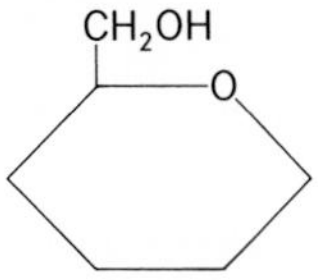

Glucose 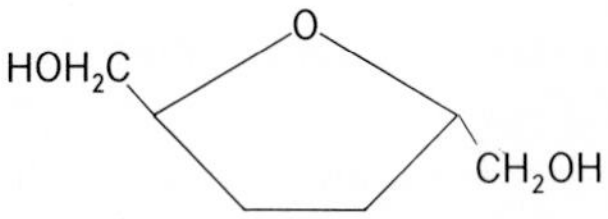        Fructose

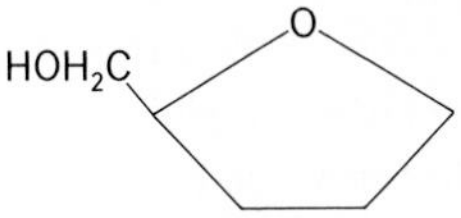

Ribose

These and some other monosaccharides are listed in Table 8.1.

A **disaccharide,** or double sugar, is formed in a reaction in which the equivalent of one molecule of water is removed from two monosaccharide molecules:

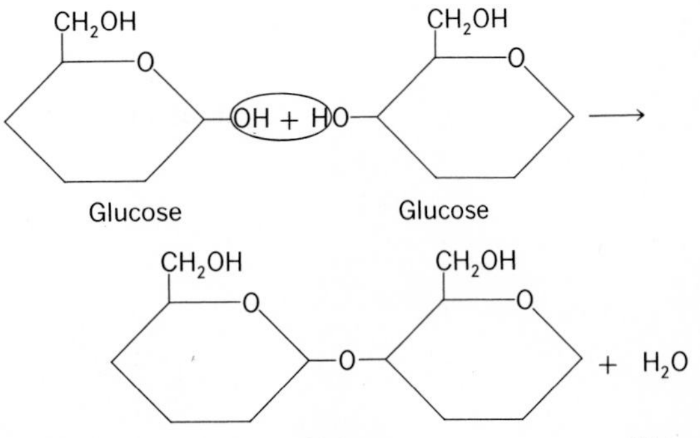

Glucose        Glucose

Maltose        Water

Such a reaction is called a **dehydration synthesis,** for a larger molecule is synthesized from smaller ones with the loss of water. A molecule of maltose is derived from two glucose molecules, and a molecule of sucrose (ordinary table sugar or cane sugar) is derived from one molecule of glucose and one of fructose:

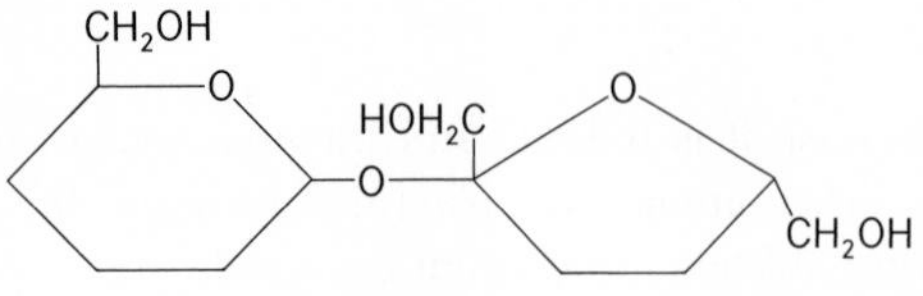

Sucrose

These and some other dissacharides are listed in Table 8.1.

Carbohydrates formed by dehydration syntheses involving three or more monosaccharides are called **polysaccharides.** In many cases polysaccharides consist of several hundred or several thousand repeating units of glucose or other monosaccharide. The formulas of starch, glycogen, and cellulose are indicated at the bottom of the page. The general formula for starch or cellulose, both of which are plant products, is $(C_6H_{10}O_5)_x$. In the case of starch $x$ is approximately 1,000; in the case of cellulose $x$ is approximately 2,500. Notice also the different orientation of the glucose units in starch

**Table 8.1**
Some Carbohydrates in Living Things

Monosaccharides (simple sugars)
  Glyceraldehyde, $C_3H_6O_3$
  Erythrose, $C_4H_8O_4$
  Ribose, $C_5H_{10}O_5$
  Glucose (dextrose, grape sugar), $C_6H_{12}O_6$
  Galactose, $C_6H_{12}O_6$
  Fructose (levulose, fruit sugar), $C_6H_{12}O_6$
  Sedoheptulose, $C_7H_{14}O_7$

Disaccharides (double sugars), $C_{12}H_{22}O_{11}$
  Sucrose [cane sugar (glucose + fructose)]
  Maltose (glucose + glucose)
  Lactose [milk sugar (glucose + galactose)]
  Cellobiose (glucose + glucose)

Polysaccharides, $(C_6H_{10}O_5)_x$
  Starch (repeating glucose units)
  Glycogen (repeating glucose units)
  Cellulose (repeating glucose units)
  Agar (repeating galactose units)

and cellulose. Glycogen, produced in the animal body, resembles starch but is not identical with it.

# Fatty Acids and Fats

Fats, like carbohydrates, contain the elements carbon, hydrogen, and oxygen, but they contain

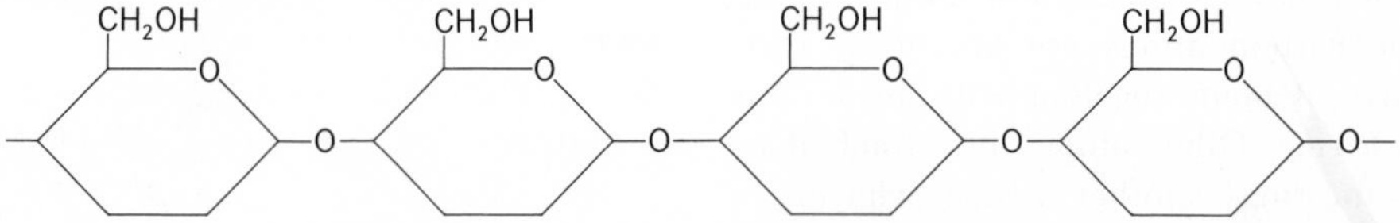

Part of a starch molecule (glycogen is similar)

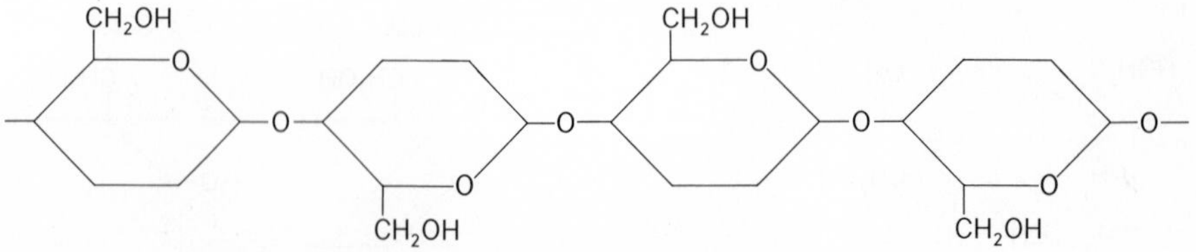

Part of a cellulose molecule

relatively less oxygen than do carbohydrates. A fat molecule is formed by the reaction of a molecule of glycerol with three fatty acid molecules. Glycerol has the formula $C_3H_8O_3$:

$$H_2-C-OH$$
$$H-C-OH$$
$$H_2-C-OH$$

Glycerol

A **fatty acid** molecule is a chain of carbon atoms of variable length, but the chain always ends in

an acid group: $-\overset{\overset{\displaystyle O}{\|}}{C}-OH$. A few fatty acids are

Acetic acid, $C_2H_4O_2$    $CH_3-\overset{\overset{\displaystyle O}{\|}}{C}-OH$

Butyric acid, $C_4H_8O_2$    $CH_3-CH_2-CH_2-\overset{\overset{\displaystyle O}{\|}}{C}-OH$

Palmitic acid, $C_{16}H_{32}O_2$    $CH_3-(CH_2)_{14}-\overset{\overset{\displaystyle O}{\|}}{C}-OH$

Stearic acid, $C_{18}H_{36}O_2$    $CH_3-(CH_2)_{16}-\overset{\overset{\displaystyle O}{\|}}{C}-OH$

The general formula for any fatty acid is $R-\overset{\overset{\displaystyle O}{\|}}{C}-OH$, in which $R$ stands for the variable part of the molecule.

One glycerol molecule and three fatty acid molecules eliminate three molecules of water when they react in a dehydration synthesis:

$$H_2-C-(OH + H)O-\overset{\overset{\displaystyle O}{\|}}{C}-R \qquad H_2-C-O-\overset{\overset{\displaystyle O}{\|}}{C}-R$$
$$H-C-(OH + H)O-\overset{\overset{\displaystyle O}{\|}}{C}-R \longrightarrow H-C-O-\overset{\overset{\displaystyle O}{\|}}{C}-R + 3H_2O$$
$$H_2-C-(OH + H)O-\overset{\overset{\displaystyle O}{\|}}{C}-R \qquad H_2-C-O-\overset{\overset{\displaystyle O}{\|}}{C}-R$$

Glycerol     Fatty acids          Fat          Water

The three fatty acids contributing to the formation of a fat may all be different, or two or three may

be the same. Most naturally occurring fats contain two or three different fatty acids.

Unsaturated fatty acids are those that have fewer hydrogen atoms than the carbon atoms potentially could hold; that is, they are not saturated with respect to hydrogen. Between two adjacent unsaturated carbon atoms is a double bond. Oleic acid ($C_{17}H_{33}COOH$), which is common in many fats and oils in living things, is an example:

Unsaturated carbon atoms
$$CH_3-(CH_2)_7-\underset{\underset{\displaystyle H}{|}}{C}=\underset{\underset{\displaystyle H}{|}}{C}-(CH_2)_7-COOH$$

Fatty acids with two or more pairs of hydrogen atoms missing and with corresponding numbers of double bonds are polyunsaturated. Linoleic acid ($C_{17}H_{29}COOH$), another common fatty acid, is doubly unsaturated.

$$CH_3-(CH_2)_4-\underset{\underset{\displaystyle H}{|}}{C}=\underset{\underset{\displaystyle H}{|}}{C}-CH_2-\underset{\underset{\displaystyle H}{|}}{C}=\underset{\underset{\displaystyle H}{|}}{C}-(CH_2)_7-COOH$$

There is considerable evidence that saturated fats (fats consisting of saturated fatty acids) may contribute to hardening of the arteries and heart attacks. For this reason, polyunsaturated fats have increased in popularity. Beef, pork, cheese, and butter are rich in saturated fats. Margarine made from corn oil is high in unsaturated fats; this is one reason for its increased use as a butter substitute.

Fats and oils are similar chemically, but fats are solid at room temperature and oils are liquid at this temperature. When heated sufficiently fats become liquid, and oils become solid when cooled enough. Butter melts in a heated frying pan, and olive oil solidifies in the refrigerator.

## Amino Acids and Proteins

69

Proteins contain the elements carbon, hydrogen, oxygen, nitrogen, and sulfur; some contain addi-

| NAME, WITH ABBREVIATION | STRUCTURAL FORMULA | NAME, WITH ABBREVIATION | STRUCTURAL FORMULA |
|---|---|---|---|
| Alanine (ala) | $CH_3-\overset{\overset{H}{\vert}}{\underset{\underset{NH_2}{\vert}}{C}}-COOH$ | Leucine (leu) | $CH_3-\overset{\overset{H}{\vert}}{\underset{\underset{CH_3}{\vert}}{C}}-CH_2-\overset{\overset{H}{\vert}}{\underset{\underset{NH_2}{\vert}}{C}}-COOH$ |
| Arginine (arg) | $\underset{HN}{\overset{H_2N}{>}}C-N-CH_2-CH_2-CH_2-\overset{\overset{H}{\vert}}{\underset{\underset{NH_2}{\vert}}{C}}-COOH$ | Lysine (lys) | $H_2N-CH_2-CH_2-CH_2-CH_2-\overset{\overset{H}{\vert}}{\underset{\underset{NH_2}{\vert}}{C}}-COOH$ |
| Aspartic acid (asp) | $HOOC-CH_2-\overset{\overset{H}{\vert}}{\underset{\underset{NH_2}{\vert}}{C}}-COOH$ | Methionine (met) | $CH_3-S-CH_2-CH_2-\overset{\overset{H}{\vert}}{\underset{\underset{NH_2}{\vert}}{C}}-COOH$ |
| Asparagine (asn or asp·NH$_2$) | $\overset{H_2N}{\underset{O}{>}}C-CH_2-\overset{\overset{H}{\vert}}{\underset{\underset{NH_2}{\vert}}{C}}-COOH$ | Phenylalanine (phe) | (benzene ring)$-CH_2-\overset{\overset{H}{\vert}}{\underset{\underset{NH_2}{\vert}}{C}}-COOH$ |
| Cysteine (cys) | $HS-CH_2-\overset{\overset{H}{\vert}}{\underset{\underset{NH_2}{\vert}}{C}}-COOH$ | Proline (pro) | $H_2C$ (pyrrolidine ring) $-COOH$ |
| Glutamic acid (glu) | $HOOC-CH_2-CH_2-\overset{\overset{H}{\vert}}{\underset{\underset{NH_2}{\vert}}{C}}-COOH$ | Serine (ser) | $HO-\overset{\overset{H}{\vert}}{\underset{\underset{H}{\vert}}{C}}-\overset{\overset{H}{\vert}}{\underset{\underset{NH_2}{\vert}}{C}}-COOH$ |
| Glutamine (gln or glu·NH$_2$) | $\overset{H_2N}{\underset{O}{>}}C-CH_2-CH_2-\overset{\overset{H}{\vert}}{\underset{\underset{NH_2}{\vert}}{C}}-COOH$ | Threonine (thr) | $CH_3-\overset{\overset{H}{\vert}}{\underset{\underset{OH}{\vert}}{C}}-\overset{\overset{H}{\vert}}{\underset{\underset{NH_2}{\vert}}{C}}-COOH$ |
| Glycine (gly) | $H-\overset{\overset{H}{\vert}}{\underset{\underset{NH_2}{\vert}}{C}}-COOH$ | Tryptophane (tyr) | (indole ring)$-CH_2-\overset{\overset{H}{\vert}}{\underset{\underset{NH_2}{\vert}}{C}}-COOH$ |
| Histidine (his) | (imidazole ring)$-CH_2-\overset{\overset{H}{\vert}}{\underset{\underset{NH_2}{\vert}}{C}}-COOH$ | Tyrosine (tyr) | $HO-$(benzene ring)$-CH_2-\overset{\overset{H}{\vert}}{\underset{\underset{NH_2}{\vert}}{C}}-COOH$ |
| Isoleucine (ileu) | $CH_3-CH_2-\overset{\overset{H}{\vert}}{\underset{\underset{CH_3}{\vert}}{C}}-\overset{\overset{H}{\vert}}{\underset{\underset{NH_2}{\vert}}{C}}-COOH$ | Valine (val) | $CH_3-\overset{\overset{H}{\vert}}{\underset{\underset{CH_3}{\vert}}{C}}-\overset{\overset{H}{\vert}}{\underset{\underset{NH_2}{\vert}}{C}}-COOH$ |

8.1 Twenty naturally occurring amino acids.

tional elements. Proteins are formed by chemical reactions among amino acids. Twenty amino acids are common in the living world. Amino acids all contain the elements carbon, hydrogen, oxygen, and nitrogen; a few contain sulfur. They are characterized by the following terminal group:

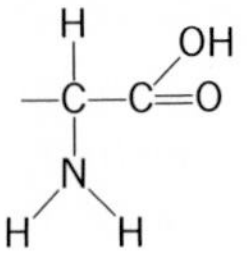

Again, the —C=O—OH is the acid group; the —NH$_2$ group is the amino group. Hence the name amino acid.

The general formula for an amino acid is

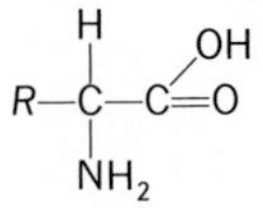

in which R stands for the variable part of the molecule. The structural formulas of the most common naturally occurring amino acids are in Fig. 8.1.

Two amino acid molecules may unite through the acid group of one molecule and the amino group of the other, forming a dipeptide and water in a dehydration synthesis:

Amino acids          Dipeptide          Water

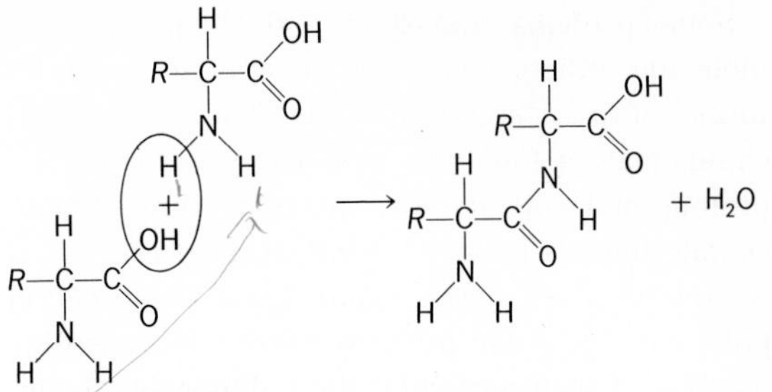

Notice that the dipeptide still has one amino group at one end that can react with the acid group of another amino in another dehydration synthesis. By the sequential addition of amino acids, a chain called a polypeptide is formed:

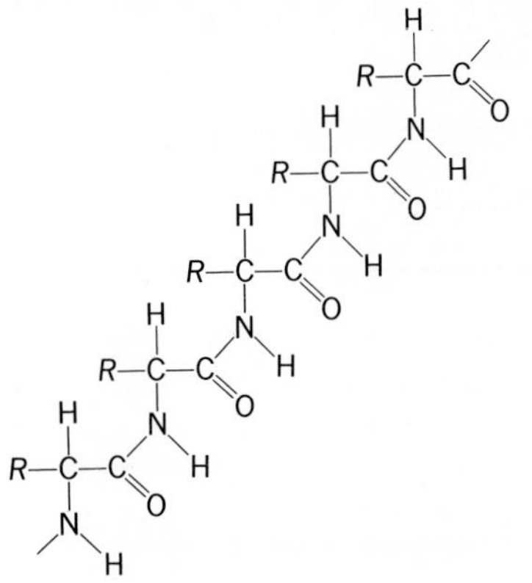

Part of a polypeptide molecule

**8.2** Structure of the insulin molecule in cattle.

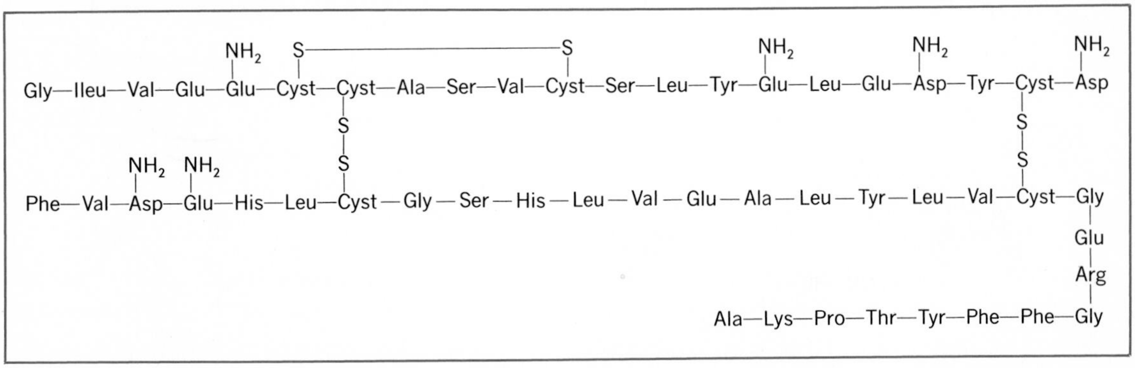

71

Some proteins consist of a single polypeptide molecule; others consist of several polypeptides united into a larger molecule. The polypeptide chains may be folded in such a way as to give the protein molecule its own peculiar three-dimensional shape. Chemical reactions between the R groups of some of the amino acids hold several polypeptides or the folds of a single polypeptide together in their particular three-dimensional configuration (Fig. 8.2).

Most proteins consist of several hundred amino acid units. Because there are 20 amino acids common to living things and because these can occur in almost any combination, the number of possible proteins is astronomical. As we shall see later in this book, the great variety of proteins is partially responsible for the many variations that life takes on this planet.

## Digestion

Macromolecules are large organic molecules. Polysaccharides, fats, and proteins are only some of them. As a general rule (but a rule with some significant exceptions) macromolecules do not move readily across plasma membranes, and therefore they do not pass freely from cell to cell within the body of a multicellular plant or animal. Instead, macromolecules must be broken down, or digested, to their smaller components, which can move to other cells where they are used. Moreover, macromolecules cannot enter directly into respiration (or many other chemical reactions). Starch and glycogen, common stored foods in plants and animals, respectively, must be digested to glucose before they can be respired. Fats, also, must be digested to fatty acids and glycerol before they can be respired. Most of the individual cells of plants and animals have the abilities both to synthesize macromolecules and to digest them.

**Digestion** of a macromolecule is the reverse of its synthesis. For every linkage broken between two adjacent subunits of a macromolecule, one molecule of water is required. For this reason digestion is also called hydrolysis (from *hydro*, meaning "water," and *lysis*, "breaking down"). The water becomes incorporated into the products.

Molecules of intermediate size, such as disaccharides and dipeptides, are also digested to their subunits.

DIGESTION OF STARCH:

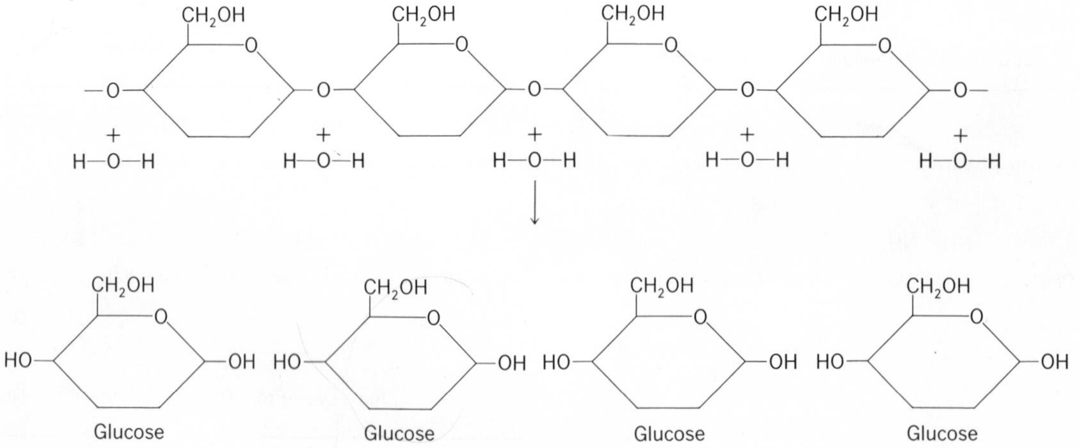

72

Glucose          Glucose          Glucose          Glucose

## DIGESTION OF A FAT

$$
\begin{array}{c}
H_2C\!-\!O\!-\!\overset{O}{\overset{\|}{C}}\!-\!R \quad H\!-\!O\!-\!H \\
H\!-\!\overset{|}{C}\!-\!O\!-\!\overset{O}{\overset{\|}{C}}\!-\!R + H\!-\!O\!-\!H \longrightarrow \\
H_2C\!-\!O\!-\!\overset{O}{\overset{\|}{C}}\!-\!R \quad H\!-\!O\!-\!H
\end{array}
\qquad
\begin{array}{c}
H_2C\!-\!OH \quad HO\!-\!\overset{O}{\overset{\|}{C}}\!-\!R \\
HC\!-\!OH + HO\!-\!\overset{O}{\overset{\|}{C}}\!-\!R \\
H_2C\!-\!OH \quad HO\!-\!\overset{O}{\overset{\|}{C}}\!-\!R
\end{array}
$$

Fat  Water  Glycerol  Fatty acids

## DIGESTION OF A DIPEPTIDE

Dipeptide  Water  Amino acids

The chemical reactions mentioned in this chapter are controlled by enzymes, which are discussed in more detail in Chapter 10.

Before proceeding to the next chapter, we might consider briefly some of the functions of foods. First, some carbohydrates, fats, and proteins serve as structural elements in cells, that is, as parts of cell membranes, cell walls, and other cellular constituents (Chapter 17). Second, some proteins function as enzymes (Chapter 10), substances that speed the rates of individual reactions in living cells. Nearly all chemical reactions in cells are controlled by enzymes; thus the presence or absence of particular enzymes has a great deal to do with the characteristics of a particular cell. Third, living things use the energy in certain carbohydrates and fats, and, to a lesser extent, the energy in proteins. The energy present in these foods is the energy source for nearly all biological activi-

ties: growth, movement, repair of wounded tissue, and reproduction. Finally, these substances can be converted to many other organic compounds essential for living things: vitamins, hormones, nucleic acids, and others.

## Questions

**1.** How do you distinguish between an inorganic compound and an organic compound? *Carbon*

**2.** Are the following compounds organic or inorganic?

| | |
|---|---|
| $H_2O$ ✓ | $MgSO_4$ |
| $NaCl$ ○ | $CH_3COOH$ ○ |
| $C_6H_{12}O_6$ ○ | $NH_4NO_3$ ✓ |
| | $CH_2NH_2COOH$ ○ |

**3.** Describe a dehydration synthesis and a digestion. What relationship do these two reactions have to each other?

**4.** Identify each of the following as a carbohydrate, fatty acid, fat, amino acid, or protein.

$R\!-\!\overset{NH_2}{\underset{H}{\overset{|}{C}}}\!-\!COOH$ *amino acid*

$C_6H_{12}O_6$ *carbohydrate*
glucose

stearic acid

alanine

glutamic acid

disaccharide

$CH_3\!-\!CH_2\!-\!CH_2\!-\!COOH$ *fatty acid*

$C_{12}H_{22}O_{11}$ *(Disaccharide) carbohydrate*

$(C_6H_{10}O_5)_x$ *(Polysaccharide) carbohydrate*

$(-\!\overset{H}{\overset{|}{N}}\!-\!\overset{R}{\underset{H}{\overset{|}{C}}}\!-\!\overset{O}{\overset{\|}{C}}\!-\!)_{50}$ *Protein*

$\begin{array}{c}
C\!-\!O\!-\!\overset{O}{\overset{\|}{C}}\!-\!R \\
C\!-\!O\!-\!\overset{O}{\overset{\|}{C}}\!-\!R \\
C\!-\!O\!-\!\overset{O}{\overset{\|}{C}}\!-\!R
\end{array}$ *fat*

*glycerol + fatty acids*

**5.** What are the end products of the complete digestion of a polysaccharide? Of a fat? Of a protein? *amino acids / glucose*

**6.** What is the difference between a saturated and an unsaturated fatty acid? *fewer hydrogen atoms*

**7.** What characteristic of carbon makes organic compounds so different from inorganic ones? *formed by organisms / carbon*

**8.** What portion of the molecule do all organic acids have in common? *acid group (variable part)*

73

*organic compound* *Chain of carbon atoms terminated by acid group* *organic acid containing an amino group on the carbon atom, adjacent to terminal acid group. Proteins are synthesized from amino acids*

**9.** Differentiate between a fatty acid and an amino acid.

## Suggested Readings

Baker, J. J. W., and G. E. Allen. *Matter, Energy, and Life*. Reading, Mass.: Addison-Wesley Publishing Co., Inc., 1965.

Green, D. E. "The Synthesis of Fats," *Scientific American* 202(2): 46–51 (February 1960).

Kendrew, J. C. "The Three-dimensional Structure of a Protein Molecule," *Scientific American* 205(6): 96–110 (December 1961).

*Scientific American* 197(3). September issue devoted to giant molecules.

Stein, W. H., and S. Moore. "The Chemical Structure of Proteins," *Scientific American* 204(2): 81–92 (February 1961).

We all have some concept of what energy is, for we hear it mentioned frequently. A young boy running around a baseball diamond and scoring a home run is said to be expending energy, and a breakfast food commercial claims that its product is full of energy. A physicist defines energy as the ability to do work. By this he means that energy is the ability to bring about some change in matter. The energy spent by our young baseball player has sent a ball soaring through the air, tossed a bat to the ground, carried his body around the bases, and kicked up a lot of dust; it probably also added some smudges and rents to his clothes and perhaps some cuts and bruises to his person. Now the energy in the breakfast cereal is not of itself capable of doing these things, but if the food is eaten by the boy, his metabolism is able to utilize this energy in just such activities. In fact, all his energy comes from the food he eats.

## Energy and Chemical Reactions

The bonds between atoms of a molecule represent chemical energy, or **bond energy.** Different chemical bonds have different amounts of energy; that is, the energy of a single bond between a carbon atom and an oxygen atom is different from the energy of a double bond between two such atoms, and both of these are different from the energy of a bond between a nitrogen atom and a hydrogen atom, and so on. In a chemical reaction some bonds are broken and/or some are formed. If the total amount of bond energy in the reactant(s) is greater than the total amount of bond energy in the product(s) of the reaction, then the reaction releases energy. If the total amount of bond energy in the reactant(s) is less than the total amount of bond energy in the product(s), then the reaction requires energy.

# 9

# A Little Physics: Energy

1 In an ordinary chemical reaction energy is neither created nor destroyed.

2 ATP is an energy-transferring compound in living things.

3 One form of energy may be converted into another form of energy.

4 Food contains potential energy that can be stored for a long time before organisms use it.

75

# First Law of Thermodynamics

One of the basic principles of physics, the first law of thermodynamics, states that in ordinary chemical reactions energy can be neither created nor destroyed.[1] Let us consider a chemical reaction in which the substances $A$ and $B$ react and form $C$ and $D$:

$$A + B \longrightarrow C + D$$

If the total bond energy in compounds $A$ and $B$ is greater than the total bond energy in compounds $C$ and $D$, then energy is released. Sometimes the energy is released spectacularly, as in the explosion of a fuel dump or in a Fourth of July fireworks display. In other cases the release of energy is not ordinarily noticeable, as when a nail rusts.

Energy-releasing reactions often are written in this form:

$$A + B \longrightarrow C + D + \text{energy}$$

This must not be construed to indicate the creation of energy. The energy indicated on the right side of the equation just accounts for the difference in energy content of the reactants and the products:

Energy in $C$ + energy in $B$ = energy in $C$
    + energy in $D$ + energy released

All chemical reactions are, in theory at least, reversible. If we wish to form $A$ and $B$ from $C$ and $D$,

$$C + D \longrightarrow A + B$$

then we must supply energy equal to the amount released by the first reaction:

Energy in $C$ + energy in $D$ + energy supplied
    = energy in $A$ + energy in $B$

In the equation of an energy-requiring reaction, the required energy is usually written over the arrow:

$$C + D \xrightarrow{\text{energy}} A + B$$

# Energy Transfer and ATP

There are many reactions in living cells that require energy; among them are the synthesis of polysaccharides from monosaccharides, the synthesis of fats from glycerol and fatty acids, and the synthesis of proteins from amino acids. Before these syntheses occur, energy must be extracted from some food, often glucose, by the process of respiration (Chapter 12), which is an energy-releasing reaction. The energy released is transferred to the energy-requiring reaction by a special group of organic compounds, the adenosine phosphates.

Adenosine monophosphate (AMP), which is not an energy-transferring compound, is formed from three simpler compounds:

Adenine          Ribose

Phosphoric acid

[1] In a nuclear reaction (such as the explosion of an atomic bomb) a small amount of matter is converted into a great deal of energy. This type of reaction does not concern us in this book, but it does illustrate the modern concept that matter and energy are really two forms of the same thing.

Together adenine and ribose are called adenosine, hence the name adenosine monophosphate:

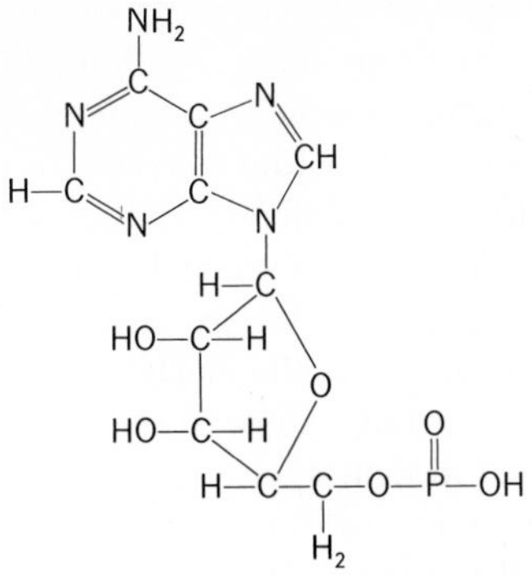

Adenosine monophosphate (AMP)

or more briefly,

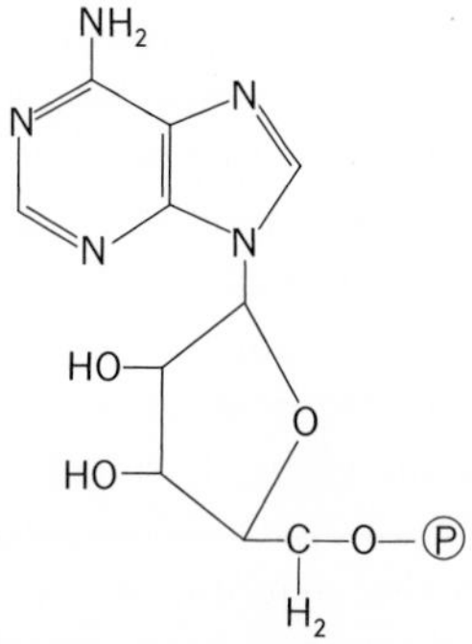

AMP

The addition of a second phosphate to the one already in the molecule results in the formation of adenosine diphosphate (ADP or AMP~ⓅⓅ):

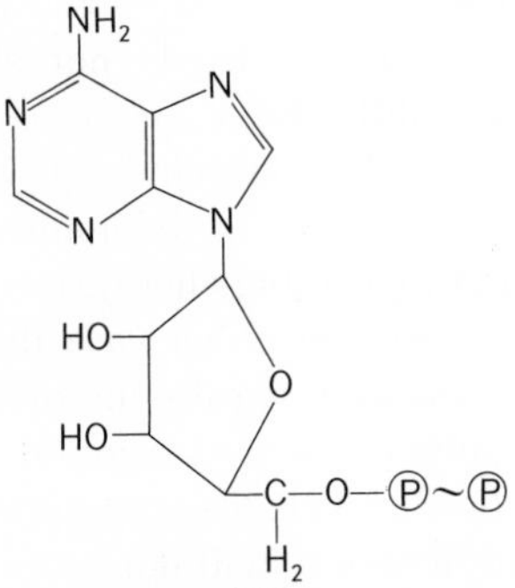

ADP

The bond between the two phosphates is a **high-energy bond** that contains more energy than most chemical bonds. It is indicated by a wavy line rather than by a straight line.

The addition of a third phosphate to the molecule results in the formation of adenosine triphosphate (ATP or ADP~Ⓟ or AMP~Ⓟ~Ⓟ):

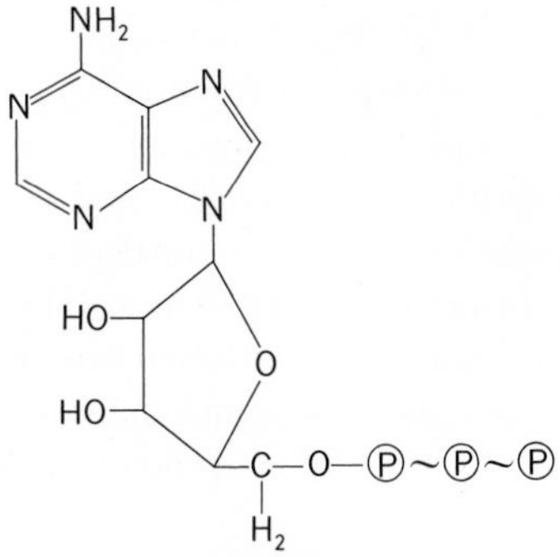

ATP

The bond between the last two phosphates is also a high-energy bond. A great deal of energy is required to form these high-energy bonds, and a great deal of energy is released when they are broken.

ATP is the most common energy-transferring compound in living cells. It transfers energy from energy-releasing reactions to energy-requiring reactions. For example, when glucose is used in respiration, it is broken down to simpler substances, and much of the energy released by this reaction is used in the synthesis of ATP from ADP and phosphate:

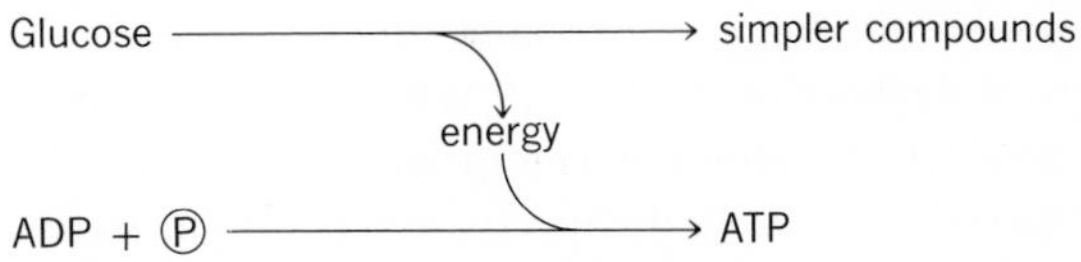

The energy that goes into the formation of the last phosphate bond may be utilized in the synthesis of polysaccharides, fats, proteins, or other complex molecules when that bond is broken by the conversion of ATP to ADP and phosphate:

77

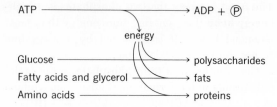

The ADP and (P) that are formed can be used again in the synthesis of more ATP whenever more energy is released by respiration of glucose. The complete respiration of one glucose molecule releases enough energy for the formation of at least 38 ATP molecules. Energy released by one ATP molecule is enough to unite two monosaccharide molecules, two amino acid molecules, or a glycerol molecule and one fatty acid molecule in a dehydration synthesis.

## Energy Conversions

Energy exists in several forms. There are two general kinds of energy: **kinetic energy** and **potential energy.** Kinetic energy is the energy of moving particles; some examples are electricity (movement of electrons along an electric circuit), heat (the never-ending motion of atoms and molecules in all substances), and mechanical energy (the motion of larger objects). Potential energy is the energy present in matter because of its location or its configuration. A rock resting at the top of a hill has potential energy; it has more than does a similar rock at the base of the hill. When a spring is compressed, its potential energy increases; when the spring is released, the energy is released as kinetic energy. Another form of potential energy of great importance in biology is the energy in chemical bonds, especially those in food molecules.

Although energy cannot be created or destroyed, one form of energy may be converted into another. When you turn on a light switch, you make use

of several conversions of energy. Like the rock on the hilltop, water above a waterfall has a certain potential energy because of its elevated position. As the water drops down the falls, the potential energy becomes mechanical energy that can be used to turn a turbine that converts the mechanical energy into electrical energy. Where waterfalls are not available, the burning of fuel may be used to release the potential energy in the fuel, and this turns the turbine. As the electrical current passes through a light bulb, some of the electrical energy is converted to light energy. As the light shines in the room, it is absorbed by the objects that it strikes and is converted to heat.

Energy conversions also occur in living things. When glucose is oxidized in living things, some of the bond energy of the glucose molecules is converted into the bond energy of ATP molecules; this energy may be converted into bond energy in some cellular constituents (for example, polysaccharides, fats, or proteins), or, in muscles, it may be converted into mechanical energy as the organism moves. Thus we see that the energy in the breakfast cereal, after undergoing some changes, can indeed propel our young baseball player to home plate.

Energy-requiring reactions do not occur spontaneously; they happen only if enough of the right kind of energy is available. Let us return to the rock at the bottom of a hill; for it to acquire as much energy as the rock at the top, it must be raised to the top of the hill (or to some other place of equal height). No one would expect this to happen spontaneously. Rocks do not of their own accord jump uphill. But a person could carry it to the top of the hill, thereby expending energy. A properly placed stick of dynamite might, upon exploding and expending energy, toss the rock to the top of the hill. In either case the expending of energy is necessary to raise the rock to a higher level and results in the possession of more potential energy by the rock. To accomplish this work not any kind of energy will do: for example, light may shine on the rock and deliver to it many millions times as much energy as would be needed to raise it, but because light is not the kind of energy that can cause rocks to rise, the rock will continue to lie in place.

The energy-requiring reactions of living things proceed only when the right kind of energy is supplied to them. In nearly all of these reactions the right kind of energy is the energy in the phosphate bonds of ATP. In some of these reactions heat theoretically might supply the necessary energy, but so much heat would be required that the cell well might be incinerated.

We have seen that the energy in the ATP molecule comes from the food molecules in the cell. We might look at the source of energy in foods. Green plants have the ability to manufacture food from carbon dioxide and water by the process of photosynthesis. This is an energy-requiring reaction that utilizes the energy in sunlight (and here light is the only kind of energy that will do). Thus we find that the energy our baseball player expends is actually converted sunlight. With the exception of a few chemosynthetic bacteria, this is true of energy expended by all organisms. Thus practically all organisms are totally dependent, directly or indirectly, on sunlight. All organisms are continually utilizing energy by converting one form of it to another.

One of the significant features of foods is that they are a store of potential energy that can be kept more or less indefinitely before they are used by organisms. In some cases potential energy may remain little changed for millions of years. Coal is the compacted remains of plants that lived about 300 million years ago. During all this time potential energy in compounds formed in photosynthesis by these plants has remained in the coal.

It is released only when the coal is burned. In a similar way the potential energy in staples such as sugar, flour, and rice and in many canned and frozen foods may be stored for at least a few years. It may also be easily transported. Thus the energy that shone as sunlight five years ago on a wheat field in Montana or on a sugar cane field in Hawaii may be moving your body today. Because your body also can store food, some of the energy you expend today may have entered your body in a meal you ate last week or last year.

## Questions

1. Define energy. List some forms of energy.
2. How many forms of energy did you use directly in the last 24 hours? What are they?
3. How many forms of energy did you make indirect use of in the last 24 hours? What are they?
4. What is the role of ATP in living things? What kind of reaction generates ATP? What kinds of reactions require ATP?
5. Differentiate between kinetic energy and potential energy. What is the significance of food in this respect?

## Suggested Readings

Baker, J. J. W., and G. E. Allen. *Matter, Energy, and Life.* Reading, Mass.: Addison-Wesley Publishing Co., Inc., 1965.

Goldsby, R. A. *Cells and Energy.* New York: Macmillan Publishing Co., Inc., 1967.

Stumpf, P. K. "ATP," *Scientific American* 188(4): 85–92 (April 1953).

# 10

# Enzymes

1 **Enzymes are proteins that catalyze (affect the rate of) chemical reactions in cells.**

2 **Enzymes are specific; each enzyme catalyzes only one chemical reaction or one group of similar reactions.**

3 **Each enzyme has its optimum temperature and optimum pH at which it functions best. Extremes of temperature or pH can inactivate enzymes.**

4 **Enzyme activity can be inhibited by other chemical compounds called enzyme inhibitors.**

5 **As far as we know, life is impossible without enzymes.**

In the preceding chapter we saw that energy-requiring reactions do not occur spontaneously. By contrast, one might anticipate that energy-releasing reactions would occur spontaneously. In some cases this is essentially true. A rock teetering on the top of a hill may require such a small push to send it crashing downhill and expending its energy that we might ignore this little investment of energy. So, too, if some highly reactive compounds come into contact with each other, an explosion may result; the heat present in the air (even on a cold day) supplies enough energy, and we say that for all practical purposes the reaction is spontaneous.

Some other energy-releasing reactions will not occur unless a relatively large amount of energy, called **energy of activation,** is available. We can draw an analogy with a rock lying in a depression some distance from the top of a hill. Someone or something must expend energy to push the rock to the top before it can fall.

If, in a chemical reaction, the reactants C and D lie behind an "energy hill," then energy of activation must be supplied to raise them to the top of the hill from which they can "fall" down and release energy as they become the products A and B (Fig. 10.1).

**10.1** Energy of activation. Although a chemical reaction may release energy, a certain amount of energy, called energy of activation, is necessary to initiate the reaction.

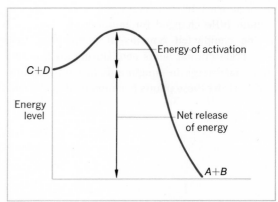

Wood burning — glucose reacts with oxygen (most of the energy is
~~produces~~ carbon dioxide ~~oxygen~~ given off in the form of heat)
energy released very quickly —

ENZYMES

One of the most common energy-releasing reactions in cells is respiration in which glucose reacts with oxygen and produces carbon dioxide and water. Now this reaction requires a fairly large amount of energy of activation. Because this is so, glucose is very stable and can be stored for long periods of time even while exposed to the oxygen of the air. Glucose and oxygen in a bottle or beaker at ordinary temperatures simply will not react. If, however, enough heat is applied, the two compounds do react. But in this case the glucose burns, and its energy is released very quickly, much of it as heat. (This is the sort of thing that happens when logs are burned in a fireplace. An initial piece of burning kindling supplies enough heat for the logs to catch fire. The wood then releases a great deal more heat.) If this were to happen in a living cell, the cell would be killed.

That great quantities of heat are not required as energy of activation and that such violent reactions do not occur in living cells are due to the presence of a group of compounds called **enzymes.** Enzymes are organic catalysts naturally produced only in living cells. Let us look first at the nature of catalysts and then proceed specifically to enzymes.

## Catalysts

A **catalyst** is a substance that accelerates (or, very rarely, slows down) the rate of a chemical reaction; we say that a catalyst catalyzes a reaction. In industry catalysts ordinarily are used to speed up chemical reactions. Catalysts cannot initiate a reaction that otherwise would be impossible, nor do they contribute energy; but they can speed up reactions that without them ordinarily would proceed so slowly as to be nearly imperceptible or that would require the application of impractical amounts of heat or other forms of energy. Industries that depend on chemical reactions often use catalysts—powdered platinum or nickel, for example—to make these reactions economically feasible. Catalysts are known to enter somehow into the reactions that they catalyze, but they remain unchanged and can be used over and over. For this reason only a very small amount of a catalyst is necessary to complete a reaction in a large batch of reactants.

Catalysts may be either inorganic or organic substances. Inorganic catalysts are relatively nonspecific; that is, a given catalyst may catalyze several different kinds of reactions. Organic catalysts, which are specific in the reactions they catalyze, are important in living things.

## Enzymes

Enzymes, the catalysts of living things, are all proteins, although the proteins often are associated more or less intimately with nonprotein molecules. In some cases the nonprotein portion is a metal (such as iron, copper, manganese, or zinc) or a coenzyme. A coenzyme is an organic substance consisting, in part, of a vitamin (such as thiamin or riboflavin). Like inorganic catalysts, enzymes enter into the reactions they catalyze, and they emerge unchanged and are utilized again and again. Therefore the concentration of any one enzyme in a given cell can be extremely low compared to the concentrations of many other organic materials.

### Enzyme Specificity

Enzymes differ from inorganic catalysts in being specific, that is, a given enzyme usually catalyzes only one reaction or one group of similar reactions. For example, an enzyme that catalyzes the synthesis of starch cannot catalyze the synthesis of sucrose or a fat. The compound on which an enzyme acts is called its **substrate.** The specificity of an

81

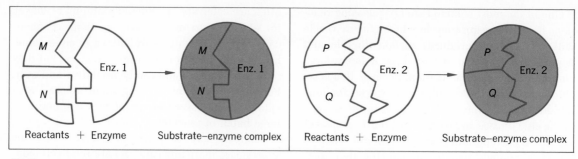

**10.2** Enzyme specificity. Geometrical relationship between enzymes and reactants resembles that of a lock and a key. Enzyme 1 fits only reactants $M$ and $N$. Enzyme 2 fits only reactants $P$ and $Q$. All figures are fanciful and are not intended to represent the actual shapes of any molecules.

enzyme is believed to be due in part to its three-dimensional shape (caused by the folding of the protein molecule) which permits a lock-and-key relationship with the reactant molecules with which it forms a temporary complex. This is illustrated in the fanciful diagram in Fig. 10.2 in which it is obvious that enzyme 1 makes a good "fit" with reactants $M$ and $N$ but not with reactants $P$ and $Q$; therefore it could catalyze a reaction involving reactants $M$ and $N$ but not one involving $P$ and $Q$. Enzyme 2, on the other hand, could catalyze a reaction with $P$ and $Q$ but not one with $M$ and $N$. Enzyme molecules also bear positive and negative electrical charges at various locations on their surfaces (because portions of the molecule may ionize). The attraction of positive charges on the enzyme molecule to negative charges on the substrate molecule or vice versa further serves to enhance the specificity of enzymes.

Nearly all the chemical reactions in cells are catalyzed by enzymes. Because enzymes are proteins, theoretically there is a very large number of possible enzymes; these can account easily for the catalysis of the many reactions in cells.

## How Enzymes Function

Enzymes seem to function by lowering the energy hill that must be overcome in some reactions. Like an inorganic catalyst, an enzyme acts by for-

ming a complex with the reactants. For instance, if the reaction $M + N \longrightarrow S + T$ is catalyzed by an enzyme, then a temporary intermediate complex, called an enzyme–substrate complex, is formed between $M$ and $N$ and the enzyme. When the products $S$ and $T$ are formed, the enzyme molecule is released unchanged:

$$M + N + \text{enzyme} \longrightarrow M\text{-}N\text{-enzyme} \longrightarrow S + T + \text{enzyme}$$

The intermediate complex has a lower energy state than that required in an uncatalyzed reaction. The enzyme effectively lowers the energy hill (Fig. 10.3). This makes it much more likely that the reaction will occur. When it is released the enzyme molecule can be used again. A single enzyme molecule may be able to catalyze the same reaction several million times in a minute.

Notice that the energy level of the reactants is the same whether or not a given reaction is catalyzed, and the energy level of the products is the same in both cases, too. Therefore, providing that a reaction does occur, the net amount of energy released per unit of reactants is the same with or without the assistance of enzymes. Many reactions in living things occur in series or metabolic pathways, $A$ forming $B$, $B$ forming $C$, and so on, with a different enzyme catalyzing each reaction in the pathway:

$$A \xrightarrow{\text{enzyme 1}} B \xrightarrow{\text{enzyme 2}} C \xrightarrow{\text{enzyme 3}} D \xrightarrow{\text{enzyme 4}} E$$

82

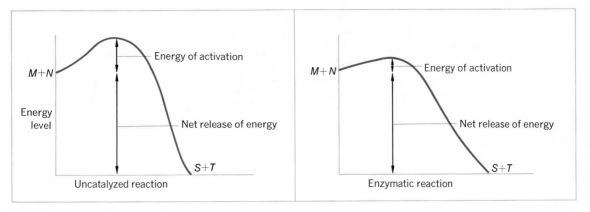

**10.3** Enzymes reduce the amount of energy of activation required to make a reaction proceed.

Each enzyme lowers the energy hill of its specific reaction, and, in the case of energy-releasing reactions, the overall release of energy is stepwise (Fig. 10.4).

Oxidation of organic compounds by burning, as in a fireplace, releases a great deal of heat and light in a single step, as in Fig. 10.1. The oxidation of glucose by respiration (Chapter 12) in living cells

**10.4** Stepwise energy release in a metabolic pathway. The hypothetical pathway shown here consists of four chemical reactions, A → B, B → C, C → D, and D → E, each catalyzed by a different enzyme. Each reaction requires its own energy of activation, and each releases a small amount of energy. Compare with Fig. 10.1 in which energy is released in one step.

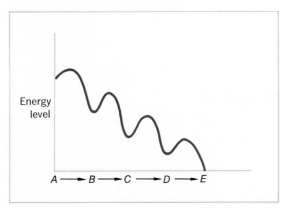

is a stepwise reaction involving many intermediate reactions that release energy gradually, as in Fig. 10.4. Although burning and respiration release the same amount of energy by the oxidation of identical substrates (glucose, for instance), respiration releases no light and less heat than burning. Some of the bond energy in the glucose molecules remains as chemical bond energy, for it is used in the synthesis of ATP from ADP. Therefore, respiration not only occurs at a temperature comfortable for living cells, it also retains some of the energy in a form immediately useful to the cell.

We have mentioned already that one of the significant features of foods is that they are a store of potential energy that can be kept indefinitely. This is because they lie behind an energy hill that cannot be surmounted at ordinary temperatures without the presence of appropriate enzymes. In the flour sack or sugar bowl, food remains unreactive and can be stored against the future. So, too, in some living tissues, such as the liver of animals or the roots of plants, food can be stored until it is needed; at this time the appropriate enzymes are synthesized.

This does not mean that foods cannot spoil; it is an everyday experience that they do, but this is just another example of enzyme activity. For many foods to spoil, enzyme-synthesizing micro-

83

organisms such as bacteria and fungi must be present; these organisms use the food much the same way that you do. Also, enough water must be present to support the growth of these organisms. The sugar in an open sugar bowl probably has received plenty of bacterial and fungal cells from the air, but it lacks water. A can of soup has plenty of water, but in the canning process it was sterilized and so lacks microorganisms. Given both the organisms and water, most foods will spoil as the microorganisms digest, oxidize, or otherwise alter the organic compounds.

### Effects of Temperature

Enzymatic reactions are sensitive to temperature changes. Within limits, the rate of an enzymatic reaction, like other reactions, is increased by an increase in temperature. However, the shapes of many protein molecules, including enzymes, are distorted quickly by high temperatures. Such distortion is called **denaturing** or **coagulation** and often is irreversible. (The white of an egg is nearly all protein; when the egg is cooked the three-dimensional configuration of the protein molecules is changed in such a way that the white changes from a transparent fluid to an opaque solid; chilling will not restore the egg white to its original state.) Because the activity of an enzyme depends on its geometry, denaturing almost inevitably destroys its enzymatic activities. The rate of an enzymatic reaction is increased by an increase in temperature only until the temperature that denatures the enzyme is reached; then the rate of the reaction falls abruptly (Fig. 10.5). The temperature at which an enzyme is denatured varies with the particular enzyme. Many enzymes are denatured at about 40°C, which is only a little higher than normal human body temperature (37°C). Others require higher temperatures for inactivation; enzymes in the few organisms that grow in hot springs remain active at 80°C or higher.

Because of its denaturing effect, heat can be used to sterilize foods, medical equipment, and other

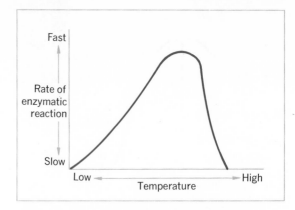

**10.5** Effect of heat on the rate of an enzymatic reaction. At low temperatures an increase in temperature increases the rate of the reaction, but eventually a temperature is reached that inactivates the enzyme. At this point the reaction rate drops rapidly. Different enzymes are inactivated at different temperatures.

materials. Application of an appropriate degree of heat for a sufficient length of time inactivates the enzymes of microorganisms, and the microorganisms die.

Most enzymes are not denatured by freezing; some can function well below the freezing point. But low temperatures do reduce the rates of most reactions in living things. This is the reason for using refrigerators and freezers for preserving food. At 5°C, the temperature of most refrigerators, the chemical activities of bacteria and fungi are slowed enough that most foods remain fresh for several days. At the below-freezing temperatures of a freezer, the reactions are slowed down to near zero, and the food may be kept for much longer periods of time.

### Effects of pH

Enzymes also are sensitive to changes in pH (Chapter 7). Each enzyme has its own optimum pH at which it functions best; at a slightly higher or a slightly lower pH the enzyme functions, but not so well; at a more extreme pH the enzyme may be inactivated completely. The enzymes that func-

84

tion in the acid contents of the human stomach have a low (acid) optimum pH; those that function in the alkaline contents of the small intestine have a high optimum pH. Most enzymes function best, however, in neutral or near neutral solutions.

## Enzyme Inhibition

**Enzyme inhibitors** are substances that are able to block the action of enzymes. Mercury, arsenic, and some other poisons exert their harmful effects by combining with enzymes at their active sites and so inactivating them. In many enzymes these active sites contain sulfhydryl groups (-SH) on their side chains:

$$\begin{array}{l} -SH \\ -SH \end{array}$$

Mercury compounds react with the sulfhydryl groups:

$$\begin{array}{l} -SH \\ -SH \end{array} + HgCl_2 \longrightarrow \begin{array}{l} -S \\ -S \end{array} Hg + 2HCl$$

Portion of     Mercuric
protein molecule     chloride

The sulfhydryl groups thus are rendered ineffective, the enzymes are inhibited, and the reactions for which the enzymes are specific do not occur. Interference with many reactions almost certainly kills a cell. Mercuric chloride is a powerful poison; in low concentrations it can be used to disinfect laboratory equipment and tabletops if other methods are inconvenient. Mercurochrome and merthiolate, used externally to disinfect wounds, are somewhat less potent mercury compounds; but they still are poisonous if taken internally.

Competitive inhibitors represent another group of enzyme inhibitors. A competitive inhibitor molecule has a shape that resembles that of a substrate

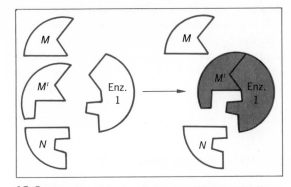

**10.6** Enzyme inhibition. Substance $M'$ resembles $M$ sufficiently to combine with its enzyme, but it differs enough from $M$ to prevent the approach of $N$. It also blocks the approach of $M$. Therefore the enzyme cannot catalyze its specific reaction between $M$ and $N$. Compare with $M$-$N$-$E$ complex in Fig. 10.2 (Figures are fanciful.)

molecule. It is able to occupy the substrate's site on the enzyme molecule. However, its shape differs enough from that of the substrate molecule that it cannot enter into the chemical reaction catalyzed by the enzyme (Fig. 10.6). It continues to occupy the site on the enzyme molecule and prevents the approach of the reactants. Thus it effectively blocks the reaction.

Sulfanilimide, one of the sulfa drugs discovered in the 1930's, was found to be effective in the treatment of some diseases caused by bacteria that require the substance para-aminobenzoic acid, which is used in the formation of folic acid, one of the vitamin B group. The sulfanilimide molecule has a structure similar to that of the para-aminobenzoic acid molecule and competes with it for a place on an enzyme molecule, but sulfanilimide is sufficiently different from para-aminobenzoic acid that it cannot be used in the synthesis of folic acid.

Para-aminobenzoic acid      Sulfanilimide

85

Administration of sulfanilimide in a high concentration renders the specific enzyme ineffective and kills the bacteria. Because human beings do not require para-aminobenzoic acid, administration of sulfanilimide is less harmful to them than it is to bacteria. (Sulfa drugs do have undesirable side effects, and after World War II they began to be replaced by newly discovered antibiotics.)

## Substrate Concentration

Up to a certain point the rate of an enzymatic reaction depends on the concentration of the substrate molecules. At very low concentrations of the substrate, even a relatively few enzyme molecules catalyze their specific reaction quickly because in relation to the substrate concentration the enzyme is present in excess. The addition of more enzyme molecules does little to speed the reaction, but the addition of more substrate molecules to the reaction mixture does increase the rate of the reaction (Fig. 10.7). At very high substrate concentrations,

**10.7** The effect of changing substrate concentration on the rate of an enzymatic reaction; enzyme concentration is assumed to remain constant. At low concentrations of the substrate, an increase in substrate concentration increases the rate of the reaction; this is reflected in the rising portion (A) of the curve. At high substrate concentrations, the addition of more substrate molecules has no effect on the reaction rate, for all the enzyme molecules are occupied fully and cannot accommodate more substrate molecules at any moment; this is represented by the horizontal portion (B) of the curve.

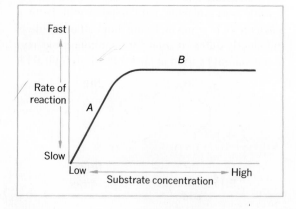

the enzyme molecules become saturated with substrate molecules; the addition of more substrate does not increase the rate of the reaction.

## Naming of Enzymes

Most enzymes are named for the types of reactions they catalyze or for the substrates on which they act. The ending -*ase* is used for most enzyme names. **Dehydrogenases** remove hydrogen from a wide variety of substrates. **Decarboxylases** remove carbon dioxide from some organic acids. **Phosphatases** remove phosphates from phosphorylated compounds. **Hydrolases** digest (hydrolyze) certain compounds. Some hydrolases are **sucrase,** which catalyzes the digestion of sucrose; **maltase,** which catalyzes the digestion of maltose; **amylase,** which catalyzes the digestion of starch (*amylum* is the Latin word for starch); **lipases,** which catalyze the digestion of lipids (fats and related substances); and **proteases,** which catalyze the digestion of proteins. Some of the enzymes discovered before this system of nomenclature was accepted have names that do not agree with the system. Two protein-digesting enzymes, pepsin and trypsin, are examples.

Because an enzyme is neither a reactant nor a product, in writing the equation for an enzymatic reaction we place the name of the enzyme over the arrow. Thus the digestion of maltose to glucose by maltase is written:

$$C_{12}H_{22}O_{11} + H_2O \xrightarrow{\text{maltase}} 2C_6H_{12}O_6$$

maltose    water      glucose

## Cells and Enzymes

Life without enzymes does not seem possible. Without enzymes so many of the reactions occurring in living cells would proceed much too slowly to support life as we know it. Furthermore, a cell is not a little box in which all reactions of which it is capable occur simultaneously and continuously. Rather, only certain enzymes are present and active in a cell at any one time; one set may function only during reproduction, another set during active growth, another when a particular food is present, and so on. This pattern of activity ensures that the chemical reactions in the cell are coordinated and contribute efficiently to the life of the cell.

A cell synthesizes its own enzymes, generally only when they are needed, and frequently the enzymatic reaction occurs within that cell. In other cases, cells secrete enzymes, and the enzymatic reaction occurs outside the cell. Digestion of macromolecules in the food we eat takes place largely in our digestive tracts. Because the digestive tract is a tube open at both ends, anything in the tract is, in one sense, outside the body. Generally, macromolecules cannot pass across plasma membranes (Chapter 17) and so cannot be absorbed by our cells. They must be digested to smaller, soluble molecules which can be absorbed. Several parts of the digestive system (Chapter 31) secrete enzymes into the digestive tract where digestion occurs. The products then enter the cells lining the digestive tract. From these cells the circulatory system transports them to other cells where they can be used in a variety of ways, including the synthesis of new macromolecules. (It may be interesting to note here that the secretion of enzymes—which are proteins, and therefore macromolecules themselves—is one of the exceptions to the generalization that macromolecules do not cross plasma membranes.)

For these reasons human beings and other animals cannot obtain their enzymes in their food, for they are destroyed before they could be absorbed by the body. But this is no loss to us. To a large extent a chicken is a chicken because of some of its peculiar enzymes (see Chapter 43 for the genetic control of this), and a carrot is a carrot for the same reason. We might be altogether different creatures if our bodies absorbed the enzymes in our food and incorporated them into our cells.

# Questions

1. Define the following terms.
catalyst *= substance that accelerates or rarely slows down the rate of a chemical reaction*
enzyme *= catalysts of living things — are proteins*
energy of activation *— energy required for initiation of chemical reaction*
substrate *— compound on which the enzyme acts.*

2. How does an enzyme resemble an inorganic catalyst? How do they differ? *energy released very quickly —*

3. Contrast the burning of wood with the biological oxidation of glucose. How do enzymes account for the difference? *they catalyze rate of chemical reaction — energy released as needed — form CO₂ + H₂O — otherwise the cell would burn up. (rate of)*

4. Compare the amount of the energy released by the burning of a substance with that released by its biological oxidation.

5. How are enzymes affected by temperature? By pH? *denatures   See back pg.*

6. Give examples of chemical reactions catalyzed by enzymes.

7. Chemically, what kind of a substance is an enzyme? *Proteins*

8. Of what significance is the fact that an enzyme is not used up in a chemical reaction? *The cell would die (burns up)*

9. Why must we synthesize our own enzymes rather than obtain them from our food? *Macromolecules — membranes @ Digested cannot pass plasma*

10. Why can some foods, like starch, be stored for long periods of time whereas fresh peaches spoil quickly? *must have water to support growth of bacteria*

11. Why are extreme heat and cold useful in preserving food? *inactivates the enzymes of bacteria   slows down chemical reactions of bacteria & fungi*

12. Give arguments for and against the possibility of life without enzymes.

# Suggested Readings

Baker, J. J. W., and G. E. Allen. *Matter, Energy, and Life.* Reading, Mass.: Addison-Wesley Publishing Co., Inc., 1965.

Bernhard, S. *The Structure and Function of Enzymes.* New York: W. A. Benjamin, Inc., 1968.

Devlin, R. M. *Plant Physiology,* 2nd ed. New York: Van Nostrand Reinhold Company, 1969.

Goldsby, R. A. *Cells and Energy.* New York: Macmillan Publishing Co., Inc., 1967.

Koshland, D. E., Jr. "Correlation of Structure and Function in Enzyme Action," *Science* 142: 1533–1541 (1963).

Locke, D. M. *Enzymes—The Agents of Life.* New York: Crown Publishers, 1969.

87

Neurath, H. "Protein-digesting Enzymes," *Scientific American* 211(6): 68–70 (December 1964).

Pfeiffer, J. E. "Enzymes," *Scientific American* 179(6): 28–40 (December 1948).

Phillips, D. C. "The Three-dimensional Structure of an Enzyme Molecule," *Scientific American* 215(5): 78–90 (November 1966).

(From Lock + key complex with reactants)

②. They both enter into the reactions they catalyze & emerge unchanged + are utilized again & again.

Different — Enzymes catalyze specific reactions — Only one reaction or one group of similar reactions.

Inorganic — are not specific — They catalyze several different kinds of reactions.

5. Temperature denatures protein molecules, incl. enzymes, that is their shape is changed & they cannot have a key & lock relation with the reactants, thereby they are inactivated.

PH — each enzyme has its own optimum PH at which it function best. An extreme in the PH. caused inactivate them

⑧ They emerge from reactions unchanged
The enzyme can be utilized over & over again.
Therefore only small amts are needed to complete a reaction in a large batch of reactants.

③④ Both release the same amt of energy int —

④ Burning of a substance — oxidation of organic compounds by burning releases a great deal of heat & light in a single step.
the oxidation of glucose by respiration in living cells is a stepwise reaction that release energy gradually.

**88** Respiration releases no light & less heat than burning
Some of the bond energy of the glucose molecules is retained for the synthesis of ATP from ADP. Retain some of the energy in a form immediately useful to the cell.

③ Enzymes — require many intermediate reactions that release energy gradually
The intermediate complex has a lower energy state than that required in uncatalyzed reactions. (Lock & key with substrate)

If one were asked to name the most important chemical reaction carried on by living things, he might find it difficult to select a single reaction. Indeed he might complain that the question is not fair, for there are many reactions known to be essential for the maintenance of life, and all of these well could rank first in a many-way tie for first place. Among these many important reactions that he might consider are photosynthesis and other synthetic reactions, respiration, and digestion. Respiration, digestion, and some other reactions are carried on by all living organisms. Photosynthesis is not. But photosynthesis possesses a unique significance: it is the only chemical process that utilizes light and that makes this energy generally available to living things; it is, for all practical purposes, the only chemical reaction that supplies the energy on which human life and the lives of all plants and animals depend.[1] In photosynthesis the light energy is converted into the chemical bond energy in food molecules. Another unique feature of photosynthesis is that it is the only process that replenishes our atmosphere with oxygen, a substance essential for many organisms, including humans.

**Photosynthesis** is carried on by most kinds of plants; these include algae, liverworts, mosses, ferns, conifers, and flowering plants. Only the fungi and most of the bacteria are not photosynthetic; these plants lack chlorophyll. Most of man's food comes directly or indirectly from land plants, especially the flowering plants. His carbohydrates are derived primarily from the cereal grains (rice, wheat, corn, barley, sorghum) and to a lesser extent from starchy tubers and roots (white potatoes, sweet potatoes, yams, cassava, taro). His protein sources are primarily beans and peas if he is poor, and domestic animals (cattle, pigs, sheep, poultry) if he is richer; in the latter

[1]Chemosynthesis, a process by which a few species of bacteria synthesize food by utilizing the energy in the chemical bonds of some inorganic compounds, is an exception to this statement; but the amount of energy chemosynthesis contributes to the biological world as a whole seems insignificant.

# 11
# Photosynthesis

1 Photosynthesis is the synthesis of glucose (or other food) from carbon dioxide and water in the presence of light by plant cells possessing chlorophyll.

2 Photosynthesis is the source of nearly all the food of the organisms of the world.

3 Photosynthesis is virtually the only source of gaseous oxygen in the atmosphere.

4 The light reactions of photosynthesis convert light energy into the chemical bond energy of ATP and $NADPH_2$.

5 The dark reactions of photosynthesis utilize the chemical bond energy of ATP and $NADPH_2$ in the reduction of carbon dioxide to glucose.

case the animals have fed upon plants. The fats in his diet also may come from either plant or animal sources, depending on his affluence. His vitamins come from an assortment of fruits, vegetables, meats, milk, and eggs.

In spite of the fact that the total amount of food synthesized in the sea far exceeds that formed on land, relatively little of man's food comes from the ocean. Most of the photosynthetic plants of the ocean are microscopic algae; and although they produce more food than do terrestrial plants, no economically feasible way of harvesting them for direct consumption by humans has been devised. Neither does there seem to be a great demand for algae on the dinner table; this may reflect a real or supposed lack of mouth-watering flavor in most algae. However, these plants do represent the basic menu for many small marine animals, which in turn are fed upon by other animals. Among these are tuna, salmon, lobster, shrimp, and clams. Some seashore communities, such as those on small oceanic islands or where climate or soil are poor for agriculture, depend heavily on marine animals for their food, but the world's human population as a whole utilizes little of the food produced by photosynthesis in the sea.

## Factors Affecting Photosynthesis

The overall equation for photosynthesis as it is performed by most plants may be written:

$$6CO_2 + 6H_2O \xrightarrow{\text{light, chlorophyll}} C_6H_{12}O_6 + 6O_2$$

This equation indicates that at least four things are necessary for photosynthesis: light as the energy source, chlorophyll as the light-absorbing compound, and carbon dioxide and water as reactants. Actually, the process of photosynthesis is a much more complex one than this equation would indicate. It consists of many individual chemical reac-

90

tions, most of which are catalyzed by their own specific enzymes; some of them are discussed later in this chapter.

### Light

The light source used naturally by plants is the sun, but some artificial lights are capable of supporting photosynthesis adequately enough to maintain healthy plants. Visible light is one of several forms of radiation that physicists call electromagnetic radiation (Fig. 6.1). Some other forms are ultraviolet and infrared radiation, X-rays, and radio waves; these cannot be seen by humans. Visible light from the sun appears white to us, but if it passes through a prism, it separates into its component parts, which are called the rainbow colors: violet, blue, green, yellow, orange, and red. A display of these colors spread out is called a spectrum. Of the several colors of visible light, violet, blue, and red are the most efficient in supporting photosynthesis, and green, yellow, and orange are the least efficient. For this reason lamps rich in green, yellow, or orange are very poor choices for anyone wishing to raise plants in a basement or other place lacking sunlight. Lamps producing a better imitation of sunlight are preferred.

Because light is necessary for photosynthesis, this process will not occur in the complete absence of light. In general, an increase in intensity of light causes an increase in the rate of photosynthesis, but very high light intensities will decrease the rate. The light intensity at which photosynthesis occurs most rapidly is its optimum light intensity. The optimum intensity varies with different species of plants; generally it is lower for shade-loving species than for plants of fields and meadows.

### Chlorophyll

It is a basic rule in photochemical reactions that in order for light to be utilized in a reaction it must first be absorbed by a pigment. The light-absorbing pigment in photosynthesis is **chlorophyll.** Chlorophyll exists in many plants in a mixture of four pigments: chlorophyll a, chlorophyll b, carotene, and xanthophyll. Chlorophylls a and b are green; their structural formulas are shown in Fig. 11.1. The main portion of the molecule is a porphyrin ring, a ring composed of four smaller rings linked together; the center is occupied by an atom of magnesium. The porphyrin ring bears several short side chains and a longer chain of phytol alcohol. The entire molecule is linked to a protein (not shown in the illustration). Chlorophyll a is the pigment directly involved in photosynthesis. Any energy absorbed by chlorophyll b must be passed to chlorophyll a before it can be used. Carotene, which is orange, and xanthophyll, which is yellow, have very different molecular structures. In most cases they do not seem to be directly involved in photosynthesis, but they may transfer the energy they absorb from light to chlorophyll a as chlorophyll b does.

When white light shines on chlorophyll, violet, blue, and most of the red light are absorbed; this is the reason these colors are most efficient in photosynthesis. Very little of the green, yellow, and orange is absorbed by chlorophyll, and this is why they are so inefficient. Green, yellow, and orange light are transmitted or reflected by chlorophyll; because green predominates in the transmitted and reflected light, chlorophyll appears green to the human eye. If a vial of chlorophyll is placed in a beam of light that passes through a prism, the spectrum displays only the transmitted colors; the areas corresponding to the colors absorbed by chlorophyll are black.

Because chlorophyll is necessary for photosynthesis, only those cells possessing it are photosynthetic. In most terrestrial plants, chlorophyll resides mostly in the leaves, which are typically thought of as *the* photosynthetic organs. Young stems and fruits often have an appreciable quantity of chlorophyll, and other organs may also possess some. In algae all or nearly all of the plant is photosynthetic.

**11.1** Structure of chlorophyll a. In the living plant this molecule occurs in combination with a protein.

### Carbon Dioxide

The carbon dioxide necessary for photosynthesis is present as a gas in the atmosphere, and it is dissolved in lakes, oceans, and other natural waters. Its concentration in the atmosphere is about 0.03 per cent by volume. Carbon dioxide is absorbed by the leaves of terrestrial plants (see Chapter 22). The concentration of carbon dioxide available to the plant has a direct effect on the rate

91

of photosynthesis. No photosynthesis occurs in a complete absence of this gas. An increase in its concentration up to an optimum increases the rate of photosynthesis, but above this it becomes inhibitory. The optimum concentration varies from species to species and also with several environmental factors; it may range from about 0.2 to 0.5 per cent. Because of its ordinarily low concentration in the atmosphere, carbon dioxide usually is the limiting factor in photosynthesis during daylight hours.

## Water

The water used in photosynthesis is absorbed by roots and transported to the leaves. Active, living cells usually have ample water for photosynthesis.

## Temperature

Temperature has an effect on photosynthesis as it does on other enzymatically controlled reactions. Photosynthesis generally occurs most actively at temperatures within the range of about 0° to 35°C. Within this general range an increase in temperature increases the rate of photosynthesis, but when the temperature is reached at which the photosynthetic enzymes are inactivated, the rate of photosynthesis decreases. The optimum temperature for photosynthesis varies with the species of plant.

# The Chemistry of Photosynthesis

If we look at the structural formulas of the reactants and the products of photosynthesis, we see that the carbon atom in carbon dioxide undergoes a change in the kinds of elements to which it is bonded. In carbon dioxide all four of its bonds are with oxygen: $O{=}C{=}O$; but in glucose most of the

carbon atoms have only one bond with oxygen and one with hydrogen (and the rest with other carbon atoms):

When an atom (or molecule) has lost some bonds with oxygen and/or has gained some with hydrogen, that substance is said to be reduced. Thus in photosynthesis the carbon becomes reduced. The opposite of reduction is oxidation, in which a substance gains some bonds with oxygen and/or loses some with hydrogen. (Chemists define reduction and oxidation in somewhat different terms; they prefer to say that a substance becomes reduced when it gains an electron and becomes oxidized when it loses an electron. This definition has the advantage of covering more situations than reactions involving only bonds with oxygen and hydrogen, but for persons with little chemical background it often is easier to follow hydrogen and oxygen atoms in reactions, and it will be all we will require here.)

The reduction of carbon requires energy. This is where the light comes in. But light itself is not a form of energy that is capable of directly reducing carbon dioxide to glucose. (If it were, the end of every day probably would find the ground coated with a frosting of sugar.) Rather, the light is utilized in the formation of ATP from ADP and phosphate (Chapter 9), with the light energy being converted into the chemical energy of the high-energy phosphate bond. This bond energy is part of the energy used in the reduction of carbon dioxide to glucose as the ATP is converted back to ADP and phosphate.

The hydrogen required for the reduction of carbon dioxide comes from water. Water molecules are split, releasing some of their hydrogen to a special hydrogen-transporting compound called nicotinamide adenine dinucleotide phosphate (NADP); light is necessary for this process, too. NADP is another of the adenosine phosphates. It is one of two common hydrogen-transferring compounds common to cells throughout the living world; nicotinamide adenine dinucleotide (NAD) is the other. NAD and NADP are coenzymes of several dehydrogenase enzymes. The structural formulas for NAD and NADP are given in Fig. 11.2; note that they both consist in part of one of the B vitamins, nicotinic acid (niacin). Each of these compounds is capable of accepting two hydrogen atoms. When NADP accepts two hydrogen atoms it becomes reduced to $NADPH_2$. The two hydrogen atoms are transferred to carbon dioxide by a series of chemical reactions, and the $NADPH_2$ is oxidized to NADP, which can be used again. Ultimately the carbon dioxide is reduced to glucose.

These reactions of photosynthesis are divided into two groups called the light reactions and the dark reactions. They are summarized briefly in Fig. 11.3. The light reactions are those which are immediately dependent on light; they include the synthesis of ATP from ADP, the splitting of water into hydrogen and oxygen, and the synthesis of $NADPH_2$ from NADP and the hydrogen from the water. The dark reactions are those that can occur in the dark, but because they require the products of the light reactions, they are not completely independent of light and ordinarily occur in the light. The dark reactions include the reduction of

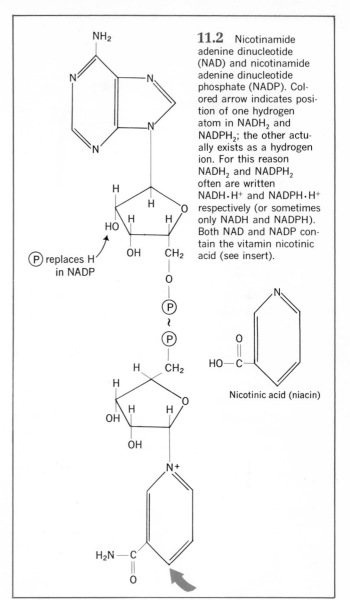

**11.2** Nicotinamide adenine dinucleotide (NAD) and nicotinamide adenine dinucleotide phosphate (NADP). Colored arrow indicates position of one hydrogen atom in $NADH_2$ and $NADPH_2$; the other actually exists as a hydrogen ion. For this reason $NADH_2$ and $NADPH_2$ often are written $NADH \cdot H^+$ and $NADPH \cdot H^+$ respectively (or sometimes only NADH and NADPH). Both NAD and NADP contain the vitamin nicotinic acid (see insert).

carbon dioxide to glucose with the energy of the ATP and the hydrogen of $NADPH_2$; ATP and $NADPH_2$ are converted back to ADP and NADP.

We might look at these reactions in a little more detail, especially with reference to the role of

93

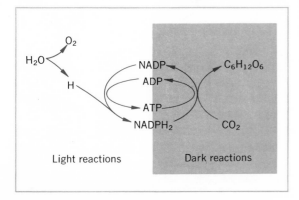

Light reactions    Dark reactions

**11.3** Summary of photosynthesis showing relationship of light and dark reactions.

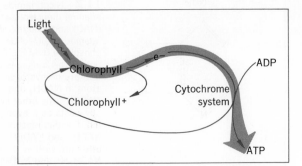

**11.4** Cyclic photophosphorylation. Energy pathway is indicated by colored arrow.

chlorophyll and the energy pathways in photosynthesis.

## Light Reactions

The light reactions are of two types: cyclic photophosphorylation and noncyclic photophosphorylation. These two names seem a little formidable at first, but we will see that they are descriptive of what happens in each set of reactions. They are photophosphorylations because something is phosphorylated with light energy (ADP is phos-

phorylated to ATP). In the one case an electron follows a cyclic path, leaving a chlorophyll molecule but eventually returning to it; in the other case an electron follows a noncyclic path, leaving a chlorophyll molecule and ending up in a different substance. We will look more closely at both of these sets of reactions.

In **cyclic photophosphorylation** (Fig. 11.4) when a molecule of chlorophyll a is struck by light, one of the electrons in the molecule absorbs its energy and becomes a high-energy electron. Such an electron has sufficient energy to escape from the chlorophyll molecule; when it does so, the molecule has a positive charge. The electron is accepted by a series of cytochrome molecules. **Cytochromes** are pigments with a porphyrin ring similar to that of chlorophyll, but the metal in the center is iron rather than magnesium. Like chlorophyll, cytochromes are linked with proteins. Cytochromes are common electron-transporting compounds in cells of nearly all living things. When an electron is transferred from one cytochrome molecule to another, the electron loses a little energy. The cytochrome system has the important characteristic of shunting some of this energy to the formation of

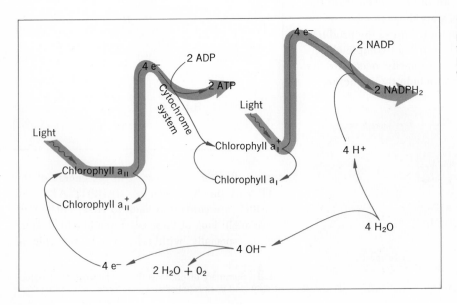

**11.5** Noncyclic photophosphorylation, somewhat simplified. Energy pathways indicated by colored arrows.

94

ATP from ADP. Thus we see that in cyclic photo-phosphorylation the energy of light is carried by a high-energy electron from the chlorophyll molecule to ATP by way of the cytochromes. This is the most important aspect of cyclic photophosphorylation. The only product of cyclic photophosphorylation is ATP. When the electron leaves the cytochromes, it returns to the chlorophyll molecule. By this time the electron has lost all its extra energy and has fallen back to its original low energy state.

In **noncyclic photophosphorylation** (Fig. 11.5) two separate chlorophyll molecules are involved. These are both molecules of chlorophyll a; but they are slightly different, and they absorb slightly different wavelengths of red light. They are called chlorophyll $a_I$ and chlorophyll $a_{II}$. Each of these chlorophyll molecules releases a high-energy electron after absorbing light. The electron released from chlorophyll $a_{II}$ passes along the cytochrome system, and the energy lost by the electron is utilized in the synthesis of more ATP. The ATP formed is one of the products of noncyclic photophosphorylation. The electron does not return to the molecule of chlorophyll $a_{II}$ from which it came; rather it replaces the electron lost by chlorophyll $a_I$.

In any sample of water, a few molecules are ionized into hydrogen ions ($H^+$) and hydroxyl ions ($OH^-$). Two hydrogen ions from water and two electrons released from two molecules of chlorophyll $a_I$ react with one molecule of NADP, reducing it to $NADPH_2$. The $NADPH_2$ is another product of noncyclic photophosphorylation.

The hydroxyl ions from the water react with each other, producing electrons, oxygen, and some water. Each of the electrons reacts with a molecule of chlorophyll $a_{II}$, replacing the electron that it lost.

We should also note here that the oxygen, another product of noncyclic photophosphorylation, comes from water (rather than from carbon dioxide as was once assumed). For this reason it has become popular to write the photosynthetic equation

$$6CO_2 + 12H_2O \longrightarrow C_6H_{12}O_6 + 6O_2 + 6H_2O$$

By adding the six molecules of water to each side of the older, traditional equation, we indicate enough water on the left side to account for all the oxygen produced on the right side.

To summarize, the products of noncyclic photophosphorylation are ATP, $NADPH_2$, oxygen, and water. For every four molecules of water originally ionized, one molecule of oxygen and two molecules each of ATP, $NADPH_2$, and water are formed. The newly formed water can be used in the reaction again.

## Dark Reactions

Although we speak of photosynthesis as the reduction of carbon dioxide to glucose, carbon dioxide itself is not directly reduced. In the dark reactions (Fig. 11.6) each molecule of carbon dioxide reacts with one molecule of ribulose diphosphate, a 5-carbon sugar, forming two molecules of phosphoglyceric acid. It is with these that two molecules of ATP from the light reactions react, forming two molecules of diphosphoglyceric acid. ADP is also formed. One molecule of $NADPH_2$ reacts with each molecule of diphosphoglyceric acid and reduces it to phosphoglyceraldehyde; it is at this point that the actual reduction of carbon occurs as the third carbon atom in diphosphoglyceric acid loses one bond with oxygen and gains one with hydrogen. Phosphoglyceraldehyde is a sugar—a phosphorylated, 3-carbon sugar. A pair of phosphoglyceraldehyde molecules can react, forming one molecule of glucose. Several other molecules of phosphoglyceraldehyde enter a complex series of reactions that regenerate ribulose diphosphate.

The dark reactions can be seen to be cyclic, with carbon dioxide, ATP, and $NADPH_2$ entering the cycle. NADP and ADP, which are produced in the cycle, can be used again in the light reactions. The other product, glucose, represents a net gain of energy to the plant; the light energy that was con-

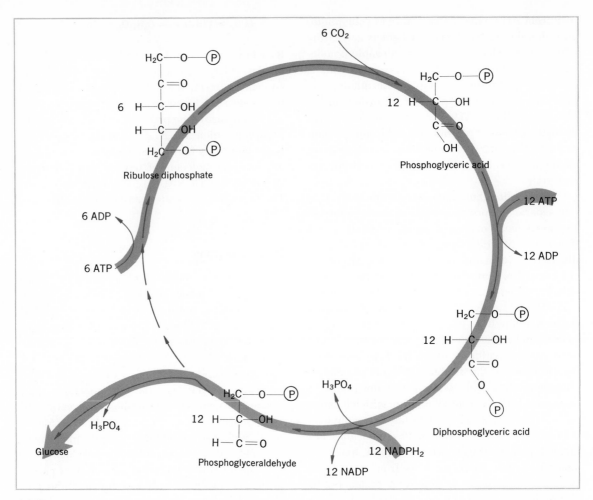

**11.6** Dark reactions of photosynthesis, somewhat simplified. Colored arrows indicate energy pathway.

verted to chemical bond energy in ATP and $NADPH_2$ in the light reactions is transferred to glucose in the dark reactions.

Although glucose is most commonly thought of as the sole product of photosynthesis, it is now known that amino acids and fatty acids also can be formed in photosynthesis. Fatty acids (and fats) require approximately 1.8 times as much energy (per carbon atom reduced) as does glucose. For this reason fats are more concentrated stores of energy than are sugars and other carbohydrates.

## The Significance of Photosynthesis

Glucose may be used either as an energy source or as a reactant in the synthesis of other organic compounds, such as fatty acids, fats, amino acids, proteins, vitamins, and hormones. To maintain itself, a healthy, actively growing plant must produce enough glucose to provide enough energy and

96

enough organic matter for its needs; otherwise it will die. Actually, most plants produce a great deal more than they require. The extra food is stored in roots, seeds, fruits, or other organs. Some of the food in seeds is utilized by young seedlings before they develop their own photosynthetic capacity, but much of the excess food produced by plants is eaten by human beings and other animals, and some decays when attacked by bacteria and fungi. The organic compounds consumed by these organisms become assimilated into their own bodies and serve as energy sources and as reactants in the synthesis of new cellular material. Thus an important feature of photosynthesis is that it converts light, carbon dioxide, and water into chemical bond energy and organic compounds that can be used by and that are required by all the living things of the world.

Man has learned to make some additional uses of the indirect products of photosynthesis. For example, he uses wood for houses, furniture, and paper; cotton, wool, fur, and leather for clothing; and a wide variety of materials for flavoring foods, for medicinal uses, and for a number of private vices ranging from the use of hard narcotics to drinking cocktails to chewing gum. In the labor of domestic animals, man utilizes the energy once trapped in photosynthesis; this energy is also released in the burning of certain fossil remains—coal, gas, and oil. An idea of the magnitude of the contributions of plants to life on earth can be obtained from the fact that every year they synthesize at least 200 billion tons of glucose and other organic compounds and somewhat more of oxygen. Our complete dependence on plants cannot be overemphasized.

# Questions

1. Describe photosynthesis from the point of view of the energy changes involved. Of what significance are these changes to the biological world?

2. Of what significance are the chemical products of photosynthesis to the biological world?

3. Summarize the light reactions of photosynthesis.

4. Summarize the dark reactions. What products of the light reactions do the dark reactions use?

5. In what compound or compounds does useful energy reside at the completion of the light reactions? The dark reactions?

6. What are the roles of the following in photosynthesis?

  chlorophyll
  water
  ATP
  NADP
  cytochrome pigments

7. Why are violet, blue, and red light more efficient in photosynthesis than yellow, orange, and green light?

8. What substance is the source of the oxygen released in photosynthesis? *water*

9. Why is it considered more accurate to write the photosynthetic equation with water on both sides of the equation?

10. We usually call photosynthesis a reduction. What element is reduced in photosynthesis, and what element is correspondingly oxidized?

11. Consider your food, your clothes, your books, the heating of your home, your transportation. How do these depend upon photosynthesis? Can you think of anything that you use in your daily life that does not derive directly or indirectly from photosynthetic plants?

# Suggested Readings

Arnon, D. I. "The Role of Light in Photosynthesis," *Scientific American* 203(5): 104–118 (November 1960).

Asimov, I. *Photosynthesis.* New York: Basic Books, Inc., Publishers, 1968.

Bassham, J. A. "The Path of Carbon in Photosynthesis," *Scientific American* 206(6): 88–100 (June 1962).

Clayton, R. K. *Light and Living Matter,* Vol. 2. New York: McGraw-Hill Book Company, 1971.

Devlin, R. M. *Plant Physiology,* 2nd ed. New York: Van Nostrand Reinhold Company, 1969.

Fogg, G. E. *Photosynthesis.* New York: American Elsevier Publishing Co., Inc., 1968.

Goldsby, R. A. *Cells and Energy.* New York: Macmillan Publishing Co., Inc., 1967.

Lehninger, A. L. "How Cells Transform Energy," *Scientific American* 205(3): 62–82 (September 1961).

Levine, R. P. "The Mechanism of Photosynthesis," *Scientific American* 221(6): 58–70 (December 1969).

Rabinowitch, E. I., and Govindjee. "The Role of Chlorophyll in Photosynthesis," *Scientific American* 213(1): 74–83 (July 1965).

In some respects, **respiration** is the opposite, or near opposite, of photosynthesis. In photosynthesis, carbon dioxide is reduced, usually to glucose; in respiration, glucose is oxidized, often to carbon dioxide. Photosynthesis requires light energy, which it converts to the chemical bond energy of glucose; respiration releases this energy from glucose and converts some of it to heat, and some of it is used in the synthesis of ATP.

Respiration, unlike photosynthesis, is common to all living things. It is not restricted to certain organisms or certain cells, nor does it occur only in the light. Although the rate of respiration is affected by several factors, all living cells in all living things respire day and night. The rate of respiration may be so rapid as to generate enough heat to maintain the temperature of an active mammal well above that of its surroundings or to cause spontaneous combustion in oily rags containing rapidly respiring microorganisms. At the other extreme, dormant seeds and spores may have respiratory rates so low as to be barely detectable. Complete cessation of respiration results in death.

Like other biochemical reactions, respiration is enzymatically controlled, and its response to temperature is similar to that of other enzymatic reactions. Its rate increases with an increase in temperature, but only until the point is reached at which the respiratory enzymes are inactivated by the heat. The optimum temperature for respiration varies from species to species.

The concentration of oxygen may have different effects on the rate of respiration in different organisms. Two general types of respiration occur in most living things: aerobic respiration and fermentations. Aerobic respiration requires the presence of molecular oxygen and cannot occur without it. Fermentations require no molecular oxygen and thus may occur under anaerobic conditions; however, they often do occur in the presence of molecular oxygen. Of the two, aerobic respiration releases by far the most energy. Most organisms, and especially the large, active ones, are obligate aerobes; that is, they rely to such an extent on the

# 12
# Respiration

1 **Respiration is the oxidation of glucose (or other food) with the release of metabolically useful energy.**

2 **Respiration is essential for all organisms and is performed by all living cells.**

3 **Aerobic respiration oxidizes glucose completely to carbon dioxide and water and releases all the energy that was used in the synthesis of the glucose.**

4 **Fermentations oxidize glucose only partially and release relatively small quantities of energy.**

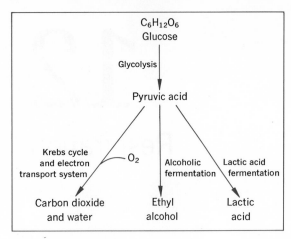

**12.1** Some common respiratory pathways.

energy released by aerobic respiration that they cannot live long in the absence of an adequate supply of molecular oxygen. Among the microorganisms there are those that are also obligate aerobes, a few that are obligate anaerobes and that cannot grow in the presence of molecular oxygen, and many facultative organisms that grow either with or without molecular oxygen.

# The Chemistry of Respiration

Like photosynthesis, aerobic respiration and fermentations are each a series of chemical reactions that may be divided into several groups. The first series of reactions in aerobic respiration and fermentations are identical and are called **glycolysis;** the remaining reactions are different for each type of respiration. In glycolysis, glucose is converted to pyruvic acid (Fig. 12.1). In aerobic respiration the pyruvic acid is oxidized completely to carbon dioxide and water by a cyclic process called the Krebs cycle and its associated electron transport system of cytochromes. In fermentations pyruvic

100

acid is converted to a product or products characteristic of the particular fermentation being carried out (ethyl alcohol and carbon dioxide in alcoholic fermentation, lactic acid in lactic acid fermentation). We will look at these processes now in some detail.

## Glycolysis

Glycolysis (Fig. 12.2) oxidizes only partially the carbon in glucose. Like other oxidations of carbon compounds, it releases energy, and like many other biochemical reactions it requires energy of activation (Chapter 10) before it can proceed. The activation energy in this case is provided by two molecules of ATP that phosphorylate each molecule of glucose and convert it into fructose diphosphate. This molecule then splits into two molecules of phosphoglyceraldehyde. These are phosphorylated further by phosphoric acid (not ATP) to diphosphoglyceraldehyde. Then the partial oxidation occurs: NAD (Fig. 11.2) removes two hydrogen atoms from each molecule of diphosphoglyceraldehyde, oxidizing them to diphosphoglyceric acid. Two molecules of ATP are now synthesized from ADP by the dephosphorylation of the two molecules of diphosphoglyceric acid to phosphoglyceric acid. At this point the original "investment" of ATP has been paid back; any additional ATP produced is a "profit" and represents energy that was stored in the glucose. Up to this point the reactions resemble some of the dark reactions of photosynthesis in reverse (Chapter 11); the remaining steps, however, bear no resemblance to those of photosynthesis. Each molecule of phosphoglyceric acid is dephosphorylated to pyruvic acid by the synthesis of ATP from ADP.

Pyruvic acid is an important compound that is an intermediate in several cellular processes and thus links them (Fig. 12.3). From this point several chemical reactions may lead into aerobic respiration, into any of several fermentations, and also into the synthesis of fatty acids and amino acids.

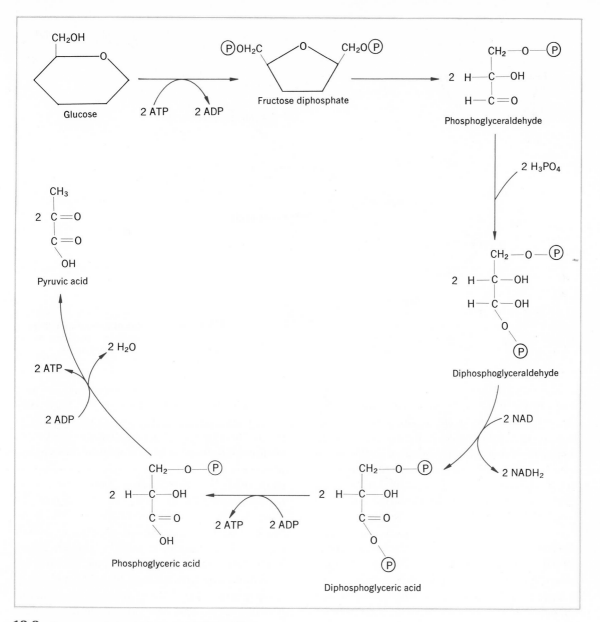

**12.2** Summary of glycolysis, somewhat simplified.

## Aerobic Respiration

Aerobic respiration begins with the glycolysis of glucose to pyruvic acid. Then the pyruvic acid undergoes a series of reactions in which it is oxidized by NAD (which becomes reduced to $NADH_2$), and also decarboxylated (Fig. 12.4). Decarboxylation is the removal of $CO_2$ from the

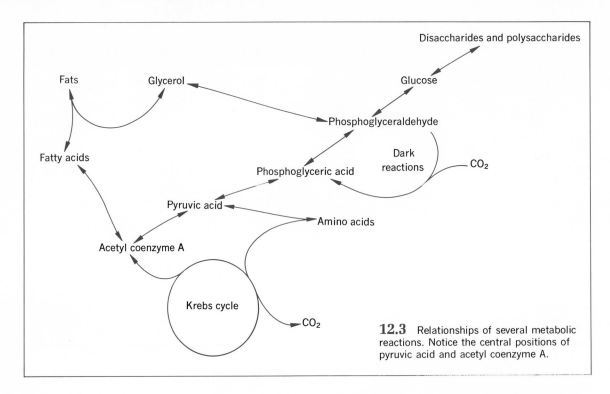

**12.3** Relationships of several metabolic reactions. Notice the central positions of pyruvic acid and acetyl coenzyme A.

acid group (—COOH) of a molecule, and it shortens the carbon chain. In this case, the 3-carbon pyruvic acid molecule becomes a 2-carbon fragment. The decarboxylation is catalyzed by cocarboxylase, a coenzyme consisting largely of the B vitamin thiamin (Fig. 12.5). The 2-carbon fragment resulting from the oxidation and decarboxylation combines with the coenzyme A (CoA Fig. 12.6), forming acetyl CoA. CoA is another of the adenosine phosphates, and it consists in part of pantothenic acid, another of the vitamin B group. Acetyl CoA is important not only because it leads into the Krebs cycle but because it is an important intermediate in both the synthesis and the oxidation of fatty acids (Fig. 12.3).

The **Krebs cycle** (also called the citric acid cycle) is one of the many important cyclic processes in living things. Acetyl CoA enters the cycle by reacting with oxaloacetic acid, a 4-carbon organic acid. The products of this reaction are CoA, which can

be used again as described above, and citric acid, a 6-carbon organic acid. The remaining steps in the Krebs cycle produce oxaloacetic acid from citric acid. Among the several intermediates in this process is alpha-ketoglutaric acid, a 5-carbon organic acid. It is formed from citric acid by decarboxylation and by oxidation; in this case the oxidation is accomplished by NADP, which removes the hydrogen. Alpha-ketoglutaric acid is converted to oxaloacetic acid by another decarboxylation and by three oxidations; two of these oxidations are accomplished by NAD, and one is accomplished by flavine adenine dinucleotide (FAD), another adenosine phosphate (Fig. 12.7).

If we pause to total up the hydrogen atoms removed in aerobic respiration, we find 12 pairs of hydrogen atoms removed for each molecule of glucose oxidized: two pairs from diphosphoglyceric acid in glycolysis, two pairs from pyruvic acid, two pairs from citric acid in the Krebs cycle, and

102

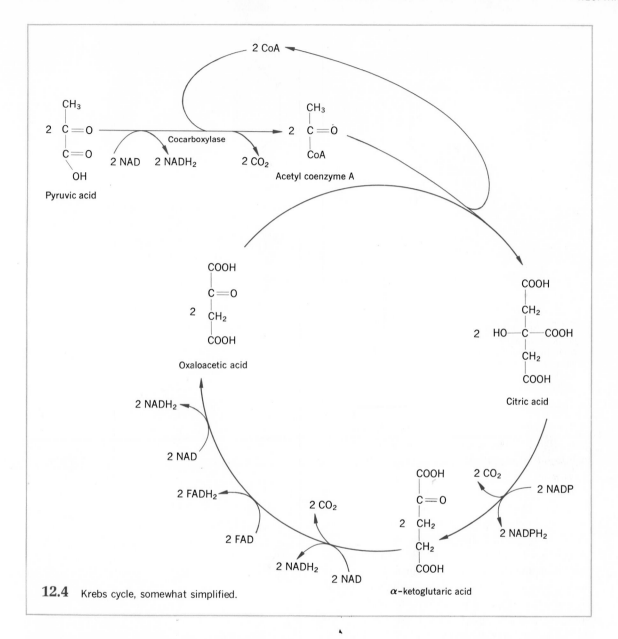

**12.4** Krebs cycle, somewhat simplified.

six pairs in the conversion of alpha-ketoglutaric acid to oxaloacetic acid in the Krebs cycle. (Because we have not indicated all the individual reactions in complete detail and therefore have not indicated the entrance of water at several points,

the student need not worry about what appears to be the impossible removal of 24 hydrogen atoms from $C_6H_{12}O_6$.)

The hydrogen atoms removed eventually combine with oxygen and form water. The first step

103

**12.5** Structure of cocarboxylase (thiamine pyrophosphate). Note that it consists largely of thiamin, one of the vitamin B group.

in this process is the transfer of hydrogen atoms from NAD to FAD (in one case from NADP to NAD to FAD). Fig. 12.8 shows that when the hydrogen is released from FAD, each atom separates into a hydrogen ion and an electron. The electron is passed along a series of cytochrome molecules. Like the cytochrome system associated with photosynthesis, this one is linked with synthesis of ATP; here the formation of ATP is called **oxidative phosphorylation** to distinguish it from the photophosphorylation of photosynthesis. As the electrons pass from one cytochrome molecule to another, they lose some of their energy, and part of this is used in the synthesis of ATP from ADP and phosphoric acid. For each pair of electrons that passes along this cytochrome system, three ATP molecules are formed. Thus, for each molecule of glucose oxidized, 38 molecules of ATP are produced (36 in oxidative phosphorylation and

104

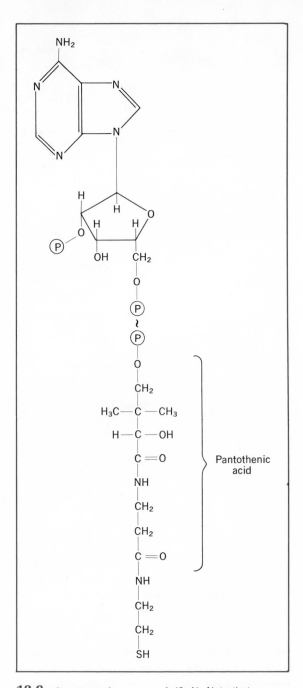

**12.6** Structure of coenzyme A (CoA). Note that pantothenic acid, one of the vitamin B group, is part of the molecule.

two in glycolysis). These then can be used in the energy-requiring processes of the cell: syntheses, movement, absorption, excretion, secretion, and other processes.

After passing along the cytochrome system, the electrons rejoin the hydrogen ions and combine with oxygen, forming water:

$$24H^+ + 24e^- + 6O_2 \longrightarrow 12H_2O$$

The overall equation for aerobic respiration is

$$C_6H_{12}O_6 + 6O_2 + 6H_2O \longrightarrow 6CO_2 + 12H_2O$$

or, subtracting six molecules of water from each side:

$$C_6H_{12}O_6 + 6O_2 \longrightarrow 6CO_2 + 6H_2O$$

Glucose is not the only substrate of respiration; other foods may be oxidized as well. Fatty acids enter the Krebs cycle by way of acetyl CoA (Fig. 12.3). Because they are a more concentrated store of energy than glucose, they release more energy than glucose. Therefore it takes less fat than glucose (or other carbohydrate) to provide the same amount of energy.

Some foods are not regularly used as energy sources. Proteins serve us largely as enzymes and structural compounds, and cellulose cannot be digested by humans (although some miroorganisms growing in our intestines can digest cellulose). All naturally occurring organic compounds are utilizable by some organisms, however. The utilization

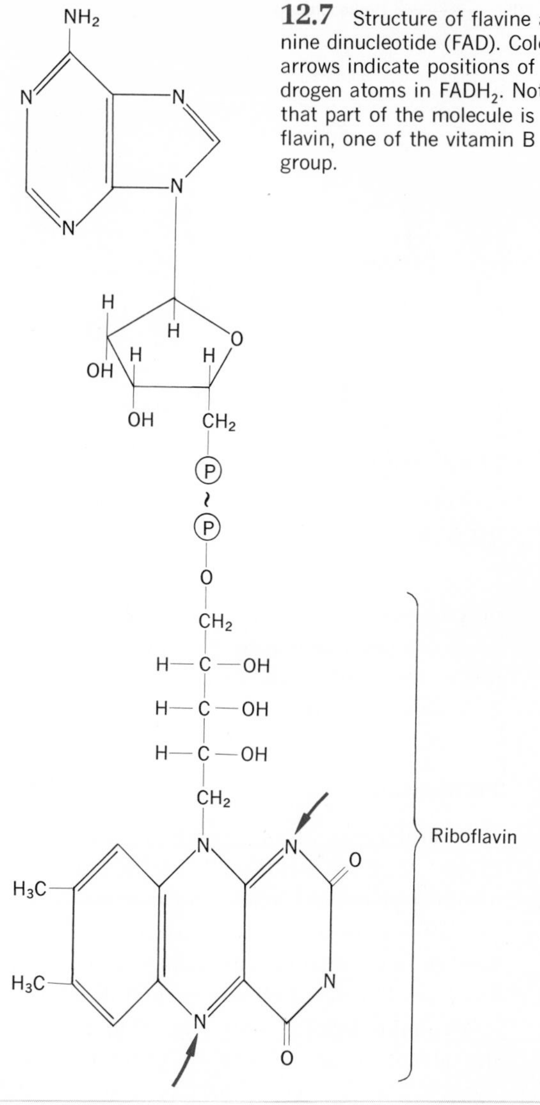

**12.7** Structure of flavine adenine dinucleotide (FAD). Colored arrows indicate positions of hydrogen atoms in FADH$_2$. Note that part of the molecule is riboflavin, one of the vitamin B group.

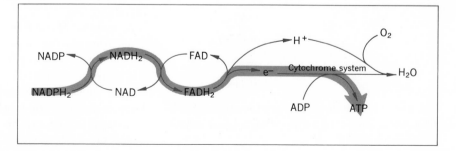

**12.8** Oxidative phosphorylation, somewhat simplified. Colored arrow indicates energy pathway.

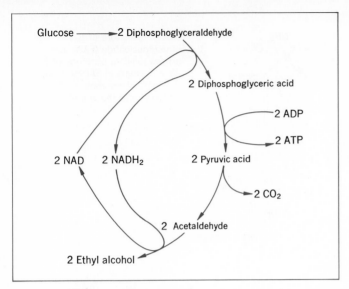

Glucose ──────▶ 2 Diphosphoglyceraldehyde

2 Diphosphoglyceric acid

2 ADP

2 ATP

2 NAD    2 NADH$_2$    2 Pyruvic acid

2 CO$_2$

2 Acetaldehyde

2 Ethyl alcohol

**12.9**  Summary of ethyl alcohol fermentation.

may have to wait until the death of the organism possessing it, at which time decay microorganisms make use of it; or in some cases, parasites may utilize compounds obtained from a living host.

## Fermentation

Several types of fermentation occur in living things. Most of them result in the formation of alcohols or organic acids, and in some cases carbon dioxide or other gas is produced. One of the best-known fermentations is the one that produces ethyl alcohol ($C_2H_5OH$) and carbon dioxide.

**Alcoholic fermentation** (Fig. 12.9) begins with the glycolysis of glucose to pyruvic acid. The pyruvic acid is then decarboxylated to acetaldehyde, and carbon dioxide is released:

$$CH_3$$
$$C=O$$
$$C=O \longrightarrow H-C=O$$
$$\phantom{C}OH \qquad CO_2$$

Pyruvic acid            Acetaldehyde

The acetaldehyde then is reduced to ethyl alcohol by NADH$_2$ (produced earlier in glycolysis), which becomes oxidized to NAD:

CH$_3$        NADH$_2$  NAD        CH$_3$
H—C=O  ──────────▶  H$_2$C—OH
Acetaldehyde                    Ethyl alcohol

Lest the student be confused by the statement that reduction occurs in an oxidation, we might mention that whenever something is oxidized, something else is reduced, and vice versa. In the case of photosynthesis, the reduction of carbon dioxide to glucose is accompanied by the oxidation of the oxygen in water to molecular oxygen. In aerobic respiration the opposite events occur: glucose is oxidized to carbon dioxide, and molecular oxygen is reduced to the oxygen in water. In fermentations, where no molecular oxygen is involved, the oxidation of some carbon atoms in the original glucose is accompanied by the reduction of others:

Reduced          CH$_3$
carbon atom ⟶  H$_2$C—OH
                 Ethyl alcohol

Oxidized                 O
carbon atom ⟶      C
                       O
                 Carbon dioxide

The overall equation for the alcoholic fermentation of glucose is

$$C_6H_{12}O_6 \longrightarrow 2C_2H_5OH + 2CO_2$$

and there is a net production of two ATP molecules (formed in the glycolysis portion of fermentation).

A number of microorganisms are capable of fermenting glucose to alcohol, but the one most commonly used commercially is a yeast called *Saccharomyces cerevisiae*. This yeast is a microscopic, single-celled fungus. Like other fungi it lacks chlorophyll and is incapable of making food. It obtains its energy by fermenting glucose. It grows on the surfaces of fruits and other plant parts that are rich in sugar. Wine can be made from the fruits or other parts of many plants, but the most popular natural source of fermentable sugar is the grape.

Ripe grapes are picked when their sugar content is at its highest and are then crushed. The more primitive method, still in use in some places today, is stamping on them. This breaks the grapes and mixes the yeast growing on the skins with the juice. Grape leaves, which usually carry large yeast populations, sometimes are mixed in. A disadvantage of this method is that other wild microorganisms can be introduced; many of them are capable of producing undesirable compounds that spoil the flavor and the aroma of the wine. In modern wineries the juice is extracted by machinery and then is sterilized to kill the microorganisms that were introduced on the grapes. The juice is then inoculated with a pure culture of *Saccharomyces*. Regardless of the method used, the mixture is permitted to stand for a week or two at a temperature suitable for fermentation. For most of this time, conditions in the mixture are kept anaerobic to discourage aerobic respiration by the yeast; this would produce carbon dioxide and water, but little or no alcohol. Fermentation continues until the yeast cells are killed by the rising concentration of alcohol. The wine is bottled and permitted to age up to several years. Most wineries maintain their own cultures of *Saccharomyces* which they have developed for their ability not only to produce a good flavor in the final product but also to withstand as high a concentration of alcohol as possible. These cultures are jealously guarded from competitors.

Production of beer is similar to that of wine, but malt (sprouting barley grains) is used as the source of fermentable sugar. Brandies, whiskeys, and other beverages with very high alcoholic content are produced by distilling a product of fermentation.

The baking industry also makes use of alcoholic fermentation, but in this case the desired product is the carbon dioxide. Flour, water, and other ingredients are mixed with *Saccharomyces*. When the resulting dough is kept in a warm place, the growing yeast cells produce ethyl alcohol and carbon dioxide, and bubbles of this gas become trapped in the dough. The warmed gas expands, the bubbles increase in size, and the dough rises. In the oven the bubbles expand still more, giving the bread a light texture. The alcohol evaporates during the baking, and virtually none is present in the final product. The heat also kills the yeast. Bakeries maintain their own strains of yeast, and packages of baker's yeast can be purchased in groceries and supermarkets for home use. In earlier days housewives used to retain a small portion of unbaked dough from one baking day to the next. This dough contained living yeast cells and was used to inoculate the next batch of dough.

Ethyl alcohol contains a considerable amount of bond energy and is capable of further oxidation with the release of more energy. The acetic acid bacteria oxidize ethyl alcohol to acetic acid. Acetic acid is the main component of vinegar, which is made commercially by adding cultures of acetic acid bacteria to wine or hard cider. These and other bacteria can be bothersome in the wine industry, for their presence can quickly spoil wine by souring it.

**Lactic acid fermentation** (Fig. 12.10) begins with glycolysis of glucose to pyruvic acid. The pyruvic

**12.10** Summary of lactic acid fermentation.

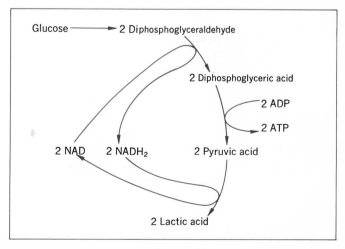

acid is then reduced to lactic acid by the $NADH_2$ produced in glycolysis.

Pyruvic acid        Lactic acid

*oxidize = increase the valence of an element.* (handwritten marginal note)

Here, again, as in alcoholic fermentation, there is both oxidation and reduction of some of the carbon atoms from the original glucose molecule:

Reduced carbon atom $\longrightarrow$

Oxidized carbon atom $\longrightarrow$

Lactic acid

The overall equation for lactic acid fermentation is

$$C_6H_{12}O_6 \longrightarrow 2C_3H_6O_3$$

Glucose        Lactic acid

Lactic acid fermentation also results in the net production of two ATP molecules (those produced in glycolysis) for each molecule of glucose fermented.

Souring of milk is caused by the lactic acid bacteria, which ferment milk sugar (lactose) to lactic acid; these bacteria can also cause the spoiling of wine, but they are used in the commercial production of sour cream and many cheeses. Pickles and sauerkraut are produced by bacterial fermentation of the sugars in cucumbers and cabbage, respectively, to lactic acid. The lactic acid not only flavors these foods, but it also provides some preservative qualities.

Lactic acid fermentation is also carried on by muscles. Ordinarily aerobic respiration occurs in muscle cells, but during prolonged, strenuous exercise, the circulatory system may not be able to bring oxygen to the muscles fast enough to support enough aerobic respiration for continued activity.

108

In these circumstances, lactic acid fermentation occurs. Although it releases only a fraction of the energy that aerobic respiration does, it has the advantage of being independent of the presence of free oxygen and thus can release a slow but steady supply of ATP to the muscles. When oxygen again becomes available, some of the lactic acid is oxidized completely to carbon dioxide and water, and some is used in forming more glucose.

# The Significance of Respiration

An important feature of respiration is that it is a stepwise process. The complete oxidation of glucose by burning releases exactly the same amount of energy as does complete aerobic respiration of glucose. Burning releases all of this energy quickly, and a great deal of it is released as heat; as a consequence, the temperature in the vicinity of the flame becomes very high. In aerobic respiration each of the several steps releases a small amount of energy; some of it is released as heat, but a considerable portion of it is retained in ATP. The heat produced by respiration is not always wasted. In some cases it protects the organism against cold. In warm-blooded animals a balance between heat-producing processes and heat-dissipating processes warms the animal in cold weather and cools it in hot weather. For example, when a person becomes uncomfortably cold, he shivers; this involuntary motion raises the rate of respiration and increases the amount of heat produced.

In all forms of respiration a certain amount of ATP is produced. Because ATP is synthesized in the light reactions of photosynthesis, one well might ask why plants do not merely perform cyclic photophosphorylation and produce ATP as the only product of photosynthesis. Why go to the trouble of performing the dark reactions which transfer the energy of ATP to glucose and then have to engage in some form of respiration to transfer it back to ATP? Why could plants not merely accumulate the photosynthetically produced ATP and use it whenever the occasion demands? Plants then would not require the metabolic machinery for the dark reactions, and no organisms, neither plant nor animal, would require the machinery for any type of respiration. However, the difference between storing energy in ATP molecules and storing it in glucose molecules is something like the difference between storing one's wealth in pennies or dimes. Pennies are bulkier than dimes, and each is worth only one tenth of a dime; it takes much more storage space to keep a certain amount of money as pennies than it does if it is kept as dimes. Because it is only the high-energy phosphate bond of ATP that ordinarily is used, from the point of view of energy an ATP molecule is worth only one thirty-eighth of a glucose molecule; it is therefore more efficient to use glucose (or other food derived from it) for long-term storage. Furthermore, glucose (and many other foods) are less reactive than ATP and are less likely to be squandered in unnecessary reactions. The controlled release of energy through respiration results in its expenditure only when that is needed. Another advantage of glucose as an energy-storing compound is that it can pass easily from cell to cell; nonphotosynthetic cells can thus receive food from photosynthetic cells. ATP does not pass readily from cell to cell; it must ordinarily be synthesized in every cell by the respiration of food in that cell.

In addition to releasing energy from stored food, respiration serves another function: some of its intermediates are reactants in the synthesis of other compounds (Fig. 12.3). The 2-carbon fragments of acetyl CoA unite, forming the long chains of fatty acid molecules; glycerol is formed from phosphoglyceraldehyde; and fats are formed by reactions between fatty acids and glycerol (Chapter 8). At least three amino acids are derived from three organic acids that are intermediate com-

pounds in respiration. In reactions with ammonia or ammonium compounds, alanine is formed from pyruvic acid, glutamic acid from alpha-ketoglutaric acid, and aspartic acid from oxalo-acetic acid. Proteins are formed from amino acids. Most of these reactions require energy, which usually is derived from the oxidation of other glucose molecules.

The efficient release of energy by aerobic respiration has permitted the evolution of large, complex, and active organisms. The small amount of energy released by fermentations is sufficient to sustain life, but there is every reason to believe that fermentation cannot sustain any but the simplest organisms. The only living things able to rely exclusively on anaerobic processes are single-celled ones such as bacteria and yeasts.

The difference between aerobic respiration and anaerobic processes has another aspect that has recently taken on a new importance. The complete decomposition of animal carcasses and dead plants, fallen leaves and the excrement of animals, garbage and sewage, and many industrial wastes requires aerobic conditions. If the amount of organic matter to be decomposed is large, aerobic respiration may deplete the immediately available oxygen supply before decomposition is complete. As conditions become anaerobic, only those organisms capable of living anaerobically can grow in the material. Many of these organisms produce disagreeable odors, as a brief trip to a garbage dump or a manure pile will confirm. Some of the compounds they produce are poisonous.

The dumping of massive quantities of raw sewage, garbage, and industrial wastes has had similar effects on our lakes and streams. The high organic content of these wastes soon makes the waters anaerobic. This kills fish and other aerobic aquatic organisms and thereby adds more organic matter to the already overburdened water. The effect is not only on the fish, however, for animals that feed on fish, whether they be bears, eagles, or human fishermen, soon feel the effects of a dwindling food supply. Chapter 16 has more material on pollution.

109

# Questions

1. Compare and contrast photosynthesis and aerobic respiration.
2. How do fermentations differ from aerobic respiration? *conversion glucose to pyruvic acid*
3. Differentiate between glycolysis and the Krebs cycle. ✓ *Kreb. Series of chemical reactions where pyruvic acid is oxidized to carbon dioxide*
4. What is the function of the following in aerobic respiration?
   NAD – *Hydrogen carrier. It is a coenzyme*
   FAD – *separates atom into hydrogen ion + electrons*
   *Leads to Krebs Cycle*
   coenzyme A – *the synthesis of fatty acids*
   cytochrome pigments *function as Electron carriers in photosynthesis + aerobic resp*
   oxygen – *Electrons combine with oxygen to form water*
5. Give some examples of how vitamins function. – *coenzymes – (Protein + Vitamin) Synthesis of fatty acids*
6. We usually call respiration an oxidation. What element is oxidized in aerobic respiration, and what element correspondingly is reduced? *oxygen (molecular)* *Carbon in glucose*
   *Glucose is oxidized to Carbon Dioxide*
7. From a chemical standpoint, what is the function of yeast in brewing and baking? *alcoholic fermentation*
8. Why does alcoholic fermentation release much less energy than aerobic respiration? *No molecular oxygen involved*
9. Name several ways that lactic acid fermentation is of use to us.
10. Respiration and photosynthesis take place by a series of stepwise chemical reactions rather than as a single reaction. Of what benefit is this to cells?
11. Compare and contrast photophosphorylation with oxidative phosphorylation.
12. If you were deprived of molecular oxygen for ten minutes, you would die. Yeast can live indefinitely without molecular oxygen. Explain the difference.
13. Do you think that the amount of food respired by a green plant in its lifetime is more than, less than, or equal to the amount of food it produces in photosynthesis?

# Suggested Readings

Amerine, M. A. "Wine," *Scientific American* 211(2): 46–56 (August 1964).

Goldsby, R. A. *Cells and Energy.* New York: Macmillan Publishing Co., Inc., 1967.

Lehninger, A. L. "Energy Transformation in the Cell," *Scientific American* 202(5): 102–114 (May 1960).

_____. "How Cells Transform Energy," *Scientific American* 205(3): 62–82 (September 1961).

Margaria, R. "The Sources of Muscular Energy," *Scientific American* 226(3): 84–91 (March 1972).

Rose, A. H. "Beer," *Scientific American* 200(6): 90–100 (June 1959).

(1) *Photosynthesis = the reduction of carbon dioxide to glucose is accompanied by the oxidation of the oxygen in the water to molecular oxygen.*

*Aerobic respiration = the opposite events occur, glucose is oxidized to carbon dioxide, and molecular oxygen is reduced to the oxygen in the water.*

(2) *Fermentation = there is no molecular oxygen involved, the oxidation of some carbon atoms in the original glucose is accompanied by the reduction of others.*

(9) *Does not require molecular oxygen –*
*Releases a slow but steady supply of A T P. to the muscles.*
*Provides some preservative qualities in foods – Cheeses, pickles + etc.*

# More About the Environment

# Part IV

Unlike modern industrial societies, nature produces no real wastes. The waste products of one organism are recycled back into the system as the nutrients of another. In accordance with the law of conservation of matter, no element ever is consumed. During its passage through a cycle, an element enters into several different compounds, some of which can be utilized by only certain organisms. The cycles of only three elements are considered here: carbon, nitrogen, and sulfur.

# 13

# Cycles of Matter

## Cycles of Matter

### Carbon Cycle

Figure 13.1 illustrates the carbon cycle. Green plants utilize carbon in an inorganic, highly oxidized form: carbon dioxide. In photosynthesis they reduce it to glucose, some of which is converted to other organic compounds. When animals eat plants, some of these organic compounds remain in their reduced form and become incorporated into the animal's tissues. After death the bodies of both plants and animals decay under the metabolic activities of nonphotosynthetic microorganisms of soil and water, especially bacteria and fungi. These organisms incorporate into their own bodies some of the organic compounds with the carbon still in reduced form; and when they die, other microorganisms utilize their bodies similarly. Decay microorganisms also utilize urine and feces of animals, the hair that is shed from animals, and the leaves that drop from plants. In their lifetimes all living things respire, and in so doing they oxidize organic compounds to carbon dioxide, which then becomes part of the general pool of carbon dioxide available to photosynthetic plants.

The passage of the element carbon from the nonliving world to the living world and back again thus involves two main types of reactions.

1 **Elements used by living things are cycled from organism to organism and are used over and over again.**

2 **Green plants utilize oxidized, inorganic forms of carbon, nitrogen, and sulfur and reduce them to the organic compounds used by all living things.**

3 **Microorganisms decompose dead bodies and wastes and oxidize their organic compounds to inorganic forms used by green plants.**

113

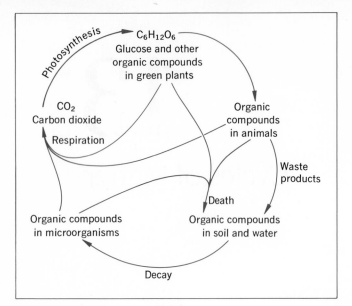

**13.1** The carbon cycle.

For every complete turn of the cycle, carbon alternately is reduced and oxidized—reduced by photosynthetic plants and oxidized by all living things.

The essential roles of nonphotosynthetic microorganisms in an ecological system can be seen in this cycle. Even though the respiration of plants and animals returns some carbon to the air and water, without decay more and more carbon would become sequestered in dead bodies which never would decompose. Not only would the accumulation of corpses become an inconvenience, but the concentration of carbon dioxide in the air gradually would decline, thus reducing the ability of plants to produce food and feed animals. All living things depend on the nonphotosynthetic microorganisms of decay as much as they do on photosynthetic plants. The death of all decay microorganisms would mean the eventual death of all other organisms as well.

**114**

## Nitrogen Cycle

The nitrogen cycle (Fig. 13.2) has some steps with counterparts in the carbon cycle. From the soil green plants absorb nitrates ($-NO_3$), which are oxidized inorganic forms of nitrogen, and reduce them to the amino group ($-NH_2$) of amino acids, proteins, and other nitrogenous organic compounds. When animals eat plants, the amino groups in these compounds remain in essentially the same reduced form, even though the proteins are digested to amino acids which may be used in the synthesis of other proteins. When plants and animals die, their proteins are digested to amino acids by microorganisms many of which incorporate the amino acids into their own bodies. Others remove the amino group and release it as the gas ammonia ($NH_3$), another reduced form of nitrogen. Microorganisms also remove the amino groups from urea ($NH_2-\overset{\overset{\text{O}}{\|}}{C}-NH_2$), one of the major components of urine.

Ammonia can be utilized directly by green plants and by many soil microorganisms, but special bacteria, the **nitrifying bacteria,** oxidize ammonia by a two-step process, first to nitrites ($-NO_2$) and then to nitrates. Nitrates are highly soluble in water, and plants absorb them readily. Nitrifying bacteria require oxygen for nitrification, and so they function best in well-aerated soil. Plowing soil aerates it and contributes to its fertility partly by encouraging the growth of nitrifying bacteria.

Under anaerobic conditions, **denitrifying bacteria** reduce nitrates to nitrites and then to gaseous nitrogen ($N_2$) which cannot be used by plants or animals. The soil becomes depleted of its nitrogen as the gas diffuses into the atmosphere, and its fertility decreases. Ammonia is another possible product of reduction, but it can be utilized by some organisms. Most farmers dread floods not only because they leach valuable minerals from his soil and carry them downstream, but also because they create anaerobic conditions in the soil and encourage the growth of denitrifying bacteria. (On the other hand, the yearly flooding of the Nile River always has been desired by Egyptian farmers

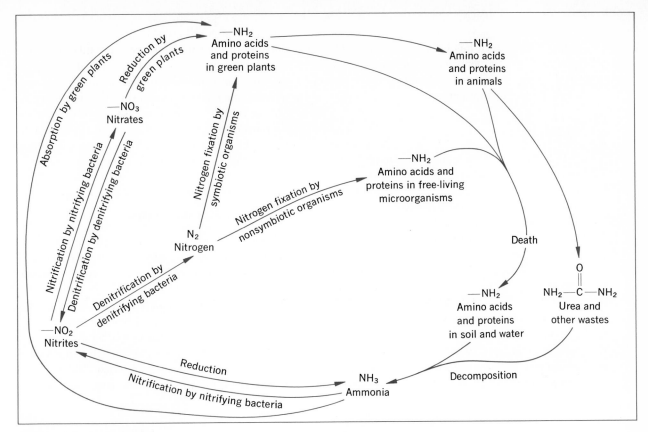

**13.2** The nitrogen cycle.

because its mineral-rich waters fertilize their fields.)

Fortunately a few bacteria and a few algae can utilize gaseous nitrogen. By a process called **nitrogen fixation** (Chapter 4), they reduce the gaseous nitrogen further to amino acids or other organic forms that can be utilized by plants. *Rhizobium,* a symbiotic nitrogen-fixing bacterium, grows in the roots of leguminous plants (which include beans, peas, clover, soybeans, and alfalfa). Farmers rotate legumes with nonleguminous plants to enrich their fields with nitrogen. Often, to insure an ample population of this bacterium, they purchase packages of *Rhizobium* with which to dust the seeds before planting, for although the legume can grow without *Rhizobium,* no nitrogen fixation occurs without it. An acre of legumes can fix more than 100 pounds of nitrogen in a growing season. Of

this, some is incorporated into the plant and some is excreted into the soil. When the farmer removes the edible portions of the legume crops and sells them, some of the nitrogen is lost from his field; but when he plows the remaining portions of the crop into the ground, the nitrogen in these parts is added to the soil as they decay. For each acre of leguminous crop raised in a season, there may be a net gain of about 60 pounds of nitrogen or more.

Nonsymbiotic nitrogen fixers include some bacteria and some blue-green algae. The latter are especially valuable in southeastern Asia, where so many farmers are poor and cannot afford even to use the feces of animals as fertilizer. Blue-green algae grow well in flooded rice paddies, where they supply nitrogen to the fields. The amount of nitrogen fixed by nonsymbiotic organisms varies a great deal, but it has been estimated at 20 to 50 pounds per acre per year in the temperate zone. In the tropics, blue-green algae may fix 70 pounds of nitrogen per acre per year.

## Sulfur Cycle

Sulfur is another valuable element used by green plants primarily in the oxidized, inorganic form (Fig. 13.3). Green plants absorb sulfates ($-SO_4$) and reduce them to the sulfhydryl group ($-SH$) of amino acids, proteins, and some other organic compounds. The sulfur remains in reduced form after being eaten by animals. Microorganisms decomposing the dead bodies of plants and animals digest the proteins to amino acids and remove the sulfhydryl groups as the gas hydrogen sulfide ($H_2S$). Under aerobic conditions, several species of aerobic sulfur-metabolizing bacteria oxidize hydrogen sulfide to sulfates, thereby increasing the fertility of the soil. Under anaerobic conditions, anaerobic sulfur bacteria reduce sulfates and decrease the fertility of the soil. Here, as in the nitrogen cycle, green plants use primarily the oxidized form of the element, and microorganisms complete the cycle by oxidizing the reduced form. Here, too, as in the nitrogen cycle, floods decrease the fertility of the soil by encouraging the growth of anaerobic bacteria, whereas plowing the soil encourages the growth of aerobic bacteria.

## Significance of the Cycles

Carbon, nitrogen, and sulfur are only three of the elements that cycle from the living to the non-living world and back again, but their cycles are sufficient to illustrate that in nature there are no dead ends where elements might become permanently lost to the biological world.

It is not so with many synthetic substances made by man. Many plastics, though organic compounds, cannot be decomposed by microorganisms. A plastic bag may become torn and

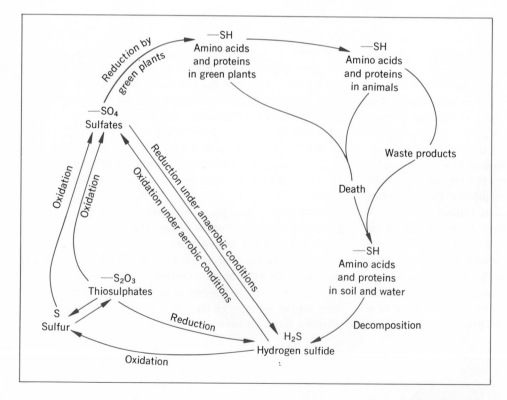

**13.3** The sulfur cycle.

shredded by winds or water currents, but its many little pieces equal the total volume of the original plastic bag. Burning quickly disposes of the bags, but in the case of some types of plastic, burning releases poisonous gases into the air.

The total number of new synthetic organic compounds increases every year. Some of these are degradable by microorganisms, but others, like the early nonbiodegradable detergents and pesticides, are virtually indestructible, and we are destined to have them with us for a long time. The harmful aspect of this is not so much the loss of carbon and other elements from the great cycles of nature, but the damage that these substances may do during their lifetimes.

Some of the customs of civilized man tend, if not to break the great cycles of nature, at least to dislocate some parts of them. Consider what happens to the minerals in a bunch of carrots or a chicken that a farmer sends to market. In the natural course of events, the minerals would have returned to the soil in the general area where the carrots and chicken lived, for it is likely that they would have died there. But these particular organisms were raised for the dinner table—possibly a table hundreds of miles away from the farm. Some parts of these organisms are not eaten. The carrot tops, the carrot parings, the chicken bones, and probably the chicken's head, feet, and a few internal organs that the average middle-class citizen does not relish are ground up in the garbage disposal or collected by the local garbage truck; the minerals they contain are likely to be sent flowing down the most convenient stream or to be burned or buried in the nearest garbage dump. Of the parts that are eaten, a certain amount becomes part of the tissues of the diner and is interred with him when he dies; but by far the greatest amount is excreted and eliminated in urine and feces daily throughout his lifetime. In most cities these waste products, with their rich supply of minerals, are conducted to a sewage disposal plant, and from there they are likely to follow the garbage downstream. Both are lost to the farm, and the streams

become overburdened with an excess of minerals.

In recent years, ranchers and farmers have found it increasingly convenient and economical to raise cattle in feedlots rather than on the range or in meadows. To prevent undesirable accumulations of feces and urine, these materials are disposed of in the nearest stream. Again, minerals are lost from the land and added to the water.

It has been the practice of farmers to replace the lost minerals with inorganic chemical fertilizers, but this does not replace the humus that no longer forms in the soil. With little or no humus, mineral components of the inorganic fertilizers wash out of the soil more quickly than they otherwise would. Nitrates, which are so highly soluble in water, move into bodies of water that are used for drinking purposes. In some parts of the United States the nitrate content of municipal water supplies already has become high enough to be considered a health hazard.

Well-kept lawns surround millions of American houses. Each home gardener who discards his lawn clippings and then spreads chemical fertilizers on the grass adds to the problem in his own small way—a way multiplied a few million times.

Garbage, sewage, and other organic wastes need not be sent on an inexorable voyage to the sea. Some of these wastes can be, and in some cities are, converted into fertilizer. One problem not yet completely surmounted is that of the presence of disease-causing bacteria in sludge (the solid product of sewage disposal plants) and to a lesser extent in other organic wastes. Intestinal diseases are likely to be contracted by eating food contaminated by improperly processed sludge. At present, sludge is not used to fertilize root crops or other crops whose edible portions are likely to touch the ground. Eventually we must learn to salvage all materials so that they can be recycled again.

An aside about *recycling*, a word sometimes misused today: Returning beverage bottles for refilling is an example of recycling. Crushing them to incorporate them into paving materials is not, unless the paving were to be used again when a

117

**13.4** Green alga growing along the shoreline in a major metropolitan area. This alga has become more abundant as the human population has increased along this shore. The discharge of more and more mineral-laden sewage into the lake nurtures the growth of the alga.

road is resurfaced again. Use of biodegradable detergents or insecticides that are converted by wild microorganisms of soils or streams to harmless compounds is another form of recycling that generally is acceptable, but it is not without its pitfalls. The early nonbiodegradable detergents formed such heads of foam on rivers and lakes that the new biodegradable phosphate detergents were hailed as an excellent solution to the problem. However, their unprecedented phosphate burdens encouraged the growth of algae, the death and decomposition of which depleted the oxygen supply (Fig. 13.4). The fact that a substance is biodegradable is no assurance that its products are harmless.

Another aspect of the cycles of matter is that all organisms are dependent on the products of others. We and other animals cannot utilize many inorganic compounds (carbon dioxide and most minerals) directly. Instead we depend on plants to remove them from air and water and to convert them into the protoplasm we use as food.

We are also dependent on the activities of decay microorganisms to return minerals to the soil and on the activities of other microorganisms, especially aerobic ones, for oxidizing the products of decomposition and so converting them to the most useful state for green plants. Without microorganisms, none of the great cycles of nature would be completed. This is true not only in cultivated fields, but in forests and oceans as well. The natural waters are almost never sterile. The largest of them may be considered as dilute mixtures of many species of organisms living together in ecological balance. So, too, is it with soils, where bacteria, fungi, algae, protozoa, insects, and worms mingle with the roots of higher plants. Rich soils of forest or grassland literally are teeming with life. Half of some of them consists of living things.

Because we are dependent on microorganisms in an indirect way for our food and oxygen, it is imperative that these organisms be protected. Careless use of chemical sprays, especially the defoliants (2,4-D, 2,4,5-T, and others) that have become so popular in recent years for the killing of weeds, may be a threat to bacteria, algae, and fungi. These microorganisms have plantlike characteristics and often are classified as plants. They well might be expected to be harmed by compounds that kill plants.

DDT was produced to kill harmful insects, and it was spread lavishly around the world without much thought of its effects on other animals. Gradually it gained notoriety not only as a killer of beneficial insects but as a threat to nearly all animal life as well. Then in 1968 it was discovered that DDT has a serious effect on plants also. When present in concentrations as low as a few parts per billion, it reduces the rate of photosynthesis in marine algae. Marine animals are essentially completely dependent on algae for their food, and the reduction of photosynthesis in the sea could wreak havoc in the oceans.

These pesticides and others that never have been tested adequately for their effects on the biological world are a potential threat not only to the soil microorganisms in sprayed areas, but also to the microorganisms in the bodies of water to which the pesticides are carried by runoff water. Wastes from factories manufacturing pesticides add to the danger. The accidental sinking of ships laden with pesticides is a possibility not pleasant to consider.

It has been argued that the use of pesticides and fertilizers and the dumping of sewage, manure, and garbage is the most economical way of raising our food and disposing of our wastes. In the long run such an economy is likely to prove a false one and one that well may cost us more than we now save. As the amount of life in the soil and in the oceans decreases, so will the amount of life that walks on the land.

## Questions

1. Follow the element carbon through the carbon cycle. What kinds of changes does it undergo? What important function do green plants perform in the cycle? What functions do microorganisms perform? How do these complement each other?
2. Why does not all the carbon in the world become sequestered in just one form such as carbon dioxide or organic compounds?
3. Define the following terms:
   nitrogen fixation
   nitrification
   denitrification
4. From the following list select those organisms that are helpful to farmers.
   nitrogen fixers
   nitrifying bacteria
   denitrifying bacteria
   aerobic sulfur bacteria
   anaerobic sulfur bacteria
5. Compare the role of green plants in the nitrogen and sulfur cycles with their role in the carbon cycle. Do the same for microorganisms.
6. In what way does man interfere with the cycles of nature?
7. Explain why short-term economies in our disposal of wastes are likely to result in eventual long-term losses to society.

## Suggested Readings

Bolin, B. "The Carbon Cycle," *Scientific American* 223(3): 124–132.

Bormann, F. H., and G. E. Likens. "Nutrient Cycling," *Science* 155: 424–429 (1967).

_____. "The Nutrient Cycles of an Ecosystem," *Scientific American* 223(4): 92–101 (October 1970).

Deevey, E. S., Jr. "Mineral Cycles," *Scientific American* 223(3): 148–158 (September 1970).

Delwiche, C. C. "The Nitrogen Cycle," *Scientific American* 223(3): 136–146 (September 1970).

Kamen, M. D. "Discoveries in Nitrogen Fixation," *Scientific American* 188(3): 38–42 (March 1953).

Pratt, C. J. "Chemical Fertilizers," *Scientific American* 212(6): 62–72 (June 1965).

Stewart, W. D. P. *Nitrogen Fixation in Plants.* London: Athlone Press, 1966.

# 14

# Food Chains and Food Webs

1 **Most food chains consist of producers, consumers, and decomposers.**
   **a Producers are plants that manufacture food.**
   **b Consumers are animals that eat plants or other animals.**
   **c Decomposers cause the decomposition of the bodies and wastes of plants and animals.**

2 **The energy flow in a food chain is from producers to consumers and from both producers and consumers to decomposers.**

3 **Complex food webs composed of many interlinked food chains are much more stable than are simple communities with only a few organisms.**

4 **Unlike the elements, energy is not recycled; the amount of useful energy decreases greatly with each succeeding trophic level.**

5 **Substances, both beneficial and harmful, tend to accumulate in the bodies of animals at high trophic levels.**

In the abundance of today's America, food seems to spring forth directly from the supermarket shelves. The purchase of one bag of potato chips somehow generates the appearance of another bag. As long as the shelves are full, the average American does not look beyond them to the real sources of his meals, which is the plants of the world. Directly or indirectly, plants supply him with the energy his body requires.

The chemical bond energy in the chunks of tuna sharing a luncheon plate with a quartered tomato arrives there through a longer chain of biological events than does the energy in the tomato. In life the tuna feeds on smaller fish like mackerel that feed on still smaller fish like herring. The herring eat smaller aquatic animals, most of which graze on algae, minute floating plants of the ocean. The line of organisms leading from the algae to the tuna constitute part of a **food chain.** The tomato is part of a food chain, too, but in this case a much shorter one (Fig. 14.1).

## The Concept of Food Chains and Food Webs

The simplest food chain consists of only two kinds of organism: producers and decomposers. Most food chains contain three kinds: producers, consumers, and decomposers. The **producers** are green plants, synthesizers of food from the inorganic compounds they absorb from their environment. They form the base of all food chains. The **consumers** are animals, and they rely directly or indirectly on the producers. **Herbivores** are consumers that feed on plants (Fig. 14.2), and **carnivores** are consumers that eat animals (Fig. 14.3). Primary carnivores eat herbivores, secondary carnivores eat primary carnivores, tertiary carnivores eat secondary carnivores, and so on. **Decomposers** are the decay organisms that consume the bodies of dead plants and animals and the waste products

| | | |
|---|---|---|
| Man | Quaternary carnivore | |
| Tuna | Tertiary carnivore | |
| Mackerel | Secondary carnivore | |
| Herring | Primary carnivore | |
| Crustacean | Herbivore | Man |
| Algae | Producer | Tomato |

**14.1**  Two of many possible food chains leading to man. Decomposers are not indicated.

**14.2**  These beef cattle are herbivores feeding on grasses and other pasture plants.

of living organisms. Decomposers include bacteria, fungi, and some protozoans; most of these are microscopic organisms, but some fungi, such as mushrooms (Fig. 14.4) and bracket fungi, become quite large. Like consumers, decomposers obtain their food directly or indirectly from green plants.

The simplest food chain, then, might consist of only a green plant and the bacteria that dispose of its remains when it dies. A longer food chain might consist of wheat, the herbivorous insects that feed on the wheat, the mice (primary carnivores) that eat the herbivorous insects, the snakes (secondary carnivores) that eat the mice, the hawks (tertiary carnivores) that eat the snakes, and the variety of decay organisms that decompose the wastes, and, eventually, the bodies of all these organisms when they die. Each link in a food chain is called a **trophic level.**

Food chains do not exist independently of each other. Rather, the many food chains are linked with each other in a complex fashion forming a **food web** (Fig. 14.5), most species eating more than one species and being eaten by more than one. Some organisms can occupy more than one trophic level. Mice are omnivorous animals that fill the role of herbivore in eating wheat and that of car-

**14.3**  This cat is a carnivore, here feeding on a chipmunk that it caught. Chipmunks are herbivorous and feed largely on seeds.

**14.4**  Mushrooms are decomposers utilizing the food in dead plants, their fallen leaves and twigs, and other dead organic material. They often can be found growing on the trunks or roots of dead trees.

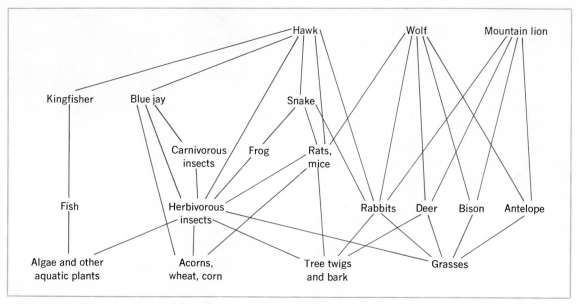

**14.5** A fragment of a food web. The entire food web of a community would be much too complex to show in a small space. Only a few of the larger organisms are indicated. No attempt has been made to show the many microorganisms that may parasitize the plants and animals when they are alive or that dispose of their wastes or their bodies when they die.

nivore in eating insects. The omnivorous blue jay lives as herbivore, primary carnivore, or secondary carnivore, depending on the food available. Similarly, the hawk can be either primary, secondary, or tertiary carnivore, depending on what animals it eats and what they, in turn, have eaten. Man is an omnivore that occupies primarily the herbivorous level but occasionally enjoys the several levels of carnivorous existence.

## The Passage of Energy Along a Food Chain

It must be remembered that food is a store of energy that is essential for life. Every animal obtains its energy from organisms in the trophic level immediately below it, and the photosynthetic plants obtain energy from sunlight. The first law of thermodynamics (the law of conservation of energy, Chapter 9) makes it obvious that the amount of energy at any one trophic level cannot exceed the amount of energy at the trophic level below. Indeed, the **second law of thermodynamics** says that the amount of useful energy available at any trophic level must be less than that at the trophic level below. In practice, it is much less.

The second law of thermodynamics has been stated in several forms. One form is the following: heat never moves of its own accord from a colder body to a warmer one. It can be made to move in this direction, however, by the proper application of energy from an external source. A refrigerator removes heat from itself and transfers it into a warmer room, and an air conditioner takes heat from a warm room or building and conducts it outside to where the temperature is higher. Both the refrigerator and the air conditioner can perform their functions only if they are supplied with appropriate energy, usually from an electric current or the burning of a gas flame.

A consequence of the second law of thermodynamics is that for every transfer of energy, some of the energy is converted to heat. The function of an electric lamp is the production of light from electricity, but only a portion of the electrical energy flowing through the circuit in the bulb is converted to light energy; some of it becomes heat—a fact that can be checked easily by touching a bulb that has been lit for a while. The heat becomes dissipated in the environment and ordinarily is not used again. (There are, of course, processes in which heat, if properly channeled, performs a useful function. The heat from a burning fuel can be used to produce steam from water; the steam then runs a steam engine, which in turn performs some useful task. But even in this case, much of the heat from the fire escapes into the environment and does not contribute to the boiling of the water.) Ironically, refrigerators and air conditioners increase the total amount of heat in the environment, even though their function is the cooling of a local area.

No process, then, is 100 per cent efficient. Different processes and different devices vary in their efficiency—a neon light bulb loses proportionately less heat than an incandescent one, and a firefly's tail lantern loses even less—but all produce some waste heat. It becomes obvious, then, that there is no cycle of energy comparable to the cycles of matter. Although energy is neither created nor destroyed and the total amount of energy remains the same, the amount of useful energy decreases with time. This explains the failure of men to invent a perpetual motion machine. No machine supplied with a finite amount of energy can recycle it indefinitely; eventually all the useful energy is converted to heat, and the machine ceases to operate. That life does not disappear from the earth as a consequence of using up the useful energy is due to the constant replenishment of energy from the sun. As long as sunlight shines on green plants, photosynthesis channels this energy into food chains.

The second law of thermodynamics does impose some important limitations on food chains, however. Animals vary in their efficiency for utilizing the energy in their food, but on the average, about 10 per cent of the energy available at one trophic level reaches the next highest trophic level, and 90 per cent is lost to the environment. At the lowest trophic level, a photosynthetic plant converts a given amount of light into the chemical bond energy of organic compounds. Much of this dissipates in the respiration of the living tissues of the plant. Of the remaining substance, some may be useless as an energy source to the herbivore that consumes it; cellulose, for instance, cannot be digested by some animals. At the next trophic level, herbivorous animals respire away most of the digestible food that they consume, and some of it is converted to tissues that are not useful to their predators; hair, bones, and the hard exoskeletons of insects, for example, may not be digestible by the predators. The amount of energy available at the top levels is very small indeed. Of the total energy available at the producer level, about 10 per cent is available to herbivores, about 1 per cent to primary carnivores, about 0.1 per cent to secondary carnivores, about 0.01 per cent to tertiary carnivores, and correspondingly less to any additional carnivores that may exist. The amount available to decay organisms and parasites depends on what organisms they use as a food supply.

The amount of useful energy at all trophic levels often is pictured graphically as a pyramid (Fig. 14.6), with the base representing the useful energy at the producer level and the top of the pyramid representing that at the top trophic level. It becomes obvious, then, that if life requires energy, the higher the trophic level, the less abundant can life be. Sometimes this is reflected in the absolute numbers of individuals at the various trophic levels; millions of pasture plants are required to raise to marketable size the several beef steers that one human being may eat in his life time. It is also

123

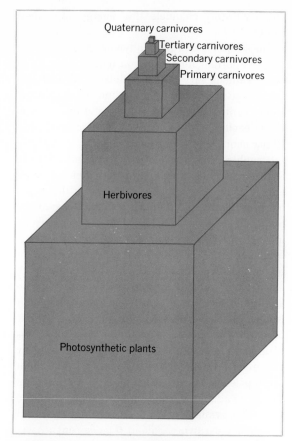

**14.6** An energy pyramid. The energy available at each trophic level is approximately one-tenth that available at the trophic level next below. For this reason the biomass at each level is about one-tenth of the biomass at the level next below. This illustration may be used to represent the biomass pyramid in a food chain.

obvious that relatively few trophic levels are possible, for at very high trophic levels there would not be enough energy to support life.

It is sometimes more convenient and often more significant to consider **biomass** rather than numbers of individuals. The biomass is the total amount of living material within any prescribed limits. Because the available energy decreases with the height of the trophic level, the biomass must decrease correspondingly, but the number of decomposers usually is many times larger than the

number of individuals they decompose. Billions of bacteria and other microorganisms dispose of the body of a dead elephant, but their total biomass is much less than that of the single elephant. Similarly, the total biomass of thousands or millions of insects in a tree and eating its leaves or twigs is less than that of the tree itself. Figure 14.6 can serve as a biomass pyramid as well as an energy pyramid.

It must become apparent that the human biomass that can be supported in a given area depends in part on what trophic level the people occupy. If they are vegetarians, the population can be much larger than if they eat primarily meat. Conversely, the more people living in an area, the less meat and animal products can there be in their diets. In densely populated areas like India, the average person eats rice as the main item of every meal. Usually it is garnished with a few vegetables, rarely with meat. Many of the world's people of necessity must eat less meat in a year than does the average American in a week. As the numbers of persons in the world continue to grow, more and more of them must relinquish the luxury of steak, bacon and eggs, and milk.

Diets in most societies are based primarily on cereal grains or other starchy foods supplemented with fruits and vegetables. Meat, the proportion of which varies greatly in the diets of the several peoples of the world, comes mostly from herbivores like cattle and sheep. Only rarely is man a secondary or tertiary carnivore; that is, he usually does not eat carnivores. Wolves and tigers ordinarily do not find themselves on the dinner table. It would be economically unfeasible to subsist solely on primary carnivores, for the amount of land required to support a single person would be about 100 times that required for a vegetarian. Correspondingly more and more land would be needed to support anyone feeding solely on secondary carnivores, tertiary carnivores, or higher carnivores.

On the other hand, in harvesting food from the sea, humans occupy higher trophic levels, sometimes being tertiary or even quaternary carnivores. The diner feeding on tuna is a quaternary carnivore. If, instead, he eats mackerel, he shares the role of tertiary carnivore with the tuna, and if he eats herring he is a secondary carnivore. Although some of the larger algae of the intertidal zones are eaten by coastal dwellers, man usually is not a herbivore when he draws sustenance from the sea.

The plant life of the vast expanses of open ocean and of many bodies of fresh water consists primarily of single-celled algae. Although the total amount of food present in the algae of a given area of sea compares favorably with that of the terrestrial plants on an equal area of land, these cells are distributed throughout the upper few feet of water, and their concentration is very low in a given sample. The energy that must be expended in harvesting such a scattering of microscopic plants has made it uneconomical to garner from the bottom of the food chain in the open seas, whereas a single line or net can take several pounds of concentrated food in the form of fish. Probably because of this, humans never have learned to relish diatoms or green algae as they do tuna, salmon, lobster, or caviar. In view of the exigencies of the food chain, it is not surprising that fish are a major item of diet among relatively few peoples of the world, and it is doubtful if any untapped reserves of fish in the sea will assist in the solution of more than a small fraction of the food problems of the world.

We might consider still another consequence of the second law of thermodynamics. Recall our example of the air conditioner heating the environment as it cools a room. Because heat from the environment moves back through the walls into the room, the air conditioner must work continuously or at least intermittently if it is to keep the room cool. If the air conditioner were to stop working after it had cooled the room to the desired temperature, the temperature in the room immediately would begin to approach that of the environment. In other words, just to maintain a given temperature, the air conditioner must continue to expend energy. The same is true of a furnace in winter. The furnace cannot stop working once it has achieved the desired temperature but must continue to burn fuel to keep the house at a comfortable temperature.

Similarly, a living organism must expend energy continuously just to maintain itself in a living state. This energy is in addition to that spent on such obvious energy-requiring processes as growth, repairing wounded tissue, or playing baseball. A living cell is a complex thing. Often it must absorb substances against concentration gradients or excrete other substances against concentration gradients. A cell cannot maintain itself without food to supply it with energy. Thus all organisms require food to keep themselves alive. With light, a green plant makes its own food. All other organisms obtain their food from organisms below them in the food web.

# Food Webs and the Stability of Communities

A complex food web is a stable thing, whereas a single isolated food chain always is in danger of being destroyed. Imagine an island populated by only a few kinds of organisms: grasses, deer, and wolves. If all the deer should die, the wolves must starve to death. If a food web had existed, however, the extinction of the deer might not be so serious for the wolves, for other prey such as bison or antelope might be available (Fig. 14.5). On the other hand, if all the wolves were to die first, the deer population would begin to increase tremendously. Then, as they ate more and more of the grass, the vegetation would become insufficient to support them, and large numbers would die. The community might stabilize eventually, probably with much lower numbers of each species than originally; or both deer and grass might become extinct leaving the island with no inhabitants.

125

Were there other predators, such as mountain lions, these carnivores could take the place of the missing wolves and prevent a devastating population explosion among the deer.

The loss of only one species from a complex, stable community rarely destroys the entire community, but with the loss of each succeeding species the community loses a little more of its stability. As the community becomes very simple, it is more likely to be severely altered by any additional changes. It surprises some persons that the tundra of Alaska can be called fragile when many of its plants and animals survive the rigors of some of the most severe winters on earth. Yet, compared with grasslands and deciduous forests, the tundra community has relatively few species and so is more susceptible to disruption. For this reason, conservationists are concerned that the exploitation of the natural resources of the Arctic tundra be preceded by thorough study of its ecology so that only a minimum amount of damage may be done.

In the past, many wild communities were destroyed by human activities that reduced the numbers of species. This was done in part by the systematic killing of some predator species that occasionally supplemented their wild diets with domestic cattle or sheep, by hunting them for sport, or by depriving them of suitable habitats, as by cutting down trees in which many animals find shelter. Pesticides and industrial wastes have taken their tolls as well.

In the interests of efficient food production, modern farmers seek to support only a very simple food web:

Man
Cow, sheep, or other edible
herbivorous animal
Plants edible to man
or his farm animals

Not wishing to share the fruits of his labor with other animals, man protects his crops and meat animals from competitors with pesticides and other poisons, traps, and guns. Weeds, fungi, insects, mice, rats, hawks, owls, bison, antelope, wolves, and coyotes are only some of the organisms he eliminates from the environment.

In its ecological simplicity, modern agriculture makes itself more susceptible to epidemics of disease or insect infestations. Thousands of acres covered only by corn plants are highly susceptible to rapidly spreading epidemics of fungus infections that occasionally threaten to destroy the entire corn crop of a large geographical region (Fig. 14.7). Primitive vegetable gardens of American Indians and other people without modern methods of agriculture consisted of mixtures of several types of plants—perhaps corn, beans, squash, and even a few trees. Although such gardens are less efficient in terms of the ability of the farmer to weed, prune, or harvest, and although they do not lend themselves to the use of machinery, they are not so likely to suffer from epidemics, for an infestation of insects or fungi on an isolated group of corn plants is not likely to spread to another group of corn plants when beans and squash occupy the intervening area. Even should an epidemic occur, it would spread much more slowly than in a monoculture (cultivation of only one crop), and the loss of one crop is not likely to result in starvation or economic disaster when the other crops still stand.

In terms of the total amount of energy required to produce a given quantity of food, the modern farm is less efficient than the primitive farm, for machinery requires that fuel be burned, and, of course, energy is expended in the manufacturing of the machinery itself. The modern farm, however, is more efficient than the primitive garden in terms of the amount of food raised per person. In primitive agriculture the labor of one person generally produces enough food for only slightly more than one person. On a modern mechanized farm, one person produces enough food for many persons.

The use of machinery dictates the adoption of monoculture. Were rows of tomatoes to alternate with rows of wheat, the mechanized harvesting of

126

**14.7** An example of a monoculture. This is a large field planted only to corn. In 1970 much of the United States corn crop was threatened by corn leaf blight.

the wheat would wreak havoc among the tomatoes. Modern farming thus demands the planting of vast acreages to a single crop, but some of the efficiency of monoculture is lost wherever a widespread epidemic sweeps a crop. The Irish potato famine of the nineteenth century is an excellent example of the tragedy resulting from dependence on a monoculture.

After the potato, a native of the Western Hemisphere, was discovered by European explorers, it was widely planted in several parts of Europe, especially in Ireland, where it became a staple item in the diet. The potato is subject to a disease called late blight of potato, which is caused by a fungus. This fungus often is present sporadically on potatoes and usually causes only minor damage, but it thrives in prolonged damp weather such as is common in the British Isles. The years 1846 and 1847 brought weather particularly favorable for the blight, and the potato crops failed. Almost completely dependent on the potato, thousands of persons starved to death, and others emigrated to America in search of a new life. By 1851 the population of Ireland had dropped from about 8 million to about 6 million. Most of those remaining lived in dire poverty.

## Pollution and Food Chains

The possibility of mass starvation always exists among those who live on marginal diets should a particularly virulent parasite spread rapidly through one of the major food crops of the world. But disease organisms have no planned place in man's scheme for a simple agricultural food chain, and he uses pesticides lavishly to kill the unwelcome guests. Unfortunately the pesticides enter the food chains and do damage that they were not intended to do. One consequence of the workings of food chains is that substances move upward through them and tend to accumulate in higher trophic levels. Each animal in its lifetime must eat many times its own weight, and from the food is extracted substances that the animal needs. Because of the great quantity of food that the animal eats relative to its own size, some substances accumulate in its body in high concentrations. The process repeats when this animal and its fellows become the food of animals at the next higher trophic level. These latter animals concentrate the substance even more. Top carnivores at the highest

127

trophic levels acquire the highest concentrations (Fig. 14.8).

Some of the pollutants released to the environment by man contaminate the leaves of plants or are absorbed by their roots. Herbivores eat these contaminants, which then work their way up through trophic levels. Mercury and DDT are only two such poisons that have been killing the birds of prey that occupy the highest trophic levels.

Mercury, which is a highly toxic poison, has been discharged by industry into waterways in the mistaken belief that because of its weight and its insolubility in water mercury would sink to the bottom and remain there inert and harmless. However, some of the bacteria living in the mud of stream and lake bottoms convert the mercury to dimethyl mercury, which is soluble in water. The dimethyl mercury enters the food chain by being absorbed by microorganisms and then passing upward to herbivorous fish and then carnivorous fish. It also may enter the fish directly through their gill membranes. Bald eagles, fish-eating birds at the top of a food chain, have died of mercury poisoning acquired through feeding. Some mercury compounds used as fungicides on seeds intended for planting have accumulated in wild pheasants to such an extent that the birds may be dangerous for human consumption. Mercury affects primarily the nervous system and in humans it causes numbness, blurred vision, tremors, mental derangement, and, in sufficient quantity, even death.

DDT, which can kill outright some birds, such as robins, has reduced the populations of brown pelicans, eagles, and peregrine falcons by causing eggs with unusually thin shells to be laid. The eggs break under the weight of the parent birds, and the embryos die. Again it is the top carnivores that suffer the most, for their bodies contain the greatest concentration of DDT. Populations of these carnivores have been declining rapidly in areas where pollution is most prevalent. Some of these species are in danger of extinction; their demise will make the natural communities less stable.

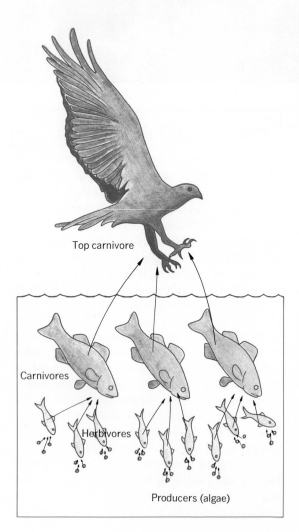

Top carnivore

Carnivores

Herbivores

Producers (algae)

**14.8** Increasing concentration of pollutants at high levels of a food chain. Algae absorb pollutants from the water in which they live and concentrate some of them. Herbivorous aquatic animals eat many times their own weight in algae, and the concentration of pollutants becomes higher in the herbivores than in the algae. Each carnivorous fish eats many times its own weight in herbivores, and so the concentration of pollutants becomes higher in the primary carnivores. Top carnivores, like the osprey, have still higher concentrations. Millions of algae form the basis of the food chain of a single osprey, which receives much of the pollutants that the algae absorbed from the water.

That most persons have not obviously suffered from exposure to DDT and thousands of other toxic substances carelessly disseminated by man may be due in part to higher resistance of the human body to some of these substances, but man surely must be protected somewhat now by his relatively low position in the food chain. Given sufficient time and the ingestion of sufficient quantities of these substances, the human species well may be adversely affected by the wastes of its civilization. Some deaths have occurred among those directly exposed to high concentrations of some of the potent pesticides and among those known to have eaten mercury-contaminated fish.

Several radioactive substances produced by nuclear bomb explosions enter the food chains and become concentrated at high trophic levels. The radioactive substance causing the most concern among biologists is strontium-90, which becomes incorporated into bone in the place of calcium. Because many blood cells are formed in the marrow of bones, these cells, which perform so many vital functions for the body, become subject to damage by unusually high doses of radiation. Two other radioactive substances that are of concern are iodine-131, which accumulates mostly in the thyroid gland in the place of nonradioactive iodine, and cesium-137, which replaces phosphorus in some tissues, especially muscle. Many Eskimos have unusually high burdens of radioactivity; relying heavily on caribou for food, they occupy a higher trophic level than most other peoples, who are partially or completely vegetarians.

The potential danger of nuclear reactors generating electrical power lies probably not so much in a nuclear explosion—an event claimed to be of remote possibility by their designers—but in the small amount of radioactivity that leaves the reactor in the cooling water or that might leak out of storage containers for radioactive wastes. Although the absolute amounts of radioactive materials escaping to the environment may be small, they can accumulate to dangerous concentrations high in the food chain.

Radioactive substances endanger both the individuals radiated and their offspring not yet born and not yet conceived. Radioactivity is known to cause malignant tumors. Before its dangers were understood, radium, a radioactive element, was incorporated in the paint on the numbers on the faces of luminous clocks and watches. The workers habitually pointed their paintbrushes by licking them, and thus ingested radium. Many of these workers have died of cancer. Among the survivors of the atomic bombs dropped on Hiroshima and Nagasaki in 1945, there has been a high incidence of leukemia, a cancer of white blood cells.

In 1954 an unexpected change in the direction of the wind carried radioactive fallout from the detonation of a thermonuclear bomb on Bikini Island to Rongelap Island in the Marshall Islands. Many of the natives of Rongelap were exposed to the fallout for two days before they were evacuated. Within a few days or weeks the people suffered from nausea, vomiting, radiation burns on the skin, loss of hair, and a depression in the numbers of some cells of the blood. Recovery from most of these conditions occurred within a few days or weeks, but it took 11 years for blood cell counts of the victims to return to normal.

Not only were the Rongelap natives exposed directly to fallout, but the radioactive material entered the food chain, and in 1962 Rongelapese, who since had returned to their island, had strontium-90 levels in their bodies about 24 times higher than those of persons living in the United States and cesium-137 levels 300 times higher. That most of this material comes from their food and is not a residual effect of the original exposure is indicated by the fact that Rongelapese who were not directly exposed to the 1954 fallout carry concentrations of cesium-137 similar to those of persons who were so exposed.

Iodine-131 accumulated in the thyroid glands of the Rongelapese. Among the 19 children who were younger than 10 years of age when they were first exposed, 15 had developed nodules of the thyroid

129

gland sometime between 1963 and 1968. None of these thyroid nodules has yet proved to be malignant, but thyroid cancers developed in survivors of the atomic bombings of Japan, and the incidence of the cancer corresponds to the intensity of the radiation received.

Exposure to radioactivity also alters the genetic material of the cell (Chapter 46), and genetic damage can be passed from generation to generation. Because humans are not suitable experimental animals and because the human species has been exposed to more than the natural background of radiation for only about one generation, relatively little is known about radiation-induced genetic variations in people. For fruit flies and other rapidly reproducing laboratory animals a large body of information is available, and radiation-induced genetic changes range from modification of superficial traits that have little effect on the individual inheriting them to crippling and debilitating diseases. Still others are lethal and may cause death at almost any stage of life.

Not all mutations are caused by exposures to external sources of radiation. The incorporation of even small amounts of radioactive substances into the body, as by way of the food chain, exposes the body to low-level, but continuous, radiation which, in the long run, may have just as serious effects.

## Questions

1. Define the following terms.
   trophic level
   herbivore
   carnivore
   omnivore
   producer
   consumer
   biomass

2. Differentiate between a food chain and a food web.
3. How does a food chain illustrate the consequences of the second law of thermodynamics?
4. Why is it unlikely that a food chain would have a dozen trophic levels?
5. Why is there less biomass at the top of a food chain than at the bottom?
6. Why does hamburger cost more per pound than the bun you put it on?
7. Arrange the following in order from the top of a food chain to the bottom.
   herbivores
   primary carnivores
   producers
   tertiary carnivores
   secondary carnivores
   At how many places would it be logical to add decomposers?
8. Try to reconstruct the food chains that produced the items in your last meal.
9. Space flights of several months or years would require the raising of food in a space capsule. Do you think that a space capsule community consisting of protein-rich algae as a food source, astronauts as consumers, and bacteria as decomposers would be adequate? Explain.
10. Explain why top carnivores generally are more adversely affected by pollution than organisms at lower trophic levels.
11. Why is there no great energy cycle in nature comparable to the carbon and nitrogen cycles?

## Suggested Readings

Gates, D. M. "The Flow of Energy in the Earth," *Scientific American* 224(3): 88–100 (September 1971).

Kemp, W. B. "The Flow of Energy in a Hunting Society," *Scientific American* 224(3): 104–115 (September 1971).

Odum, E. P. *Ecology.* New York: Holt, Rinehart and Winston, Inc., 1963.

Rappaport, R. A. "The Flow of Energy in an Agricultural Society," *Scientific American* 224(3): 116–132 (September 1971).

Woodswell, G. M. "The Energy Cycle of the Biosphere," *Scientific American* 223(3): 64–74 (September 1970).

Many an inquisitive child has tried to solve the problem of how much money he would have at the end of one month if he had a job that paid him 1 cent the first day, 2 cents the second day, and 4 cents the third day and continued to double the amount each succeeding day. The calculations for the first week are done quickly, and those for the second week are not difficult. By the end of the third week calculations become tedious, but the amount of money received each day is surprising. At the end of the month the hypothetical recipient is a millionaire several times over.

Anyone who takes the time to perform this exercise learns something about compound interest: the original amount increases slowly at first, then rapidly, indeed startlingly to those for whom this is the first experience with logarithmic growth.

Perhaps the example of logarithmic increase familiar to most adults is the growth of a savings account collecting compound interest. This is an example of a **positive feedback** mechanism. This is a system with two components, each component growing and its growth causing still further growth of the other component. This mutual effect results in a greater amount of growth with each succeeding unit of time. The savings account interest of one year increases the size of the account, which in turn increases the size of the interest the next year (we assume a constant interest rate). This increases the size of the account again, and this increases the size of the interest again. As long as the money is left in the bank, each component continues to grow by virtue of the other's growth.

## Arithmetic and Logarithmic Progressions

We should digress to compare arithmetic and logarithmic progressions. An **arithmetic progression** is familiar to nearly everyone: it is found on

# 15
# Growth Curves

1 An arithmetic progression grows by equal increments—for example, 1–2–3–4–5 . . . .

2 A logarithmic progression increases by some constant multiple—for example, 1–2–4–8–16 . . . .

3 Biotic potential is the maximum possible reproductive rate of which a species is capable; unrestricted biotic potential produces a logarithmic increase in numbers.

4 Environmental resistance includes those environmental factors that prevent organisms from realizing their full biotic potential.

5 The interaction between biotic potential and environmental resistance generally keeps population levels constant.

131

ordinary rulers, thermometers, and bathroom scales:

In an arithmetic progression equal spaces have equal values. The distance between 1 and 2 is the same as that between 6 and 7, each of these spaces having a value of 1. The distance between 3 and 5 is the same as that between 8 and 10, each having a value of 2. Similarly, the distance between 2 and 5 is the same as that between 6.5 and 9.5, and so on.

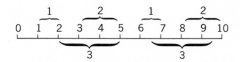

An arithmetic progression need not begin with zero. The lowest temperature registered on an outdoor thermometer may be 20° or 30° below zero, whereas a fever thermometer may begin at 94° (F) above zero.

Arithmetic progressions need not proceed by increments of one. A few possible arithmetic progressions are: 2, 4, 6, 8, 10, . . . ; 5, 10, 15, 20, 25, . . . ; and $\frac{1}{4}, \frac{1}{2}, \frac{3}{4}, 1, 1\frac{1}{4}, \ldots$ .

A list of the amount of money received each day in the exercise described at the opening of this chapter is an example of a **logarithmic progression:**

```
 1    2    4    8    16    32    64    128
```

Each number is double the preceding number, and therefore equal spaces do not have equal values. The space between 4 and 8 has a value of 4, and the space between 32 and 64 has a value of 32 even though it is the same size. What the two spaces do have in common is the increase in value by a certain multiple, in this case, by 2. Similarly, the spaces between 4 and 16 and between 32 and 128

are equal in length and represent a multiplication by 4.

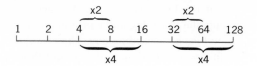

Often it is convenient to use a logarithmic progression in which each number is 10 times larger than the number preceding it:

```
1    10    100    1,000  10,000  100,000  1,000,000
```

Or the same progression may be written:

$$10^0 \quad 10^1 \quad 10^2 \quad 10^3 \quad 10^4 \quad 10^5 \quad 10^6$$

The number zero does not appear in a logarithmic progression, for if the progression is continued toward lower numbers, only smaller and smaller decimals (or fractions) are encountered. They approach zero more and more closely, but they do not reach it:

```
0.00001  0.0001  0.001   0.01    0.1     1        10
```

Or

$$10^{-5} \quad 10^{-4} \quad 10^{-3} \quad 10^{-2} \quad 10^{-1} \quad 10^0 \quad 10^1$$

### Plotting Logarithmic Progressions on Graph Paper

A logarithmic progression on the Y axis (vertical scale) of a graph can be useful in plotting growths of populations for several reasons. One reason is that they permit the convenient plotting of data that range from extremely small numbers to very large ones. Let us assume that a single bacterium lives in a test tube of nutrient broth, a solution

containing all the substances the bacterium requires to grow and divide. Let us also suppose that
this cell divides into two cells at the end of one
hour, and that every hour thereafter each progeny
cell divides:

| | |
|---|---|
| 0 hour | 1 cell |
| 1 hour | 2 cells |
| 2 hours | 4 cells |
| 3 hours | 8 cells |
| 4 hours | 16 cells |
| 5 hours | 32 cells |
| 6 hours | 64 cells |
| 7 hours | 128 cells |
| 8 hours | 256 cells |
| 9 hours | 512 cells |
| 10 hours | 1,024 cells |

**15.1**  Curve of a population growing at a logarithmic
rate and plotted on arithmetic graph paper. See text.

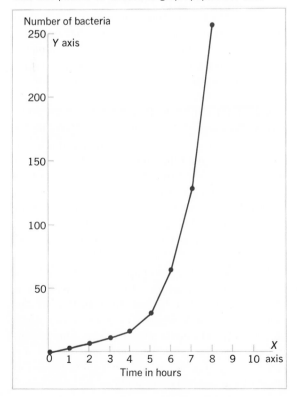

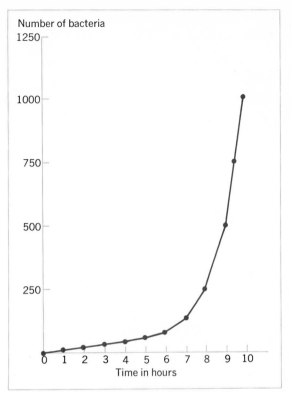

**15.2**  Curve of a population growing at a logarithmic
rate and plotted on arithmetic graph paper. See text.
Compare the shape of this curve with that in Fig. 15.3,
in which the same data are plotted.

If we try to plot these data on ordinary arithmetic
graph paper in which both the X axis (horizontal
scale) and Y axis (vertical scale) are arithmetic, a
problem becomes apparent immediately. If the
scale of the Y axis is so arranged as to enable one
to distinguish easily the increase in numbers in the
first few hours, the curve soon runs off the top of
any graph paper of convenient size (Fig. 15.1). If,
on the other hand, the highest number encountered is placed near the top of the Y axis, the
smallest numbers are crowded too closely together
to distinguish them clearly (Fig. 15.2). The rapid
increase of logarithmic growth is evident, however,
in the dramatic sweep of the curve which bends
ever more steeply upward.

The problem can be avoided to a large extent

133

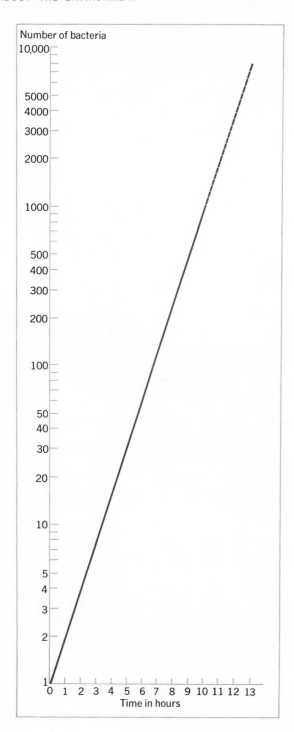

**15.3** Curve of a population growing at a logarithmic rate and plotted on semilogarithmic graph paper. The Y axis is a logarithmic progression. See text. Compare the shape of the curve with that in Fig. 15.2, in which the same data are plotted.

by plotting the data on semilogarithmic graph paper (Fig. 15.3). The X axis is an ordinary arithmetic progression, but the Y axis is a logarithmic one. When the data for the preceding example are plotted, the curve takes on a different aspect. Whenever the growth rate remains constant, as it does in this example of doubling every hour, the curve is a straight line. Because of this, another advantage of semilogarithmic graph paper becomes apparent. With a constant growth rate it is possible to extrapolate or interpolate without making any mathematical calculations. One needs only a straightedge with which to extend the curve to any desired length. Extending the curve in Fig. 15.3 indicates that in three more hours the bacterial count will be approximately 8,200 (actually 8,192). With arithmetic graph paper it is not possible to extend the curve accurately without first calculating the next point (or points) to be plotted.

Given only two points to plot—perhaps the population at the end of the fourth hour and the population at the end of the seventh hour—one can plot these on semilogarithmic graph paper and join the points by a straight line that extends beyond them in both directions. From this line one can read off the population at any given time—before the fourth hour, between the fourth and seventh hours, or after the seventh hour. These manipulations are possible *only if the rate of growth remains constant.*

Semilogarithmic graph paper has a third advantage of allowing one to perceive immediately a change in growth rate and the direction of the change. It is not always easy to detect such changes on arithmetic graph paper. Should the growth rate change, it becomes apparent on the semilogarithmic graph paper as an alteration in the slope of the curve. Let us suppose we begin again

134

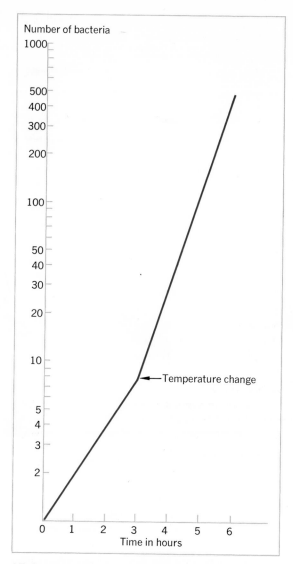

15.4 This curve plotted on semilogarithmic graph paper shows an abrupt increase in the rate of growth.

| 0 hour | 1 cell |
|---|---|
| 1 hour | 2 cells |
| 2 hours | 4 cells |
| Temperature change → 3 hours | 8 cells |
| 3½ hours | 16 cells |
| 4 hours | 32 cells |
| 4½ hours | 64 cells |
| 5 hours | 128 cells |
| 5½ hours | 256 cells |
| 6 hours | 512 cells |

These data are plotted in Fig. 15.4. At the end of the third hour, when the growth rate changes, the curve slopes more steeply upward. Note that the second part of the curve is a straight line also, for the growth rate here is constant (doubling every half-hour).

Let us return to our original bacterium again and suppose that at the end of the third hour the con-

15.5 This curve plotted on semilogarithmic graph paper shows an abrupt decrease in the rate of growth.

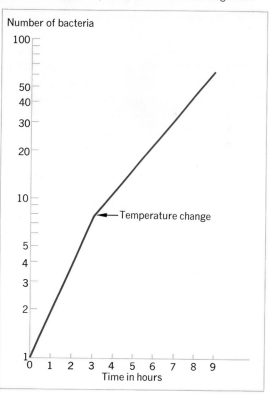

with a single bacterium in nutrient broth in a test tube. For the first three hours the bacterium and its offspring behave as before and double every hour, but let us suppose that the test tube is warmed somewhat, improving conditions for the bacteria, and then the cells double every half-hour:

135

ditions in the test tube had become cooler and less favorable for the bacteria and the cells divided only once every two hours:

| | | |
|---|---|---|
| | 0 hour | 1 cell |
| | 1 hour | 2 cells |
| | 2 hours | 4 cells |
| Temperature change | 3 hours | 8 cells |
| | 5 hours | 16 cells |
| | 7 hours | 32 cells |
| | 9 hours | 64 cells |

The curve in Fig. 15.5 shows a change in slope of the curve at three hours, this time to a less steep slope because the growth rate has changed to a less rapid one. It should be noted that after the third hour the population is still growing in size; it is only the rate of growth that has slowed.

A population that is decreasing in size has a negative growth rate, and its curve slopes downward. If the decrease is a logarithmic one, for example 64–32–16–8 . . . or 1,000,000–100,000–10,000 . . . , the curve is a straight line with downward slant on semilogarithmic graph paper.

## Growth Curves

### Growth Curve of a Pure Culture

The growth curve of a culture of bacteria introduced into a test tube of fresh nutrient media may resemble the curve in Fig. 15.6. The culture grows slowly at first as the cells become adjusted to their new environment. Then the growth rate increases and reaches a constant logarithmic rate. As the population grows, it uses up the supply of substances that it needs, the new individuals occupy more and more space, and the waste products of their metabolism accumulate in the environment. The change of conditions from favorable to unfavorable gradually slows down the growth rate to zero. At this point the maximum number of live individuals is reached, and as conditions become less and less favorable the deaths of cells exceeds the production of new ones. The curve falls slowly at first and then reaches a phase of logarithmic decrease. The curve may then fall gradually to zero, leaving no living cells in the tube.

This growth curve is typical of a pure culture of organisms confined to a finite space and provided with a finite supply of food and other necessities. Natural populations rarely consist of a single species, and because the wastes of one are the food of another, populations in mixed communities are not so likely to run out of food or to be poisoned by their own waste products. In a balanced community the population levels remain fairly constant, varying within a relatively narrow range over a long period of time.

Under certain conditions logarithmic increases occur in natural populations, but such an increase usually does not last for very long. The introduc-

**15.6** The growth of a single species confined to a finite space and provided with a finite amount of food. Plotted on semilogarithmic graph paper, it shows very slow growth (*A* to *B*), a gradual increase in the rate of growth (*B* to *C*), growth at a constant rate (*C* to *D*), a gradual decrease in the rate of growth which is still positive (*D* to *E*), growth rate of zero (*E* to *F*), a gradual increase in the rate of decrease as the growth rate becomes negative (*F* to *G*), decrease at a constant rate (*G* to *H*), and a decrease in the rate of decrease (beyond *H*).

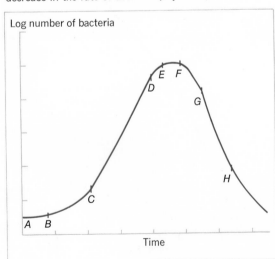

tion of a species into a new area, an island, for instance, can result in logarithmic growth if conditions there favor reproduction of the species (the cactuses and rabbits introduced into Australia are examples, Chapter 4). But as growth of the population deprives it of some of the conditions it requires, growth slows down. Assuming that the new species can live in harmony with the rest of the community, the population levels off to a nearly constant level which, with minor fluctuations, continues indefinitely.

## Natural Populations

The rate at which a population grows or decreases in size depends on two factors: the birth rate and the death rate. For a natural population, over a long period of time, the birth rate and the death rate usually equal each other, and there is no net growth; that is, the growth rate is zero. The minor fluctuations that always are present in populations are due to temporary inequalities in the birth and death rates. If the birth rate exceeds the death rate, the growth rate becomes positive, and the population increases in size. Conversely, when the death rate exceeds the birth rate, the growth rate falls below zero, and the population size decreases.

The **biotic potential** of a species is its maximum possible reproductive rate. In most species the biotic potential is enormous, adults having the biological capability of reproducing themselves tens, hundreds, thousands, or even millions of times over (Fig. 15.7). Unchecked, biotic potential produces a logarithmic growth of a population. This is achieved rarely under natural conditions, and never for very long, because **environmental resistance,** which includes those factors in the environment that increase the death rate or decrease the birth rate, serves as a balance to the biotic potential. Exhaustion of the food supply, accumulation of wastes, grazing or predation, competition for mates or living spaces, and epidemics of disease all serve to reduce the numbers

**15.7** Female hog-nosed snake with her eggs. With this clutch of 24 eggs, this snake and her mate would appear to have reproduced themselves 12 times over. If such biotic potential were to remain unchecked, the world soon would be overrun with hog-nosed snakes. Most of these offspring will not survive to maturity. Many of the eggs or the young snakes that hatch from them will fall victim to predators, parasites, or unfavorable physical factors in the environment. Only a few will survive long enough to produce their own offspring. (Courtesy of The American Museum of Natural History.)

of surviving offspring. In the long run, each parent leaves, on the average, one offspring that survives and reproduces (this is equal to two offspring per pair of parents in monogamous species).

## Questions

**1.** Are the following arithmetic or logarithmic progressions?
    1, 2, 3, 4, 5
    3, 9, 27, 81
    0.5, 1.0, 1.5, 2.0
    0.001, 0.01, 0.1, 1.0
**2.** If a population doubles in size every day, is it growing arithmetically or logarithmically?
**3.** Is a positive feedback mechanism likely to maintain a steady state or a logarithmic increase?
**4.** What does each of the following indicate in a population curve drawn on semilogarithmic graph paper?
    a. Straight line with an upward slant.
    b. Curve that slants more and more steeply upward.
    c. Curve that continues to rise but becomes less and less steep.

5. On semilogarithmic paper, what would be the appearance of a curve of population with zero growth?

6. Why do logarithmic increases not occur in natural populations very often? When they do occur, what types of factors are likely to stop such growth?

7. There is an old problem about a farmer whose pond has been invaded by an aquatic weed. Each day the weed covers twice as much area of the pond as it did the day before. At this rate of growth it will cover the pond completely in 30 days. The farmer decides to kill the weed when it covers half of the pond. On what day will he act? Suppose he became ill with a 24-hour flu on that day and decided to postpone the job; how would this affect the pond?

8. Define biotic potential and environmental resistance. What relation do these have to each other in natural populations?

There are too many people in the world today. This is the opinion of many biologists who have studied the problem of overpopulation and its sister problem, pollution.

## The Overpopulation Problem

The human population problem had its origins primarily in the lowering of the death rate by decreasing environmental resistance rather than in the raising of the birth rate, although a decrease in environmental resistance may have the effect of raising the birth rate. No other species has managed so to manipulate its environment. In prehistoric times, environmental resistance very nearly equaled the human biotic potential. Once the human species had evolved and established itself throughout the world, its population leveled off at an estimated 5 million individuals. The total world population then was less than that of any of several of the major metropolitan areas today.

### History of the Problem

The people first were hunters and gatherers of food. This meant that they could have no permanent residence, for they had to follow the game herds and travel to wherever fruits, seeds, or tubers were ripe. Not only did their nomadic existence prevent them from storing more food than they could carry with them, but most of their foods did not lend themselves well to more than a few days or weeks of preservation. An abundance of food in the summer or autumn did not ward off the specter of starvation in the winter. Shelters were crude and temporary. Exposure to the elements must have taken its toll, especially in colder climates. The necessity to roam through woods or meadows to find food and clothing exposed humans to attacks by carnivorous animals from whom they could protect themselves only by the

# 16
# Overpopulation and Pollution

1 The human species is experiencing an unprecedented population problem.

2 The problem arises primarily from man's ability to decrease environmental resistance rather than from an ability to increase his biotic potential.

3 If the human population does not voluntarily reduce its reproductive rate (preferably by birth control), environmental resistance (in the form of famine, pollution, or disease) once again will limit population growth.

139

sticks or stones at hand or whatever crude implements they had fashioned.

Death rates were high, and the average life expectancy was less than 25 years. It is not surprising

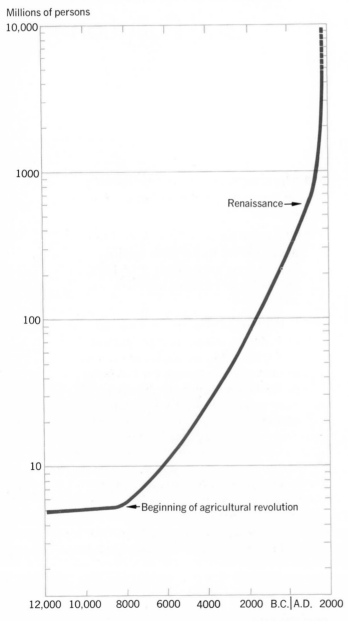

**16.1** Estimates of world population during the last 14,000 years. These figures are plotted on semilogarithmic graph paper. Note two obvious changes in the rate of growth.

Millions of persons

Renaissance→

←Beginning of agricultural revolution

12,000  10,000  8000  6000  4000  2000  B.C.|A.D.  2000

that the injunction to be fruitful and multiply is well rooted in human culture. If the prehistoric birth rate had not equaled the high death rate, the human species eventually would have become extinct. As it was, the birth rate exceeded the death rate so slightly that the population growth rate in prehistoric times probably was only a minute fraction of 1 per cent. Plotted on a graph, our best estimates of growth in this period produce a line that appears to be essentially horizontal, that is, a growth rate only slightly above zero (Figs. 16.1 and 16.2).

The first major lowering of the death rate came with the development of agriculture, especially the raising of cereal grains. Unlike hunting and gathering, most forms of agriculture permit the establishment of permanent residences. Villages sprang up in which not only a common store of food could be saved against lean times, but ideas could grow, be preserved, and exchanged among the residents. The importance of the farming of cereal grains cannot be overemphasized. Not only do these crops lend themselves to relatively long storage, thus decreasing the danger of famine, but they also are highly concentrated energy sources. The amount of human effort required to raise a given amount of food in the form of cereals was much less than that required to hunt wild game animals or to gather wild fruits. Animals were domesticated; their meat, milk, and hides were immediately available and did not need to be hunted down. Some leisure was permitted the farmer, and clever minds devised more efficient ways of farming and the means of extending agriculture into desert or semidesert areas. Plows and other implements were invented. The labor of large domestic animals, such as cattle, supplemented that of humans. Canals were built to bring water to areas otherwise too dry to farm. Not only did production of essentials on the farm reduce the number of dangerous trips into forest and field, but as agriculture spread and wild areas were reduced, large predators became less common in settled areas. The new agricultural revolution grew like a posi-

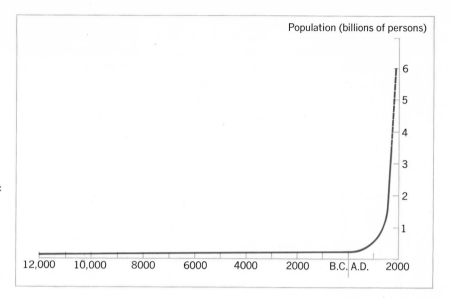

Population (billions of persons)

6
5
4
3
2
1

12,000   10,000   8000   6000   4000   2000   B.C. | A.D.   2000

**16.2** World population figures plotted on arithmetic graph paper. The rise in numbers during recent history is displayed dramatically. The increase caused by the agricultural revolution is not apparent because of the scale of the Y axis.

tive feedback system, with new improvements producing more leisure time that could be reinvested by using it to make more improvements. All of these reduced the environmental resistance to the human biotic potential, and as the death rate fell, the growth rate rose. More babies were likely to survive to adulthood and to produce a new crop of babies with equal or better chances of survival.

Fortunately the logarithmic growth of the ancient population was at a low rate; probably it did not exceed 0.05 per cent (which corresponds to doubling every 1,400 years), yet it was sufficient to bring about a population of nearly 200 million at the time of Christ. From then until the Renaissance the growth rate was an estimated 0.07 per cent (doubling every 1,000 years). During all this time food was more abundant than it had been before the ancient agricultural revolution, but widespread famines were not unknown from time to time. Medical knowledge had not advanced much beyond what it was in the ancient world. Wherever population densities were high, epidemics were more likely to take high tolls and to reduce the population again. One of the most spectacular epidemics of this time was the black

plague that ravaged Europe during the middle of the fourteenth century, killing one out of every three persons and wiping out some villages entirely.

The Renaissance (fourteenth to sixteenth centuries) was, as its name suggests, a rebirth—a cultural rebirth. Knowledge increased, and learning was held in great respect. The groundwork was laid for new discoveries and inventions. New interests arose in the sciences as well as in the arts, and information could be disseminated by means of the newly invented printing press.

One of the products of the post-Renaissance period, when perhaps 500 million (a half billion) persons lived on the earth, was the microscope. It made visible to the human eye organisms whose existence previously had been unsuspected. Among them were the bacteria and other agents of disease. After many years of study, their full significance was realized, and new methods of treating disease could be devised and tested. The effect of these began to be felt in the nineteenth century and continues today. Immunizations, quarantines (practiced in earlier times, but now used with greater understanding), antibiotics,

141

transfusions, transplants, and pesticides are only a few of the factors that have decreased the death rate of the human species, increased its life expectancy, and increased its growth rate.

Over the last few hundred years, improved methods of transportation and communication have made their contributions to population growth. Medicines now can be rushed within a few hours to almost any place on earth where they are needed. Physicians may diagnose diseases and prescribe treatment by telephone, thereby saving lives of people that they might not otherwise be able to reach. Fruits, vegetables, meats, and seafoods can be transported to almost any market in fresh condition; this permits a more varied diet and reduces the likelihood of a mineral or vitamin deficiency. Previously goiters were seen in persons living inland far from the seacoasts where seafoods rich in iodine could be found, and scurvy was common where citrus fruits were rare.

The present rise in the growth rate began slowly in the sixteenth and seventeenth centuries, then gathered momentum in the eighteenth and nineteenth centuries, and reached explosive proportions in the present century. Frighteningly, the rate of growth still is increasing. The present rate is an unprecedented 1.9 per cent for the world as a whole; at this rate the population doubles every 37 years. The United Nations' most optimistic and most pessimistic projections into the future estimate the world of the year 2000 to have a population of from 5.5 to 7 billion persons.

## Natural Resources for the Human Population

For all its intelligence the human race still is dependent for its survival on the resources of the earth. These may be divided into two groups: renewable and nonrenewable. Plants and animals are renewable resources; they replace themselves by producing offspring before they die. We are dependent on them for food, fibers, lumber, paper, and other things. Nonrenewable resources include

142

mineral deposits, such as iron and copper, and the fossil fuels, such as coal and oil. With care, renewable resources and some nonrenewable resources can provide the human race with certain of its requirements almost indefinitely. Used carelessly, both types can be depleted to the point that the planet can support few or no humans.

"Renewable resource" does not imply an infinite abundance. A farmer raising cattle cannot sell every animal he has, or he will go out of business. To ensure a steady income, he must at least keep the size of his herd constant. Each year he can sell a number of animals equivalent to the number that were born minus the number that died. If he sells more than that, his breeding stock dwindles, and the number of offspring produced each year falls accordingly. For several years he may enjoy a large income, but he is living on his capital, and the day will come when he has no herd left at all and must face the prospect of life with no income. He is not unlike a person whose sole income is that derived from $100,000 invested at an annual rate of 5 per cent. Its $5,000 annual income is a modest sum on which to live, but adequate to stave off starvation if properly husbanded. As long as he spends no more than $5,000 each year, he is assured of an income for life. This person, however, prefers to live in a grander manner and in the first year spends $15,000, thus reducing his capital by $10,000. The second year, his $90,000 capital yields only $4,500 in income. If he again spends $15,000, he reduces his capital by $10,500 to $79,500. For a few years he lives comfortably, but it is not difficult to see that he is on his way to bankruptcy within a few years.

Throughout much of its history mankind has lived on its capital of renewable resources. In some cases the condition of bankruptcy has been reached. Fertile farmlands have become deserts, and many species have become extinct. Whalers have been living on their capital (the world's breeding stock of whales) for so long that several species, including the blue whale, the largest animal that ever lived, are on the verge of extinction.

Yet the whaling industry continues to deplete the whale stock as if it were unaware that it must be putting itself out of business.

Such practices, which would not be tolerated in a privately owned business or farm, are common in the case of public property or where there is no owner (such as on the high seas). The prevailing attitude here is basically a selfish one. The philosophy is that if I don't reap my share of the harvest, the other fellow will. Unfortunately, the size of "my share" almost inevitably is overestimated, and a portion of the capital is harvested. (We must remember, too, that the larger the human population becomes, the smaller the individual's share in the world's resources becomes.) Solutions probably lie in governing agencies (international, national, or local, as a particular situation demands) that have the same concern for our renewable resources that the private owner has for his breeding stock. Such agencies must have the power to enforce their regulations.

Nonrenewable resources may be divided into those that are consumed in the process of being used and those that are not. Theoretically the latter can be recycled, but there always is some loss, and the cost of recycling sometimes is prohibitive. The metal in old coins, for example, can be melted down and used to make new ones. It will not make the same number of coins, for years of handling will have worn a small amount of the metal away, and the lost metal is essentially irretrievable. Today it is considered too costly to salvage old automobiles or to sort out the odds and ends of a city's rubbish and garbage for recycling. This may not always be the case.

Fuels are nonrenewable resources that cannot be replaced once they have been used. We always can hope that another vein of coal or another oil deposit will be found, but eventually all that can be economically exploited will be consumed. Even the radioactive substances used as fuels in nuclear reactors and once hailed as almost endless sources of energy are expected to be depleted by the turn of the century if all of the presently planned nuclear reactors are completed as scheduled. Two new types of reactor, breeder reactors and fusion reactors, may solve this problem. If they are perfected and become a reality, breeder reactors will produce more fuel than they use and so provide fuel for other reactors, and fusion reactors may use hydrogen as fuel which is in plentiful supply. Breeder reactors, however, will produce plutonium, a highly radioactive waste, with its attendant storage problems. Fusion reactors, which are expected to produce significantly less radioactive wastes than present-day reactors, will require the maintenance of such high temperatures that the technological problems will not be solved soon.

One hope may lie in the discovery of a way to harness solar energy economically. Sunlight is a resource that we cannot turn off (except perhaps by smog). Its energy pours on the earth in such abundance that, properly channelled, it could provide all the power needed in the foreseeable future. Unfortunately, we still await news of an economically feasible way to harness its energy on a broad scale.

In his understandable effort to make life easier, more comfortable, and more secure, man has created a problem unlike that of any other species. By artificial means he has made the earth to yield its resources at such a rate that they cannot be replaced. The problem is not new, but it is of unprecedented proportions. Over the ages unwise agricultural practices have ruined farmland. Parts of the Fertile Crescent, the cradle of civilization in the Middle East, a land "flowing with milk and honey," are today largely desert (Fig. 16.3 and Plate VIB), its low, sparse vegetation capable of supporting little more than a few hardy sheep and goats. Over many years of antiquity the great forests of the Mediterranean areas were cut down, leaving denuded hills in Greece and Italy. The floods that ravage Florence with some regularity are only one consequence of this. Ruins of once prosperous cities lie baking in the sands of North Africa.

Today, in an effort to increase crop yields for

143

**16.3** Once fertile, these denuded hills in the Middle East are now desert. Without vegetation the soil washes away in rain. Notice erosion lines in the hills.

the ever-growing human population, forests and lands better suited to grazing have been turned into croplands, often with disastrous effects. The cereal grains are tall grasses that grow best in the tall-grass prairies where the rainfall is about 30 to 40 inches per year. The short-grass prairies receive only 10 to 20 inches of rain, and mixed-grass prairies are intermediate. Attempts to farm short-grass prairies sometimes are successful, especially if there is a succession of years with higher than average rainfall. Eventually there come years with rainfall less than average, and the crops fail. The dry soil, no longer protected by its natural prairie grasses, is exposed to the air and blows away with the winds (Fig. 16.4). In later rains, it washes away with the runoff. A decade of agriculture can de-

**16.4** Dust storm in Colorado in the 1930's. The dry soil of short-grass prairies planted to corn and other tall cereal grains blew away in great clouds. (From *The Prairie World* by David F. Costello. Copyright © 1969 by David F. Costello. With permission of Thomas Y. Crowell Company, Inc., publisher.)

144

**16.5** An abandoned farm in the panhandle of Oklahoma. This is short-grass prairie, photographed in 1937 when it had lost its fertility after an unfortunate attempt at agriculture. Notice the remains of several corn plants in the field. (USDA Photo.)

stroy a layer of topsoil that required hundreds of years to form. It is not just a matter of a few seasons of crop failure during droughts, but the complete ruin of the land for years to come (Fig. 16.5). The famous Dust Bowl of the American plains is only one example of grazing land that was put to crops; in the 1930's gigantic dust storms carried the dry topsoil as far as the east coast of the United States. As the world population increases there will be more and more pressure to farm such marginal lands and to run the risk of converting them to deserts. The result can only be less arable land to feed a larger population.

A second method of increasing the food production of the world has been the breeding of new varieties of crops that give higher yields. However, the greater the yield, the greater the loss of minerals from the soil and the greater the need for replacement with chemical fertilizers. Chapter 13 discusses the contamination of water supplies with runoff from chemically fertilized fields. Many high-yield varieties are more susceptible to disease than older varieties. Monoculture of such high-yield varieties sets the stage for possible widespread epidemics of crop diseases.

Liberal use of pesticides is a third method of

145

increasing yields. Although government regulations limit the time of applications and their permissible concentrations, detectable quantities of pesticides wash into streams or remain on harvested crops. Tremendous fish kills in the Mississippi River and elsewhere have been attributed to the use of pesticides. The effects of long-term exposures to low concentrations of pesticides on human beings never have been determined, but they are not expected to be beneficial.

Where water is limiting, irrigation is a fourth factor that can increase yields. Irrigated fields often produce bountifully at first, but then they tend to become salty. The salts are those that arrive in the irrigation water and that remain in the soil when the water evaporates. Eventually the salt content of the soil becomes so great that it can no longer support crops and it reverts to desert. Presently, the Imperial Valley, a former desert in southern California, which now produces much of the winter fruits and vegetables of the United States, is beginning to show signs of salting as a result of irrigation.

Although these and other techniques have increased the amount of food available, they have not, in general, raised the standard of living of most of the world's population. (The reader must not use his own condition as a standard; the file clerk in almost any American office lives better than most of the people of the world.) Rather, they have allowed the survival of more people in the same, often pitiful, conditions that their ancestors lived in. This is the danger of the so-called Green Revolution, which seeks to raise crop yields by using new varieties of crop plants and by increased use of fertilizers. It has been reported to increase crop yields in Mexico, the Philippine Islands, India, and Pakistan. The Green Revolution has been hailed in some quarters as a potential solution to the overpopulation problem, but without population control, it is likely to set the stage for a newer, larger version of the Irish potato famine (Chapter 14).

In 1900 the world's population was about 1.6

billion persons. In 1970, it was over 3.5 billion, and it continues to rise. More than half of the persons of today's world are undernourished, malnourished, or starving. In other words, the number of persons today with substandard diets exceeds the entire world population at the turn of the century. Presumably the total misery has increased.

The problem is not merely one of distribution, for if the world's food supply were to be evenly shared among all its people, no one would have an adequate diet. It has been estimated that all the food raised today could nourish no more than 1.2 billion persons at the dietary standards presently prevailing in the United States. Tripling food production to feed everyone adequately within the next few years is an impossibility. Moreover, by the time such a tripling could be accomplished, unchecked population growth would have kept ahead of the new food supply.

This was essentially the message of the clergyman Malthus, who late in the eighteenth century wrote a treatise in which he stated, "Population, when unchecked, increases in a geometrical ratio. Subsistence increases only in an arithmetical ratio. . . . That population does invariably increase where there are the means of subsistence, the history of every people that have ever existed will abundantly prove. . . . In short it is difficult to conceive any check to population which does not come under the description of some species of misery or vice."

Although Malthus frequently is considered the father of our present concern with population problems, a search of older literature will reveal that the problem was recognized by ancient thinkers in various parts of the world. Among them were Tertullian, a church father of the late second and early third centuries, and Han Fei-tzu, a Chinese philosopher of the third century B.C. As long ago as the eighth century B.C., the Old Testament prophet Isaiah wrote, "Shame on you! you who add house to house and join field to field, until not an acre remains, and you are left to dwell alone in the land. The Lord of Hosts has sworn in my hearing: Many houses shall go to ruin, fine large houses shall be uninhabited. Five acres of vineyard shall yield only a gallon, and ten bushels of seed return only a peck." (Isaiah 5: 8–10, NEB.) Isaiah understood the principle of diminishing returns; trying to obtain more and more produce from a given parcel of land eventually results in reduced yields.

Most of the inhabited areas of the world probably approached the condition of overpopulation relatively soon after their colonization. Some, like the Middle East and the American Dust Bowl, have suffered extensive damage because of this. Others have managed to escape so dire a fate because unoccupied lands were available to be colonized by the excess population. Today nearly all the habitable portions of the world have been occupied. There is no place left to go. Two thirds of this world's people are experiencing the inevitable consequences of growth as predicted by Malthus, Isaiah, and many others. Modern "prophets of doom" warn us of worldwide famines in the 1970's and 1980's, and the possibilities of global epidemics are high in a time of high population densities and easy, rapid international travel.

To solve a problem, one first must be able to state it clearly. Too often the problem we are dealing with has been stated, "There is not enough food for the people of the world." The obvious solution to such a problem is an attempt to increase the food supply, but such attempts have not met with success except locally. It is the author's contention that the problem really is, "There are too many people for the amount of food the world can produce." This statement of the problem leads to a different approach, that of lowering the world's population.

Two theoretical possibilities present themselves: raising the death rate and lowering the birth rate. The former has little to commend itself to a civilized society, although infanticide has been practiced sporadically throughout history, and tribal peoples living a precarious hand-to-mouth existence have abandoned their aged and infirm in times of stress. It should be emphasized that if the human population is allowed to continue to follow its present tendencies of increase, the day will come when death control—by starvation, disease, or pollution—will solve the problem. It is unpleasant to contemplate the amount of misery involved in this laissez-faire "solution."

Far more attractive is the solution by means of birth control (Chapter 40). A decrease in the number of births to the point at which they equal the number of deaths eventually would serve to stabilize the population at a growth rate of zero (although, if the proportion of young persons in the population is high, as it is in much of the world today, the population will continue to grow for some time because the number of births will exceed the number of deaths, the latter being largely the deaths of old persons who represent only a small part of the population.) A further decrease in the number of births would bring about an ultimate decline in the world's population.

What is the optimum size of the human population of the world? It is difficult to come up with an absolute figure, for the answer to the question depends on the standard of living people wish to maintain. The world well might support five, ten, or even more times its present population if everyone were willing to live at a mere subsistence level. Considering only food supplies, we might estimate an optimum number of no more than 1.5 billion persons if everyone wants a diet similar to the present American diet. If everyone were to maintain the standard of living of the average American (with central heating, automobiles, television sets, college educations, elaborately equipped hospitals, and numerous other items we take for granted every day), the world might support no more than 0.6 billion (600 million) persons. This figure is based on the fact that slightly more than 200 million United States citizens consume at least one third of the world's resources, but limited supplies of specific resources may reduce this number even further. Whatever we decide is the optimum number of persons inhabiting the world, that number

147

almost certainly must be well below the number of persons presently alive—at least if we assume everyone's right to a nutritious diet, reasonably good health, and some degree of personal dignity.

## The Pollution Problem

Limited supplies of food and other resources are not the only problem in an overpopulated world. We have seen that the accumulation of waste products can cause a decrease in the size of a popu-

lation, but that because waste products are recycled when they are used by other organisms (Chapter 13), natural populations rarely are poisoned by their own wastes, though this may happen locally where an imbalance in the relationship among species exists. (An extreme example is the pure cultures of yeast mentioned in Chapter 12; used in the brewing industry, they produce ethyl alcohol until the high alcohol concentration kills them.) Humans, on the other hand, produce industrial wastes that have no use or that presently are considered too difficult or too expensive to recycle. Many are essentially impossible to retrieve once they have been released into the environment: the

**16.6** Smoke from factories near Pittsburgh blots out the sky. (Courtesy of Environmental Protection Agency.)

soot and gases from smokestacks (Fig. 16.6), pesticides used in farms and gardens, and mercury and many other wastes dumped into streams, lakes, and oceans. Some remain virtually unchanged for years in the environment. These are the substances that are not biodegradable; no organisms can metabolize them and convert them into compounds useful to themselves or other living things. Chapter 13 discusses the effects of the dumping of some biodegradable substances, the products of which have undesirable effects. Only a few of the many nonbiodegradable or very slowly biodegradable substances produced by man will be discussed here.

The chlorinated hydrocarbons (DDT, dieldrin, aldrin, endrin, chlordane, and others) persist for years in the soil. Of these widely used insecticides, DDT is perhaps the best known. Because it quickly kills insects and has no immediate effect on humans as it ordinarily is used, DDT was long considered harmless to humans. Since World War II it has been applied liberally throughout much of the world. India, once the most malaria-ridden country in the world, had a plan to spray every building in the country with DDT. Projects like this greatly reduced the incidence of malaria and some other insect-borne diseases in tropical and semi-tropical regions. On the other hand, DDT has been less than an outstanding success in controlling Dutch elm disease (which is spread by beetles). Its indiscriminate use has led to its worldwide presence in air, water, land, and living organisms. Fatty tissues of humans over the entire world now store DDT, and human milk presently contains twice the upper limit recommended for consumption by babies. Relatively little is known about the effects of DDT on humans, but it is suspected of causing cancers, cirrhosis of the liver, interference with sex hormone metabolism, and nervous disorders. It is known to cause liver cancer in mice. It has certainly caused disastrous ecological upsets in predator–prey relationships (Chapter 46).

The chlorophenoxy acid group of herbicides includes 2,4-D, 2,4,5-T, and picloram. When ap-plied in proper concentrations they cause defoliation of plants. The chlorophenoxy acids achieved notoriety for their use by the United States government in Vietnam, where they were sprayed to defoliate forests and to destroy crops that might feed the enemy. Ecological damage to defoliated mangrove swamps is expected to persist for at least 20 years. These compounds also are used in the United States to kill weeds. Evidence that they may harm animals, including humans, has been accumulating.

A source of numerous pollutants is the internal combustion engine, which moves essentially all the automobile-centered cultures of today. The complete combustion of a fuel such as gasoline results in the production of only carbon dioxide and water, two compounds that recycle back into the environment (although they do not reconstitute new gasoline). Complete combustion requires a plentiful supply of oxygen, which is not available in an internal combustion engine. Without it carbon monoxide and a wide variety of hydrocarbon compounds are produced. Additives or impurities in the fuel produce other emissions such as lead compounds and sulfur oxides. Nitrogen and oxygen from the air react in the engine, producing nitrogen oxides. In 1965 the automobile was responsible for 86 million of the 142 million tons of air pollutants produced in the United States. In the presence of sunlight, these compounds react, forming a host of dangerous compounds that form the photochemical smog for which Los Angeles is so famous but which other cities endure as well (Fig. 16.7). As a final insult to the environment, most old automobiles find their final resting places in junkyards rather than being recycled.

Throughout history, large bodies of water have been considered suitable disposal sites for all manner of refuse. "Running water purifies itself," was a comforting saying, and in areas of low population densities it often was true. Immediately below a sewage outfall, water was not fit to drink; but during its trip 10 or 20 miles downstream, the sewage became diluted, the pathogenic bacteria

149

**16.7** New York City in the grip of air pollution. (Courtesy of CBS News.)

died, and the activities of other microorganisms in the water degraded the sewage to harmless compounds. The next town downstream could drink the water in safety. This was also true of other biodegradable wastes, such as garbage. Today towns are larger, producing a greater quantity of wastes; they are closer together, often contiguous, thus not permitting sufficient time for wild microorganisms to degrade substances present in the sewage or for pathogenic microorganisms to die. Sewage now contains a vast array of nonbiodegradable substances. Increasingly, water supplies of most towns and cities require more treatment to render them potable. Some rivers, like the Cuyahoga River in Cleveland, Ohio, are not even fit for fish to live in. The wastes of a variety of industries has made it one of the few rivers that can catch fire and burn.

Carbon dioxide usually is not considered a pollutant in the ordinary sense of the word. It is a natural waste product of animals and an essential substance for green plants. But any substance might be considered a pollutant if it is produced faster than the environment can cope with it, and this is the case with carbon dioxide (so, too, with the phosphates released by biodegradable phos-

phate detergents and most of the degradation products of sewage mentioned in Chapter 13).

Carbon dioxide is an invisible gas. Visible light from the sun passes more or less unimpeded through it and the other gases of the atmosphere (most of the longer and shorter wavelengths are absorbed by components of the upper atmosphere) and reaches the surface of the earth. Here much of it is absorbed by the soil, bodies of water, and living things. Some of it is used in photosynthesis and other photochemical reactions. Much of it—as a result of the second law of thermodynamics—eventually is converted into heat (infrared radiation of wavelength somewhat longer than that of visible light; see Fig. 6.1) that is radiated back to space. Carbon dioxide, which is transparent to visible light, absorbs heat and so slows its radiation to space. The higher the carbon dioxide content of the air, the slower the loss of heat to space, and the higher the temperature of the earth becomes.

So well balanced are the earth's natural annual heat gains and losses that over a period of one or a few years there is no perceptible change in the mean global temperature. However, the burning of fuels, whether gasoline, wood, coal, gas, or oil, introduces carbon dioxide into the air. It is estimated that an increase of 10 per cent in the carbon dioxide concentration of the atmosphere could raise the mean global temperature 0.5°C if no other factors counteracted the effect. A rise of 3.0°C is estimated to be sufficient to melt the polar icecaps. Within this century the concentration of carbon dioxide rose 7.5 per cent. From 1880 to 1940 the mean global temperature rose 0.4°C, but since 1940 it fell about 0.2°C. It well may be that the introduction of small particles (such as soot, dust, and aerosols) into the air has reduced the amount of sunlight reaching the earth and thus has reduced the warming effect of the sun. Indeed, scientists are uncertain whether to expect a rise or a fall in the world's average temperature in the future. The warming effects of some factors such as carbon dioxide and thermal pollution may more than compensate for the cooling effects of others (such as particulates in the upper atmosphere), or the opposite may occur. The simultaneous action of opposing factors may put off the day when we will have to face an unpleasant climate of one extreme or the other, but it is unlikely that the several types of pollution will balance each other's effects exactly.

This leads us to a completely different form of pollution—thermal pollution. Whereas all material wastes can, in theory, at least, be recycled back into the environment, the heating of the environment is the inevitable result of the second law of thermodynamics (Chapter 14). Essentially every activity, whether it be turning a single page in this book or generating the electricity required to keep New York City functioning, produces waste heat. Stand near the exhaust of an automobile and you will be aware of the phenomenon. Or touch the tires immediately after a long drive. The warmth of a kitchen comes from the heat that didn't cook the food. Most persons are not aware that their use of any electrical equipment, whether it be air conditioner or stove, television set or phonograph, lamp or doorbell, requires the generation of heat at a power plant that must burn some fuel or employ nuclear reactions to supply the electricity to run the equipment. Heating and cooling devices, such as stoves, irons, and air conditioners, are more wasteful in this way than are radios or phonographs. Doorbells and electric toothbrushes, which are used for only a few seconds or minutes during the day, contribute relatively little waste heat to the environment, as do the movements of the human body. Every activity, however, adds something to the total heating of the environment. The demand for energy in the United States is growing at an annual rate of 7 per cent. This means that the amount of energy used and the amount of heat generated double every ten years. It also means an eight-fold increase from 1970 to 2000.

Presently, water is used to conduct waste heat away from industrial equipment that becomes excessively hot with use. The water taken from a stream or lake returns to it warmer than it was upstream from the point of removal. Because hot water holds lower concentrations of gases than does cold water, the oxygen supply in heated water falls. At the same time the need for oxygen by aquatic organisms increases, for chemical reactions generally occur faster at higher temperatures. When the temperature rises high enough, fish and other organisms die. Nuclear power plants require a great deal of water for cooling; one such plant in Connecticut uses 370,000 gallons of water per minute. Concern has been expressed for the fate of our lakes and streams, especially those that will have several nuclear power plants along their shores. An alternative is the construction of cooling towers in which the heated water is permitted to evaporate into the air, but this has the disadvantage of increasing fog, rainfall, and snowfall in the surrounding area. A second type of tower, which cools water confined in pipes and thus produces no evaporation, is much more costly.

Regardless of the methods used for disseminating waste heat, the environment inevitably is heated. If the amount of heat produced by human activities combined with the effect of carbon dioxide is enough to counteract the cooling effect of the reflection of sunlight away from the earth by a possible future layer of global smog, the gradual rise in temperature could have disastrous effects. Melting of the polar icecaps could raise sea level and drown all the coastal cities of the world; this would include many of the major metropolitan centers. Most of Florida, the haven of the retired and the mecca of winter vacationers, would be under water. The amount of land available to the world's growing population would be even less than it is now. Unhappy as this prospect is, its "cure" by a cooling layer of perpetual global smog is even less felicitous. The resulting decline in crop production must spell mass starvation.

An exhaustive discussion of thousands of man-made pollutants released to the environment is not possible here. The reader is referred to current issues of almost any periodical of general interest.

**151**

He can hardly fail to find a great deal more information on the subject. What is more important here is to emphasize that pollutants are dangerous to both humans and wildlife, and that the increased abundance of pollutants provides a real threat to life on earth. The free use of air and water as public dumps has been taken for granted throughout history, but just as we cannot harvest our capital without depleting it (as discussed earlier), so we cannot discard wastes faster than the natural world can dispose of them without losing something as well. A small amount of sewage dumped into a lake produces insignificant consequences, for the wastes are recycled within the framework of the existing community of the lake. When the concentration of sewage becomes great enough to change the population of organisms living in the lake, the quality of the water declines from the point of view of both aesthetics and public health. The same sort of thing is true for industrial and agricultural wastes, whether disposed of in water or air or on the land. We might look on the capacity of the environment to absorb our wastes as a form of capital. As more and more of this capacity of the environment is destroyed by too great a burden of our wastes, we lose still more of this form of capital. The problem of large cities to find places to dispose of garbage is only one of the more readily visible aspects of the problem. Somewhat more serious is the accumulation of invisible poisons in air, food, and water.

In the final analysis, use of the environment as a dump is not free. Someone pays for it: in increased costs of purifying drinking water, in more frequent cleaning of clothes and painting of homes, in loss of wildlife, in increased unpleasantness of surroundings, in poor health, in doctor bills, in premature death. Industries have complained that they cannot afford to reclaim materials or to treat them so as to render them harmless. But the cost of pollution *is* part of the cost of doing business. The only question is who will pay the price and how?

Businesses are conducted today with the aim not only of making a profit, but usually of making a profit larger than last year's. Whether by built-in obsolescence or by advertising campaigns to make us all feel embarrassed wearing last year's clothes and driving a three-year-old automobile, the consumer is urged, even coerced, to buy products the manufacturing of which depletes our resources and pollutes our environment. It is questionable whether such a system yields more than a paper profit to the population as a whole.

Let us consider only one small aspect of conducting business: packaging. To increase the attractiveness of a product or for the customer's convenience, more and more products come encased in one, two, or more layers of plastic, paper, cardboard, steel, or aluminum. Perhaps the ultimate absurdity is a single grapefruit, already naturally packaged in its own rind, nestled on a bed of tissue in a cardboard tray, the whole tightly wrapped in transparent plastic. As an exercise, let the reader separate his family's kitchen wastes for one week into two piles: every bit of packaging in which the food arrived and the real garbage (carrot tops, bones, and so on). He may be surprised at the proportions of the two piles. Not only does the consumer pay for all that packaging at the time the food is purchased, when he is through with it he must pay to have it hauled away and disposed of. Depending on the method of disposal, he may pay for the packaging once again by inhaling smoke when it is burned or through the loss of valuable land as another dump is created. The food business's profit on all that packaging is everyone's loss.

A society prospers only when all or nearly all its members work for the good of the society. Of course, this implies benefits to the individual, but carried to the extreme, an obsession with personal or corporate profits without concern for the welfare of the entire system works to the detriment of the whole. If profit-induced pollution causes us all misery or even causes the extinction of the human species, wherein lies the profit?

The reader may be becoming aware that a solu-

152

tion to the "popullution" problem requires a three-fold attack:

1. Reduction in the human population of the world, preferably by birth control. The average citizen can help here. Indeed, he *must* help here. Without voluntary limiting of individual family sizes, no program short of some sort of coercion such as compulsory sterilization or abortion is likely to help. Reduction of our population will not solve all our problems, but it will make many of their solutions easier. Without population reduction, some problems may have no solution.

2. Reducing the amount of wastes produced by reducing the amount of unnecessary goods and power produced and consumed. The individual can contribute greatly here, also. The less he purchases, the less demand he puts on the environment. Let the consumer not be misled by clever advertising that urges him to purchase more electrical gadgets and so seduces him into using more power, that implies that melting down old glass bottles to make new ones is just as good a job of recycling as reusing the old ones intact, that implies that clothing washed in soap rather than in detergents is unsanitary, and that urges him to buy a hundred other unnecessary goods and services. Old-fashioned frugality benefits the environment as well as the individual.

3. Recycling all wastes the production of which cannot be avoided. Short of collecting newspapers and returning glass bottles there is relatively little the individual can do directly. But he can agitate for recycling by industry and government. These organizations must be made willing to do it, and the individual's refusal to purchase products of major polluters and his vote can wield considerable power. A contribution from science is required, too, for research will be necessary to find a way to recycle some materials. Already some industries collect their wastes and process them into salable products from which they derive profit. Domestic sewage and feedlot wastes should be converted to fertilizer whenever possible and returned to the fields from which they came.

Implicit in these three recommendations is a fourth: the cultivation of a spirit of selflessness that considers carefully the far-reaching consequences of a proposed action before that action is taken. The human brain is clever enough to devise ways to make a fast buck. If that brain is not clever enough to see that many short-term profits for individuals, corporations, or governments lead to larger long-term losses for society as a whole, then man is not the intelligent, superior animal he fancies himself to be.

## Nine Misconceptions

Those who feel that there is no serious population or pollution problem often support their views with arguments based on misunderstood or poorly understood biological principles. Discussions of several of these misconceptions follow.

*Misconception No. 1. The surplus food produced in the United States can feed the starving of the world.* The United States presently produces more food than its citizens require, but there is no known way that it can supply the tremendous world food deficit. As we have seen earlier in this chapter, the total *world* food production can feed adequately only about one-third of its population.

*Misconception No. 2. The presently untapped food reserves of the ocean someday will feed the starving of the world.* Because the edible foods of the oceans are mostly fish, many of which are high in the food chains, the contribution that seafoods will make to the prevention of starvation is likely to be small. However, because fish are largely protein, they can be valuable supplements to protein-poor diets. The seemingly endless bounty of the sea may be less abundant than expected, for presently the oceans are becoming adversely

153

affected by disposal of wastes. As deep as they are, the oceans are not bottomless pits. If their productivity is reduced, they will not live up to any expectations.

Communities as large and as stable as the oceans can absorb a lot of abuse before suffering significant damage, but once such damage is done, its reversal will require a great deal of time. A polluted river can be made clean in a relatively short time merely by refraining from polluting it. The rain waters it receives wash old pollutants seaward. No such solution can cleanse the oceans so quickly. They are the final recipients of pollution disposed of in lakes and streams, and they have no outlets. Only the action of marine organisms can clean the oceans and in its cold, oxygen-poor depths this action is slow. If the pollution reaches such proportions that it kills these organisms, the oceans will become as "dead" as Lake Erie with even less hope of recovery.

*Misconception No. 3: If the lush vegetation of the tropics is cut down, the land can be farmed to produce the food the world needs.* Attempts to transplant modern agriculture to the tropics have met with frequent failures. After tropical rain forests (popularly called jungles) are cut down, minerals leach quickly from the soil. Usually one or two crops can be raised, and then the land must be abandoned. Scattered, rather primitive tribes, moving from place to place, can be supported this way, but there seems to be no way that the tropics could assume the burden of feeding the rest of the world. In some areas, the tropical rain forests gradually regenerate themselves in clearings. Other tropical forests grow on lateritic soils, which, once denuded, harden into a bricklike consistency that supports no vegetation and is lost indefinitely to agriculture.

*Misconception No. 4: There is plenty of sparsely inhabited land to which excess population can move—many parts of the western United States or northern Canada, for instance.* Overpopulation is a matter of the amount of resources available per person and not the number of persons occupying a given area. Today most sparsely inhabited areas are that way because they cannot support a dense population. Where water is scarce, summers are short, or other conditions prevent an abundant crop yield, relatively few persons per unit area can be supported. The rich farmlands of the Midwest can provide for several hundred persons per square mile; ten persons or even one person per square mile may be more than some deserts can support.

We should note also that cities are not self-supporting in terms of food, lumber, and other natural resources. New York City inhabitants depend on Midwest prairies for meat and cereal grains, on California and Florida farms for winter fruits and vegetables, on the forests of the western United States and foreign countries for lumber, and on many other areas for metals and other resources. In a sense, then, these areas are already "occupied" by city dwellers even though no one lives visibly in field or forest. Furthermore, should these farms and forests be converted into new cities or suburbs, their productivity will be lost and their pollutants will affect adjacent areas.

*Misconception No. 5: When interplanetary or interstellar travel is perfected, the excess human population can be exported to those celestial bodies that are hospitable to human life.* Other than the planet earth, no body in our solar system holds much promise as a home for man. Temperatures too high or too low, the lack of water or oxygen, or the presence of poisonous gases face prospective settlers on our sister planets and their satellites.

Beyond the solar system is a vast expanse of space with no celestial bodies closer than the star Alpha Centauri, 4.2 light-years[1] away, and there is some reason to believe that it does not possess a planet suitable for human habitation. Even if it did, present-day spaceships would require more than 100,000 years to reach Alpha Centauri; this

154

---

[1] A light-year is the distance light travels in one year, almost 6 trillion miles.

is more than ten times the length of recorded history. Should any optimists still desire to start the journey, they must make their spaceship self-sufficient. No stocks of food could be laid in to provide for so long a voyage. All the food needed would have to be raised on the ship. The waste products of the inhabitants could not be dumped overboard but would have to be recycled. In other words, a miniature planet earth would have to be created. Because space would be at a premium in such a spaceship, presumably no more species would be taken on board than absolutely necessary. The instability of such communities is well established (Chapter 14). The failure of any one component could well cause the deaths of all occupants.

The population of the earth today is growing at the rate of more than 67 million persons per year. The cost of manufacturing and launching enough miniature "planets" to remove 67 million persons every year from the earth would impoverish all those remaining behind. Finally, the utter futility of such a project is made apparent by the fact that the travelers would face the same population problems in their spaceship that we do today on earth. If we cannot solve our population problems on earth without exporting people, it is unlikely that those exported could do so either.

*Misconception No. 6: Because humans have proved themselves adaptable to a wide variety of conditions, they will adapt to life in a polluted environment.* If this means the adaptation of the individual to pollutants, it would be difficult to see how anyone ever would die of anything except old age or physical violence. If individuals had the ability to adapt to any chemical compounds, no one would succumb to lung cancer caused by smoking cigarettes, the black lung disease of miners who inhale coal dust for many years, or the habitual use of narcotics. Sewage plants and water purification plants would be unknown if individuals could adapt to the assortment of materials found in water supplies, and the Pure Food and Drug Administration never would have been formed.

A second possible meaning of this proposition is that over several generations the human species will evolve an immunity to its pollutants. Theoretically this is possible. Evolution is accomplished through the survival of those individuals genetically adapted to new environments. This, of course, implies premature deaths of all those not so adapted, the lowering of their reproductive rate, or both. The more drastic the change in the environment, the smaller will be the number of those who are adapted, and the larger will be the number of deaths, or the smaller will be the number of births. If the change is severe enough, there will be none who are sufficiently well adapted to survive and reproduce. Survival of the human species by evolutionary adaptation to severe pollution can be accomplished only at the tremendous price of the deaths of many persons.

*Misconception No. 7: Nature produces pollutants; therefore man's addition of pollutants to the environment is of little importance.* This seems to be a favorite argument of industries unwilling to reduce the levels of their pollution. Sometimes cited are volcanic eruptions. It has been stated that three eruptions (Hekla in Iceland, in 1947; Katmai in Alaska, in 1912; and Krakatoa near Java, in 1883) ejected more particulates and gases into the air than all of man's activities throughout history. What never is mentioned is that if a sufficient quantity of fine particulates is carried into the upper atmosphere, the reflection of the sunlight back to outer space results in cooler weather. Cold, hard winters were recorded after major eruptions of Hekla, Krakatoa, Vesuvius (in Italy), and Pelée (in Martinique). The 1815 eruption of Tambora near Java produced a year "without a summer"; in 1816 Boston received snow even in the summer. A succession of a few such years could so reduce the world's food supply that the starving would no longer care whether the cause was an unusual renewal of volcanic activity, continuing proliferation of power companies, or a fleet of supersonic transports. As recently as 1963, 25,000 acres of fertile Balinese riceland suffered permanent de-

155

struction in the lava, ashes, and glowing clouds emitted by an eruption of Agung. The fate of the residents of Pompeii in A.D. 79 is well known— death by suffocation from the noxious gases emitted by Vesuvius and burial under ashes from the same source. None of these incidents argues for increased pollution just because some occurs naturally.

*Misconception No. 8: The more persons there are on the earth, the more minds and the more money will be available to combat the pollution produced by all these people.* This is circular reasoning. One might just as well argue that heart attacks are valuable because they provide the opportunity to find cures for them. As a population increases in size, the problems it generates tend to multiply faster than the population itself. The extra minds and money must be spread ever more thinly. We might draw an analogy with a traffic intersection. Let us imagine two automobiles arriving there simultaneously. Each driver need worry only about the other, and only one possible interaction (accident) can occur:

Now let us imagine four automobiles arriving simultaneously at the intersection. The population has doubled, but each driver now has three other drivers to consider, and the number of two-car interactions has increased to six:

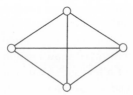

Furthermore, three-car accidents now become possible, and there are four ways they can occur. And, of course, all four cars could pile up. The possibilities for accidents have increased way out of proportion to the increase in numbers of automobiles.

It might be helpful to imagine today's world as a 4-billion-way intersection. With the possibility of thousands of different pollutants arriving from any direction, the problems seem insurmountable without reduction in the number of polluters and the quantity of pollutants they produce.

*Misconception No. 9. The wealthy are entitled to have large families, for they can support them comfortably, but the poor should have fewer children.* On the face of it, this proposition may seem reasonable, but the wealthy add more than their share of burdens to the environment. A child of affluent parents well may have his own telephone, radio, television set, phonograph with accompanying stacks of records, tape recorder and tapes, bicycle, and, eventually, an automobile. The most affluent may acquire second homes, boats, and even private airplanes. The pollution caused by the manufacturing and use of these luxuries and the demands they make on the natural resources of the world cannot be afforded by society. Rich and poor, we all are affected by lead in our air, mercury in our water, and DDT in our food; but these are generated at a higher rate by the affluent. Today, the United States, the richest nation in the world, with a little less than 6 per cent of the world's population, uses at least 33 per cent of its resources and produces a similar share of its pollution.

Actually this proposition is a self-defeating one. If a rich family permits itself 12 children, and if each child has a similar number, and if this were to continue for six generations, there would be nearly 3 million great-great-great-great-grandchildren (Table 16.1). It is unlikely that all these offspring would find themselves at the top income levels no matter how rich their ancestors were. Only a few could be executives. Many would be blue-collar workers, and some would find only menial tasks like scrubbing floors, collecting garbage, or cleaning cattle feed lots. Certainly children do not come cheaper by the dozen today.

| | Numbers of Offspring with | |
|---|---|---|
| | 2 Children per Family | 12 Children per Family |
| Children | 2 | 12 |
| Grandchildren | 4* | 144 |
| Great-grandchildren | 8* | 1,728 |
| Great-great-grandchildren | 16* | 20,736 |
| Great-great-great-grandchildren | 32* | 248,832 |
| Great-great-great-great-grandchildren | 64* | 2,985,984 |

**Table 16.1**

*These numbers do not represent increases in population size, for each grandchild is the descendant of four grandparents, each great-grandchild is the descendant of eight great-grandparents, and so on.

# Questions

**1.** What is the major reason for the present human population explosion?

**2.** The famous Netherlands fallacy runs thus: The population of the Netherlands is about 1,000 persons per square mile. If people can live at this concentration, then the world with its land area of about 57.5 million square miles should be able to support about 57.5 billion persons. Wherein lies the fallacy?

**3.** What were the ultimate sources of the things you used in the last 24 hours (bananas from Brazil, cotton from Georgia, and so on)? How much space in the world do you effectively occupy?

**4.** Differentiate between renewable and nonrenewable sources. How could our supply of renewable resources become exhausted? Can you give any examples of this happening?

**5.** Persons who are unconcerned about the possible extinction of endangered species often argue, "Is the world for people or for animals?" Can you find the fallacy in this reaction?

**6.** The summer of 1972 brought death by floods in several parts of the world. To what extent were these natural disasters and to what extent were they the result of overpopulation?

**7.** If some catastrophe were to deprive us of the use of fossil and nuclear fuels, which of the following would be most likely to survive?
   a. Boy Scout from Cleveland.
   b. Modern corn farmer from Iowa.
   c. New York City secretary.
   d. Indian from the upper reaches of the Amazon River.

**8.** If pollutants are considered to be those substances that the environment cannot degrade or assimilate as fast as they form, how many of the following would you consider pollutants?
   a. Carbon dioxide exhaled by an animal.
   b. Oxygen produced by a green plant.
   c. Carbon monoxide produced by an automobile in the Holland Tunnel in New York City.
   d. Carbon monoxide produced by a pickup truck used on a Wyoming ranch.
   e. Sewage from a Brazilian Indian village 50 miles upstream from the next village.
   f. Sewage from a San Francisco suburb.
   g. Manure from a steer in a feed lot.
   h. Manure from a steer on a Texas range.

**9.** Many persons think of the cleaning of the environment as benefiting those who are rich enough to have summer cabins in the wilderness, but the persons who benefit from the reduction of foul smoke emitted by steel mills are the persons who work in the mills or live nearby and not the stockholders. Name some other sources of pollution the removal of which would benefit the common worker.

**10.** In the list of suggested readings at the end of this chapter is an article, "The Tragedy of the Commons," by Hardin. Find this article in your school library and read it. In how many ways do you see the Tragedy of the Commons re-enacted on your campus, in your home town, and on national and international levels? (Read newspapers and watch television for examples.)

**11.** Rainfall continuously replenishes the water in our streams and lakes. Why, then, is there concern about water shortages in the future?

**12.** Someone once said we pay manufacturers for the goods they produce but we don't charge them for the bads they produce. Can you explain what he meant?

**13.** Examine several industries (construction, cosmetics, education, medicine, landscaping, food, clothing, fashion, or any other that interests you) in this respect:

　　a. Are they totally useful and operating in an ecologically sound manner?

　　b. Are they useful but partially abusive? If so, how could their operation be improved?

　　c. Are they totally useless and therefore abusive of the environment? If so, examine your conscience. Do you use the products or services of this industry? If your answer is yes, would you consider a total personal boycott of that industry, or are you too fond of its services or products to give them up?

**14.** Consider your activities for the last 24 hours. What did you do that increases pollution? What did you do to reduce it?

# Suggested Readings

Adams, M. W., A. H. Ellingboe, and E. C. Rossman. "Biological Uniformity and Disease Epidemics," *BioScience* 21: 1067–1070 (1971).

Appleman, P. *The Silent Explosion.* Boston: Beacon Press, 1965.

Boerma, A. H. "A World Agricultural Plan," *Scientific American* 223(2): 54–69 (August 1970).

Brown, H. S. *The Challenge of Man's Future.* New York: The Viking Press, Inc., 1954.

Cailliet, G., P. Setzer, and M. Love. *Everyman's Guide to Ecological Living.* New York: Macmillan Publishing Co., Inc., 1971.

Calhoun, J. R. "Population Density and Social Pathology," *Scientific American* 206(2): 139–148 (February 1962).

Callahan, D. "Ethics and Population Limitation," *Science* 175: 487–494 (1972).

Carson, R. *Silent Spring.* Boston: Houghton Mifflin Company, 1962.

Chisholm, J. J., Jr. "Lead Poisoning," *Scientific American* 224(2): 15–23 (February 1971).

Clement, R. C. "Pesticide DOs and DON'Ts," *Audubon* 72(2): 50–51 (1970).

Coale, A. J. "Man and His Environment," *Science* 170: 132–136 (1970).

Commoner, B. *Science and Survival.* New York: The Viking Press, Inc., 1966.

Cook, E. "The Flow of Energy in an Industrial Society," *Scientific American* 224(3): 134–144 (September 1971).

Croat, T. B. "The Role of Overpopulation and Agricultural Methods in the Destruction of Tropical Ecosystems," *BioScience* 22: 465–467 (1972).

Darlington, C. D. "The Origins of Agriculture," *Natural History* 79(5): 46–57 (May 1970).

Ehrenfeld, D. W. *Biological Conservation.* New York: Holt, Rinehart and Winston, Inc., 1970.

Ehrlich, P. *The Population Bomb.* New York: Ballantine Books, Inc., 1968.

_____, and A. H. Ehrlich. *Population, Resources, Environment.* San Francisco: W. H. Freeman & Co., Publishers, 1970.

Gough, W. C., and B. J. Eastlund. "The Prospects of Fusion Power," *Scientific American* 224(2): 51–64 (February 1971).

Hardin, G. "The Tragedy of the Commons," *Science* 162: 1243–1248 (1968).

_____, ed. *Population Evolution and Birth Control,* 2nd ed. San Francisco: W. H. Freeman & Co., Publishers, 1969.

_____, ed. *Exploring New Ethics for Survival.* New York: The Viking Press, Inc., 1972.

Hays, H., and R. W. Risebrough. "The Early Warning of the Terns," *Natural History* 80(9): 38–47 (November 1971).

Hubbert, M. K. "The Energy Resources of the Earth," *Scientific American* 224(3): 60–70 (September 1971).

Hulett, H. R. "Optimum World Population," *BioScience* 20: 160–161 (1970).

Jacobsen, T., and R. M. Adams. "Salt and Silt in Ancient Mesopotanian Agriculture," *Science* 128: 1251–1258 (1958).

Klein, R. "The Florence Floods," *Natural History* 78(7): 46–55 (August–September 1969).

Langer, W. "Checks on Population Growth: 1750–1850," *Scientific American* 226(2): 92–99 (February 1972).

Marx, W. *The Frail Ocean.* New York: Ballantine Books, Inc., 1967.

Odum, H. T. "A Print-out of the Future Systems of Man," *Natural History* 80(5): 24–29 (May 1971).

Peakall, D. B. "Pesticides and the Reproduction of Birds," *Scientific American* 222(4): 72–78 (April 1970).

Rienow, R., and L. T. Rienow. *Moment in the Sun.* New York: Ballantine Books, Inc., 1967.

Sax, K. *Standing Room Only,* new ed. Boston: Beacon Press, 1960.

"The State of the Species," Special Supplement, *Natural History* 79(1): 43–74 (January 1970).

Tietz, C., and S. Lewit. "Abortion," *Scientific American* 220(1): 21–27 (January 1969).

Turk, A., J. Turk, and J. Wittes. *Ecology Pollution Environment.* Philadelphia: W. B. Saunders Company, 1972.

"The Unforeseen International Ecological Boomerang," Special Supplement, *Natural History* 78(2): 41–72 (February 1969).

Wagar, J. A. "Growth Versus the Quality of Life," *Science* 168: 1179–1184 (1970).

Westing, A. H. "Ecocide in Indochina," *Natural History* 80(3): 56–61 (March 1971).

Woodwell, G. M. "Effects of Pollution on the Structure and Physiology of Ecosystems," *Science* 168: 429–433 (1970).

# The Cell

# Part V

We saw in earlier chapters that one feature of all living things is that they constantly use energy and that they do so by transforming one kind of energy into another. The machinery for utilizing these energy transformations lies in the cell. Every living cell has this function—indeed, if it lacked it, it would no longer be living.

# 17

# The Cell

## The Cell Theory

One of the basic principles in biology is the cell theory, which was developed and elaborated by several of the great biologists of the nineteenth century. The cell theory states that the cell is the fundamental unit of living things, and that all living things (except viruses) are composed of cells. This is true not only in a structural sense—multicellular organisms being composed of many cells somewhat as a wall is made of bricks—but also in a functional sense. All cells carry on certain essential functions, and within an organism each cell has the potential (which is not always actually expressed) of performing all the chemical reactions characteristic of that individual.

We may say also that the cell is the smallest unit of life. It has all the characteristics of living things; it grows, reproduces, responds to its environment, and so on. Like the atom, it cannot be fragmented into its component parts and still retain its identity, nor will it retain its quality of being alive under such conditions. The individual parts of the cell are not living entities in themselves; instead, the entire system—that is, the cell with its properly organized parts and its integrated reactions—is alive.

One corollary of the cell theory is that living cells arise only from pre-existing cells; each generation of cells is produced from an earlier generation of cells. No one has yet observed a cell arising from nonliving material, nor is he likely to do so. Today life arises only from life. (This does not

**1** The cell is a fundamental unit of life; all living things except viruses consist of cells.

**2** Cells arise from pre-existing cells by the process of cell division.

**3** The several parts of the cell perform different but coordinated functions.

**4** A eucaryotic cell possesses a nucleus and cytoplasm; the nucleus controls many chemical reactions occurring in the cytoplasm, which in turn regulates some activities of the nucleus.

**5** Membranes are an essential part of many cell parts. These membranes are the sites of many vital chemical reactions, and some membranes, by their differential permeability, control the entrance and exit of a variety of substances.

163

explain the origin of life, which will be discussed in Chapter 47.) The machinery for utilizing energy varies only slightly from one cell type to another, and the basic similarity of all cells is so remarkable that one is constrained to believe that they must have a common origin. It seems that if nature ever evolved cells built on other, fundamentally different plans, these must have been inferior and so died out, leaving only the kind we know today.

Even today we have only two major variations on the basic plan: the **procaryotic** cell and the **eucaryotic** cell. The procaryotic cell is the more primitive of the two and is possessed only by the bacteria and the blue-green algae; procaryotic cells lack true nuclei and several other parts we consider typical of most cells. All other organisms have eucaryotic cells, built on the more advanced pattern which we call typical. Viruses, which are variously described as living and nonliving by different biologists, are not considered in this chapter, for they do not possess cellular structure; they are obligate parasites and make use of their hosts' cellular machinery.

The shapes and sizes of cells vary a great deal. The "ideal" isolated cell is spherical, and many single cells do approach that shape. Others, whether isolated or in tissues, are nearly cylindrical, like many wood cells, or branched, like nerve cells. Cells in tissues frequently become polyhedral because of the pressure exerted by their neighbors. When cells are observed under a microscope they often appear to be flat, and it is easy to consider them merely as two-dimensional objects. The student must remember that cells and their **organelles** (subcellular units) all have three dimensions. An experienced microscopist repeatedly focuses up and down through the entire depth of the cell he is studying in order to be sure of observing all he can. Most cells are microscopic, with some bacteria being so small as to be barely visible with a light microscope. Some, like the yolks of eggs of birds and reptiles, are large enough to be seen easily by the unaided eye; their size is due mainly to the large amount of stored food they contain,

for the quantity of living matter is quite small. Others, although small and easily overlooked, can be seen by the keen eye; some woods, especially oak and chestnut, have cells large enough to appear like little holes just big enough to be seen without a lens.

## The Cell

### The Basic Plan

The living portion of a cell is called **protoplasm.** Protoplasm is not a single compound, but rather a complex substance containing many organic and inorganic compounds in solution and in suspension. Not only is it complex, but it is constantly changing. The protoplasm of a cell is not necessarily the same in light as in darkness, in heat as in cold, or in dampness as in drought; its very vitality permits it to adjust to changing conditions by changing itself. The protoplasm of any cell seems to have at least a few minimum parts. When properly stained and examined under a microscope, protoplasm displays two obvious parts: **nucleus** (or **nucleoid**) and **cytoplasm** (Figs. 17.1 and 17.2).

The nucleus of eucaryotic cells ordinarily is a spherical, membrane-bound body that contains most of the hereditary material of the cell and that controls many of the chemical reactions in the cytoplasm. Procaryotic cells do not have a true nucleus; usually their hereditary material is confined to the center of the cell in a region called the **nucleoid,** or **central body.** The nucleoid has no bounding membrane and often has a very irregular shape, but it performs many of the same functions that a true nucleus does in a eucaryotic cell.

The cytoplasm is the remaining portion of the protoplasm; it surrounds the nucleus. In it occur many vital processes, not the least of which are respiration, by which the cell utilizes the energy

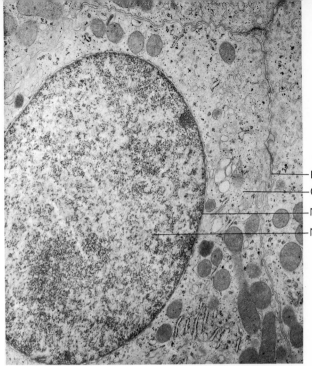

—Plasma membrane
—Cytoplasm
—Nuclear membrane
—Nucleus

**17.1** Electron micrograph of a portion of a rat liver cell. The nucleus is the largest object visible here. The double nature of the nuclear membrane is obvious. Small portions of three other cells can be seen. A very narrow space separates the plasma membranes bounding adjacent cells. The cells have no cell walls. Mitochondria, endoplasmic reticulum, and ribosomes can be seen in the cytoplasm. (Courtesy of Dr. Frederic A. Giere.)

**17.2** A plant cell. This plant cell shares many features with animal cells, but it also has a cell wall. This cell is from a root and has no chloroplasts. (From W. G. Whaley, H. H. Mollenhauer, and J. H. Leech. 1960. The ultrastructure of the meristematic cell. *American Journal of Botany* 47: 401–459; courtesy of W. Gordon Whaley.)

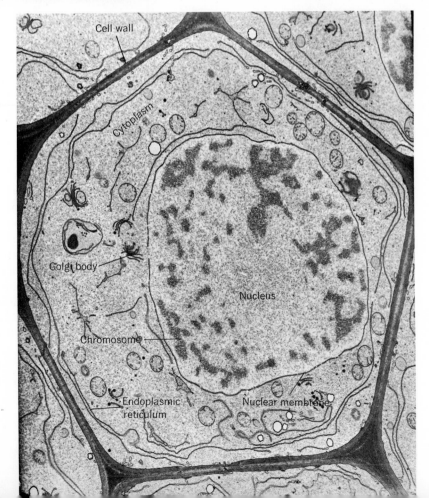

Cell wall
Cytoplasm
Golgi body
Chromosome
Nucleus
Endoplasmic reticulum
Nuclear membrane

165

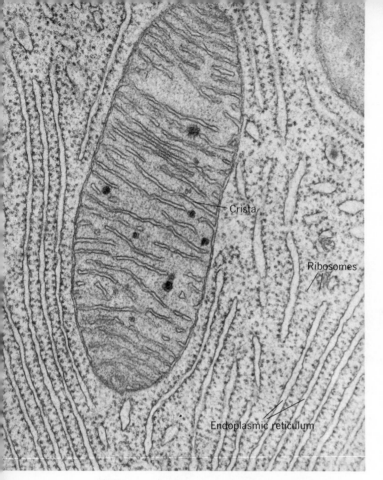

Crista

Ribosomes

Endoplasmic reticulum

**17.3** A portion of the cytoplasm of an animal cell, showing a mitochondrion, endoplasmic reticulum, and ribosomes. The mitochondrion is bounded by two membranes; the inner membrane is folded into cristae. Some ribosomes are attached to the endoplasmic reticulum; others are free in the cytoplasm. (Dr. Keith R. Porter.)

in its food, and protein synthesis, by which the cell manufactures the enzymes that catalyze the reactions of the cell. Most of the cytoplasm has a fluid or semifluid consistency and is called the **matrix,** or **ground substance.** Embedded in it may be any of several organelles. Protein synthesis occurs on one type of organelle, the ribosome (Fig. 17.3). **Ribosomes** are minute, submicroscopic bodies often present in great numbers in the cell. The outer boundary of the cytoplasm, which makes contact with the environment, is a membrane called the **plasma** or **cell membrane.** Materials

entering or leaving the cell must pass across the plasma membrane, which exercises some control over the passage of these materials.

We may pause here to say that a minimum number of parts seem to be necessary for the integration of processes in a living, functioning cell. These parts seem to be the cytoplasm, in which energy is used in many metabolic reactions; its plasma membrane, which controls the entrance of materials used in these reactions and the exit of waste materials produced by these reactions; its ribosomes, on which enzymes catalyzing these reactions are synthesized; and either a nucleus or nucleoid that exercises some control over the entire cell. These are parts of all cells—whether procaryotic or eucaryotic—but they do not exhaust all the subcellular parts known to exist in living things. In the rest of this chapter we will consider in some detail the parts of eucaryotic cells, with only brief mention of procaryotic cells.

## Detail of Some Cell Parts

The **plasma membrane,** which is the external boundary of the cytoplasm, is such a thin layer that it is invisible even to an eye aided by an ordinary light microscope, which can magnify about 1,000 times. It can be seen only by electron microscopy, which permits magnifications of many thousand times. The plasma membrane is composed of four layers: the two in the center of the membrane are monomolecular layers of phospholipids (fatty acids in chemical combination with other organic compounds and phosphates), and these are surrounded by two layers of protein, each probably also one molecule thick. This set of four layers is called a **unit membrane.** We will see that many parts of eucaryotic cells are composed of similar unit membranes. The structure of the plasma membrane allows it a great deal of flexibility. Any student who has watched an ameba assume a variety of shapes as it moves across a microscope field can appreciate the utility of flexibility of the outer membrane. Despite its thinness, the plasma membrane is fairly strong and is capable of repairing minor damage to itself.

Perhaps its most important feature is its differential permeability, that is, its ability to permit only certain substances to pass through it or to permit different substances to pass at different rates. This feature has several consequences. For one thing, it allows the entrance of required substances and permits the escape of waste materials that would otherwise become toxic. Not only does the membrane allow some substances to enter the cell, but it is also involved in their accumulation inside the cell to much higher concentrations than they exist outside. This means that the cell must expend energy, and it is believed that the ATP and the various enzymes necessary for this process are associated with the membrane. Unfortunately we cannot say that the membrane always differentiates between useful and useless substances; not only do some cells permit extraordinary accumulation of substances for which there seems to be no use, but they also permit entrance of some poisons.

The turgidity (a condition of being swollen because of internal pressure) of living cells is due to a process called osmosis (Chapter 18) which depends in part on the differential permeability of the plasma membrane. Major damage to the plasma membrane—that is, damage that is not repaired by the cell—destroys its differential permeability; this almost certainly results in the death of the cell.

The **cytoplasm** of procaryotic cells is relatively undifferentiated, but that of eucaryotic cells contains a complex of folded membranes called the **endoplasmic reticulum,** often abbreviated ER. It is visible only with an electron microscope. In many preparations it appears to be irregular and discontinuous, but one must recall that cells and their contents are three-dimensional and that cells prepared for electron microscope examination are sliced into very thin sections; one section cannot show the three-dimensional relationships of the entire endoplasmic reticulum (or of many other organelles either). It has been determined, however, that the endoplasmic reticulum is continuous with the plasma membrane and appears to be an invagination of it; that is, it seems as if the outer membrane has folded inward to become an inner membrane.

At any place along its length the endoplasmic reticulum is a double unit membrane, the space between the two unit membranes being continuous with the exterior of the cell. In most organisms the cytoplasm is in more or less constant movement as it circulates around the cell. How this affects the endoplasmic reticulum is not fully understood, for one could easily imagine it twisting endlessly around the nucleus like spaghetti around the fork of a nervous diner. Presumably this does not happen, because the endoplasmic reticulum can break and mend itself as does the plasma membrane when the occasion arises. Movement of the cytoplasm is called **cyclosis** or **cytoplasmic streaming.**

In eucaryotic cells, **ribosomes** are located along the interior or matrix side of the endoplasmic reticulum or are free in the matrix. Ribosomes are rich in nucleic acids (especially ribonucleic acid, or RNA) and proteins. Together these are known as nucleoprotein; both function in the synthesis of enzymes (Chapter 43). Ribosomes occur in large numbers in active cells, the number varying approximately with the amount of protein synthesis occurring in the cell. Ribosomes are also known to be present in the nucleus.

The **nucleus** is always located in the cytoplasm. It consists of four parts: **nuclear membrane, nuclear sap,** several **chromosomes,** and at least one **nucleolus.** The nuclear sap is a fluid substance in which are found the chromosomes and nucleoli. The nuclear membrane controls the passage of materials into and out of the nucleus. Like the endoplasmic reticulum, it is a double unit membrane; in fact, in places it is continuous with the endoplasmic reticulum (Fig. 17.4). In other places the nuclear membrane is broken by pores that permit communication between the cytoplasm and

167

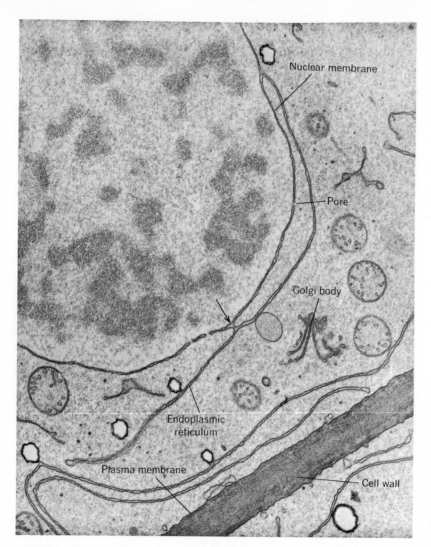

Nuclear membrane

Pore

Golgi body

Endoplasmic reticulum

Plasma membrane

Cell wall

**17.4** Continuity of endoplasmic reticulum and nuclear membrane (arrow). (From W. G. Whaley, H. H. Mollenhauer, and J. H. Leech. 1960. The ultrastructure of the meristematic cell. *American Journal of Botany* 47: 401–459; courtesy of W. Gordon Whaley.)

the interior of the nucleus. The **genes,** or hereditary material of the cell, consist of nucleic acid (in this case deoxyribonucleic acid, or DNA) in close association with protein. The genes are arranged in linear order along the lengths of the elongate, threadlike chromosomes. Chromosomes are confined to the nucleus, where they are so much entwined among each other that they are ordinarily difficult to see. They do become shorter and thicker and therefore much easier to observe in dividing cells (Chapter 19). The number of chromosomes varies from species to species. The DNA is not only capable of self-duplication, thus ensuring that two sets of hereditary material are available for distribution to the daughter cells when a cell divides, but it also can direct the synthesis of a particular type of RNA, called messenger RNA, that moves through the nuclear membrane and travels to the ribosomes in the cytoplasm and there directs protein synthesis. The nucleolus is a spherical body rich in RNA. The function of nucleoli is not completely understood, but they appear to be involved in RNA metabolism and the synthesis of ribosomes.

The nucleus does not control the cell autonomously. Although it regulates events occurring in the cytoplasm, these events in turn affect the activities of the nucleus. The entire cell then operates as a unit, with the nucleus functioning something like an administrative office that makes decisions based on information coming from other members of the organization of which it is a part. See Chapter 43 for more information about this process on the molecular level.

**Mitochondria** are spherical or rod-shaped bodies located in the cytoplasm of eucaryotic cells. They are visible with an ordinary microscope, but usually they are so small that they seem merely to give the cytoplasm a granular appearance. Their internal structure is visible only by means of an electron microscope. Each mitochondrion is bounded by a double unit membrane. The outer membrane is fairly smooth, but the inner membrane is thrown into folds called **cristae** (Fig. 17.3). The cristae contain several of the enzymes of aerobic respiration, specifically those of the Krebs cycle and the cytochrome system. Mitochondria are essential for aerobic respiration in eucaryotic cells, and the number of mitochondria in a cell corresponds approximately to the level of respiratory activity of the cell. Muscle cells, which expend energy in movement, and kidney cells, which expend energy in excretion, often have large energy demands, and they contain large numbers of mitochondria. The number of cristae per mitochondrion is also higher in active cells than in inactive ones.

**Golgi bodies** are flattened, membrane-bound sacs; several are usually stacked together like a pile of plates. They often appear as extensions of the endoplasmic reticulum, and they seem to pinch off small vesicles into the cytoplasm (Figs. 17.2 and 17.4). Their exact function is not known with certainty, but they are especially numerous in glandular cells and are believed to be involved in se-

cretion; there is also some evidence that they may function in synthetic activities.

Endoplasmic reticulum, mitochondria, and Golgi bodies are found in both plant and animal cells. Some other structures typically are associated with only plant cells or only animal cells.

A plant cell typically is surrounded by a **cell wall** that lies outside the plasma membrane with which it ordinarily is in direct contact (Figs. 17.2 and 17.4). Nearly all plant cells have a thin **primary cell wall** composed primarily of cellulose. The wall of an adult cell has a limited elasticity and keeps the cell from swelling past a certain size when it absorbs water. Unlike the plasma membrane, the primary cell wall is permeable to nearly all substances with which it might ordinarily come into contact. Cell walls frequently have openings called **pits** through which strands of cytoplasm extend from one cell to the next. As some plant cells mature, they develop a secondary cell wall that lies to the inside of the primary cell wall; usually these cells die, leaving a hollow cell wall devoid of protoplasm. The **secondary cell wall** can become very thick; in addition to cellulose it may contain such compounds as lignin (in wood cells) or suberin (in cork cells).

An organelle unique to plants and found in nearly all of them is the **plastid.** Most plastids are large enough to be seen readily under the light microscope. Three types of plastids are chloroplasts, chromoplasts, and leucoplasts. Like mitochondria, **chloroplasts** are bounded by a double unit membrane. Within each chloroplast are smaller units called **grana.** Each granum appears to consist of a number of discs piled on each other much like a pile of coins. Each disc consists of a membrane folded back on itself (Fig. 17.5). Similar membranes connect the grana. Chlorophyll is confined to the membranes of the grana and the intergranal membranes. The remaining portion of the chloroplast, the nonmembranous part, is the **stroma.** The grana contain the enzymes of the light reactions of photosynthesis, and the stroma contains the dark reaction enzymes.

169

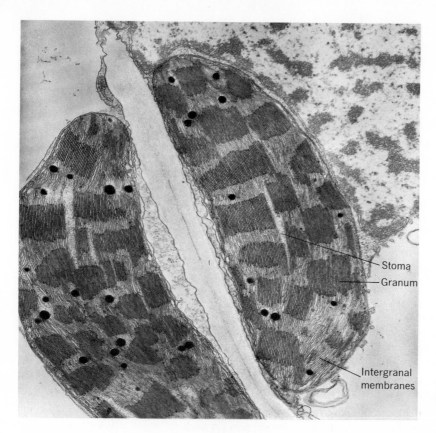

Stoma
Granum

Intergranal
membranes

**17.5** Two chloroplasts
from a corn plant. The grana
are composed of membranes.
(Courtesy of Howard J.
Arnott, The Department of
Botany and The Cell Research
Institute, The University of
Texas at Austin.)

Chloroplasts are found in plant parts exposed to light, especially leaves, and also in young stems and fruits. **Chromoplasts** are plastids of some color other than green—usually yellow, orange, red, or brown. In higher plants these are found largely in flowers and fruits, where they serve no known metabolic function, but the colors probably attract insects and other animals that transfer pollen or disseminate seeds. In some algae the chromoplasts are photosynthetic. **Leucoplasts** are colorless plastids. They store foods—especially starch, but sometimes oils—that represent the excess of food manufactured over food currently required. Starch deposited in the leucoplast forms a **starch grain.** Starch grains may grow so large that they break out of their leucoplasts and lie free in the cytoplasm. Leucoplasts are found mostly in plant parts that are not exposed to light, such as roots and tubers, or in some seeds and fruits. Certain interconversions among plastids are possible. Leucoplasts may become chloroplasts if they receive enough light; chloroplasts in some green fruits become chromoplasts as the fruits ripen.

**Vacuoles** are found in many types of plant cells. A vacuole is a drop of water surrounded by a single unit membrane. In plants the water contains a variety of dissolved materials including gases, various minerals, sugars, organic acids, and other water-soluble compounds. This solution is called the **cell sap;** it is the juice of a crisp apple or other similar plant material. Young plant cells have many small vacuoles; mature plant cells usually have one large vacuole that occupies nearly the entire volume of the cell and is surrounded by a very thin layer of cytoplasm. Animal cells typically do not have vacuoles, but some of the protozoa, such as amebas and paramecia, have food vacuoles (Fig. 17.6) in which their food is obtained, digested, and eliminated, and contractile vacuoles from which excess water is eliminated.

**Pinocytotic vesicles,** by which liquids enter some animal cells, resemble small vacuoles. Like food vacuoles, they form as invaginations of the plasma membrane.

**Lysosomes** are small membrane-bound sacs that contain digestive enzymes. Thus far they have been found with certainty only in animal cells. Presumably they release appropriate enzymes when the need for them arises. When a cell dies and the membrane around the lysosome disintegrates, the enzymes catalyze the digestion of much of the cell contents and thus contribute to its final decomposition.

Some cells can swim by means of hairlike processes called **flagella** and **cilia.** These structures lash about and propel the cell through water. Flagella and cilia of eucaryotic cells are structurally very much alike, flagella usually being longer and occurring singly or in pairs or small tufts on the cell, whereas cilia are short and often cover the entire cell. Both have their origin within the cytoplasm and extend through the plasma membrane and the cell wall, if there is one.

Flagella and cilia are not confined to aquatic organisms. The sperms of many terrestrial plants and animals are motile by means of flagella. Nor is the function of these structures confined to locomotion. Sponges are aquatic animals that live attached to rocks or other firm objects; they cannot move around in search of food, but the beating of their flagella creates water currents that bring food to them. Parts of the human respiratory tract are lined with ciliated cells. Dust and bacteria that are inhaled become trapped in mucus that is pushed upward by the action of the cilia. This material is sneezed or coughed out, thus ridding the host of potentially dangerous organisms.

Although eucaryotic flagella and cilia appear under a light microscope to be only thin, simple strands, they are much more complex, consisting of 11 strands that are surrounded by a sheath for part of their length. The main constituent of these strands is a contractile protein; the movement of the flagellum or cilium is due to the alternate contraction and relaxation of this protein. The movements of the several flagella or cilia of a cell are beautifully coordinated. Bacterial flagella also consist of protein, but they are somewhat simpler than those of eucaryotic cells; they consist of three or more strands either intertwined or lying parallel to each other.

Animal cells and the cells of some of the lower

**17.6** Food vacuole of an ameba. As a portion of the cell engulfs the food particle, a section of the plasma membrane becomes the bounding membrane of the newly formed food vacuole. Digestion of the food occurs in the vacuole. Undigested food is eliminated as the vacuole opens to the outside and merges with the plasma membrane again.

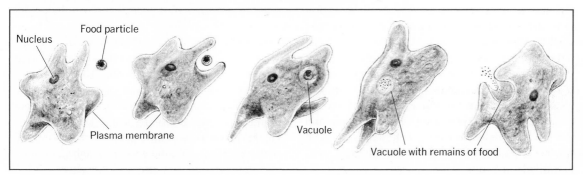

Nucleus  Food particle  Plasma membrane  Vacuole  Vacuole with remains of food

171

plants—notably the fungi—have two small bodies called **centrioles** in the cytoplasm near the nucleus. They play a role in cell division, but apparently they are not essential for this process, because they are lacking in the cells of most plants which do very well without them. Centrioles also are associated with flagella and cilia in eucaryotic cells.

Some cells contain various globules, crystals, and granules. Fats often are stored in globules in the fatty tissues of animals. Many crystals in plant cells are waste products that are stored in solid, and therefore nontoxic, form. Material to be secreted or excreted by cells sometimes is stored temporarily in granular form in cells.

The cells of multicellular organisms secrete cementing substances that hold the cells together. These substances are derivatives of polysaccharides. In animal tissues they lie outside the plasma membrane; in plants, where they form a distinct layer called the **middle lamella,** they lie outside the primary cell wall.

## Some Variations on the General Plan

Several variations exist on the general organization of "typical cells." Procaryotic cells lack plastids, mitochondria, Golgi bodies, endoplasmic reticulum, and lysosomes. Their nucleoids have no membranes and no nucleoli, and the chromosomes are distinctly different from those of eucaryotic cells. Their flagella, if they have any, are simpler than those of eucaryotes.

Some eucaryotic cells have two nuclei rather than the usual one; these include the paramecia, some human liver cells, and the cells of some fungi. Other fungi are not divided into separate cells by cell walls, and they are thus multinucleate, as are human skeletal muscle cells. A very few types of cells manage to function for a while without nuclei. Among these are human red blood cells and the sieve tube elements (food-conducting cells) of flowering plants; however, neither of these cells can divide or produce new cells.

## Interrelationships Among Cell Parts

When one looks at a cell under an ordinary light microscope, one sees discrete particles—the nucleus is here, the plastids are there, and so on. But we have learned from electron microscopy that many parts of the cell are membranous, and the interconnection of at least several kinds of membranes has been well established; the nuclear membrane is continuous with the endoplasmic reticulum, the endoplasmic reticulum with the plasma membrane and with the Golgi bodies. One is almost tempted to believe that much of the cell is just one membrane that has undergone numerous invaginations, some of which have become pinched off, forming what then become discrete units such as vacuoles, mitochondria, or plastids. Just this sort of thing happens when an ameba engulfs food: as two outgrowths of the ameba flow around and then completely surround its prey, a small part of the plasma membrane is pinched off and becomes the membrane of a food vacuole. One theory states that the endoplasmic reticulum, mitochondria, and plastids arise from the nuclear membrane.

Another theory is that mitochondria and plastids represent the descendants of primitive bacterialike organisms that entered other cells with which they gradually evolved an interdependence. If this is so, then the outer membranes of these two cell parts represent the plasma membranes of the ancestral forms. Supporting this theory is the fact that both mitochondria and chloroplasts contain the genetic material DNA and can divide and reproduce themselves.

These theories are mentioned not because they are proved, but to emphasize the abundance and similarity of membranes in a cell. The presence of the several membranes greatly increases the surfaces on which metabolic reactions can occur and therefore greatly raises the level of activity possible in living cells. Although they have a basic structural similarity, the various membranes of the cell possess different groups of enzymes all of which contribute in their own peculiar ways to the coordination of activities in the cell. These coordinated metabolic activities are an essential part of what we call life.

## Questions

1. Define the following terms:
   protoplasm *cytoplasm + nucleus*
   cytoplasm *the portion of the protoplasm exclusive of the nucleus*
   organelle *— subcellular unit in the cell*
   unit membrane *plasma membrane made of 4 layers*
   endoplasmic reticulum *complex of folded or double membrane continuous with plasma + nuclear membrane*
2. Give the function of each of the following:
   chromosomes *located in nucleus — contain most of genetic info. of an organism (Genes)*
   plasma membrane *outer Davy-Davy surrounding the cytoplasm*
   ribosomes *cytoplasmic organelle that is site of protein synthesis — they make enzymes take info. from nucleus*
   mitochondria *rod shaped bodies power house site of aerobic respiration in cytoplasm*
   chloroplasts *site of photosynthesis where energy demands are large*
   Golgi bodies *flattened membrane sacs in cytoplasm. Plant cell involved in secretion*
3. What structures are common to all cells? Differentiate between a procaryotic and a eucaryotic cell. *1st—notebook Protoplasm, plasma membrane, Mitochondria, Golgi Bodies, endoplasm reticulum*
4. Describe some differences between plant and animal cells. How is this correlated with their different modes of living? *notebook*
5. How many organelles are composed mostly of membranes? What processes occur on or in these membranes?

## Suggested Readings

Brachet, J. "The Living Cell," *Scientific American* 205(3): 50–61 (September 1961).

deDuve, C. "The Lysosome," *Scientific American* 208(5): 64–72 (May 1963).

Fox, C. F. "The Structure of Cell Membranes," *Scientific American* 226(2): 30–38 (February 1972).

Green, D. E., and Y. Hatefi. "The Mitochondrion and Biochemical Machines," *Science* 133: 13–19 (1961).

Jensen, W. A., and R. B. Park. *Cell Ultrastructure.* Belmont, Calif.: Wadsworth Publishing Co., Inc., 1967.

Kerr, N. S. *Principles of Development.* Dubuque, Iowa: William C. Brown Company, Publishers, 1967.

Loewy, A. B., and P. Siekevitz. *Cell Structure and Function,* 2nd ed. New York: Holt, Rinehart and Winston, Inc., 1969.

Margulis, L. "Symbiosis and Evolution," *Scientific American* 207(6): 56–62 (December 1962).

Morrison, J. H. *Functional Organelles.* New York: Reinhold Publishing Corporation, 1966.

Neutra, M., and C. P. Leblond. "The Golgi Apparatus," *Scientific American* 220(2): 100–107 (February 1969).

Racker, E. "The Membrane of the Mitochondrion," *Scientific American* 218(2): 32–39 (February 1968).

Robertson, J. D. "The Membrane of the Living Cell," *Scientific American* 206(4): 64–72 (April 1962).

Swanson, C. P. *The Cell,* 3rd ed. Englewood Cliffs, N.J.: Prentice-Hall, Inc., 1969.

*Both Plant cell + animal*
*(3) Protoplasm, Plasma membrane, Golgi, mitochondria, Endoplasm reticulum*

*5 — plasma membrane, lysosomes, endoplasmic reticulum, Golgi bodies, mitochondria (matrix or ground substance)*

*(3) cytoplasm — in which energy is used in many metabolic reactions*

*How do Cell functioning (eucaryotic)*

*plasma membrane — controls the entrance of materials into + out of the cell*

*Ribosomes — makes enzymes — produces other proteins (protein synthesis) takes info from nucleus.*

*Nucleus — exercises some control over the entire cell. or nucleoid — (procaryotic cell)*

*3 a Procaryotic cell has no true nucleus with nuclear membrane. the nucleoid contains the genetic material of the cell. the nucleoid is not bound by a membrane.*

*4 Plant cells contain plastids — a cytoplasmic organelle.
chloroplasts — photosynthesis — (grana, the light reaction enzymes in photosynthesis — Stroma, the dark reaction enzymes)
chromoplasts — a cytoplasmic organelle that contain pigments other than chlorophyll. (red-tomato)
leucoplasts — storage area for starch — can turn into chloroplasts.
vacuole — drop of water surrounded by membrane — as plant matures, vacuoles become one large vacuole surrounded by cytoplasm; nucleus, plastids
cell wall — made of cellulose (glucose — gives plant strength)*

*Animal cells —
vesicles —
lysosomes —
centrioles —*

173

# 18

## Diffusion, Osmosis, and Active Transport

1 **Diffusion, osmosis, and active transport are three methods by which cells absorb substances.**

2 **A substance diffuses only in the direction of its diffusion pressure gradient.**

3 **Osmosis is the diffusion of a solvent across a differentially permeable membrane.**

4 **The energy necessary for diffusion and osmosis is the kinetic energy of the molecules of the diffusing substance.**

5 **Active transport is the movement of substances against a concentration gradient—from a region of low diffusion pressure of that substance to one of high diffusion pressure of that substance.**

6 **Active transport of substances into or out of cells requires the expenditure of metabolic energy by the cells.**

No cell exists in a vacuum, and no cell is sufficient unto itself. A cell is a living, metabolizing thing that requires materials and that forms waste products. The needed materials come from outside the cell and must enter by passing across its plasma membrane; the waste materials are disposed of back across the same route. Before examining the movement of materials into and out of cells, we should look at some of the physical processes involved, especially molecular motion, diffusion, and osmosis.

## Diffusion

Molecules of all substances possess a certain amount of energy, called **kinetic energy,** that keeps them always in motion. This motion is rather restricted in solids, somewhat freer in liquids, and still freer in gases. **Molecular motion** can be speeded up by applying heat and slowed down by cooling. Theoretically, molecular motion can be stopped by exposure to $-273°C$ $(-460°F)$, but as this temperature has not been achieved, we may consider molecules as being in never-ending motion. The direction of movement of molecules is random, and at any moment approximately equal numbers of molecules in a given sample are traveling in any direction. This random movement, which keeps them bumping into each other and bouncing back, tends to keep them evenly distributed throughout whatever space is available to them. For instance, in a sealed bottle containing only the gas carbon dioxide, the molecules are virtually evenly distributed throughout the container and are not clustered to one side or at the top or at the bottom; similarly, in a solution of glucose in water the glucose molecules are evenly distributed among the water molecules.

## Diffusion Pressure

Molecular motion results in a pressure called **diffusion pressure.** If a gas or a liquid is confined to a container, it exerts pressure against the walls of the container. The diffusion pressure of a substance can be affected by several factors; among them are the concentration of the molecules, the application of pressure from without, and the temperature.

Diffusion pressure increases with increase in concentration of the substance. Every child who has blown up balloons has experienced this. Even if he is not familiar with the terminology, he knows that the increased diffusion pressure of more and more gas molecules being forced into the balloon causes it to expand. Part of the excitement of blowing up a balloon is the knowledge that, as the limit of extensibility of the balloon is approached, any additional increase in diffusion pressure on the inside increases the imminent danger of its bursting. Application of pressure from without also increases diffusion pressure within—as any child who has squeezed an inflated balloon too enthusiastically can testify.

Because an increase in temperature increases the speed at which molecules move, it also increases the diffusion pressure. This is why we are cau-tioned never to heat a sealed container, for the increased diffusion pressure of the gases inside the container may cause it to burst with explosive force.

## Diffusion

If a bottle of carbon dioxide gas were opened inside a larger container—let us say a sealed room filled with another gas, perhaps nitrogen—then the amount of space available to the carbon dioxide would suddenly increase (Fig. 18.1). At the moment the bottle is opened, the diffusion pressure of carbon dioxide is high inside the bottle, where the carbon dioxide concentration is high; and its diffusion pressure is low in the rest of the room, where its concentration is low. We say then that a **diffusion pressure gradient** (or slope) exists.

The paths of some of the carbon dioxide molecules take them out of the bottle; as more and more of them enter the room a few move back into the bottle; but for some time more of them move out of the bottle than move back in. This movement of the gas is called diffusion.

**Diffusion** is the net movement of the molecules of any substance from a region of high diffusion pressure of that substance to a region of low diffusion pressure of the same substance. Diffusion

**18.1** Diffusion of gases. In (a) carbon dioxide molecules are confined to the sealed bottle, and nitrogen molecules are confined to the room. In (b) carbon dioxide diffuses into the room; nitrogen diffuses into the bottle. In (c) diffusion has ceased because equilibrium has been reached.

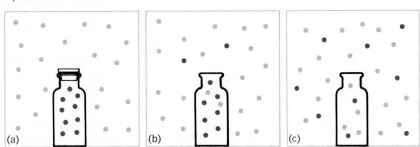

(a)          (b)          (c)

● Carbon dioxide molecule
● Nitrogen molecule

continues to occur until an equilibrium is reached, and the diffusion pressure of that substance is the same everywhere throughout the available space. Although diffusion is then impossible, molecular movement continues.

It is possible for two substances to diffuse simultaneously in opposite directions through the same general space. In our example the diffusion pressure of nitrogen is initially high in the room and low in the bottle. Nitrogen therefore diffuses from the room to the bottle until its concentration is uniform throughout the available space.

Several factors influence the rate of diffusion. Among them are temperature, the nature of the diffusing substance, the nature of the medium through which it diffuses, and the diffusion pressure gradient.

Because an increase in temperature increases the rate of molecular motion and of diffusion pressure, it also increases the rate of diffusion. In general each 10°C rise in temperature increases the rate of diffusion about 1.25 times. Diffusion of materials in air and water is therefore faster in hot weather than in cold; the higher the body temperature of an animal, the higher the rate of diffusion is likely to be within it, other things being equal.

Particles of light weight diffuse rapidly; heavy particles diffuse more slowly. For instance, oxygen (molecular weight 32) diffuses faster than carbon dioxide (molecular weight 44) but slower than hydrogen (molecular weight 2).

Diffusion is quite rapid in gases, slower in liquids, and very slow in solids. One need open the door or window of a crowded, stuffy room for only a few minutes to get a noticeable improvement in the quality of the air as oxygen diffuses into the room and carbon dioxide diffuses out, even on a windless day. A drop of ink placed carefully in a shallow dish of still water may require days or weeks to color the entire dish uniformly. Diffusion through solids is so slow that we use some solids as containers—an automobile tire, for example, is expected to hold air for a considerable length of time without going flat, and a coffee pot holds

coffee without leaking. The cellular membranes, which partake of both solid and liquid properties, vary in their permeability to various substances, and diffusion may occur relatively rapidly or slowly across them.

The steeper the diffusion pressure gradient is, the faster diffusion occurs. The steepness of the gradient is influenced by several factors; among them are the difference in the concentrations of a substance in two different places and the distance between these two places. A diffusion pressure gradient becomes steeper as the difference in the two concentrations increases. Decreasing the distance over which diffusion occurs also makes the gradient steeper. The diffusion pressure gradient across a cell membrane may be relatively steep because cell membranes are so thin.

The metabolism of living things, which increases the need for movement of materials into and out of the organism, steepens the diffusion pressure gradient of some of these substances by changing their concentrations in certain places. An actively photosynthesizing leaf, for example, lowers its carbon dioxide concentration relative to that of the environment and raises its oxygen concentration; these gradients keep carbon dioxide diffusing into the leaf and oxygen diffusing out. In animals and in any nonphotosynthetic tissue (including leaves in the dark), respiration lowers the concentration of oxygen and raises the concentration of carbon dioxide in the tissue relative to that of the environment; then oxygen diffuses from the air into the organism, and carbon dioxide diffuses away.

Absorption of a few substances by cells depends on simple diffusion across the plasma membrane. The membrane is believed to have ultramicroscopic pores through which very small molecules and ions pass; among them are oxygen, carbon dioxide, water, and some water-soluble minerals. Because the plasma membrane is composed in part of lipids, it is assumed that some fat-soluble compounds of low molecular weight dissolve in the lipid and diffuse across the membrane. In general, very small molecules pass rather freely through membranes; medium-sized molecules (like sugar, fatty acids, and amino acids) pass less easily; and most macromolecules do not cross membranes. There are, however, several exceptions to these generalities. The role of the plasma membrane in simple diffusion is a passive one, and the cell expends no energy in this process, the kinetic energy of the moving molecules being all the energy that is necessary.

# Osmosis

The movement of water into and out of cells is such a special case that it warrants a separate discussion, for it involves not only the water content of cells, but also their turgidity. Water enters the cell by a diffusion process called osmosis.

Osmosis is the diffusion of a solvent across a differentially permeable membrane. For osmosis to occur, two solutions of different diffusion pressures must be separated from each other by a differentially permeable membrane; this membrane is permeable to the solvent (water in the case of living cells) and is impermeable to the dissolved substances. Several differentially permeable membranes are found in cells; one of them is the plasma membrane.

To simplify our discussion, let us for the moment consider only the following: the plasma membrane of a cell, the water and dissolved sugars and mineral salts inside the cell, and the water in which the cell is immersed. Let us also assume, for simplicity's sake, that the membrane is completely impermeable to sugars and salts but freely permeable to water (not completely true in the living cell).

Let us place this cell in pure water, that is, in water that contains no dissolved substances (Fig. 18.2a). We may consider the water inside the cell to be diluted by its dissolved substances; the con-

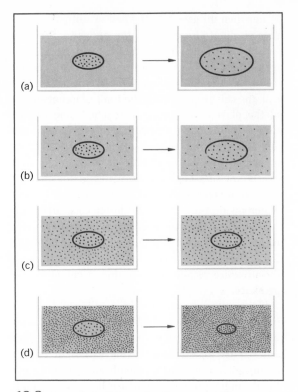

**18.2** Osmosis in living cells. Clear area: 100 per cent water. Stippled area: water containing dissolved substances; density of stippling indicates relative concentration of dissolved matter. (a) Cell immersed in pure water. Water diffuses into cell causing the cell to swell. (b) Cell immersed in a solution containing some dissolved materials but not so much as the cell itself. Water diffuses into the cell, and the cell swells but not so much as in (a). (c) Cell immersed in a solution containing exactly the same concentration of dissolved materials as the cell itself. There is no diffusion pressure gradient and no diffusion occurs. The cell does not change in size. (d) Cell immersed in a solution containing a higher concentration of dissolved materials. Water diffuses out of the cell, causing the cell to shrink.

centration of water within the cell then is something less than 100 per cent. In this case the plasma membrane separates a low concentration of water (inside the cell) from a high concentration of water (outside). A diffusion pressure gradient exists across the membrane, the diffusion pressure of the water being greater outside the cell than inside.

Water then diffuses into the cell. A diffusion pressure gradient also exists in the opposite direction for the dissolved materials, but because the plasma membrane is impermeable to them, they cannot diffuse out of the cell. As water moves into the cell, the cell swells. When the limit of extensibility of the plasma membrane is reached, one of two things may happen. Either the membrane breaks and spills out the contents of the cell, thus killing it, or the plasma membrane remains intact and exerts pressure back on the water inside the cell and so increases the diffusion pressure of the water inside the cell. In the latter case, diffusion ceases when the diffusion pressure of the water inside the cell reaches that of the water outside the cell. The cells become plump and turgid, and the pressure responsible for this is called **turgor pressure.**

Plant cells are protected by a cell wall that lies immediately to the outside of the plasma membrane (Fig. 17.2). In mature cells the cell wall does not stretch to any appreciable extent; it is also stronger than the plasma membrane and does not break under pressures that are likely to be built up in cells. If plant tissues are placed in pure water, they can develop a high degree of turgidity rarely found in animal cells. The crispness of fresh fruits and vegetables is due largely to the turgidity achieved by osmosis. In fact, turgidity is the main support of herbaceous (nonwoody) plants. The turgor pressure of growing plants sometimes manifests itself more or less spectacularly by breaking up pavements recently laid over live roots or seeds.

In nature, cells rarely are surrounded by pure water. Rainwater, soil water, and the water of streams, lakes, and oceans contain various amounts of dissolved substances; so do the body fluids bathing cells of multicellular organisms. Because the diffusion pressure of water in these solutions is lower than that of pure water, an equilibrium is reached between the cells and the external solution before enough pressure develops inside to cause the cell to burst (Fig. 8.2b).

If a cell is immersed in a solution with exactly the same concentration of dissolved particles as the water inside the cell (Fig. 8.2c), then equilibrium already exists, for the water has no diffusion pressure gradient. Osmosis does not occur, and there is no change in the size of the cell. The blood of animals and their blood cells exist in such an osmotic equilibrium. Algae and some aquatic animals stay in osmotic equilibrium with their environment by maintaining a salt concentration the same as that of the surrounding waters. Those living in fresh water have lower salt concentrations than those living in marine waters.

If a cell is placed in a solution containing a higher concentration of dissolved materials than the cell has (Fig. 18.2d), then the diffusion pressure of the water inside the cell is greater than that outside. Water diffuses out of the cell, and the cell shrinks in size. As the cell becomes more and more dehydrated, it may die, although if fresh water is supplied before that happens the cell will revive. Fresh-water organisms transferred to marine water and marine organisms transferred to fresh water often experience osmotic difficulties and may even die. Human beings stranded on rafts in the ocean cannot survive by drinking sea water, for its salt concentration (3.5 per cent) is higher than that of human body fluids (0.9 per cent) and has a dehydrating effect.

Throwing salt around the roots of weeds is an easy way to kill them, but it is best practiced in parking lots and not in flower beds or on lawns, where the salt does not discriminate between wanted and unwanted plants. Many a person has been disappointed to find a brown lawn next to the sidewalk he kept free of ice by throwing salt on it all winter. Too liberal a use of fertilizer on plants has the same effect. Candies, jams, and jellies are their own preservatives, for their high sugar content prevents most bacteria and many fungi from growing in them. Salting of meat also tends to inhibit growth of microorganisms and so preserves the meat.

# Active Transport

If diffusion and osmosis were the only mechanisms by which substances entered cells, then the concentrations of these materials would never become appreciably higher than they are in the immediate environment, for diffusion and osmosis cease when equilibrium is reached. Yet it is known that cells do accumulate materials to concentrations several times higher than that in their environment; conversely, they can also secrete some substances against a concentration gradient. Such movement requires the expenditure of energy. Here again we find use for the chemical bond energy of ATP. The mechanism of **active transport** is not completely understood, but it is believed that the plasma membrane contains the enzymes necessary for the dephosphorylation of ATP and for the coupling of this reaction with the transfer of substances across the membrane.

Because respiration is necessary for the production of ATP, only living cells are capable of active transport. Cyanides and other poisons that uncouple ATP synthesis from respiration cause cessation of active transport and secretion. Depriving cells of oxygen or glucose also lowers the rates of these processes or stops them completely.

Roots of many plants absorb mineral salts in such large quantities that the plants contain concentrations of the compounds many times that of the surrounding soil water. The value of this to the plant is not always understood, especially in cases where the absorbed mineral has no known function in the plant.

The cells lining the small intestine are known to engage in active transport, absorbing food against a concentration gradient from the cavity of the digestive tract. The value of this must be obvious, for it enables the organism to absorb and utilize foods that otherwise would be wasted because they happened to be present in concen-

trations lower than their concentrations in the tissues of the organism.

Some kidney cells are involved in active absorption. When urine is being formed it contains several useful compounds, including glucose, water, and minerals. Certain kidney cells partially reabsorb these compounds from the urine against a concentration gradient. These cells are among the most metabolically active cells in the animal body and are known to have large numbers of mitochondria that supply the ATP necessary for active absorption.

Marine fish live in ocean waters containing a higher concentration of salts than exists in their tissues. These fish drink ocean water, and the gills excrete the excess salt against a concentration gradient. Fresh-water fish, which maintain salt concentrations higher than that of the surrounding waters, have the opposite problem, and their kidneys must excrete water against a concentration gradient.

The nectar secreted by some nectar glands in plants contains such a high concentration of sugars that it is thick and viscous; yet the cells from which it continues to exude have a less sugary sap.

A variation on the entrance of materials into the cell might be mentioned here. Some cells form extensions that surround a solid food particle (Fig. 17.6). By completely engulfing the particle, the cell forms a food vacuole containing the particle. Enzymes from the cytoplasm pass through the membrane of the vacuole where digestion occurs, and the digested food is absorbed back across the membrane. Many of the protozoa, the single-celled animals, obtain their food in this way, and the white blood cells of the human body dispose of bacteria in the same way. Water containing dissolved substances is similarly taken into cells by pinocytotic vesicles (Chapter 17).

Although we know relatively little about how the plasma membrane controls the movement of substances across itself, we do know that it certainly is an active, vital part of the cell. It exercises

179

a certain amount of selective control over the passage of these materials, and the permeability of the membrane to different substances changes from time to time.

*Oxidize*

## Questions

**1.** Define diffusion and osmosis.

**2.** What is a diffusion pressure gradient, and of what significance is it in osmosis?

**3.** What kind of energy is responsible for diffusion? What is the source of this energy?

**4.** Explain why too much fertilizer can kill a plant or why jelly does not spoil readily.

**5.** Explain how the cell wall and the process of osmosis maintain the turgidity of a plant cell.

**6.** How does active transport differ from osmosis?

**7.** What kind of energy is necessary for active transport? What is the source of this energy?

*A.T.P Chemical bonds between atoms.*
*Glucose Plasma membrane contains enzymes for dephosphorylation of A.T.P.*

## Suggested Readings

Devlin, R. M. *Plant Physiology,* 2nd ed. New York: Van Nostrand Reinhold Company, 1969.

Holter, H. "How Things Get into Cells," *Scientific American* 205(3): 167–180 (September 1961).

Larimer, J. *An Introduction to Animal Physiology.* Dubuque, Iowa: William C. Brown Company, Publishers, 1968.

Rustad, R. C. "Pinocytosis," *Scientific American* 204(4): 120–130 (April 1961).

Steward, F. C. *Plants at Work.* Reading, Mass.: Addison-Wesley Publishing Co., Inc., 1964.

(2). *Sign. Maintains an equilibrium. Concentration is less inside the cell, because of the dissolved substances that pass into it thru the differentially permeable membrane.*

(3). *Kinetic — molecular motion.*

(4) *High sugar content prevents bacteria & algae from growing in them. Jams & jellies are their own preservatives.*

*Fertilizer - The diffusion pressure inside the cell is greater than on the outside (due to the dissolved substances of fertilizer on the outside) Water diffuses out of the cell — The cells shrink & die.*

(5). *The cell wall is very strong & does not break easily; able to maintain an equilibrium with the pressure on the outside of the cell. The cell become smaller with the water until equilibrium is maintained with the pressure on the outside.*

Cells that have not reached their maximum size, that are supplied adequately with food, minerals, water, and oxygen (or are protected from oxygen in the case of obligate anaerobes), that live at their optimum temperature and optimum pH, and that are not exposed to poisons almost invariably grow. They increase in size and weight as they synthesize more protoplasmic material. There is a limit, however, to the size that a cell reaches. The limit varies with the type of cell. Some bacterial cells reach maximum size at diameters of a micron (one millionth of a meter) or even half of a micron. Many cells of multicellular plants and animals are larger, with dimensions of 5, 10, 25, 50, or even more microns, and some cells, such as the yolks of the eggs of birds and reptiles, are easily seen with the naked eye. None, however, grows indefinitely.

When a cell reaches its maximum size it may divide into two daughter cells, each of which is about half the size of the parent cell. These grow, and when they have reached their maximum size, they, too, divide. New cells are produced only by division of pre-existing cells. Except for occasional abnormal divisions, two daughter cells are the result of any one division.

Cell division is essential for the reproduction of unicellular organisms, the parent cell giving rise in this way to new cells that are separate individuals. Multicellular organisms grow by a combination of cell growth and cell division; young cells grow, thereby increasing the size of the plant or animal, and then these cells divide, producing more cells that grow in turn. Repair of wounded tissues and replacement of worn-out and dying cells also depend on cell divisions. Reproduction, growth, and repair thus possess certain similarities and are really different aspects of each other.

Division of eucaryotic cells involves two events: the division of the nucleus and the division of the cytoplasm. These two events usually occur simultaneously, or nearly so, and each daughter cell receives one of the daughter nuclei. The division of the nucleus is called **mitosis** (a term that is used

# 19

# Mitosis and Cell Division

1 New cells are produced from pre-existing cells by the process of cell division; one cell divides into two cells.

2 Mitosis is the process of nuclear division that ordinarily accompanies cell division.

3 Mitosis ensures that the genetic content of the nucleus remains the same from one cell generation to another. This is accomplished by
   a Duplication of every chromosome in the nucleus prior to mitosis.
   b The distribution during mitosis of one set of duplicates to one of the new daughter nuclei and the other set of duplicates to the other daughter nucleus.

181

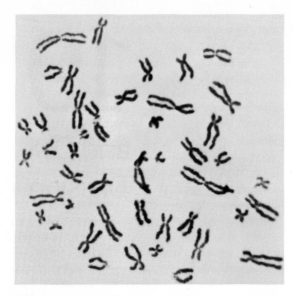

**19.1** Human chromosomes. The chromosomes shown here are from a cell of a human female. Each of the 46 chromosomes has duplicated, but the two chromatids are still held together at the centromere. (Courtesy Carolina Biological Supply Company.)

somewhat loosely and incorrectly as a synonym for cell division).

In Chapter 17 we learned that the nucleus contains the chromosomes, nuclear sap, and one or several nucleoli and is bounded by a nuclear membrane. Of these several parts, the chromosomes are most important from a genetic standpoint, for they contain nearly all the cell's genes or genetic material.

The chromosome number is usually constant for all members of a species. Humans, for example, have 46 chromosomes in the nucleus of each cell (Fig. 19.1); corn has 20, fruit flies 8, and onions 16. Chromosome numbers of a few species go as high as 500 or more, but numbers under 50 are most common. One important feature of mitosis is that it keeps the number of chromosomes constant from one cell generation to another. A cell dividing in a growing human embryo produces two cells each with 46 chromosomes, and their daughter cells each have 46 chromosomes, and so on. Simi-

larly, a corn root cell with 20 chromosomes divides into two daughter cells each with 20 chromosomes, and so on. In sexually reproducing organisms the chromosomes occur in pairs, the two members of the pair being **homologous chromosomes.** One homologue of each pair came from the male parent of the organism, and the other homologue from the female parent. These chromosomes are called paternal and maternal chromosomes, respectively.

Figure 19.2 is a diagrammatic representation of a parent cell and its daughter cells; the two chromosomes of each pair are distinguished by arbitrarily chosen colors to indicate maternal and paternal chromosomes. Not only do the two daughter cells have the same number of chromosomes as their parent cell, they also have the same *kinds* of chromosomes: two long ones, two medium ones, and two short ones, and every chromosome of the parent cell is represented in each daughter cell. This comes about because, as later paragraphs show, every chromosome of the parent cell is duplicated, forming two daughter chromosomes, one of which passes to one daughter cell and the other to the other daughter cell. By repeated mitotic divisions any number of cell generations can be produced, with all the cells genetically identical. With the exception of certain reproductive cells (which are formed by a different kind of division), all cells of the body of an organism are genetically identical because they are produced by mitotic divisions.

We now should examine in some detail the steps in the process of mitosis that bring about these results. Mitosis has been divided somewhat arbi-

**19.2** Diagrammatic representation of a parent cell and its two daughter cells. The daughter cells have chromosomes identical with those of the parent cell. Maternal and paternal chromosomes are colored differently.

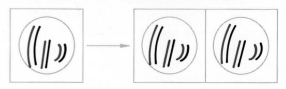

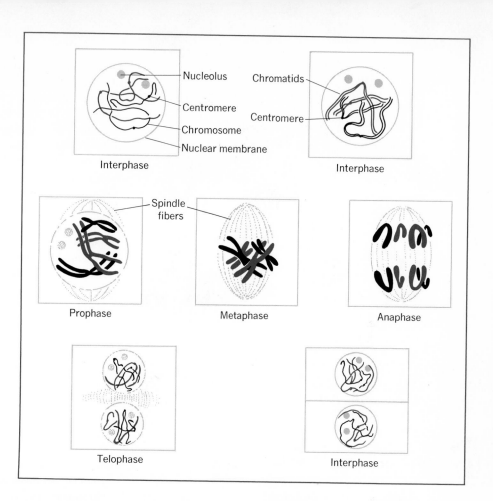

Nucleolus

Centromere

Chromosome

Nuclear membrane

Interphase

Chromatids

Centromere

Interphase

Spindle fibers

Prophase

Metaphase

Anaphase

Telophase

Interphase

**19.3** Mitosis in plants. For clarity the chromosomes have been shortened greatly in interphase and prophase, and there is no attempt to show the coiling of chromatids around each other. Maternal and paternal chromosomes are colored differently. See text for details.

trarily into four steps, called, in order of their occurrence, prophase, metaphase, anaphase, and telophase; these are illustrated in Fig. 19.3. Figure 19.4 is a photograph of several phases. Each phase merges with the phase that follows, and there is no sharp dividing line between them. A nucleus that is not dividing is said to be in interphase.

During **interphase** the nucleus is much as it is described in Chapter 17. The **chromosomes** are long, thin threads. They are so much entangled among themselves that even with the best microscopes it is practically impossible to follow the length of a single interphase chromosome. Late in interphase, as the nucleus approaches prophase, each chromosome duplicates itself. The two duplicates are called **chromatids.** At one place along

their lengths they are not duplicated; at this point is the **centromere,** which holds the two identical chromatids together. The location of the centromere may be anywhere along the length of the chromosome but is constant for any chromosome. Because of the extreme fineness of the chromosomes, it is not possible actually to see their duplication, but there is chemical evidence of it. Late in interphase the amount of the nucleic acid DNA (Chapter 43) in the nucleus doubles; and because the chromosomes are the only constituents of the nucleus that contain DNA, we infer that this is the time of formation of new chromosomal material.

During **prophase** the chromosomes gradually become shorter and thicker, and late in prophase it becomes possible to distinguish individual chro-

183

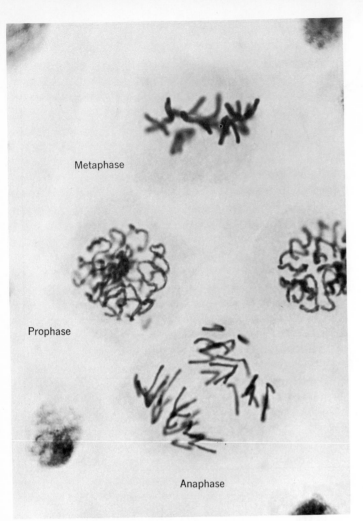

Metaphase

Prophase

Anaphase

**19.4** Mitosis in cells of an onion root tip. The prophase nuclei have long, thin chromosomes, the double nature of their chromosomes is visible at several places. (Courtesy Carolina Biological Supply Company.)

mosomes—at least in those species that have low chromosome numbers. Close examination reveals that the two chromatids of a chromosome are tightly coiled around each other. Late in prophase the chromosomes approach the center of the cell.

Other events of prophase are the disappearance of the nucleoli and the nuclear membrane and the formation of spindle fibers. The **spindle fibers** are extremely fine fibers, some of which extend from one end (or pole) of the cell to the opposite end.

Other spindle fibers extend only half the length of the cell—from one pole of the cell to the center, where each one is attached to the centromere of a chromosome. The two identical chromatids of a chromosome are connected to opposite poles by different spindle fibers.

At **metaphase** the chromosomes line up at the middle of the cell. Now the chromosomes are short and thick and are more clearly seen than during any other phase of mitosis. They still consist of two chromatids, but the chromatids are less tightly coiled around each other than before.

By **anaphase** the centromeres of the chromosomes have duplicated, and the chromatids separate from each other and move toward opposite poles of the cells. Once separated, the chromatids are called chromosomes. Because the centromere is the leading part of each chromosome and because the spindle fibers are attached to the centromeres, it appears as if the spindle fibers pull the chromosomes to the poles. In actual fact the mechanism of movement is unknown. Note that the set of chromosomes moving to one pole is identical both in number and type of chromosomes with the set of chromosomes moving to the other pole and that each of these sets is identical with the set of chromosomes possessed by the original cell.

**Telophase** begins as the two sets of chromosomes approach opposite poles. A nuclear membrane forms around each set of chromosomes, and a nucleolus or nucleoli form in the new nuclei. The chromosomes gradually become long and thin again, and the daughter nuclei enter interphase.

The division of the cytoplasm usually occurs during telophase. In plants the long spindle fibers condense toward the middle of the cell, where a middle lamella (the cementing substance between plant cells) forms across the cell. A new primary cell wall is laid down on either side of the middle lamella. Cell division is then complete.

In animal cells, which have no cell walls, the spindle fibers disappear, and the cell constricts near its middle (Fig. 19.5). As the constriction increases, the cell pinches into two cells, and cell

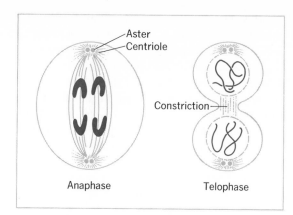

Aster
Centriole

Constriction

Anaphase          Telophase

**19.5**  Two phases in mitosis in an animal cell.

division is complete. Another difference between cell division in animals and in most plants is that the **centrioles** of animal cells play a role in division. In prophase the pair of centrioles divides and the two pairs move to opposite ends of the cell and there act as the poles. The spindle fibers emanate from the centrioles, and another system of short fibers called **asters** radiates from the centrioles. Most plants do not have centrioles and asters.

The final result of mitosis is the production of two cells with identical sets of chromosomes. The division of the cytoplasm is not necessarily equal, though it may be nearly so. Most cells are divided approximately across the middle, and the organelles of the cytoplasm are distributed in nearly equal numbers to the daughter cells. Recently it has been discovered that mitochondria and chloroplasts contain their own DNA. These organelles are now known to divide and to produce new mitochondria and chloroplasts. This keeps the numbers of these organelles per cell more or less constant from one cell generation to another.

# Questions

**1.** Diagram a cell in the several stages of mitosis. Pay special attention to what happens to the chromosomes.
**2.** Using these diagrams, explain how mitosis ensures that each daughter cell has a complete copy of the genetic material of the parent cell.
**3.** What are some differences in the division of plant cells and animal cells? Does this make any difference in the distribution of chromosomes to daughter cells? *no.*
**4.** What is the difference between cell division and mitosis?
**5.** At what time in the life cycle of a cell do the chromosomes duplicate?
**6.** What is the function of cell division in the life of an organism?

# Suggested Readings

Mazia, D. "Cell Division," *Scientific American* 189(2): 53–63 (August 1953).
———. "How Cells Divide," *Scientific American* 205(3): 100–120 (September 1961).
McLeish, J., and B. Snoad. *Looking at Chromosomes.* London: Macmillan & Company, Ltd., 1959.
Swanson, C. P. *The Cell,* 3rd ed. Englewood Cliffs, N.J.: Prentice-Hall, Inc., 1969.
Taylor, J. H. "The Duplication of Chromosomes," *Scientific American* 198(6): 36–42 (June 1958).
Wilson, G. B. *Cell Division and the Mitotic Cycle.* New York: Reinhold Publishing Corporation, 1966.

*[handwritten margin notes:]*
*Cell wall Centrioles Spindle fibers disappear in animal cells but they form cell wall in plants*

*4. mitosis - div. of nucleus*
*cell div - div. of cell into 2 separate cells - incl. mitosis & cytokinesis (div. of cytoplasm)*
*5. Late interphase*
*6. Growth, reproduction & repair*
*Keep the number of chromosomes constant from cell generation to generation.*
*To keep contact with the environment*

185

# 20
# Size and Form

1 A large object has a smaller surface-to-volume ratio than a small object of the same shape.

2 Approximately proportional to an organism's volume are its
   a Need for gases, water, minerals, and food (in animals) or light (in plants).
   b Production of wastes and the need to eliminate them.
   c Weight.

3 Approximately proportional to an organism's surface area are its
   a Ability to absorb gases, water, minerals, and food or light.
   b Ability to eliminate wastes.
   c Strength.

4 The surface-to-volume ratio of large organisms is relatively large because the organs are branched or thrown into folds; this increases the surface area without further increasing volume.

5 The tissues of all but the smallest organisms are differentiated into different types that perform different functions.

We might inquire if there is a limit to the size of cells. Could there be unicellular organisms as large as dogs or lilac bushes or whales or redwood trees? The answers seem to lie in the relationships between the size and form of the cell.

With any change in the size of an object that is not accompanied by a change in its shape, there is a change in the relationship between its surface area and its volume. To illustrate, we might examine a familiar object like a cube. The formula for the surface area of a cube is $6x^2$, where $x$ is the length of any edge. The formula for the volume of a cube is $x^3$. These two formulas suggest that if the cube increases in size, its surface area will increase as the square of the linear dimension, and the volume will increase as the cube of the linear dimensions. For example, if the edge of a cube is 1 millimeter long, then its surface area is 6 square millimeters and its volume is 1 cubic millimeter. If the size of the cube is changed by doubling the length of the edges, then the surface area becomes 24 square millimeters and the volume becomes 8 cubic millimeters. Both surface area and volume have increased, but the volume has increased proportionately more than the surface area. Table 20.1 gives surface areas and volumes for several cubes of different sizes. Notice how much more rapidly volume increases than does surface area. This is true of objects of other shapes also. The interested reader may refer to a geometry book to find formulas for spheres and other three-dimensional objects, and in every case he will find the surface area proportional to the square of the linear dimension and the volume proportional to the cube of the linear dimension.

What, now, is the significance of these surface-to-volume relationships to living things? As the volume of actively metabolizing protoplasm increases, so does the need for food, water, minerals and gases and the need for the elimination of wastes. These substances must enter or leave the cell through the plasma membrane, and therefore the rate of their passage through the membrane is proportional to its surface area. As a cell in-

**Table 20.1**

Surface Areas and Volumes of Cubes of Several Sizes

| Edge (x), mm | Surface Area ($6x^2$), sq mm | Volume ($x^3$), cu mm |
|---|---|---|
| 1 | 6 | 1 |
| 2 | 24 | 8 |
| 3 | 54 | 27 |
| 4 | 96 | 64 |
| 5 | 150 | 125 |
| 6 | 216 | 216 |
| 7 | 294 | 343 |
| 8 | 384 | 512 |
| 9 | 486 | 729 |
| 10 | 600 | 1,000 |
| – | – | – |
| 100 | 60,000 | 1,000,000 |
| – | – | – |
| 1,000 | 6,000,000 | 1,000,000,000 |

creases in size, its need to receive nutrients and to dispose of wastes increases faster than the rate at which they can enter or leave the cell. The larger the cell, the greater is its risk of starving or of being poisoned by its own waste products.

Cell division is one way of avoiding these problems. The division of a cell into two smaller cells which separate from each other increases the surface-to-volume ratio and makes it more favorable. As the young cells grow, the surface-to-volume ratio decreases, and as it becomes unfavorable again, the cells divide once more. The actual triggering mechanism that initiates cell division is not known. It might be the accumulation of a substance within the cell, or it might be the depletion of some compounds.

Cell division is an adequate solution to size and form problems in unicellular organisms, but cell division alone cannot solve these problems in large multicellular organisms, for a large mass of similar cells packed closely together are not in a much different situation than a single giant cell. Materials reaching or leaving the central cells still have to pass through the surface cells, and the weight and strength problems still exist. It is estimated that diffusion alone cannot provide adequate exchange of materials between active organisms and their environments if the dimensions of the organisms exceed 1 millimeter.

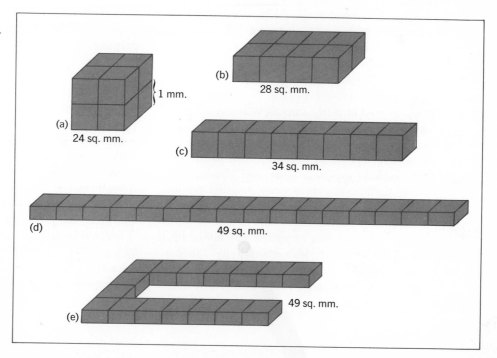

**20.1** Five objects with the same volumes but different shapes. Surface areas are given for all objects.

(a) 24 sq. mm. 1 mm.
(b) 28 sq. mm.
(c) 34 sq. mm.
(d) 49 sq. mm.
(e) 49 sq. mm.

Another type of solution to the surface-to-volume problems is the changing of shape of the object. The less compact the shape of the object becomes, the greater is its surface-to-volume ratio. In Fig. 20.1 are several objects with the same volume; all occupy 8 cubic millimeters. Object (a), which is the most compact (2 mm × 2 mm × 2 mm), has a surface area of 24 square millimeters. Object (b), which is less compact (2 mm × 1 mm × 4 mm), has a total surface area of 28 square millimeters. In objects (c) and (d) the surface areas rise to 34 and 49 square millimeters, respectively, by making the objects still longer and narrower. The object in (d) has a much more favorable surface-to-volume ratio than those in (a), (b), and (c), although its extremely long shape might be ungainly under some circumstances. The object in (e) has the same volume and the same surface area as the figure in (d), but by bending back on itself, it regains a certain compactness while retaining a high surface-to-volume ratio.

In multicellular organisms, evolution has progressed along lines that take advantage of such foldings on both the macroscopic and microscopic levels. A typical flowering plant, for example, rather than being spherical like an algal cell, has a basic cylindrical shape for its stems and roots, either of which may branch profusely; many flat organs, the leaves, occur on the branches; and each root tip bears hundreds of minute root hairs. On cursory examination animals do not seem to exhibit such extensive folding, but the increased surface area of many animals is nearly all to the inside. The many folds and minute pouches of human lungs create a surface area of about 100 square meters, or 50 times as much as the total skin surface; this area is devoted almost exclusively to the exchange of gases. Another example is the small intestine, through which food is absorbed. At 9 feet in length, the human small intestine is longer than its owner, but folded up it fits comfortably in the abdominal cavity. The inner surface of the intestine is thrown into minute fingerlike extensions called villi, which greatly multiply the absorbing surface (Figs. 31.2 and 31.3). Furthermore, the absorbing cells themselves have ultramicroscopic extensions called microvilli that contribute still more to the absorbing surface (Fig. 31.4). Without this folding of their surfaces, large organisms probably never would have existed. An active organism the size of a human being could not receive adequate oxygen or food if these substances had to enter through the skin.

Accompanying this increase of surface area by changing of shape, there usually is the development of a division of labor among the cells of an organism. These cells are specialized for certain functions, and as they mature they become differentiated in ways that adapt them to their particular roles. Cells are organized into tissues, a **tissue** being a more or less continuous mass of similar cells with the same function or functions. Several different kinds of tissue comprise an **organ.**

Gases are exchanged with the environment by leaf cells in plants and by the lungs or gills in animals. Water and minerals are absorbed by root epidermal cells of plants. The intestines of animals absorb food, water, and minerals. Because different materials are absorbed at different locations in a single organism, there is a need for these substances to be transported throughout the organism. Diffusion is fast enough to fill this need in only very small organisms, but larger animals have circulatory systems that actively pump compounds around the body in arteries and veins. Veins of a different nature distribute materials throughout higher plants (Fig. 20.2). The fact that different parts of an organism perform different functions raises the need for some mechanism of coordination that would prevent the several parts from acting antagonistically and thus decreasing the overall efficiency of the organism. Both plants and animals have developed hormonal systems that fill this need, and the higher animals have developed a nervous system as well. Some other functions assumed by specialized tissues are movement, excretion of waste products, food storage, support, and reproduction. The following chapters discuss in more detail this division of labor in plants and animals.

Heat, as well as chemical compounds, passes across the surfaces of organisms. With regard to body temperature, most species of living things are largely at the mercy of their environment. Fish and snakes, algae and maple trees, these and many others have internal temperatures scarcely different from that of their environment. But two groups of animals maintain nearly constant body temperatures regardless of external temperatures. These are the mammals and birds, the so-called warm-blooded animals. Except in very hot weather, their bodies maintain temperatures higher than that of the environment. In cold winter weather, the difference can be very great—well over 100°F (56°C). Because heat escapes through the body surface, small animals lose their heat at a higher rate (proportional to body volume) than large animals do. The heat that maintains the constant body temperature comes from respiration of living cells. Small mammals and birds, therefore, must maintain higher respiratory rates than large ones. Indeed, mice and other very small mammals spend most of their waking hours finding and eating food with which to maintain their respiration. These small rodents probably approach the lower limit of size for warm-blooded animals.

Another size-and-form problem has to do with the strength of an object and its ability to bear

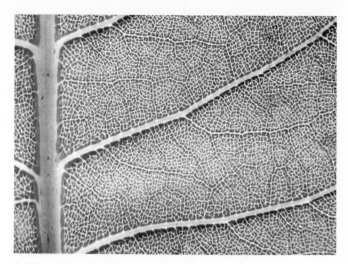

**20.2** Vein pattern of a leaf. The repeated branching of the veins greatly increases the surface area of the veins. No cell of the leaf is very distant from a vein.

weight—either its own or additional weight placed upon it. The weight of an object of given shape and composition, like its volume, increases in direct proportion to the cube of its linear dimension; its strength, like its surface area, increases in direct proportion to the square of its linear dimension. This means that as an object increases in size but maintains its shape, it becomes progressively weaker in proportion to its volume; when a certain size is reached, it can no longer support its own weight.

The problem may be solved in part by changing

**20.3** The great size of the elephant requires thick legs for support. The smaller springbok has slender bones and is dainty and agile.

the shape of the object. By increasing the cross-sectional area of a bone faster than its length, additional body weight may be supported, but eventually difficulties are encountered. As the bones become proportionately thicker, the animal becomes clumsier. Compare, for instance, the agility of the dainty antelopes with the slower, more deliberate movements of the larger, ponderous elephants (Fig. 20.3). In the case of completely terrestrial animals, the elephant probably approaches the upper limit on size. The biggest of the dinosaurs were much larger than elephants, but these giant reptiles lived in swamps where part of their weight was supported by water. The blue whale, about 75 feet long, is larger than any of the dinosaurs, but it is a completely aquatic animal.

**6.** Can you explain why small Arctic animals migrate south in winter or enter a dormant state, whereas large animals can lead active lives in the Arctic winter? Can you correlate the long ears of a Texas jackrabbit and the short, stubby ears of an Arctic fox with the climates in which these animals live? *Balancing) Shape*

**7.** Why do small animals require so much more food for their size than large animals do?

**8.** Why do large organisms require differentiation in tissues and organs whereas minute organisms have no such need?

**9.** How does the minute size of bacteria help to explain their ability to spoil food in a short period of time?

**10.** Why would a land animal much larger than an elephant probably have trouble supporting its own weight?

**11.** Review the parts of the cell and recall how many of these are membranes. Consider the internal structure of chloroplasts and mitochondria. Does this have more significance to you now?

## Questions

**1.** Explain why, as an object increases in size, its volume increases faster than its surface area.

**2.** What are some of the characteristics of an organism that depend on its surface area?

**3.** What are some of the characteristics of an organism that are proportional to its volume?

**4.** Using the answers to the previous two questions, explain why the surface-to-volume ratio of organisms is so important.

**5.** Why can a cell not grow indefinitely without dividing?

## Suggested Readings

Bonner, J. T. *Size and Cycle*. Princeton, N.J.: Princeton University Press, 1965.

———. "The Size of Life," *Natural History* 78(1): 40–45 (January 1969).

Moog, Florence. "Gulliver Was a Bad Biologist," *Scientific American* 181(5): 52–56 (November 1949).

Thompson, D'A. *On Growth and Form*, abridged ed. Edited by J. T. Bonner. New York: Cambridge University Press, 1961.

Went, F. W. "The Size of Man," *American Scientist* 56: 400–413 (1968).

(7) must maintain higher respiratory rates. Heat escapes thru body surface & In proportion to body volume small animals lose their heat at a higher rate.

(8) Diffusion is fast enough in very small organisms. Different materials are absorbed at different locations on single organism this is no need for these substances to be transported throughout the organism. In large organisms, different parts of an organism perform different functions.

(10) It would become weaker in proportion to its volume (and clumsier) (an object that increases in size but maintains its shape)

# Flowering Plants

# Part VI

Except for very small plants and animals, the cells of organisms are differentiated into several cell types that perform different functions. There exists a wide variety of plants, including algae, fungi, mosses, ferns, conifers, and flowering plants. In this and the next eight chapters we will consider primarily the tissues and organs of the flowering plants. These plants have been chosen for several reasons: they are the most highly evolved plants; they are the dominant plants over much of the land surface of the earth; and they are most familiar to us. They are the direct or indirect source of nearly all of man's food, clothing, and many other products.

## Plant Organs and Tissues

A **tissue** is a group of similar cells that perform the same function or functions. Tissues often are organized into larger structures called organs.

### Organs

The organs of a flowering plant (Fig. 21.1) are few compared with those of a highly evolved animal such as a mammal. Three vegetative organs—root, stem, and leaf—are all that is necessary to maintain a healthy plant. The flower is the reproductive organ—or more correctly, an assemblage of several organs—from which other organs, the seeds and fruits, are produced.

The **leaf** is the primary photosynthetic organ of the plant. Its flat shape provides a great deal of surface area through which sunshine can be absorbed and through which gases can be exchanged. It is also the organ from which most of the water loss of plants occurs.

The **root** is a cylindrical structure, usually branched and usually located underground. It anchors the plant in the soil, and it absorbs both water and dissolved minerals from the soil. Roots of many plants serve also as food storage organs.

# 21
# The Flowering Plant–Its Organs and Its Tissues

1  The flowering plant body consists of several organs, each of which contains several tissues. Each organ and each tissue performs one or more functions.

2  The vegetative organs of the plant are the leaf, the root, and the stem.

3  The reproductive organs of the plant are the several parts of the flower, the fruit, and the seed.

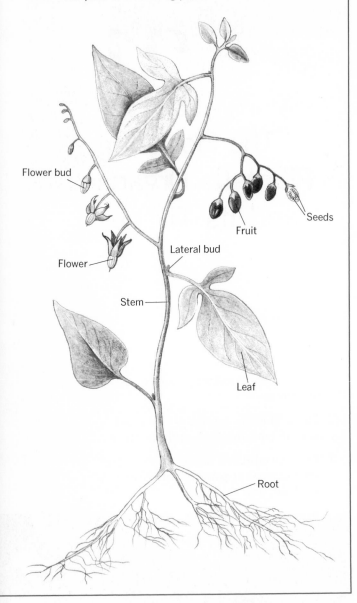

Flower bud

Flower

Stem

Lateral bud

Fruit

Seeds

Leaf

Root

pollinating agents. Another function of the stem is the transport of water and minerals absorbed by the root to the aerial parts, the transport of recently manufactured food from the leaf to the root, and the transport of stored food from the root to the tops of the plant when it is needed.

**Flowers** are not essential for the well-being of an individual plant, but fruits and seeds cannot be formed without them. A **fruit** contains one or more seeds, and each **seed** contains one embryo plant. Most flowers produce only one fruit, but some produce several one-seeded fruits.

The many families of flowering plants are divided into two groups on the basis of the number of their cotyledons (special, modified seed leaves borne by the embryo). Monocotyledonous plants (**monocots,** for short) possess one cotyledon, and dicotyledonous plants (**dicots**) possess two cotyledons.

Monocots generally have flower parts (petals, for instance) in multiples of three (Fig. 21.2) and linear (long and narrow) leaves with parallel veins running the length of the leaf (Fig. 21.3). Among the monocots are the cereals, which supply most of our carbohydrates, and some lovely ornamental plants like lilies and orchids.

**21.2**  Trillium flowers. Monocot flowers usually have their flower parts in multiples of three.

The **stem** is another cylindrical organ, and frequently it is branched. Typically it is an aerial organ, but several types of modified stem, such as bulbs, grow underground. An aerial stem and its branches support the leaves and flowers in the air where the leaves receive sunlight and the flowers are exposed to wind and insects, the most common

**21.3** Cattails. Monocot leaves typically have a linear shape.

Dicots usually have flower parts in multiples of five (Fig. 21.4) or occasionally four; their leaves are of many diverse shapes and are served by much-branched networks of veins (Fig. 21.5). Among the dicots are the legumes, which are our main plant source of proteins; the hardwood trees, such as maple and oak; and a wide variety of food and ornamental plants.

## Tissues

The plant organs are composed of several kinds of tissues the cells of which are modified in ways that enable them to perform their peculiar functions. The major cell types are described below.

Conducting cells usually are elongated cells with

**21.4** Apple flowers. Most dicot flowers have their flower parts in multiples of five, or occasionally four.

their longest axes coincident with the direction of transport. The two main types of conducting tissues are xylem and phloem.

**Xylem,** which is the woody tissue of plants, conducts water upward from the roots. Most of its cells are dead when they are functioning. As the cells mature, their protoplasm forms very thick secondary cell walls, and then they die. The cellulose of the secondary cell walls is impregnated with lignin, a rigid substance responsible for many of the characteristics of wood. **Tracheids** (Fig. 21.6)

**21.5** Buckthorn leaves. Dicots have several leaf shapes, but typically they are oval or some modification of the oval shape.

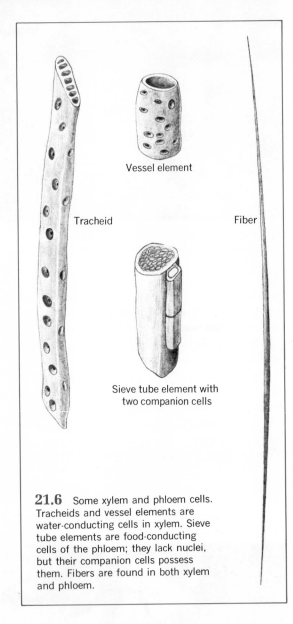

Tracheid

Vessel element

Fiber

Sieve tube element with
two companion cells

**21.6**  Some xylem and phloem cells.
Tracheids and vessel elements are
water-conducting cells in xylem. Sieve
tube elements are food-conducting
cells of the phloem; they lack nuclei,
but their companion cells possess
them. Fibers are found in both xylem
and phloem.

shorter. They do not have end walls at maturity;
because the cells fit together end to end, they form
continuous vertical hollow tubes, called **vessels,**
that run the length of the plant. Like soda straws,
vessels make excellent conductors of water. The
thick secondary cell walls of tracheids and vessel
elements make wood the major supporting tissue
of large plants.

The conducting cells of the **phloem** are **sieve
tube elements** (Fig. 21.6); these are living cells that
transport food both upward and downward in the
plant. They have thin cell walls, and they are
arranged in vertical rows called **sieve tubes.** The
end walls are perforated with pits through which
materials pass from cell to cell. Sieve tube elements have cytoplasm but no nuclei. Smaller adjacent cells called **companion cells** have nuclei that
are assumed to function for both.

**Fibers** (Fig. 21.6) are supporting cells that resemble tracheids in general shape but usually are
much longer. They have no protoplasm at maturity, and the secondary cell walls are so thick that
they sometimes fill nearly the entire cell. Fibers
occur in xylem and phloem, where they are called
xylem fibers and phloem (or bast) fibers, respectively; they may also occur in other places. Phloem
fibers of some plants can be easily removed in
strands long enough that they can be worked into
threads or rope. Phloem fibers from flax are used
in the manufacture of linen. Those from hemp
once were widely used to make cloth and rope,
but they have been replaced largely by fibers from
other plants and by synthetic fibers.

**Parenchyma** cells are thin-walled cells that are
isodiametric (having all diameters of equal size) or
nearly so. Some parenchyma cells occur as separate tissues, such as the cortex and pith of roots
and stems (Chapters 23 and 24) and the palisade
and spongy parenchymas of leaves (Chapter 22);
they also may be scattered among xylem and
phloem cells where they are called xylem parenchyma and phloem parenchyma, respectively. In
most plant organs, parenchyma cells are food-
storage cells, but those in leaves are photosyn-

are elongated xylem cells with tapered end walls
that coincide with the end walls of adjacent tracheids. The walls are interrupted by pits or other
openings that facilitate passage of water from one
cell to another. **Vessel elements** (Fig. 21.6) are
xylem cells generally wider than tracheids but

thetic. The turgor pressure of parenchyma cells is the main support of organs not well supplied with woody or fibrous tissues.

The entire plant body is surrounded by tissues that protect the plant from excessive loss of water by evaporation and from infection by fungi and bacteria. **Epidermis** and **cork** contain substances impermeable to water, and the cells have few or no air spaces among them.

Young plant parts—those only a few months old—are covered by a single layer of epidermal cells with cell walls of varying thickness. The external wall usually has a layer of cutin, a waxy material impermeable to water. The cutin layer is fairly thin on plants growing in damp places, but it can be very thick on plants of the desert where water conservation is important.

Stems and roots that live more than one growing season develop several layers of cork cells that replace the epidermis. The secondary cell walls of cork cells contain cellulose impregnated with suberin, another substance impermeable to water. It is this quality of cork that makes it a useful material for stopping bottles filled with beverages.

All new plant cells arise from embryonic tissues called **meristems.** Meristematic cells have thin cell walls and dense protoplasm, but their most characteristic feature is their ability to divide frequently and for indefinite periods. The meristems of some individual redwoods and the bristlecone pines of California have been functioning for hundreds or even thousands of years.

The meristems at the apices (tips) of roots and stems are called **apical meristems.** They have small isodiametric cells that contribute to growth in length of these organs. Growth in width of roots and stems is due to the activity of two other meristems called vascular cambium and cork cambium. Vascular cambium forms new xylem and phloem, and cork cambium forms cork. Apical meristems, vascular cambium, and cork cambium all produce new cells by mitotic cell divisions. The functioning of these meristems will be discussed in more detail in the following chapters.

## Questions

**1.** List the organs of a flowering plant. What are the most important functions of each organ? How do these functions contribute to the life of the plant?

**2.** List the tissues of a flowering plant. What are the functions of each tissue? How do these functions contribute to the life of the plant?

**3.** Select a few plant organs and cell types and tell how their structures are correlated with their functions.

**4.** How do meristems differ from other plant tissues?

## Suggested Readings

Biddulph, S., and O. Biddulph. "The Circulatory System of Plants," *Scientific American* 200(2): 44–49 (February 1959).

Esau, K. *Anatomy of Seed Plants.* New York: John Wiley & Sons, Inc., 1960.

————. "Explorations of the Food Conducting System in Plants," *American Scientist* 54: 141–157 (1966).

Ray, P. M. *The Living Plant.* New York: Holt, Rinehart and Winston, Inc., 1963.

Salisbury, F. B., and R. V. Parke. *Vascular Plants: Form and Function.* Belmont, Calif.: Wadsworth Publishing Co., Inc., 1964.

# 22

# The Leaf

1 The most important function of most leaves is the manufacture of food by the process of photosynthesis.

2 Photosynthesis occurs in the green, chlorophyll-containing cells of the leaf. More specifically, it occurs in the chloroplasts of these cells.

3 The light necessary for photosynthesis is absorbed by the chlorophyll in the chloroplasts.

4 The carbon dioxide necessary for photosynthesis enters the leaf through openings called stomata.

5 The water and minerals necessary for photosynthesis are absorbed by the roots and are transported to the leaves by the veins.

The most important function of most leaves is photosynthesis (Chapter 11), and the internal anatomy of the leaf is so arranged that all the things required for photosynthesis—chlorophyll, light, carbon dioxide, water, and minerals—can be brought together simultaneously in the internal cells of the leaf.

## Internal Anatomy of Leaves

Chlorophyll is present in the chloroplasts of a mature, healthy leaf (Fig. 22.1). In most leaves the chloroplasts are concentrated largely in the upper half of the leaf in the tissue called **palisade parenchyma.** Here they receive maximum exposure to sunlight, for a transparent, usually colorless, single-celled layer of **epidermis** is all that covers the palisade parenchyma.

Before it can be used in photosynthesis, light must be absorbed by chlorophyll. The flat shape of a typical leaf results in the spreading out of the palisade layer and gives the leaf a great deal of light-absorbing surface in relation to its volume. In most plants the leaves are arranged in such a way that the flat surface rather than the edge is directed toward the light. The leaves also shade each other as little as possible (Fig. 22.2). The green color you see when you look down on the upper surface of a leaf is largely the color of the palisade parenchyma.

The elongated palisade parenchyma cells stand in a relatively compact arrangement. Small air spaces do exist between the cells, however, and these air spaces receive oxygen from the cells and bring them carbon dioxide.

The **spongy parenchyma** in the lower half of the leaf where somewhat less light penetrates has more loosely arranged cells than the palisade parenchyma, and its individual cells contain fewer chloroplasts. Though these cells are also photosynthetic, an important function of the spongy

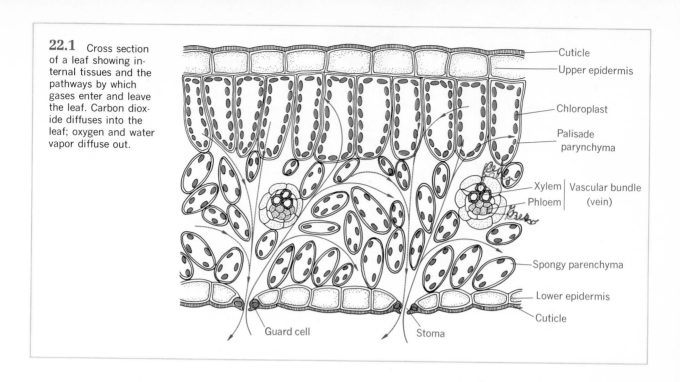

**22.1** Cross section of a leaf showing internal tissues and the pathways by which gases enter and leave the leaf. Carbon dioxide diffuses into the leaf; oxygen and water vapor diffuse out.

Cuticle
Upper epidermis
Chloroplast
Palisade parynchyma
Xylem | Vascular bundle (vein)
Phloem
Spongy parenchyma
Lower epidermis
Cuticle
Guard cell
Stoma

**22.2** A leaf mosaic. The leaves of this Boston ivy covering a wall are so arranged that they shade each other as little as possible.

**22.3** Lower surface of a red raspberry leaf. Dicot leaves typically have netted venation.

parenchyma is the distribution of gases, both carbon dioxide and oxygen, horizontally throughout the leaf. The large air spaces of this tissue make it well adapted to this function.

The lower surface of the leaf is protected by an epidermis. Both the upper and lower epidermises

**22.4** Vein pattern in a corn leaf. Monocots typically have parallel veins.

consist of cells tightly packed against each other with no air spaces between them except for certain specialized openings called **stomata.** In most species of plants, stomata either are confined to the lower epidermis or are more numerous in the lower epidermis than in the upper. Stomata serve as the principal passageways through which gases are exchanged with the environment. Carbon dioxide from the surrounding atmosphere diffuses through open stomata and enters the internal atmosphere of the leaf.

The spongy parenchyma consists of rather loosely arranged cells with large intercellular spaces that allow the carbon dioxide to diffuse laterally throughout the leaf, and small intercellular spaces extend among the palisade cells. These spaces permit carbon dioxide to reach all the photosynthetic cells of both palisade and spongy parenchymas. In many leaves the total absorbing surface inside the leaf (that is, the surfaces of palisade and spongy parenchyma cells exposed to the air passages) is 10 to 20 times the surface area of the epidermis.

Water and minerals absorbed by the roots travel upward in the stem through a system of **veins (vascular bundles).** Each leaf has at least one vein that branches off from the vascular tissue of the stem. Within the dicot leaf the vein usually branches several times, the smallest branches often being microscopic. The branches ramify through the entire leaf (Fig. 22.3), and any palisade or spongy parenchyma cell is never more than a few cells away from a vein. Monocot leaves usually have parallel veins (Fig. 22.4).

Food, the main product of photosynthesis, moves out of the leaves by way of the veins. It moves first to the stem and then may go to growing organs (such as flower buds, developing fruits, or the growing tips of stems and roots), or it may be carried to a storage organ (such as a root or tuber) where it remains until needed.

Oxygen, a by-product of photosynthesis, diffuses from the photosynthetic cells, passes through the intercellular passageways of the palisade and spongy parenchymas, and diffuses through open stomata to the outer atmosphere.

Because of the large internal surface area of the leaf, a great deal of water evaporates from the palisade and spongy cells and diffuses out through the stomata. This water loss, called transpiration (Chapter 26), is unavoidable during daylight hours when the plant must receive carbon dioxide through open stomata, but in darkness, when photosynthesis does not occur, and most plants have no need for carbon dioxide, the stomata need not be open. Many species of plants conserve water by the closing of their stomata at night; this reduces the chances of wilting or even death of the plants.

The stomata are not simply holes in the epidermis. A pair of specialized, sausage-shaped epidermal cells called **guard cells** surrounds each stoma. Modifications in the shapes of these cells can increase the size of the stoma or can decrease it until it closes completely (Fig. 22.5).

The mechanism regulating the opening and closing of stomata is not completely understood. It seems to depend in part, at least, on the possession of chloroplasts by the guard cells, whereas other epidermal cells lack chloroplasts. Therefore, guard cells are photosynthetic in light. During daylight hours the rate of photosynthesis exceeds that of respiration in the guard cells. This results in a decrease in the carbon dioxide concentration. Because some of the carbon dioxide was dissolved in water, producing carbonic acid, the pH of guard cells rises during the day as carbon dioxide is used in photosynthesis. The guard cells contain some stored starch, which tends to be converted to sugar as the pH rises. Therefore, during daylight hours, the concentration of sugar increases. The sugar is soluble in water, and so causes a decrease in the diffusion pressure of water in the guard cells relative to that of water in the neighboring epidermal cells. As a consequence, water moves by osmosis from the neighboring cells into the guard cells, which then swell. The cell walls of the guard cells are unevenly thickened, the thickest portion of the

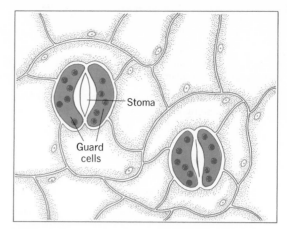

**22.5** Surface views of a leaf epidermis showing open and closed stomata.

wall being adjacent to the stoma. As the cells increase in size, the cell walls stretch, but the thickened portion does not stretch. As a consequence, each swelling guard cell assumes a sausage shape, the two ends of each cell pushing against the two ends of its partner. This results in an increased curvature of the two cells, and therefore an increase in the size of the space between them. The stoma thus opens (Fig. 22.6).

In darkness, respiration continues in the guard cells, but photosynthesis ceases. The carbon dioxide concentration increases, and therefore also the concentration of carbonic acid. As the pH falls, sugar tends to be converted to starch, and so the sugar concentration falls. This results in an increase in the diffusion pressure in the guard cells, and water moves by osmosis from the guard cells into the neighboring cells. With a loss of water, the guard cells wilt and decrease in size as they lose their turgidity. The curvature of the guard cells decreases and the opening between them becomes smaller. The stoma thus closes.

The amount of water lost by transpiration directly through the epidermal cells varies with the species. In most plants the epidermis secretes a waxy, water-impermeable substance called **cutin,** which forms a layer called **cuticle** that covers the

|  | Light | Darkness |
|--|-------|----------|
|  | Rate of photosynthesis exceeds rate of respiration in guard cells ↓ | Respiration continues in guard cells, but photosynthesis stops. ↓ |
|  | Concentration of carbon dioxide in guard cells decreases ↓ | Concentration of carbon dioxide in guard cells increases ↓ |
|  | Concentration of carbonic acid decreases ↓ | Concentration of carbonic acid increases ↓ |
|  | pH rises ↓ | pH falls ↓ |
|  | Starch is converted to sugar ↓ | Sugar is converted to starch ↓ |
|  | Concentration of sugar in guard cells increases ↓ | Concentration of sugar in guard cells decreases ↓ |
|  | Diffusion pressure of water in guard cells falls below that of water in neighboring epidermal cells ↓ | Diffusion pressure of water in guard cells rises above that of water in neighboring epidermal cells ↓ |
|  | Water moves from neighboring epidermal cells into guard cells ↓ | Water moves from guard cells into neighboring epidermal cells ↓ |
|  | Guard cells swell ↓ | Guard cells wilt ↓ |
|  | Stoma opens | Stoma closes |

**22.6** A summary of the mechanism of the opening of stomata in light and their closing in darkness.

outer surface of the epidermis. The upper epidermis has a thicker cuticle than the lower epidermis. In general, plants growing in dry climates have thicker cuticles than those growing in wet climates. Some plants, such as water lilies, which grow in water have virtually no cuticle.

## External Features

The leaves of most plants are divided into two parts: the **blade,** which is flat and usually fairly broad, and the **petiole,** which is usually a cylindrical stalk that attaches the leaf to the stem (Fig. 22.7). In some plants the petiole may be flattened, with its lower portion clasping the stem; in this

case it is called a leaf sheath. Sessile leaves are those that lack petioles and leaf sheaths.

Deciduous plants are those plants that lose all their leaves at one time of year; for instance, most broad-leaved trees and shrubs of the Temperate Zone lose their leaves in autumn, and a new crop of leaves forms the following spring. Evergreen plants retain some leaves throughout the year; individual leaves, however, do not live throughout the life of the plant. In different species a single leaf may live two years, three years, or longer.

### Leaf Modifications

Some leaves are not photosynthetic but are modified in such ways that they perform other functions. These modifications increase the ability of a plant to survive in what might otherwise be a hostile environment. The buds of many trees and

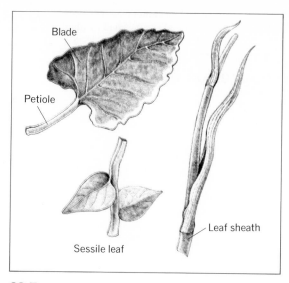

Blade

Petiole

Sessile leaf

Leaf sheath

**22.7** Three variations in leaves.

When leaves fall, they do not break off haphazardly. Rather, they break at a preformed zone, called the *abscission zone*. This zone forms in the base of the petiole, near the stem. As the zone develops, part of it becomes weak and increasingly unable to support the leaf. The leaf then breaks off in this area, leaving a scar (**leaf scar**) on the stem. The leaf scar consists of corky tissue that seals the break, protects the stem from infection, and prevents loss of water from this area.

## Economic Significance of Leaves

shrubs are covered with bud scales, which are tough leaves that protect the delicate stem tips. Leaves modified into tendrils help support climbing plants, and leaves modified into spines often protect plants from being devoured by animals. Onion bulbs consist of a series of fleshy, food-storage leaves. Desert plants sometimes have thick, water-storage leaves.

### Leaf Fall

Leaf fall in deciduous trees reduces the amount of water lost by transpiration, which otherwise would occur rapidly from flat leaves in the winter months when the air is very dry, and soil water may be frozen and so unavailable to the roots. Leaf fall also decreases the burden of snow and ice that deciduous trees bear in the winter. Nearly all the evergreen trees of the Temperate Zone have needle-shaped leaves, which, because of their more compact shape, lose less water by transpiration and hold less snow than a typical flat, deciduous leaf does.

Although the leaves of flowering plants are the ultimate source of most of our food, relatively few of them are major items in the human diet, but those that are edible are valuable as vitamin sources. Among those eaten by man are cabbage, chard, dandelion greens, lettuce, spinach, and watercress. The large, thick petioles of celery and rhubarb are edible, and the blades of the former may be consumed; but the blades of rhubarb leaves are poisonous. Leaves valued as spices or other flavoring materials include bay, marjoram, peppermint, spearmint, thyme, and parsley.

Atropine and belladona are drugs extracted from the leaves of the belladona plant; the leaves of the South American coca shrub and the ornamental foxglove yield cocaine and digitalis, respectively. The dried leaves of the tobacco plant are smoked in cigars, cigarettes, and pipes; and tea leaves are brewed to make a mildly stimulating beverage.

Leaves of grasses are not eaten by man, but when consumed by livestock, they may become transformed into milk, beefsteak, or lamb chops.

203

## Questions

1. How does the shape of a leaf contribute to its main function of photosynthesis?
2. How does the internal structure of a leaf contribute to photosynthesis? How does the leaf receive everything it needs for photosynthesis?
3. Explain the action of the guard cells in regulating the opening and closing of stomata.
4. In relation to the carbon cycle, what is the ecological significance of leaves?
5. Explain the following statement: Most of our food comes from leaves, but most of our food does not consist of leaves.
6. Explain why the removal of all the leaves from a plant may kill it.
7. List some nonphotosynthetic modifications of leaves and indicate how they aid in survival of plants.

## Suggested Readings

Baker, H. G. *Plants and Civilization.* Belmont, Calif.: Wadsworth Publishing Co., Inc., 1965.

Esau, K. *Anatomy of Seed Plants.* New York: John Wiley & Sons, Inc., 1960.

Jacobs, W. P. "What Makes Leaves Fall?" *Scientific American* 193(5): 82–89 (November 1955).

Ray, P. M. *The Living Plant.* New York: Holt, Rinehart and Winston, Inc., 1963.

Salisbury, F. B., and R. V. Parke. *Vascular Plants: Form and Function.* Belmont, Calif.: Wadsworth Publishing Co., Inc., 1964.

The root typically is the underground portion of the plant. Roots grow vertically downward and sometimes horizontally, but rarely upward. Their primary functions are anchorage and absorption of water and minerals. A secondary function is the storage of excess food synthesized in the leaf and transported down to the root by the stem.

Externally the root is fairly simple. It branches more or less extensively, the branches increasing its ability to keep the plant firmly anchored. Near the tip of each branch is a group of **root hairs** (Fig. 23.1); these are very fine cylindrical extensions of the epidermal cells. They greatly increase the surface area of the root, and they represent most of the absorbing surface of the root. One rye plant that could easily be held in one hand by a child had 14 billion root hairs and a total surface area of over 6,800 square feet, of which almost two thirds was the surface of the root hairs. As a root grows through the soil, the oldest root hairs, those that are farthest from the tip, die; and new hairs develop near the tip. In this way new hairs are formed in soil not previously occupied by older ones, and the root hair zone gradually moves through the soil, although the individual hairs are stationary.

Absorption of minerals and water from the soil is by diffusion, osmosis, and active transport (Chapter 18). The cell walls of root hairs are very thin, and the hairs are very delicate. Plants that are uprooted so roughly as to have their root hairs torn off do not transplant easily, for they have lost most of their absorbing surface. Those that are dug up carefully with the root hairs preserved intact are much more likely to survive.

It is through these root hairs and epidermal cells that minerals pass from the physical world into the biological one. Just as photosynthetic leaves represent our main source of carbohydrates, fats, and proteins, so are roots the ultimate biological source of most of our minerals. The roots absorb minerals from soil, and minerals move to other parts of the plant. Herbivorous animals obtain their minerals from the plants they eat, and the

# 23

# The Root: Primary Growth

1 The main functions of the root are the anchoring of the plant and the absorption of water and minerals.

2 Absorption occurs primarily through the root hairs.

3 Growth in length of a root is due to the activity of its apical meristem.

4 Many roots are food-storage organs.

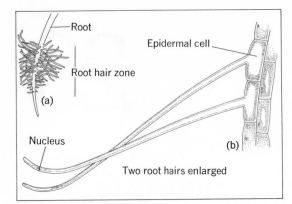

**23.1** (a) Root tip showing root hair zone. (b) Small portion of root epidermis showing two root hairs.

carnivores in turn receive them from the herbivores (Chapters 13 and 14). Nearly all the minerals we eat (except those in seafood) were extracted from the soil by the roots of flowering plants.

## Internal Anatomy of Roots

If a root tip is sectioned longitudinally (along its longest axis), four regions are visible: root cap, apical meristem, region of elongation, and region of differentiation (Fig. 23.2). These regions are not sharply delineated from each other but tend to overlap somewhat.

The **root cap** protects the delicate root tip from damage as the root grows among hard soil particles. The outermost cells of the root cap slough off as they die; they are replaced by new cells formed in the apical meristem.

The **apical meristem** is the tip of the root proper, and it contributes to growth in length of the root. Here the cells are perpetually young and embryonic. They are relatively small, have only thin primary cell walls, and have dense protoplasm with minute, almost invisible vacuoles. The cells divide frequently and regularly by mitotic cell divisions. Some of the daughter cells remain in this

area and perpetuate the apical meristem; the other daughter cells are contributed to the root cap and to the region of elongation. Although some cell divisions may occur in other parts of the root, the apical meristem is the ultimate source of all new root cells.

Above the apical meristem is the **region of elongation.** Here the cells contributed by the apical meristem elongate, the axis of elongation coinciding with the long axis of the root. There is little or no increase in width of the cells. The cell walls remain thin, but the vacuoles become so large that they merge, forming one large vacuole that fills nearly the entire cell, and squeezes the cytoplasm into a thin layer against the cell wall. Much of the growth of cells in this area is due to the increased volume of the vacuoles; and this in turn is due to

**23.2** Longitudinal section through a root tip.

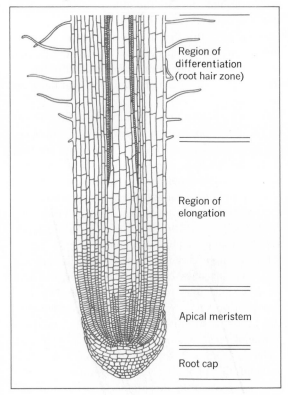

Region of differentiation (root hair zone)

Region of elongation

Apical meristem

Root cap

the absorption of water by the root. The elongation of cells in this region pushes the apical meristem deeper into the soil.

The apical meristem and the region of elongation together usually are not much more than a half inch long, but they account for all the growth in length of a root, with the apical meristem providing the cells that elongate in the region of elongation. If the apical meristem is removed, those cells in the region of elongation that have not yet reached their full size may continue to grow, but elongation of this region soon ceases if no new cells are provided. This is another reason why care should be taken to preserve roots intact when transplanting plants. The growth in width of roots is due to the activity of lateral meristems (Chapter 25).

All the cells in the apical meristem look rather similar. Except for differences in size, all those in the region of elongation are similar to each other also. But above the region of elongation, in the **region of differentiation,** cells undergo modifications into several cell types. In the center of this region is a cylinder of vascular tissue consisting of xylem and phloem cells (and sometimes pith cells). Then progressing to the outside there are, in order, pericycle, endodermis, cortex, and epidermis.

In a cross section of a dicot root in the region of differentiation (Fig. 23.3), the central mass of xylem appears to be star-shaped, the star often having four or five points, but there may be more or fewer. The phloem exists as several strands, as many as the points of the xylem and lying between them. Surrounding the xylem and phloem is a single layer of pericycle cells; these are the only cells in the region of maturation that exhibit meristematic characteristics. External to the pericycle is a single layer of endodermal cells; these cells have bands of lignin or suberin in their cell walls that force water passing across this tissue to pass across the plasma membranes. Because the membranes are differentially permeable, this provides some screening of the materials that reach the xylem and move upward to the rest of the plant. Beyond the endodermis is the cortex, which con-

**23.3** (a) Cross section of a buttercup root in the region of maturation. A dicot root typically has a central core of xylem. (Triarch Incorporated, P.O. Box 98, Ripon, Wisconsin 54971.) (b) An outline drawing of the same root with regions labeled; endodermis enlarged.

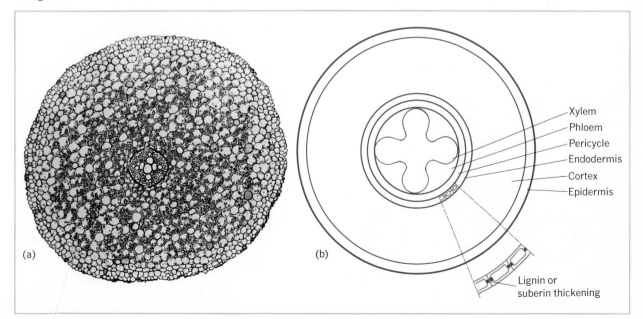

(a)

(b)

Xylem
Phloem
Pericycle
Endodermis
Cortex
Epidermis

Lignin or suberin thickening

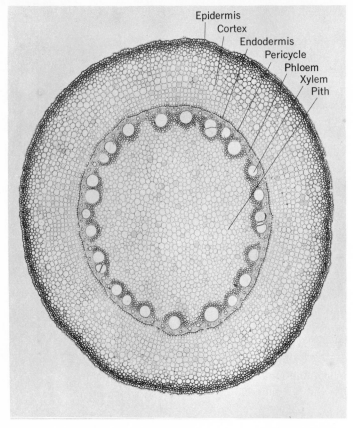

Epidermis
Cortex
Endodermis
Pericycle
Phloem
Xylem
Pith

**23.4** A cross section of a corn root. A monocot root typically has a central pith surrounded by alternating strands of xylem and phloem. (Courtesy Carolina Biological Supply Company.)

sists of several layers of cells; this is a storage tissue, and the cells may be filled with starch grains. The outermost tissue is the single layer of epidermis, through which water is absorbed. There

**23.5** Origin of branch roots. Branch roots form from the pericycle of a mature root and grow outward through the endodermis, cortex, and epidermis. Branch roots usually develop opposite the arms of the xylem.

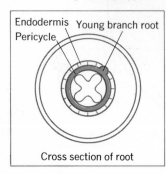

Endodermis  Young branch root
Pericycle

Cross section of root

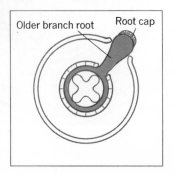

Older branch root   Root cap

are epidermal cells that fit together tightly, with no intercellular spaces among them; this provides a barrier that prevents the entrance of many microorganisms. Unlike leaf epidermal cells, they have no cutin that would impede the entrance of water.

The most conspicuous feature of epidermal cells in the region of differentiation is their root hairs. The youngest hairs, which occur low in the region, appear as mere bulges on the external surfaces of the epidermal cells. Above them, the older root hairs are much-elongated, tubular extensions of their cells. Root hairs do not occur in the region of elongation or in the apical meristem; if they did, they would undoubtedly become damaged as they were dragged downward by the cells being pushed downward by the elongation of cells above them. Root hairs are delicate and do not live long; the uppermost hairs, for example, die after only a few days of existence. The root hair zone, therefore, often is less than an inch long.

Monocot roots are similar to dicot roots, but the center of the root usually is occupied by pith, a parenchymatous tissue. Alternating strands of the xylem and phloem surround the pith (Fig. 23.4).

## Origin of Branch Roots

Branch roots arise from the pericycle. The cells of this tissue remain meristematic, and a few of them can organize a new apical meristem (Fig. 23.5). This is the apical meristem of a branch root, and it acts like the apical meristem of its parent root. It forms its own root cap and contributes cells to its own region of elongation. The new branch root grows outward across the endodermis and cortex of the parent root, and when it breaks through the epidermis it becomes visible externally. Later it may develop its own branch roots.

New roots may also arise from stems and occasionally from leaves or other organs. Such roots are called *adventitious* roots. You can see them growing from geranium or ivy cuttings placed in water.

## Symbiotic Relationships of Roots

Most flowering plants can survive without organisms in intimate contact with them. In nature, however, they are in contact with a great many organisms, especially soil organisms about their roots. Compounds produced by many soil microorganisms are absorbed by the roots of flowering plants, which may not require these substances but which flourish much better with them than without them.

Among these soil microorganisms are fungi that grow in thick layers around young roots or in them. The association is called a mycorrhiza (Chapter 4). Mycorrhizal fungi are especially valuable to plants growing in sandy soils where the mineral content is low or in acid bogs where minerals are not readily absorbed by roots. In some cases the fungi absorb minerals and water and even organic compounds from the soil and transfer them to the roots; in other cases the fungi may manufacture compounds that are absorbed by the roots. Nitrogen-fixing organisms (Chapters 4 and 13) are soil microorganisms that enrich the soil with nitrogen.

## Economic Significance of Roots

Most roots store at least a little food, but some are modified into thick, fleshy, food-storage organs that man finds edible. Among them are carrots, parsnips, turnips, beets, radishes, and sweet potatoes; these are eaten primarily as vegetables throughout much of the Temperate Zone. On the other hand, the staple foods in much of the tropics are starchy roots like sweet potatoes, cassava, and

yams; these occupy the place in the diet that cereal grains do in most of the rest of the world. Sugar beets are raised for their sugar (sucrose) content; they produce about one fourth of the world's supply.

Roots have yet another significance that is important to man: they greatly retard erosion, for the interweaving of the roots of many plants holds soil particles in place (Fig. 23.6). If all or most of the plants are removed from an area, rains easily wash away the layers of topsoil. Erosion often is encouraged by farming, grazing, and lumbering. A farmer

**23.6** Soil-holding capacity of roots. This bunch of grasses was uprooted from a gravel driveway. Notice the clump of earth surrounding the roots. This capacity of roots to bind soil prevents erosion by water or wind wherever there is a good cover of vegetation.

209

does not want weeds competing with his valuable crops for light, water, and minerals, and so he discourages the growth of as many weeds as he can and thus leaves bare soil exposed between the rows. A rancher, in the hope of getting as much meat per acre as possible, may crowd too many animals on his land, permitting them to eat plants down almost to ground level and to trample them; without their tops, the roots die. Lumbermen often cut down every tree in many acres of forest. These practices are open invitations to erosion. With no vegetation to slow its progress, falling rain strikes the bare ground with greater force and washes away soil particles no longer held by living roots. Millions of acres of topsoil have been eroded away in the United States alone. Farming, ranching, and lumbering are activities that must be carried on with careful conservation practices.

## Questions

1. What is the significance of the increased surface area contributed to a root by its root hairs?

2. Describe how a root grows in length. How do the following regions contribute to growth?

    root cap
    apical meristem
    region of elongation

3. Explain how the root hair zone can move downward in the soil when no root hair moves downward.

4. What is the advantage of having the root hairs above the region of elongation?

5. Why should a gardener be careful to keep a ball of soil around the roots of a plant he is transplanting?

6. What is the significance of roots in relation to the cycles of nitrogen, sulfur, and other minerals?

7. How do roots function in soil conservation?

8. What is the difference between a branch root and an adventitious root?

9. Explain why the removal of all the leaves of some plants does not necessarily kill them, and that they may not die even after the removal of the second crop of leaves. Do you understand why some weeds are so difficult to kill merely by cutting them off at the soil surface?

## Suggested Readings

Baker, H. G. *Plants and Civilization*. Belmont, Calif.: Wadsworth Publishing Co., Inc., 1965.

Esau, K. *Anatomy of Seed Plants*. New York: John Wiley & Sons, Inc., 1960.

Ray, P. M. *The Living Plant*. New York: Holt, Rinehart and Winston, Inc., 1963.

Salisbury, F. B., and R. V. Parke. *Vascular Plants: Form and Function*. Belmont, Calif.: Wadsworth Publishing Co., Inc., 1964.

Stewart, F. C. *Plants at Work*. Reading, Mass.: Addison-Wesley Publishing Co., Inc., 1964.

Stewart, W. D. P. "Nitrogen-Fixing Plants." *Science* 158: 1426–1432 (1967).

Torrey, J. G. *Development in Flowering Plants*. New York: Macmillan Publishing Co., Inc., 1967.

Went, F. W., and N. Stark. "Mycorrhiza," *BioScience* 18: 1035–1039 (1968).

Most stems are aerial organs that grow vertically upward, but their lateral branches may be oriented horizontally or in some intermediate position. Externally the stem is somewhat more complex than the root. It bears not only its own branches, but leaves, flowers, and fruits as well. Water and minerals absorbed by the roots are transported through the xylem of the stem to leaves, flowers, and fruits. Food synthesized in the leaves is transported by the phloem of the stem to any growing organ that may require it—stem tips, root tips, new leaves, flowers, and fruits; food may also be carried by the stem to the root or other food storage organs where it remains during dormant periods (such as winter or dry seasons), and it is transported by the stem upward to growing parts when the growing season starts.

## Internal Anatomy of Stems

The tip of the stem (Fig. 24.1), like the tip of the root, contains an apical meristem, but unlike the root tip it does not have a cap; rather, it is located deep in a bud which provides it protection. The growth in length of the stem is similar to that of the root; there is a region of elongation below the apical meristem, and a region of maturation below that. However, these regions overlap each other so much in the stem that it is even more difficult to draw lines of demarcation between them than it is in the root. The region of elongation is much longer than the one in the root, sometimes being several inches long. Together the apical meristem and the region of elongation account for growth in length of stems as they do for growth of roots. Growth in width of stems is due to the activity of lateral meristems (Chapter 25).

In the region of differentiation five kinds of tissues develop: pith, xylem, phloem, cortex, and epidermis. Aerial stems have no endodermis and no pericycle, but these are found in many under-

# 24

# The Stem: Primary Growth

1 The most important functions of the stem are the support of leaves, flowers, and fruit and the conduction of water, minerals, and food between these organs and the roots.

2 Growth in length of a stem is due to the activity of its apical meristem.

211

**24.1** Longitudinal section of the stem tip of a flax plant. Notice the small, densely staining cells in the apical meristem. Farther down, the cells have enlarged and have large vacuoles; these cells appear to be almost completely clear. The youngest leaf primordia do not have branch primordia. (Reprinted by permission from *Botanical Microtechnique*, 3rd edition, by John E. Sass, © 1958 by the Iowa State University Press, Ames, Iowa.

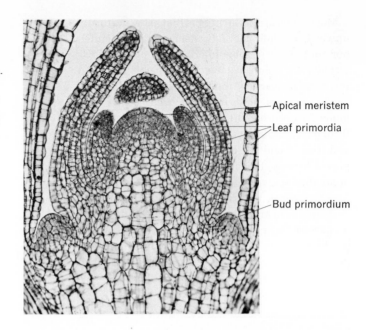

Apical meristem

Leaf primordia

Bud primordium

**24.2** Cross section of a dicot stem in the region of maturation. The section outlined is shown in detail in Fig. 24.3.

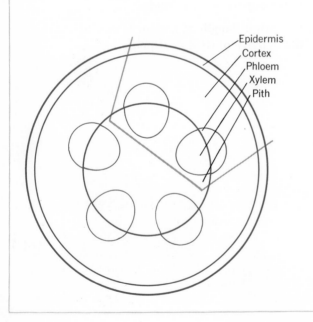

Epidermis
Cortex
Phloem
Xylem
Pith

ground stems. In the stem the vascular tissues exist as separate strands rather than as the single central cylinder of the dicot root. The vascular strands of dicot stems are arranged in a single ring (Figs. 24.2 and 24.3), but in monocots the strands are scattered throughout the stem (Fig. 24.4). Each vascular strand consists of xylem and phloem, with the xylem oriented toward the center of the stem and the phloem toward the outside.

The tissue in the center of the stem is pith, a parenchymatous tissue that resembles the cortex. In some plants the pith is a food-storage tissue; in others it may decay, leaving a hollow center in the stem. To the outside of the vascular strands is the cortex, which may store food; in young stems the outer part of the cortex often is photosynthetic. The epidermis surrounds the cortex.

Once tissues have matured they no longer elongate. Because the woody trunks of trees do not grow longer, their branches remain at a constant distance above the ground until they die and fall off. Signs nailed to trees and lover's messages carved into their bark also stay at their original heights.

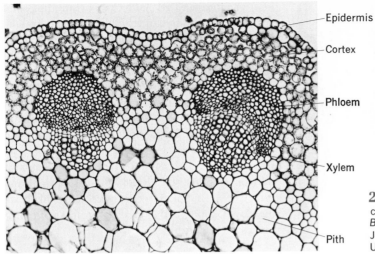

— Epidermis

— Cortex

— Phloem

— Xylem

— Pith

**24.3** A portion of a cross section of a clover stem. (Reprinted by permission from *Botanical Microtechnique,* 3rd edition, by John E. Sass © 1958 by the Iowa State University Press, Ames, Iowa.)

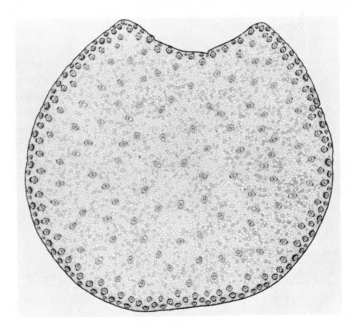

**24.4** Cross section of a monocot stem. This corn stem shows scattered vascular bundles. (Courtesy Carolina Biological Supply Company.)

### Origin of Leaves and Branches

Leaves and buds arise from the surface of the stem. A very short distance below the apical meristem, small mounds of tissue develop and grow into young leaves or leaf primordia (Fig. 24.1). The youngest primordia are those closest to the apical meristem; older ones occupy positions farther down the stem. When conditions are appropriate—such as the coming of spring—the leaf primordia expand into leaves. Too small to be viewed comfortably with the naked eye, the apical meristem and its youngest leaf usually are observed in the form of microscope slides that show

213

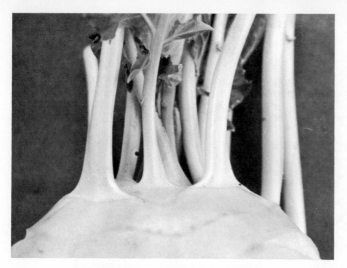

**24.5** The large stem tip of a kohlrabi plant and its leaves resemble the small apical meristem and leaf primordia of most stems. The youngest leaves are crowded near the tip. Several of the lower leaves have been removed to make some of the young ones visible. The leaves shown here are all mature leaves and not leaf primordia. For the most part, only their petioles are visible, the leaf blades extending above the edge of the photograph.

leaf primordia. This with its surrounding bud scales is a lateral bud; it is actually a small branch. The tip of the main stem terminates in a similar bud, called a terminal bud. In trees and shrubs of the Temperate Zone, the terminal bud opens in the spring and expands into a new segment of stem. In late summer or fall a new terminal bud forms; it remains dormant over the winter and opens in the spring. Many lateral buds remain dormant and never develop any further; others grow into branches. Whether or not they open depends in part on whether the terminal bud is present and intact; if it is not, branches are more likely to develop (Chapter 29). This is why pruning of plants makes them shrubbier.

## Annual Growth in Length and Bud Scale Scars

Because the dropping of each bud scale leaves a scar on the surface of the stem, and because all the bud scales of a single bud are located close together, the opening of one bud leaves a ring of bud scale scars around the stem (Fig. 24.6). The segment of stem between two successive sets of bud scale scars on a twig represents one year's growth of that twig. Variations in environmental conditions and in the heredity of the plant influ-

only two dimensions of the stem tip. The large stem of kohlrabi shown in Fig. 24.5 can serve as a three-dimensional model of an apical meristem.

Immediately above all but the youngest leaf primordia, a small mound of stem tissue becomes oriented into a new apical meristem with its own

**24.6** External features of a stem.

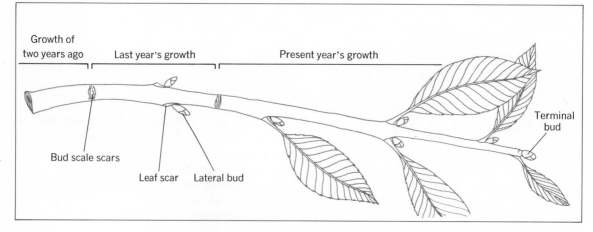

Growth of two years ago · Last year's growth · Present year's growth

Bud scale scars · Leaf scar · Lateral bud · Terminal bud

ence the amount of a year's growth. In a poor year an apple twig may grow not even a quarter inch in length, whereas willow twigs and tree-of-heaven twigs may grow 5 or 6 feet or even longer in a favorable year. Vines frequently grow even more in a single season.

# Economic Significance of Stems

Several stems are the source of important food products. Sugar cane, with the juice of its stems rich in sugar, is the major source of table sugar; it produces about 75 per cent of the world's supply of sucrose. The white potato (Irish potato) is a starchy tuber, an enlarged fleshy portion of an underground stem. A native of South America, the white potato was introduced in Europe, where it has become a staple food. Kohlrabi is an enlarged portion of an aerial stem. The stems and unopened flower buds of broccoli are eaten. Onions, garlic, chives, and leeks are bulbs; these are usually classified as underground stems, but they actually are large buds consisting mostly of leaves that arise from a very small amount of stem tissue at the base of the bulbs.

The phloem of flax, hemp, and jute stems yield commercially important fibers. Of these, flax has the highest quality and is used in the manufacture of linen cloth and linen thread for sewing and embroidery. Hemp is coarser than flax; it is used to make both cloth and rope. Jute, which is still coarser and also weaker, is used in the manufacture of burlap and other cloth of inferior quality.

Some other stem products are mentioned in Chapter 25.

# Questions

1. What are the main functions of stems?
2. Compare the growth in length of the stem with that of the root.
3. Contrast the origin of buds and leaves with those of branch roots.
4. Compare a terminal bud with a root tip. How do the differences reflect the differences between growing in air and growing in soil?
5. Describe a lateral bud in terms of its being a small branch.
6. If someone indicated a point 6 inches below the terminal bud of a twig, could you determine the age of that portion of the stem?
7. If you were presented with slides of a cross section of a one-year-old root and a cross section of a one-year-old stem, could you distinguish them? How?
8. List some stem modifications and tell how they contribute to the survival of plants.
9. Some persons think that old leaves drop from the stem because new leaves push them off. Criticize this idea.

# Suggested Readings

Baker, H. G. *Plants and Civilization.* Belmont, Calif.: Wadsworth Publishing Co., Inc., 1965.

Esau, K. *Anatomy of Seed Plants.* New York: John Wiley & Sons, Inc., 1960.

Ray, P. M. *The Living Plant.* New York: Holt, Rinehart and Winston, Inc., 1963.

Salisbury, F. B., and R. V. Parke. *Vascular Plants: Form and Function.* Belmont, Calif.: Wadsworth Publishing Co., Inc., 1964.

Stewart, F. C. *Plants at Work.* Reading, Mass.: Addison-Wesley Publishing Co., Inc., 1964.

Torrey, J. G. *Development in Flowering Plants.* New York: Macmillan Publishing Co., Inc., 1967.

# 25

# Secondary Growth of Plants

1 Growth in width of roots and stems of perennial plants is due primarily to the division of cells of the lateral or secondary meristems: vascular cambium and cork cambium.

2 The vascular cambium produces secondary xylem toward the inside and secondary phloem toward the outside.

3 The cork cambium produces cork toward the outside and may produce a small amount of phelloderm (secondary cortex) toward the inside.

The apical meristems of roots and stems described in Chapters 23 and 24 are called primary meristems and the tissues that differentiate from the cells produced by them are known as primary tissues. Primary growth increases the lengths of these organs, but it contributes little to increase in width.

Growth in width is accomplished by **lateral** or **secondary meristems:** vascular cambium and cork cambium. These are not present in the primary tissues found in the region of differentiation, but in perennial stems and roots a narrow band of cells lying between the primary xylem and primary phloem remains undifferentiated for some time. As the first growing season advances, these cells differentiate into vascular cambium, which produces secondary xylem and secondary phloem. Cork cambium, which arises from the pericycle of the root and from either the epidermis, primary cortex, or primary phloem of the stem, produces mostly cork (the outermost tissue of most stems and roots more than a year old), but it may form a small amount of cortical tissue as well. Tissues produced by secondary meristems are called secondary tissues, and their formation contributes to the growth in width of stems and roots. Like primary meristems, secondary meristems continue to function throughout the life of the plant. Most monocots are annual plants and so do not have secondary growth; palm trees are one notable exception.

## Vascular Cambium

The vascular cambium of the stem begins its differentiation between the primary xylem and the primary phloem of the vascular bundles; then more vascular cambium cells differentiate between the pith and the primary cortex. Ultimately all these sections join up, forming a continuous hollow cylinder of vascular cambium. In a cross sec-

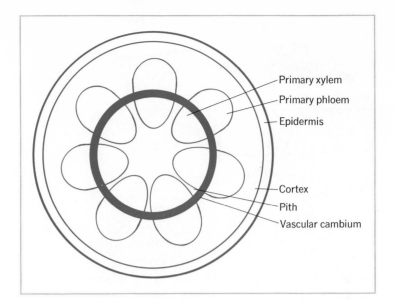

**25.1** Cross section of a stem at time of differentiation of vascular cambium. A portion of the vascular cambium is shown in Fig. 25.2.

tion of a stem, the vascular cambium appears as a ring (Fig. 25.1) only one cell thick. Like the meristematic cells of the apices of roots and stems, vascular cambium cells have only thin primary cell walls.

The vascular cambium functions by the division of its cells. The new cell wall between the two daughter cells of a division usually is oriented parallel to the outer surface of the stem (Fig. 25.2). The two daughter cells thus lie along the same radius. One of them remains a vascular cambium cell; the other becomes either a xylem or phloem

**25.2** Production of secondary tissues from vascular cambium. (a) Vascular cambium cells. (b) Division of vascular cambium cells produces secondary xylem cells and more vascular cambium cells. (c) A division like (b). (d) Division of vascular cambium cells produces secondary phloem cells and more vascular cambium cells. (e and f) After several divisions, the descendants of each vascular cambium cell form a long radial row. Young secondary xylem cells and young secondary phloem cells near the cambium may divide also, but as they age they lose the ability to divide. C = vascular cambium cell, 2°X — secondary xylem cell, 2°P = secondary phloem cell.

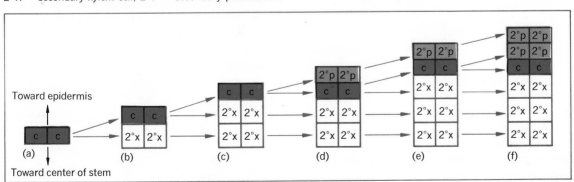

217

Vascular cambium                           Primary phloem

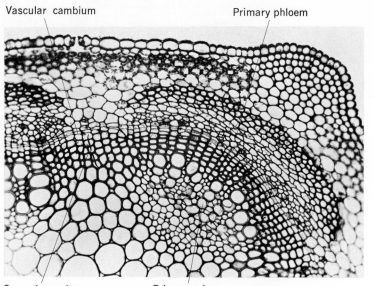

Secondary xylem                    Primary xylem

**25.3** Cross section of a portion of an alfalfa stem showing the beginning of vascular cambium activity. A continuous band of secondary xylem has been formed, but very little secondary phloem is visible. Notice the radial rows of secondary xylem cells. (Reprinted by permission from *Botanical Microtechnique,* 3rd edition, by John E. Sass, © 1958 by The Iowa State University Press, Ames, Iowa.)

cell. Most commonly the inner cell becomes a xylem cell and the other remains a vascular cambium cell; but sometimes the inner remains a vascular cambium cell, and the outer becomes a phloem cell. In this way the cambium continues to regenerate itself while producing new secondary tissues. The secondary xylem cells usually are larger than the secondary phloem cells, and much more secondary xylem forms than secondary phloem. Because the vascular cambium is a continuous tissue, the secondary xylem and secondary phloem that it forms are continuous tissues as well (Figs. 25.3 and 25.4). Unlike the primary vascular tissues of the stem, they do not occur in discrete bundles.

All the cells descending from a single cambial cell, both xylem and phloem, lie along the same radius. This radial arrangement usually is absent in primary tissues, but often it is conspicuous in secondary tissues, although it may not be so obvious where unequal growth of cells tends to obscure the rows.

The **secondary xylem** is the wood of trees and shrubs. When first formed, the tracheids and vessels have primary cell walls and living protoplasm,

but later they develop thick secondary cell walls, and the protoplasm dies, leaving the empty lumens typical of water-conducting cells. The tracheids and vessels formed in the spring are large, and their secondary cell walls are not so thick as those of the smaller tracheids and vessels produced later in summer. As the growing season advances, the activity of the cambium slows down, and late in summer or in autumn it ceases. No new secondary xylem cells form again until spring when the cambial activity is renewed. This seasonal change in the characteristics of secondary xylem cells causes the xylem tissue formed in the spring to appear light to the naked eye and the later-formed xylem to be darker. The color changes sharply between the late wood of one year and the early wood of the following year (Fig. 25.5). The alternating light and dark bands in the wood form the **growth rings** that enable one to estimate the age of a tree or one of its branches. One growth ring (consisting of both the light **spring wood** and the dark **summer wood**) ordinarily forms each year, although it is possible for two growth rings to form in a single year if a short period of unusually unfavorable environmental conditions, such as drought or an

**25.4** Cross section of a stem showing early secondary growth. The amount of phelloderm is small, and some stems may not have any. In older stems the pith, some of the secondary phloem, the primary phloem, and the primary cortex become crushed by the action of the two cambiums.

Cork
Cork cambium
Phelloderm
(secondary cortex) Bark
Primary cortex
Primary phloem
Secondary phloem

Vascular cambium
Secondary xylem (wood)
Primary xylem
Pith

**25.5** Cross section of the wood of hemlock. One growth ring and portions of two others are visible. The spring wood consists of large cells, the summer wood of small cells. Notice the abrupt change in size between the cells of the summer wood of one year and the cells of the spring wood of the following year.

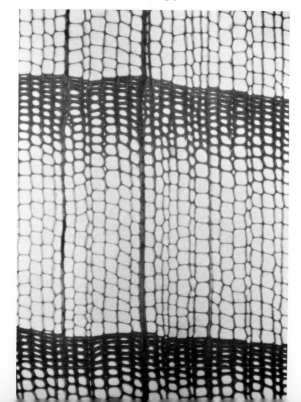

insect infestation, causes the temporary cessation of cambial activity during the growing season. The width of growth rings in secondary xylem reflects the character of the growing conditions during the time that that ring was formed, more favorable years producing wider rings. Tropical and subtropical trees also produce growth rings if conditions alternate between favorable and unfavorable for growth—wet and dry seasons, for example.

The youngest secondary xylem cells lie adjacent to the vascular cambium, the oldest near the center of the stem. As xylem cells age, their lumens often become plugged with a variety of organic materials, and the cells become nonfunctional, for they are unable to transport water. This central, nonfunctioning wood is the **heartwood** (Fig. 25.6). Usually it is noticeably darker than the surrounding functional **sapwood,** and it may rot away, leaving a hollow trunk. Some persons are surprised to learn that a hollow tree can survive for many years, but if only heartwood decays, no functional tissue is lost and the tree is not damaged.

In addition to the water-conducting tracheids

219

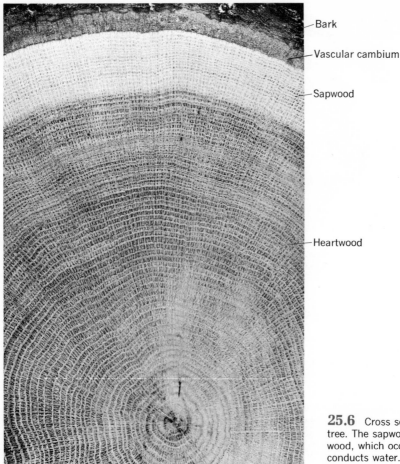

Bark

Vascular cambium

Sapwood

Heartwood

**25.6** Cross section of the trunk of a red oak tree. The sapwood appears light. The darker heartwood, which occupies most of the trunk, no longer conducts water. (Courtesy Carolina Biological Supply Company.)

220

and vessels, the wood of flowering plants contains xylem fibers, xylem parenchyma cells, and xylem ray cells. Although they are longer and thinner than tracheids, fibers otherwise resemble them in appearance; they have thick secondary cell walls and empty lumens. They do not conduct water, but they contribute to the strength of the wood. Wood parenchyma cells are small, inconspicuous living cells that store food. Arranged end to end in horizontal rows, the ray cells form the rays that are visible to the unaided eye and radiate from the center of the stem. These cells are alive, and they conduct food laterally through the wood.

To the lumberman, the woods of flowering plants (for example, oak, maple, birch, cherry, mahogany) are known as hardwoods. The softwoods come from coniferous trees (for example, pine, spruce, fir, redwood, larch), and the wood is simpler, consisting of only tracheids and ray cells. The grain of a particular wood depends on the kinds and arrangements of the several types of xylem cells. The growth rings feature largely in the grain. So, too, do the rays in some woods, such as oak, in which these are unusually large. The grain gives different woods characteristic patterns and contributes to their attractiveness.

**Secondary phloem** of the flowering plants consists of five types of cells: the sieve tube elements, which conduct food upward and downward in the stem or root; the companion cells, which are associated with the sieve tube elements; and phloem fibers, phloem parenchyma cells, and phloem ray cells, all of which have functions similar to those of their counterparts in the xylem. The phloem rays are continuous through the vascular cambium with the xylem rays.

Although the formation of secondary phloem is a seasonal process like that of secondary xylem, growth rings are not nearly so readily visible in the phloem because of several factors. The amount of secondary phloem produced each year is much less than the amount of secondary xylem, and the phloem cells are smaller, with less seasonal difference in size. Furthermore, the older phloem cells become crushed as they are caught between the activity of the vascular cambium and the cork cambium, which lies still farther out in the stem. Often it is impossible to distinguish individual cells in secondary phloem that is a few years old. The cortex similarly becomes crushed.

# Cork Cambium

The cork cambium differs from the vascular cambium in arising from a differentiated tissue. In the root it develops from the pericycle, which already has performed the function of producing branch roots. In the stem its origin is various; commonly it differentiates from the outermost row of primary cortex cells, those immediately under the epidermis. Sometimes it arises from deeper parts of the cortex, sometimes from the epidermis, and occasionally from the phloem. Regardless of its origin it resembles the vascular cambium in being a hollow cylinder only one cell thick. Its pattern of division is similar, the two daughter cells of any one division lying along the same radius. In most cases the outer daughter cell of any division of a cork cambium cell matures into a cork cell, and the inner daughter cell remains a cork cambium cell. Rarely the outer daughter cell remains a cork cambium cell, and the inner cell matures into a cortical cell (called phelloderm or secondary cortex); the amount of phelloderm produced ordinarily is very little (Fig. 25.4). As in the case of the vascular cambium, the secondary tissues produced by the cork cambium possess a radial arrangement of cells. Recently formed cork cells are alive, but the older cells have no living contents.

The activity of the cork cambium pushes outward any tissues that lie beyond it (epidermis, cortex, and endodermis of the root, and varying tissues of the stem, depending on the origin of the cork cambium). These tissues eventually die and are worn away, and the **cork** becomes the outermost tissue of stems and roots more than about a year old. Because the secondary cell walls in cork contain the waxy material suberin, cork is impermeable to water and makes an excellent outer covering that protects the plant from loss of water through the older surfaces of stems and roots.

Not only is cork tissue impermeable to water, it is impermeable to gases as well. Were it not for small openings called lenticels in the cork, this could result in the lack of oxygen and the accumulation of carbon dioxide in the interior of tree trunks. Like any living tissue, the live portion of a tree trunk requires oxygen for respiration and releases carbon dioxide. Many portions of a stem with secondary growth are distant from the stomata in leaves and therefore are not likely to exchange gases as quickly as necessary. Furthermore, in the winter even these openings are unavailable in deciduous trees, for the leaves drop in autumn and their scars remaining on the stem consist of a seal of corky tissue. The lenticels in the cork of the stem, therefore, allow for the escape of carbon dioxide and the entrance of oxygen.

Lenticels appear on the surface of tree trunks as small dots or transverse lines that look like

221

**25.7** Cambial activity resulted in the growth of this oak tree around a fence. Figure 25.7b was taken four years later than Fig. 25.7a and shows nearly complete encirclement of the horizontal bar.

(a)

(b)

small eruptions in the bark. Lenticels arise by increased activity in localized areas of the cork cambium; therefore lenticels extend only across the cork and do not penetrate to deeper tissues.

All tissues lying to the outside of the vascular cambium comprise the **bark** of the stem. These include the secondary phloem, primary phloem, primary cortex, phelloderm (if any), cork cambium, and cork. Because of its exposure to the environment, bits of bark always are being sloughed off. Sometimes these pieces include only the outermost part of the cork, but in other cases the cork cambium may be worn away as well. When this happens a new cork cambium usually regenerates from deeper tissues.

If someone removes the entire thickness of bark from a tree trunk, the vascular cambium usually comes with it. If only a small piece is lost, the tree usually is not damaged, for the lost tissue may be regenerated from the adjacent vascular cambium remaining on the trunk. If a complete ring of bark and vascular cambium is removed from the tree, the food manufactured in the leaves cannot be transported to the roots because there is a break in the phloem, and new secondary phloem cannot be formed because the vascular cambium is missing as well. For a while the roots remain alive, for they can utilize the store of food earlier transported to them by the phloem. During this time the roots absorb water, which the xylem transports upward; and the tree appears to be healthy. Eventually, the supply of food in the roots becomes exhausted and the roots starve. When the root hairs die, they no longer absorb water, and the upper parts of the plant die for lack of water. Extensive damage to forests by such girdling of trees may occur in hard winters when deer, porcupines, beavers, and other animals find bark their last source of food.

The activities of both the vascular and cork cambiums account for growth in width of stems; cambial activity may even result in the growth of stems around fences or other objects (Fig. 25.7).

## Economic Significance of Wood and Bark

Wood, which constitutes nearly the entire volume of the trunks of large trees, varies in its properties from species to species and so has been used for a variety of purposes. Woods from several species of pine, spruce, and redwood are used as structural timbers because of their strength and durability. In furniture and interior finishes the appearance of the wood is an additional consideration; the colors or patterns in the wood of oak, maple, walnut, and mahogany make them popular for interior uses. Balsa wood, which is extremely light in weight, has been used for boats and rafts, and it is familiar to any model airplane fancier. Whiskey is aged in barrels made of white oak, a wood which, because of its low porosity, holds liquids with a minimum of loss. It does not cause undesirable changes in flavor or odor of the whiskey, but the whiskey owes its color to the wood, which is charred before use. Lumber also finds use in railroad ties, fence posts, and utility poles.

Paper may be made from a variety of plant materials, but many papers contain at least some wood fibers. Coniferous trees, such as pine, spruce, and hemlock, are common sources of pulpwood in North America and Europe. Books, newspapers, paper bags, stationery supplies, and paper plates are only a few uses for paper. On paper we have preserved the discoveries, ideas, philosophies, and mistakes of earlier generations.

Other products derived directly or indirectly from wood include turpentine, wood alcohol, acetone, photographic film, cellophane, and polyethylene.

Because of its high organic content, wood is a good fuel, and once it was much used for this purpose. But many of the great forests, especially in the regions where the ancient civilizations flourished, have been destroyed, and the demand for energy in modern societies is satisfied to a great extent by the burning of coal, gas, and oil. Coal itself consists of fossilized woody plants.

The cork oak tree of the Mediterranean region is the world's commercial source of cork. Workers strip off its thick bark with care to avoid damaging the vascular cambium and killing the tree, which can reach the age of several hundred years and yield a new harvest of cork each decade. Because of its impermeability to water and many other liquids, cork is a popular material for bottle stoppers. (Corks must be cut so that the lenticels run transversely and not vertically. Examine a cork and you will see the lenticels appearing as long lines.) In life preservers, its buoyancy has saved many lives, and its many empty, dead cells make it excellent insulating material. It also finds use in bulletin boards, linoleum, and gaskets.

Rubber comes from various botanical sources throughout the world, but the Para rubber tree produces most of the world's rubber. Native to Brazil but cultivated successfully in the tropical Far East, its bark is slashed to collect the milky latex from which rubber is derived.

The traditional drug for treatment of malaria was quinine, a substance obtained from the bark of several species of quinine tree of South America. Newer drugs have replaced quinine partially.

Sassafrass and cinnamon are two flavoring materials obtained from bark. Bark on the roots of sassafras trees yields sassafras oil, which gives root beer its characteristic flavor. The highest-quality bark from the stems of cinnamon trees is cut into the small curled cinnamon "quills" of commerce. Ground cinnamon comes from bark of poorer quality.

Tannins are chemical compounds extracted from various plant sources, including wood and bark.

223

Because of their action on proteins, tannins are used in the tanning of animal hides, a process that converts them to leather. The wood of the quebracho tree of South America and the bark of oaks, hemlocks, and mangroves are important commercial sources of tannins.

# Questions

1. What contribution do lateral meristems make to the growth of plants?
2. Indicate whether the following are produced by apical or lateral meristems:
   primary xylem
   secondary xylem
   primary phloem
   secondary phloem
   epidermis
   cork
   pith
3. Describe the formation of a few secondary xylem and secondary phloem cells from vascular cambium.
4. Why is there much more wood than bark in a tree trunk?
5. What causes growth rings in trees?
6. Could you distinguish between primary and secondary tissues in a cross section of a stem? How does this correlate with their origin?
7. Distinguish between tracheids, vessel elements, xylem rays, xylem fibers, and xylem parenchyma. What are their functions?
8. Distinguish between sieve tube elements, companion cells, phloem rays, phloem fibers, and phloem parenchyma. What are their functions?
9. How much of a ten-year-old tree is ten years old? How much is one year old?
10. Why does it not matter that old secondary phloem becomes crushed in tree trunks?
11. Why does girding a tree kill it? Does death occur immediately? Why or why not?
12. Distinguish between heartwood and sapwood.
13. Why does it not matter if the center of a large tree trunk decays?
14. What is the function of lenticels?

# Suggested Readings

Baker, H. G. *Plants and Civilization.* Belmont, Calif.: Wadsworth Publishing Co., Inc., 1965.

Esau, K. *Anatomy of Seed Plants.* New York: John Wiley & Sons, Inc., 1960.

Fritts, H. C. "Tree Rings and Climate," *Scientific American* 226(5): 97–100 (May 1972).

Ray, P. M. *The Living Plant.* New York: Holt, Rinehart and Winston, Inc., 1963.

Salisbury, F. B., and R. V. Parke. *Vascular Plants: Form and Function.* Belmont, Calif.: Wadsworth Publishing Co., Inc., 1964.

Schulman, E. "Tree Rings and History in the Western United States," *Economic Botany* 8: 234–250 (1954).

Stewart, F. C. *Plants at Work.* Reading, Mass.: Addison-Wesley Publishing Co., Inc., 1964.

Williams, S. "Wood Structure," *Scientific American* 188(1): 64–67 (January 1953).

The movement of water within a flowering plant is largely a one-way affair. The soil ordinarily keeps the roots more or less well supplied with water. The leaves and other aerial organs, on the other hand, are exposed to the air, which often has a drying effect on them. As they lose water to the atmosphere, it is replaced by water absorbed by the roots and transported upward through the stem. We will consider first the loss of water by aerial parts, then the rise of water in the plant and the entrance of water into the roots.

## Transpiration

Transpiration is the loss of water in the form of water vapor from the plant. It occurs from all aerial organs, but by far the most water is lost from the leaves. Lesser amounts evaporate from flowers, fruits, and stems. Soil water enters the root, moves through the stem, and reaches the leaves. Along the way, a certain amount of water is utilized by the living cells in the several organs through which it passes, but most of it—up to 95 or even 99 per cent—evaporates from the leaves.

Transpiration involves two different processes: the evaporation of water from the wet surfaces of the internal cells of the leaf and the diffusion of the resulting water vapor out of the leaf. Some water may also be lost directly from the epidermis; the amount of this loss depends on the thickness of the cuticle. Aquatic plants, some of which have very thin cuticles on their leaves, are most likely to lose much water by cuticular transpiration if their leaves extend above the surface of the water; replacing the water presents no problem for these plants.

### Evaporation

Evaporation of water occurs whenever the surface of a body of water contacts air that is not

# 26

# Transpiration and the Movement of Water

1 Transpiration is the loss of water in the form of water vapor from the aerial portions of the plant, especially from the leaf.

2 The transpiration rate is increased by increases in temperature, wind velocity, and the area of the transpiring surface and by a decrease in the relative humidity of the surrounding air.

3 The rise of water in plants is due mostly to transpirational pull and in lesser amount to osmotic pressure developed in the roots.

4 Desert plants and evergreen plants of cold climates have adaptations that decrease the rate of transpiration and conserve water.

225

saturated with water vapor. Water molecules moving from the body of liquid water pass through its surface and into the air and become part of the water vapor of the air. Simultaneously some water molecules move in the opposite direction, but as long as the air is not saturated, the net movement of water molecules is from the liquid water to the air. This is evaporation. When the air becomes saturated with water vapor, the number of water molecules moving in one direction equals the number moving in the opposite direction, and evaporation ceases.

The rate of evaporation of water depends in part on the relative humidity of the air in contact with the evaporating surface. When the air holds all the water vapor that it can, it has a relative humidity of 100 per cent; such air is saturated. Because air can hold more water at high temperatures than at low temperatures, relative humidity is not an expression of the absolute water content of the air. At 5°C, air with a relative humidity of 90 per cent holds less water than air at 50°C with a relative humidity of only 10 per cent. The lower the relative humidity at a given temperature, the faster the rate of evaporation.

In addition to its effect on evaporation through relative humidity, temperature exerts an effect by influencing the speed of molecular motion. Only the most rapidly moving molecules in a liquid break through the surface and enter the air. A rise in temperature accelerates the rate of evaporation by increasing the speed of molecular motion, thereby increasing the number of molecules with sufficient speed to break through the surface.

Because evaporation occurs through a surface, the total amount of surface area affects the rate of evaporation as well. A certain quantity of water spread out in a shallow, uncovered tray evaporates much more rapidly than does the same quantity of water confined to a narrow tubing and exposed to the air only at the top of the tube. A thin film of water spread on a blackboard evaporates rapidly enough to give a good visual demonstration of the process.

## Diffusion

The center of a leaf is so constructed that there is a maximum of surface area exposed to the internal atmosphere of the leaf (Fig. 22.1). This permits efficient exchange of carbon dioxide and oxygen which is necessary for the process of photosynthesis, but it also allows the evaporation of a great deal of water from the walls of cells in the palisade and spongy parenchymas. During daylight hours when the stomata are open, the water vapor is free to diffuse out of the leaf to the external atmosphere. Because the internal atmosphere of a living leaf is saturated with water or nearly so, there usually exists a diffusion pressure gradient, the steepness of which depends in part on the relative humidity of the external atmosphere. On a dry day the diffusion pressure gradient is steep and the rate of diffusion is higher than on a humid day with the same temperature when the gradient is less steep. Any decrease in the relative humidity within the air spaces of the leaf resulting from diffusion of water vapor out of the leaf is only momentary, for evaporation from the internal cells keeps the relative humidity high there. When the external atmosphere is saturated, as in rain or fog, there is no diffusion pressure gradient between the air inside and outside, and so no water vapor diffuses from the leaf, and the transpiration rate falls to zero.

When moisture from transpiration accumulates in the immediate vicinity of the leaf, the diffusion pressure gradient becomes less steep and the rate of evaporation decreases. If there is a wind that blows away the moist air and replaces it with drier air, the diffusion gradient remains relatively steep, and the transpiration rate continues at a more or less constant rate. A very brisk breeze can remove a great deal of water from a plant and cause wilting.

By increasing both the rate of evaporation and the rate of diffusion, a rise in temperature increases the rate of transpiration. A hot, dry, windy day is the most dehydrating for plants. On such

226

days wilting is most likely to occur, but if wilting becomes severe, one of the many homeostatic mechanisms of plants comes into play. The guard cells themselves wilt, thus closing the stomata and conserving water by greatly reducing the subsequent transpiration rate.

Desert plants, which live in some of the most dehydrating environments known, have several adaptations that reduce water loss. Most of them have such thick cuticles on their leaves and stems that essentially no water evaporates from the epidermis. Many have stomata sunken into the leaf; the longer passageway between the internal and external atmosphere makes a less steep diffusion gradient and reduces the rate of transpiration. These features are shared with evergreen plants of the Temperate Zone, where cold winters bring extremely dry air. Many of these evergreens are conifers with compact, needle-shaped leaves that present less surface area to the air than do the broad, flat leaves of deciduous plants (Fig. 26.1). Some desert plants have physiological adaptations that enable them to withstand a certain amount of desiccation; some also drop their leaves in times of severe drought.

## The Rise of Water in Plants

The ability of plants, especially tall trees, to raise water to their uppermost parts long presented an intriguing problem to botanists who sought to understand the mechanism by which plants could support columns of water more than 300 feet high. Of the many theories advanced, one based on the strong cohesive forces among water molecules is held in greatest favor today. Water is not a very extensible substance; that is, it cannot be stretched. If suction is applied to a column of water (as, for example, when someone drinks through a soda straw), the column does not stretch,

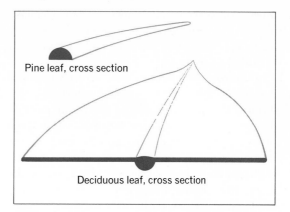

**26.1** Pine trees retain their needle-shaped leaves throughout the year, but deciduous leaves drop before winter. Partly because of its larger surface area per volume, the deciduous leaf would lose much more water in winter if it were retained.

neither does it ordinarily break. The cohesion of water molecules to each other maintains the integrity of the water column. This, together with the applied suction, draws water upward.

The evaporation of water from the mesophyll cells of the leaf reduces their water content, which is replaced by water that diffuses into them from nearby xylem cells by osmosis. The water lost by these xylem cells is replaced by water from adjacent xylem cells. Because functioning xylem cells are dead and have no protoplasm or cellular membranes, osmosis is not involved in the movement of water from xylem cell to xylem cell. As some water molecules leave the xylem cells of the leaf, adjacent water molecules take their place because of cohesion. They, in turn, are replaced by the water molecules below them. In this way, a "pull," called **transpirational pull**, develops in the xylem. It extends from the xylem of the leaf through the xylem of the stem and down to the xylem of the root, and it causes the upward flow of water.

Water moves upward through the tracheids and vessels in the xylem of flowering plants and through the tracheids of conifers. Tracheids have end walls that slow the passage of water from cell to cell; but vessels, which are essentially hollow

227

tubes, have no end walls when the vessel elements mature. For this reason the xylem of flowering plants is a much more efficient conductor of water than that of conifers.

In the root, osmosis again is involved in water movement. Water from the soil enters the root through the epidermal cells of the young parts of the root. The root hairs, which provide most of the surface area of the root, account for much of the water absorption. Water moves across the cortex, endodermis, and pericycle, and then enters the xylem. In each case, the passage is into and/or out of living cells by osmosis.

Movement of water by transpirational pull is along a diffusion pressure gradient, with the greatest diffusion pressure in the soil water surrounding the root:

Soil water.
Epidermis and its root hairs.
Cortex of root.
Endodermis of root.
Pericycle of root.
Xylem cells of root, stem, and leaf.
Mesophyll cells of leaf.
Internal atmosphere of leaf.
External atmosphere.

Freezing of soil water reduces its diffusion pressure to about zero and prevents its entrance into the root. Plants find themselves in a physiological desert in winter; the water lost to the cold, drying winds may not be replaced if the roots are in soil frozen for a long time. For this reason the water-conserving adaptations of the leaves of evergreens are especially useful to these plants. Deciduous plants adapt to winter (and to dry seasons) by dropping their leaves and thus greatly reducing their transpirational surface. The corky leaf scars do not lose water.

228 Under ordinary daylight conditions throughout much of the year, the water absorbed by roots barely replaces that lost by transpiration, or it may not even accomplish that. During this time the water in the xylem is under *tension* and is being

*pulled* up. When the amount of water absorbed exceeds that lost by transpiration, the water in the xylem is under *pressure*, called **root pressure**, and is *pushed* upward. Root pressure is responsible for the rise of relatively little water in plants, for it develops only when high humidity and/or closed stomata reduce the transpiration rate and abundant soil water is available to the roots. Such conditions are likely to occur only in rainy seasons (such as spring in much of the Temperate Zone) or at night. Plants that were in a wilted condition on a hot summer afternoon may revive by the following morning even if there was no rain, for the nightly closing of the stomata can reduce the transpiration rate below the rate of water absorption.

## Significance of Transpiration

Too many days and nights without rain can result in permanent wilting and death, for eventually the soil water becomes depleted, and nightly absorption cannot replace the daily loss. This is the annual worry of the farmer who has large plantings in unirrigated fields. A hot, dry summer can destroy his entire crop. During one growing season, an acre of corn may transpire more than 300,000 gallons of water; spread out over an acre, this water would stand 11 inches deep. When we consider that some rainfall is lost by runoff to streams and some evaporates directly from the soil, it is little wonder that short-grass prairies with only 10 to 15 inches of rain per year are damaged by attempts to farm them (Fig. 16.5).

A gardener transplanting plants often removes the lower, older leaves and any dead leaves. This reduces transpiration while the roots grow into the new soil. This is especially important if some of the roots have been damaged during transplanting, for the rate of water absorption then will be sharply but temporarily reduced.

Substances other than water move in xylem and phloem in the dissolved state. For this reason starch, fats, and proteins are good storage forms of food for they are not soluble in water and remain immobile in the plant. They may be digested to their components—sugars, fatty acids, glycerol, and amino acids—which are soluble in water and which therefore can move to a growing part requiring food.

Movement of organic compounds occurs in the phloem sieve tubes. During much of the growing season, when the leaves produce more food than the plant requires, the phloem transports to the growing parts whatever food they require, and it conducts the rest to the root (or other storage organ) where the excess accumulates and remains until it is needed. During the cold of winter a perennial does not grow, and there is little demand on the root for its stored food, but in early spring the first spurt of growth requires energy that can be provided only by the food stored from the previous year. Early in the season the plant has no green, photosynthetic leaves—only the leaf primordia that lie concealed in the leaf buds and that require an energy supply before they can expand and become photosynthetic themselves. At this time food moves upward through the sieve tubes.

The mechanism of food translocation in the phloem has not been adequately explained. It occurs too rapidly to be mere diffusion, nor is transpiration involved. Because it requires the presence of living cells, the protoplasm undoubtedly is involved in moving food.

Most minerals just absorbed by the roots move in the xylem, but those that have become incorporated into organic compounds move in the phloem.

The significance of transpiration in the life of the plant is difficult to find. Evaporation has some cooling effect on leaves, but even in hot sunshine, leaves that are not transpiring do not warm up appreciably. Transpiration increases the rate of water transport within the plant and the rate of water absorption by the roots (if it is available to them), but without transpiration such increases would not become necessary. Furthermore, water moves into water-deficient parts of the plant even though transpiration may have stopped. Transpiration does not seem to affect the rate at which minerals are absorbed, but it may facilitate their movement within the plant once they have entered. The risk of wilting, with its deleterious effects, seems to be the unavoidable consequence of having openings that permit the exchange of gases necessary for photosynthesis.

The cooling effect of transpiration may have at least as much effect on other organisms as on the transpiring plants themselves. The air around a transpiring plant is cooler than similar air not near a plant. In forests, where there are many plants, the effect can be appreciable. The discomforts of summer in a large city is due in part to the heat-generating activities of humans, and because there is little vegetation, the discomfort is not alleviated by the cooling effect of transpiration.

# Questions

1. What roles do evaporation and diffusion play in transpiration?

2. What factors affect the rate of transpiration? In what ways do they influence it?

3. Diagram a plant and show the path of water from tissue to tissue from the time the water enters the plant until it leaves. What factors are involved in its movement in each tissue?

4. What is the most widely accepted theory to explain the rise of water in plants?

5. What tissues function primarily in the transport of water and food in a plant?

6. Does transpiration seem to be beneficial or harmful to plants? Explain.

7. In what kinds of climates would it be beneficial to plants to reduce their transpiration?

8. Name some plant modifications that reduce transpiration.

9. Of what advantage is the dropping of leaves in winter to deciduous plants? What modification of evergreen plants help them to survive the winter?

10. Why is it a good idea to remove the older leaves from a plant when you transplant it?

11. What is the relationship between osmosis and root pressure?

229

**12.** If you must leave a potted plant unattended and unwatered for several days, why would covering it with a transparent plastic bag help it to survive?

**13.** Transpiration has been called a necessary evil. Explain.

# Suggested Readings

Crafts, A. S. *Translocation in Plants.* New York: Holt, Rinehart and Winston, Inc., 1961.

Devlin, R. M. *Plant Physiology,* 2nd ed. New York: Van Nostrand Reinhold Company, 1969.

Greulach, V. A. "The Rise of Water in Plants," *Scientific American* 187(4): 78–82 (October 1952).

Ray, P. M. *The Living Plant.* New York: Holt, Rinehart and Winston, Inc., 1963.

Salisbury, F. B., and R. V. Parke. *Vascular Plants: Form and Function.* Belmont, Calif.: Wadsworth Publishing Co., Inc., 1964.

Stewart, F. C. *Plants at Work.* Reading, Mass.: Addison-Wesley Publishing Co., Inc., 1964.

Zimmerman, M. H. "How Sap Moves in Trees," *Scientific American* 208(3): 132–142 (March 1963).

To many persons the most attractive features of plants are their flowers. Their beauty is equaled only by their diversity. Thousands of different kinds of flowers exist, ranging from the minute, inconspicuous flowers of willow and oak trees to the large, flamboyant blooms of orchids, but they all represent variations on a certain basic pattern. Whatever the variation, the function of flowers is sexual reproduction.

# 27
# Flowers

## Structure and Function of the Flower

A **flower,** which for convenience sometimes is referred to as a single organ, actually consists of several organs, each of which contributes in its own way to the function of sexual reproduction. Four different kinds of organs comprise most flowers: pistils, stamens, petals, and sepals (Fig. 27.1). Of these, only the first two are directly involved in sexual reproduction and for that reason are called essential organs. They produce the gametes (eggs and sperms) that fuse, forming a **zygote,** the first cell in the life of a new **embryo.** Eggs form within the pistil, the sperms within the pollen grains produced by the stamens. Sepals and petals, the accessory organs, contribute indirectly to reproduction.

Most commonly a single **pistil,** which is the female organ of the flower, occupies the center of the flower. Its lower, swollen portion is the **ovary,** which narrows into a short stalk, the **style.** The tip of the style is the **stigma,** which is receptive to pollen grains. The ovary contains one or several chambers, within each of which is one or more **ovules** attached to the ovary by a short stalk (Fig. 27.2).

The cells of a young ovule, like nearly all the other cells of the plant, have paired chromosomes. These diploid cells are formed by mitotic cell divisions (Chapter 19) which produce daughter cells

1  The flower is an assemblage of organs, the function of which is sexual reproduction.

2  The pistil, which is the female organ, produces haploid eggs in the ovules of the ovary.

3  The stamen, which is the male organ, produces pollen grains in which the haploid sperms form.

4  Fertilization of a haploid egg by a haploid sperm results in a diploid zygote.

5  As the zygote grows into an embryo plant, the ovule ripens into a seed and the ovary ripens into a fruit.

6  Fertilization is preceded by pollination—usually by either wind or insects.

7  Day length and temperature determine the time of flowering of many species.

231

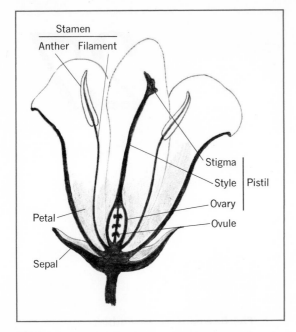

**27.1**  Section of a flower.

with the same number of chromosomes as the parent cell. This is true of the divisions that produce the pistil, but one of the cells within each ovule divides by meiosis, a division that happens

**27.2**  Cross section of an ovary showing six ovules. The number of chambers in the ovary and the number of ovules vary with the species.

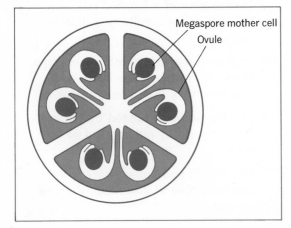

at some point in the life cycle of every sexually reproducing species. Chapter 41 describes meiosis in detail, but here we should note that one result of meiosis is the formation of cells with half the number of chromosomes of the parent cell; such cells are haploid. Meiosis actually consists of two divisions accompanied by only one duplication of the chromosomes. As a consequence, a diploid cell produces four daughter nuclei (often in four different cells), each with the reduced number of chromosomes. Meiosis is a necessary part of sexual reproduction, for it results in haploid eggs and sperms. The fusion of these cells produces a diploid zygote that develops into a diploid embryo by mitotic divisions. It grows into a mature plant by further mitotic divisions. Meiosis occurs again only in its pistils and stamens. In this way the chromosome number remains constant from generation to generation rather than doubling each generation as it would if meiosis did not alternate with fertilization.

Within each ovule is a single large diploid megaspore mother cell that divides by meiosis (Fig. 27.3). Of the four haploid daughter cells, called megaspores, three disintegrate and one remains viable. The nucleus of the surviving megaspore divides mitotically, producing two nuclei. These divide again, producing four, and the four divide once more. The result is an eight-nucleate cell called the **embryo sac,** because an embryo later develops within it. All eight nuclei are haploid because they were produced by the mitotic divisions from the haploid nucleus of the megaspore. Of the eight nuclei, one is the **egg.** Two **polar nuclei** occupy the center of the embryo sac. Only these three nuclei seem to play an important role in the life cycle. Among the five remaining nuclei are two synergid nuclei and three antipodal nuclei: they contribute little or nothing to further development. At the eight-nucleate stage the embryo sac is ready for fertilization. After fertilization, the ovule ripens into a seed. We will return to this later.

Below the pistil and surrounding it are the **stamens,** the male organs of the flower. Each consists

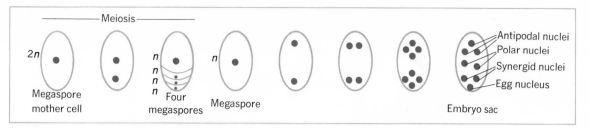

**27.3** Meiosis in megaspore mother cell and development of embryo sac from megaspore. Meiosis produces four megaspores from a single megaspore mother cell. Only one megaspore survives, and it develops into an embryo sac by mitotic divisions of its nucleus. The nuclei in the megaspores and in the developing embryo sac are monoploid. $2n$ = diploid; $n$ = monoploid.

of a slender **filament** topped by an **anther.** The four chambers of a young anther (Fig. 27.4) are packed with diploid microspore mother cells, each of which produces four haploid microspores by meiosis (Fig. 27.5). Each microspore matures into a **pollen grain** as its nucleus divides and one of the resulting nuclei divides again. The second division often occurs only after the pollen grains leave the anther and reach a receptive stigma. Of the three haploid nuclei, two are **sperm nuclei,** the male gametes of the plant; the third is the tube nucleus. The mature pollen grain is a very small, very simple structure. About the size of a speck of dust and barely visible to the unaided eye, pollen grains are produced by the dozens in a single anther, and the pollen grains produced by a single plant may number in the thousands or millions.

Mature pollen grains are released by the opening of the anthers. Wind or insects disseminate them, and some are carried to the stigmas of pistils of the same species. Here they germinate, forming long pollen tubes that grow down through the style

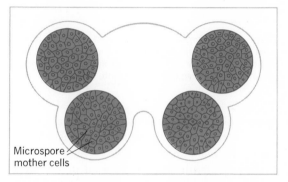

**27.4** Cross section of an anther. In most species an anther contains four groups of diploid microspore mother cells.

to the ovary. A pollen tube penetrates an ovule and discharges its sperm nuclei into it. One of these fuses with the egg nucleus, forming a diploid **zygote** (Fig. 27.6). This fusion of two gametes is called **fertilization** and is typical of sexual reproduction in both plants and animals. The zygote develops into a multicellular embryo within the

**27.5** Meiosis in microspore mother cell and development of pollen grain. Four microspores are produced by meiosis of a microspore mother cell. Each microspore matures into a pollen grain by mitotic divisions.

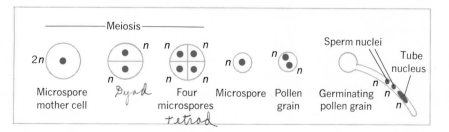

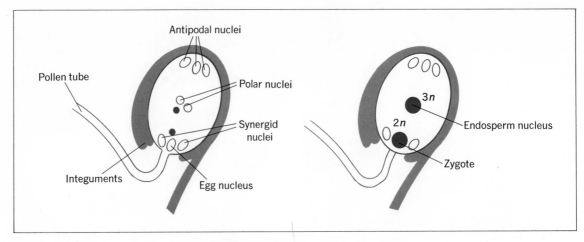

**27.6** Penetration of ovule by a pollen tube. One of the sperm nuclei (dark shading) fuses with the egg nucleus, forming a diploid zygote; the other sperm nucleus fuses with two polar nuclei, forming a triploid endosperm nucleus. When the ovule matures into a seed, the integuments ripen into seed coats.

ovule, which simultaneously ripens into a seed. In the absence of fertilization the ovule usually remains small and undeveloped.

A second fertilization, the fusion of the other sperm nucleus with the two polar nuclei, is peculiar to flowering plants and occurs in no other organisms. It produces a triploid endosperm nucleus; from this a triploid tissue called **endosperm** forms within the ripening seed. The endosperm is a nutritive tissue that in some species stores food that the seedling uses during germination. In other species the endosperm does not develop to any great extent, and food is stored within the embryo.

The **petals** grow immediately below the stamens. They have a flat, leaflike shape, and their colors range through almost endless shades and patterns. In insect-pollinated species the colors may serve as identification marks or as guides for the insects. **Sepals,** located below the petals, also have leaflike shapes, and they usually are green and photosynthetic. They contribute some food to the plant, and before the flower opens they protect the interior parts.

Flowers vary so much that there is hardly any that can be considered typical. The number of stamens, petals, and sepals often is three or five, and there usually is a pistil with five or fewer chambers within it; but the numbers of any of these flower parts may be fewer or may be in the dozens; some may even be lacking completely. Typically, monocots have their flower parts in multiples of three, whereas in dicots they are in multiples of five or four (Figs. 21.2 and 21.4). Flowers with either the stamens or pistils lacking are called imperfect; in some species both pistillate (having pistils but lacking stamens) and staminate (having stamens but lacking pistils) flowers may be on the same plant (as in corn, with pistillate flowers in the ear and staminate flowers in the tassel) or on separate plants (as in willow). The flower parts may all be separate from each other (buttercup, magnolia) or some of the parts may be fused. Any part may be fused to any other adjacent part, but the fusion of petals into a tube is conspicuous in some flowers (snapdragon, fuchsia). Petals and sepals may be so similar as to appear essentially identical (tulip, lily). Sepals, petals, and stamens originate below the ovary in some flowers (snapdragon, tulip, lily) and above it in others (apple, blueberry, orchid).

# Pollination

Wind and insects are the most common agents of pollination—the transfer of pollen from an anther to a stigma. Wind dissemination is most wasteful of pollen, for only a small fraction of it ever reaches the stigma of a flower of the same species. The dry, light pollen, which the wind blows about easily, is produced in great abundance by such species. Wind-pollinated flowers (oak, grasses) usually have small and inconspicuous petals and sepals that do not interfere with the dissemination of pollen (Plate VII). The flowers of oak open and shed their pollen early in spring before the leaves expand from their buds, and the grasses raise their flowers high above their leaves. The stamens may be long, exposing the anthers to the breeze, and the stigmas often are large—an adaptation that increases the likelihood of catching wind-blown pollen.

Insect-pollinated flowers (snapdragons, orchids), on the other hand, generally produce nectar or emit various odors that attract insects; large, conspicuously colored petals serve as guides for the insects (Plate VII). In some species the individual flowers are small, but they are massed into large inflorescences that mimic flowers (sunflower, goldenrod, milkweed). Their sticky pollen adheres to the body of the insect and then to the stigma of the flower next visited by the insect.

For genetic reasons (Chapter 42) cross-pollination usually results in the setting of a greater percentage of viable seeds than does self-pollination. It also produces greater variety among the off-

spring. Several mechanisms evolved by flowering plants ensure that pollen will be transferred from the anthers of one flower to the stigmas of another flower; often these flowers are on separate plants. Some species (date palm, willow, hemp) have imperfect flowers, the staminate and pistillate flowers being on different plants. In other species (beet, red clover, avocado) the anthers shed their pollen and the stigma of the same flower is receptive to pollen at different times; therefore, pollen that happens to reach the stigma of the same flower does not effect pollination. In still other species the pollen of a given plant is not compatible with stigmas of flowers of the same plant or even the same variety—if self-pollination occurs, the pollen does not germinate; or if it does, the pollen tube stops growing before it reaches an ovule. Several varieties of fruit trees are self-incompatible, and an orchard must have two varieties to ensure fruit set.

Insect-pollinated species have a variety of adaptations that tend to prevent self-pollination. Some orchid flowers are so constructed that once the insect enters, it cannot escape the same way it entered but must follow a one-way route that leads it first past the stigma and then past the anthers. Any pollen that it may have received on a earlier visit to another flower is brushed off onto the stigma. Then the insect picks up fresh pollen when it touches the anthers; because the insect does not pass the stigma again, self-pollination does not occur. Primroses have stamens and pistils of different lengths. Some of the flowers have long stamens and pistils with short styles; others have short stamens and pistils with long styles. An insect visiting a flower receives pollen at a position on his body that corresponds to the position of the stigma of some flower that it may later visit.

Not all species are self-incompatible. Tomatoes and some peas commonly are only self-pollinated. In many species either cross-pollination or self-pollination can occur, although the former is more likely to produce an abundance of seed.

235

# Timing of Flowering

Flowering in many plants is correlated with the time of the year. Hepaticas and violets, for example, bloom in spring, roses and lilies in the summer, and asters and witch hazel in autumn. In most species the lengths of the light and dark periods during the natural 24-hour day control the change from the vegetative to the reproductive state. Plants are classified as long-day plants if their blooming is initiated by long light periods and short dark periods. Short light periods and long dark periods stimulate flowering in short-day plants. Day-neutral plants bloom under most light regimes. Many summer-blooming plants are long-day plants, and most of those that flower in spring and fall are short-day plants.

Under the appropriate light regime at least some of the apical meristems of the stems cease elongating and producing new leaves and lateral buds and produce instead the pistils, stamens, petals, and sepals of flowers.

Knowledge of the effect of light on flowering makes it possible to bring many plants into bloom at any desired time of year merely by lengthening daylight with artificial lights or by placing plants in dark rooms several hours before sundown. Florists now can have appropriate plants at the height of their bloom in time for Christmas, Easter, and other occasions. Should it be desired, one could have poinsettias in July and chrysanthemums for Easter. Chapter 45 discusses in more detail the effects of long and short days on flowering.

Flowering is not independent of temperature, however, for the length of time elapsing between the initiation of flower buds (change of the vegetative meristem to a reproductive one) and the opening of the flower often is influenced by temperature, with a longer time being required at cold temperatures than at relatively warm temperatures. On the other hand, one or more short exposures to very low temperatures (often below freezing) are necessary to break the dormancy of some flower buds. Some fruit trees, apple and cherry, for instance, form their flower buds in autumn, and the buds remain dormant throughout the winter and open in the spring. The cold requirement for breaking of dormancy prevents flowers from opening in autumn when the resulting seeds might be killed by the cold of winter before they mature.

# Economic Significance of Flowers

The commercial uses of flowers depend largely on their beauty. They grace weddings, funerals, and other ceremonies of life. Bouquets and corsages convey sentiments to their recipients. Landscaping of private properties and public places relies to a great extent on flowers. The pleasant scents of some flowers not only enhance gardens but have commercial value; roses, orange blossoms, jasmine, violets, and other flowers are used in the manufacture of perfumes. People eat relatively few kinds of flowers, but the tips of broccoli consist of unopened flower buds, and cauliflower is a large cluster of aborted flower buds. The dried flower buds of cloves are used whole or ground as a spice in cooking, and when pickled, those of capers make a spicy condiment. The brewing industry utilizes the dried pistillate flowers of hops, which contribute the bitter flavor to beer.

More important than these, the significance of flowers lies in their function in the production of seeds and fruits from which most of our food and many other products are derived.

# Questions

1. What are the parts of a flower? How does each part contribute to the flower's function of reproduction?
2. Distinguish between pollination and fertilization.

3. What is endosperm? How is its origin related to that of the embryo? How does it differ from the embryo?

4. Compare the origins of the egg nucleus and the sperm nuclei.

5. What is the function of meiosis in the life cycle of a flowering plant?

## Suggested Readings

Davenport, D. "The Esthetics of Orchid Pollination," *Natural History* 77(4): 44–49 (April 1968).

Echlin, P. "Pollen," *Scientific American* 218(4): 80–90 (April 1968).

Esau, K. *Anatomy of Seed Plants.* New York: John Wiley & Sons, Inc., 1960.

Grant, V. "The Fertilization of Flowers," *Scientific American* 184(6): 52–55 (June 1951).

Naylor, A. "The Control of Flowering," *Scientific American* 186(5): 49–56 (May 1952).

Ray, P. M. *The Living Plant.* New York: Holt, Rinehart and Winston, Inc., 1963.

Salisbury, F. B. "The Flowering Process," *Scientific American* 198(4): 108–117 (April 1958).

———, and R. V. Parke. *Vascular Plants: Form and Function.* Belmont, Calif.: Wadsworth Publishing Co., Inc., 1964.

# 28

# Fruits, Seeds, and Seedlings

1 A fruit is a ripened ovary, and it contains one or more seeds.

2 A seed is a ripened ovule, and it contains an embryo plant. Some seeds also contain endosperm.

3 The embryo of a ripe seed consists of at least a stem tip, a root tip, one or two cotyledons, and usually a few vegetative leaves.

4 Most seeds enter a period of dormancy which must be broken by special environmental factors before the seeds can germinate.

5 Either endosperm or cotyledons contain a food supply used by the embryo during germination.

6 Some seeds, especially the cereals, are so rich in food that they have become the major food supply of the human population.

With a few exceptions, fruits and seeds develop only if pollination of the flower and fertilization of the egg and polar nuclei have occurred. As the embryo forms, the ovule ripens into a seed, its outer parts becoming the seed coat. Simultaneously the ovary ripens into a fruit. Like flowers, fruits come in many forms.

## Fruits and Seeds

Botanically, a **fruit** is a ripened ovary sometimes accompanied by other associated parts. The average person recognizes peaches, grapes, oranges, and bananas as fruit, but some other fruits are called vegetables in the supermarket—corn, peppers, string beans, watermelon, pumpkins, eggplant. The edible portions of most nuts are seeds, but the entire fruit, with its hard, dry coat, may be sold, as are peanuts with the shell. In the case of other nuts, only the seed finds its way to market, the fruit coat having been removed from almonds, Brazil nuts, and cashews. Sunflower "seeds" are fruits, too.

A few fruits consist mostly of tissues other than ripened ovary tissue. The conical stem of the strawberry flower ripens into the red, juicy edible portion, and the true fruits are the tiny flecks of "straw" that develop from the several pistils of the flower. The ripened ovary of an apple or pear is little more than the core of the fruit. The fleshy portion has been variously interpreted as stem tissue or as the fused, enlarged bases of stamens, petals, and sepals.

Not all fruits are edible for humans, or at least they are not generally acceptable in the diet (acorns, the winged fruits of maple trees, the milkweed pods, with their plumed seeds, the fruits of numerous weeds), and some fruits contain poisonous or otherwise unpleasant substances (deadly nightshade).

## Dissemination of Fruits and Seeds

The dropping of fruits and seeds in the immediate vicinity of the parent plant could lead to severe crowding and competition if all the seeds germinated in place. Most plants have adaptations that ensure dissemination of either fruits or seeds for distances ranging from only a few inches or feet to many miles. Although most of the fruits and seeds finally may come to rest in inhospitable places, the species benefit from the arrival of at least some of the embryos in distant places favorable for germination and growth. Wind, animals, water, and even the plants themselves serve as agents of dissemination.

Most wind-disseminated fruits and seeds bear structures that are likely to keep them airborne for at least a little time (Plate VIII). The fruits of maple, elm, and ash and the seeds of pine and spruce have wings that slow their descent from the parent tree and increase their chances of being blown away by the wind. Silky hairs act like parachutes and keep milkweed seeds and the fruits of dandelions airborne for long periods of time.

Some animal-disseminated seeds and fruits (Plate VIII) bear various hooks, spines, and barbs that become attached to the fur or feathers of passing animals only to drop off later. In autumn one can hardly walk through fields or woods without becoming a disseminator himself as cocklebur, beggar-ticks, or burdock fruits become attached to his clothes. Seeds of marsh plants become embedded in mud and then are transported when the mud sticks to the feet of birds or other animals. When cherries and some other edible fruits are eaten whole by an animal, the fleshy part is digested, but the seeds remain intact as they pass through the digestive tract, and may even require this passage through the animal, for the action of digestive enzymes on the hard seed coats renders them permeable to water which they must absorb before germination can occur. During the time they reside on or within the animal, seeds can be transported many miles from their point of origin. Birds

especially may carry seeds from island to island and sometimes bring new vegetation to oceanic islands. Squirrels bury walnuts, acorns, and other nuts; some of these are retrieved later, but others, left neglected, germinate where they were buried.

Water-disseminated seeds like coconuts are light enough to float. The loose fibrous fruit coat of the coconut with many air spaces makes it buoyant. Washed up on shores, the coconuts may germinate. Coconuts are not long-distance travelers, however. After a few days at sea, salt water seeping into the seed kills the embryo.

When the fruits of wild geranium dry, they break open in such a manner that the seeds are shot away from the parent plant. The ripe fruits of touch-me-not (jewelweed) snap open when they are touched, and they hurl their seeds away. The squirting cucumber develops an internal pressure that shoots out its seeds with explosive force.

Man has become an important international disseminating agent. He intentionally introduces crops or ornamental plants into new areas. Other seeds lie hidden in cargoes or cling to boots and clothing. Many of the wild flowers considered so typical of the American countryside are native in Europe and probably were introduced by settlers. Among them are Queen Anne's lace, daisy, teasel, and the ever-present dandelion.

## Dormancy and Breaking of Dormancy

Like winter buds, seeds of many species enter a dormant state in which they remain through periods unfavorable for growth. During dormancy there is essentially no growth, and the rate of respiration is almost imperceptible. The water content is low, perhaps about 10 per cent. In this dry state, the seeds are more resistant to environmental extremes than are actively growing plants. High or low temperatures that would kill photosynthesizing leaves may have little or no effect on the dormant embryo.

Some dormant seeds will not germinate if they merely receive water and a suitable temperature.

239

They require some special treatment to break dormancy, and then germination results if adequate water is available and the temperature is favorable. The factors that break dormancy are those appropriate for the particular species and its habitat. Alternate periods of hot and cold weather must precede the germination of some seeds. This adaptation prevents germination in autumn when the coming winter would kill the tender seedlings. Seeds of some annual desert plants contain a water-soluble substance that inhibits germination. A heavy rain that brings enough water to support a plant through its entire life cycle washes the inhibitor from the seeds and permits them to germinate. Small amounts of rain do not remove enough of the inhibitor to release the seed from its inhibition. This mechanism, sometimes called a "rain gauge," prevents seeds from being stimulated to germinate by only a little moisture and then dying when the meager water supply fails.

Seeds with hard coats impervious to water (honey locust, black locust) must be cracked before water can reach the embryo and stimulate its growth. Alternate freezing and thawing accomplishes this, or the seed coat may be abraded away by being tumbled among rocks. Digestion by enzymes secreted by some microorganisms may also render the seed coat permeable to water.

Lettuce seeds of the Grand Rapids variety require exposure to light before they germinate. While buried deeply in the soil they remain dormant even though all conditions necessary for germination except light may be present. Seeds very near the surface where they receive some light are likely to germinate. The amount of light required is exceedingly small, but the requirement keeps seeds from germinating if they are so deeply buried that the seedlings would be unlikely to reach the surface before they exhaust their food supply.

Rarely do all the seeds produced by a species in any one year germinate at the next appropriate season. If, by chance, an unseasonal cold or drought should kill the young seedlings, the seeds still remaining in the dormant state retain the ability to germinate when the appropriate season comes around again.

The length of time a seed remains viable while dormant varies with the species and with the individual seed. Rarely do seeds die within a few days of dissemination, but those of willow and poplar must germinate within a few weeks or they lose their viability. Most seeds remain viable for at least a few years, some for decades. Some lotus seeds determined to be 2,000 years old and lupine seeds believed to be at least 10,000 years old germinated when placed in a suitable environment, but these represent extreme cases.

Seeds of many cultivated plants have been bred for the ability to germinate quickly whenever they are planted. The highest percentages of germination result when these seeds are planted in the growing season following their production. Percentage of germination declines with succeeding years.

# Germination and Seedlings

Of the many kinds of fruits and seeds produced by flowering plants, we will consider the germination of only two in some detail: corn and bean. Corn is one of the cereal grains which are our basic food supply. Beans are legumes, the major protein source for those of the world's poor who cannot afford meat, fish, or other animal products.

## Corn

The fruit of the corn plant is a dry grain or kernel (Fig. 28.1). It contains only one seed, which is so tightly encased within it that the entire structure commonly is called a seed. The fruit coat and the seed coat are firmly attached to each other, and they surround the endosperm and the embryo which together fill the seed. In most varieties of

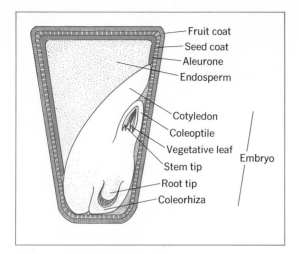

**28.1**  Longitudinal section of a corn kernel.

corn, including popcorn, the **endosperm** stores starch, but sweet corn owes its flavor to the presence of sugar in the endosperm. (Popped corn consists mostly of the white endosperm that expanded explosively under the pressure developed by heating.) The starch or sugar in the endosperm serves as a reserve food that is used by the embryo when it germinates. The outermost layer of endosperm, the **aleurone,** is rich in protein.

The very young **embryo,** formed by mitotic cell divisions of the zygote, is a small mass of undifferentiated cells, but as it grows it forms the vegetative organs typical of an adult plant. The main axis of the mature embryo in the kernel terminates in a stem tip at one end and a root tip at the other. Both contain apical meristems, and the root tip bears a root cap. The first leaf to form is the **cotyledon,** a leaf distinctly different from ordinary vegetative leaves. It arises from the side of the axis, but the primordia of the several vegetative leaves arise from the apical meristem. Protecting the root tip and its root cap until they emerge from the grain during germination is a sheath, the **coleorrhiza.** A corresponding sheath, the **coleoptile,** surrounds the stem tip and its vegetative leaves until they emerge.

Production of the starch-digesting enzyme amy-

lase in the endosperm precedes germination. The amylase catalyses the digestion of starch. The resulting sugar moves into the cotyledon and from there to the root and stem tips and the leaf primordia. Using the sugar as an energy source, these young organs resume their growth. The root emerges first from the grain, then the young leaves and stem come out (Fig. 28.2). As the growing tissues utilize the food that they drain from the endosperm, the endosperm shrinks in size; but before the food supply becomes exhausted, the leaves rise above the ground and begin to photosynthesize, producing food and making the young plant independent of the endosperm. The primary root of corn grows for only a short time and then dies. The stem, on the other hand, continues to grow to the height of several feet; it produces adventitious roots (roots that grow from some organ other than a root) and more leaves, and eventually flowers and fruits.

The starch-rich endosperm of corn and other cereals is a compact food supply. The mature

**28.2**  Young seedling emerging from a corn kernel.

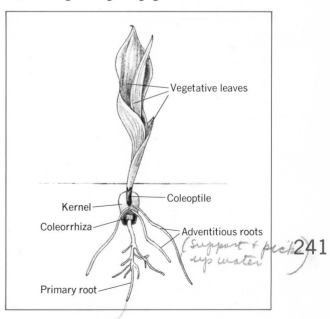

grains have little moisture, and as long as the starch is dry, it remains stable and does not digest to respirable sugars. For this reason the grains can be stored for a long time and can provide food during the winter or other hard times. It is possible that the major civilizations of the world might never have arisen if man had not learned to cultivate the cereal grains. These plants were the major food supply of the ancient civilizations—rice in the Far East, wheat and other grains in the Mediterranean area, and corn in Central and South America. Today these three plants account for about three fourths of the world's cereal production. Other important cereal grains are barley, oats, rye, millet, and sorghum. Some cereal grains, such as rice and sweet corn, may be eaten whole after cooking. The flour extracted from the endosperm of most cereals is used to make a variety of food products: breads of many types, breakfast cereals, cakes and cookies, waffles and pancakes, macaroni and spaghetti. Throughout most of the world, cereals are indeed the staff of life.

## Bean

The fruit of the bean plant is a pod containing several seeds (Fig. 28.3). Mature bean seeds lack

**28.3** Longitudinal sections of a bean fruit (a) and seed (b). The stem tip of the embryo is hidden between the vegetative leaves.

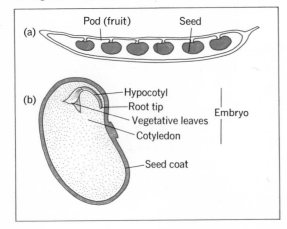

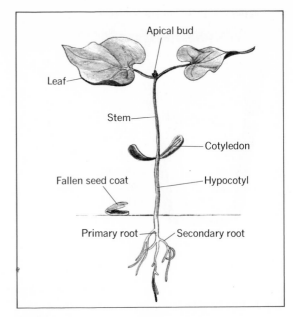

**28.4** Bean seedling.

an endosperm, and the embryo fills the entire seed. Like the corn embryo, the bean embryo consists of an axis with root and stem tips at opposite ends. The portion of the axis between the two tips is called the **hypocotyl.** Several young vegetative leaves are present. There is no coleorrhiza or coleoptile. Instead of one cotyledon, there are two attached to the main axis where the small stem and the hypocotyl join. The plump cotyledons are so large that the rest of the embryo occupies only a small space between them.

The main food supply of the bean embryo is stored in the cotyledons. Under conditions favoring germination, this food supply is mobilized and moves into the main axis of the embryo where it provides the energy necessary for the resumption of growth. Much of the early growth occurs in the hypocotyl, which elongates forcing the root tip through the seed coat. In beans the primary root persists and produces branch roots. The continued elongation of the hypocotyl raises the cotyledons and the small stem with its leaves above the ground (Fig. 28.4). The expansion of the cotyledons tears the seed coat, which may remain underground or may drop off later if it is carried upward by the cotyledons.

Gradually the cotyledons become smaller and more shriveled as food travels from them to the growing parts of the seedling. Eventually they drop off, but by this time the vegetative leaves have expanded and newer ones continue to be formed. These are green, and they assume the function of photosynthesis. Later, flowers form, and fruits and seeds are set.

Beans and other legumes have a high protein content. This is due to their possession of symbiotic nitrogen-fixing bacteria in their root nodules (Chapter 13). The bacteria incorporate the free nitrogen of the atmosphere into amino acids from which the protein is synthesized. The fruits of some legumes, such as string beans, are picked and eaten while the fruit and its seeds are still tender and not completely ripe and dry. The seeds of

others, like pinto beans, are picked dry and can be stored for long periods of time. Split pea soup and Boston baked beans are only two dishes made from legumes. Among the important legumes with edible seeds are garden peas, chick peas, peanuts, broad beans, lima (butter) beans, kidney beans, soybeans, and lentils.

Corn and beans represent only two of many variations of seeds, which range from the fine, almost microscopic orchid seeds to the large coconuts. The relative size of endosperm and embryo varies a great deal. In the ripe seed the embryo may range from the almost formless lump of cells with no distinctly differentiated tissues in many orchid seeds to the sprouting mangrove seedling with its root emerging an inch or more before the seed drops from the tree.

**Table 28.1**

| Some Seeds and Fruits Used as Food by Man* | | |
|---|---|---|
| Cereal grains | Brazil nuts | Citrus fruits |
| Corn | Macadamia nuts | Oranges |
| Wheat | Cashews | Lemons |
| Rice | Pistachio nuts | Limes |
| Barley | Chestnuts | Grapefruit |
| Rye | Almonds | Tangerines |
| Oats | Cucumbers | Mangoes |
| Millet | Watermelon | Papayas |
| Sorghum | Pumpkins | Coconuts |
| Legumes | Squash | Dates |
| String beans | Canteloupes | Figs |
| Garden peas | Avocados | Olives |
| Soybeans | Apples | Tomatoes |
| Peanuts | Pears | Pineapple |
| Kidney beans | Peaches | Guava |
| Lentils | Apricots | Buckwheat |
| Chick peas | Plums | Breadfruit |
| Walnuts | Quinces | Pomegranate |
| Butternuts | Cherries | Okra |
| Hickory nuts | Bananas | Cranberries |
| Pecans | Grapes | Blueberries |
| Hazelnuts (filberts) | Eggplant | Strawberries |

*The list does not include plants used primarily as beverages, spices, or flavoring materials.

# Economic Significance of Fruits and Seeds

The economic value of fruits and seeds lies mainly in their use as food, primarily for man himself and secondarily for his domestic animals. A few examples have been mentioned and Table 28.1 lists a few more.

Oils expressed from soybeans, peanuts, cottonseed, corn, and olives are all edible; they are used in salad dressings, as cooking oils, or in margarine. Oils extracted from flaxseed, coconuts, and castor beans are used in the manufacture of soaps, lubricants, paints, varnishes, and linoleum.

Fruits or seeds used as spices include allspice (pimento), anise, caraway, cardamom, celery seed, dill, mace, nutmeg, mustard, black pepper, red pepper, poppy seed, sesame, and vanilla. Coffee and cocoa are beverages obtained from the seeds of the coffee tree and the cacao tree, respectively.

Opium and its derivatives morphine and heroin are obtained from the fruit of the opium poppy.

243

Cotton, one of the most commonly used natural fibers, comes from the seeds of the cotton plant. The dry fruit, or "boll," contains several small seeds covered with long, single-celled fibers that not only fill the fruit but extend beyond it when the mature fruit cracks open. The strong fibers consist of almost pure cellulose and can be worked into threads that are used in the manufacture of cloth and cordage.

Aside from their immediate economic value to man, fruits and seeds are of great importance as structures that help to reproduce the species. In annual plants they are the only form in which the species exists during the winter or dry periods. In this resistant state the species manages to survive unfavorable periods.

# Questions

1. Define the following terms:
   fruit
   seed
   endosperm
   aleurone
   embryo
   cotyledon
   coleorrhiza
   coleoptile

2. Compare an embryo with a mature plant. How many resemblances do you find?

3. Why do seeds with a large endosperm usually not have embryos with large cotyledons and vice versa?

4. Why would it be a disadvantage for a corn seedling to have its cotyledon leave the seed when this does not harm a bean seedling? Explain in terms of the different functions of cotyledons in these two species.

5. Name some of the produce you would find on a fruit counter in a supermarket. How many of these are true fruits in the botanical sense? How many true fruits would you find on the vegetable counter?

6. Name some modifications of seeds and fruits that help to disseminate them.

7. Why will most seeds of the Temperate and Arctic Zones ordinarily not germinate in autumn? How is germination at inappropriate times prevented?

8. How does the economic significance of seeds and fruits compare with that of other plant organs?

9. What plant part is each of the following?
   stalk of celery with leafy top
   stalk of celery without leafy top
   bunch of celery
   peanuts with shells
   peanuts without shells but retaining red papery covering
   peanuts with papery covering removed
   beet
   rhubarb
   bean sprouts
   onion
   cherry
   broccoli
   Brussels sprouts
   white potato
   sweet potato
   sunflower seed

# Suggested Readings

Baker, H. B. *Plants and Civilization.* Belmont, Calif.: Wadsworth Publishing Co., Inc., 1965.

Esau, K. *Anatomy of Seed Plants.* New York: John Wiley & Sons, Inc., 1960.

Koller, Dov. "Germination," *Scientific American* 200(4): 75–84 (April 1959).

Ray, P. M. *The Living Plant.* New York: Holt, Rinehart and Winston, Inc., 1963.

Salisbury, F. B., and R. V. Parke. *Vascular Plants: Form and Function.* Belmont, Calif.: Wadsworth Publishing Co., Inc., 1964.

The fact that an organism as complicated as a flowering plant functions smoothly implies the existence of some regulatory mechanism that integrates the many processes going on within the plant. Hormones contribute to this integration by causing growth to start, stop, or proceed faster or slower and by regulating the differentiation of tissues and organs.

**Hormones** are defined as organic compounds that are produced in one part of an organism and that move to another part where they exert their effects. In this way they provide communication between the parts of an organism. The distance hormones travel may vary from exceedingly small distances to the length of an organ or even the entire organism. The mechanisms by which some hormones exert their effects have begun to become clear only recently, and this field undoubtedly will reveal much interesting new information in the future. Hormones act in such low concentrations that they must have enzyme-related effects. There is evidence that at least some hormones initiate the production of certain enzymes by influencing the genetic material that carries the genetic code for the synthesis of these enzymes (Chapter 43). Probably many plant hormones act in this fashion, but others well may be found to function in other ways.

Three groups of plant hormones will be discussed here in some detail: auxins, gibberellins, and cytokinins. Other plant hormones are known, and additional ones undoubtedly will be discovered.

## Auxins

The **auxins** probably are the best known of the plant hormones. They were discovered late in the 1920's and since then have been the subject of intensive research. The most common auxin in plants is indole-3-acetic acid (IAA); Figure 29.1

# 29

# Plant Hormones

1 The function of plant hormones is the integration of processes in the plant body.

2 The three main groups of plant hormones—auxins, gibberellins, and cytokinins—control cell division, cell elongation, and differentiation of tissues and organs.

3 Tropisms, which are growth movements of plants, are the result of uneven distribution of auxin in a plant organ in response to external stimuli, such as light and gravity.

4 The effects of hormones depend not so much on their absolute concentrations, but on their relative concentrations in a given tissue.

5 Homeostasis in a living organism is the maintaining of a constant internal environment despite external change.

245

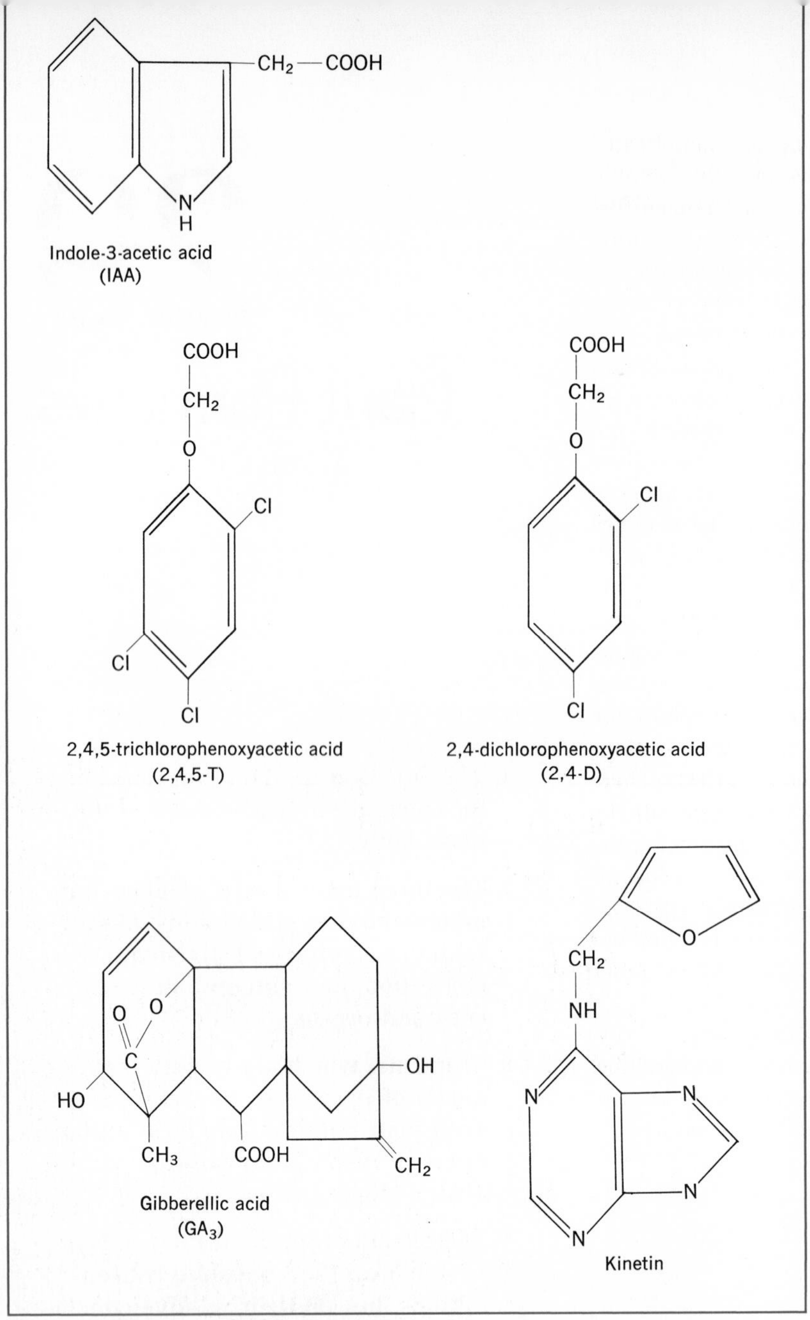
CH2—COOH
N
H
Indole-3-acetic acid
(IAA)
COOH
CH2
O
Cl
Cl
Cl
2,4,5-trichlorophenoxyacetic acid
(2,4,5-T)
COOH
CH2
O
Cl
Cl
2,4-dichlorophenoxyacetic acid
(2,4-D)
O
O
C
HO
CH3
COOH
OH
CH2
Gibberellic acid
(GA3)
O
CH2
NH
N
N
N
N
Kinetin

**29.1** Molecular structures of some plant hormones.



shows its molecular structure. Produced by active apical meristems, young leaves, and other tissues with dividing cells, this auxin moves in a basipetal direction (from stem tip or root tip to the base of the stem or root) and affects growth and differentiation in several ways.

## Auxin and Elongation

Discovered early was the ability of auxin to stimulate elongation of cells. When the apices of oat coleoptiles (which respond to auxin in much the same way as stems do) are removed, the decapitated coleoptiles soon cease to elongate, but if the apices are replaced, the coleoptiles continue to elongate for a while. Applying auxin extracted from coleoptile tips to decapitated coleoptiles also restores their ability to elongate, as does synthetic auxin.

Auxin moves downward from the apical meristem of a stem to the region of elongation where it exerts its effect. The cell walls become more plastic as the bonds between the cellulose molecules break, and the walls then can stretch when water enters the cells. It is likely that auxin exerts its effect indirectly by affecting some step in the basic metabolism rather than acting directly on cellulose molecules.

For each organ of the plant there is a concentration of auxin that promotes maximum elongation. Slightly lower or higher concentrations also promote elongation, but to a lesser extent. At very low concentrations there is no effect; at very high concentrations not only is growth inhibited, but various deformations and even death may result. The optimum concentrations for elongation are different in different organs; concentrations that stimulate growth of stems inhibit growth of roots (Fig. 29.2).

## Tropisms

Through its role in cell elongation, auxin functions in several tropisms (growth movements). **Phototropism** is the bending response of an organ to unilateral illumination. Young stems and petioles, which are positively phototropic, bend toward bright illumination (Fig. 29.3). Many roots do not respond to light, but some are negatively phototropic, bending away from bright illumination.

If a stem is more brightly illuminated from one side, a higher concentration of auxin accumulates on the shaded side than on the illuminated side. This causes the stem to bend toward the brighter light source. A petiole responds similarly if only one side of the leaf blade is illuminated. Auxin moves from the blade to the petiole, and more

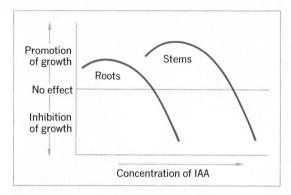

**29.2** Effect of auxin on elongation of roots and stems. Concentrations of IAA that inhibit the growth of roots promote the growth of stems.

**29.3** Positive phototropism. Tomato plants were raised in an opaque box open at only one side. The growing stems have bent toward the light source.

**29.4** Negative geotropism. This tomato plant was placed on its side two days before the photograph was taken. The elongating portion of the stem has bent returning the stem tip to a vertical position. Note that the lower portion of the stem, which has ceased elongating, did not bend.

moves from the shaded side than from the illuminated side. The higher auxin concentration on one side of the petiole causes it to bend in such a way that it moves the blade into brighter light. Inspection of most plants shows the leaves so oriented that the blades shade each other as little as possible. Boston ivy and other vines often show patterns called **leaf mosaics** when they grow on walls; the leaves are so uniformly spaced that they cover the wall with little overlapping (Fig. 22.2).

**Geotropism** is a response to gravity. Stems and petioles are negatively geotropic, bending upward away from the earth. Roots are positively geotropic, bending toward the earth. If the stems and roots of a plant are placed in a horizontal position, in only a few hours the young portion of the stem and the younger petioles bend upward, the curve in the stems usually occurring an inch or more from the tip in the region of elongation (Fig. 29.4). In the root, where the region of elongation is much closer to the tip, the downward bend will appear at a corresponding distance from the tip, about a half inch.

In a stem held in a horizontal position, the lower half contains much more auxin than does the upper half. The higher concentration in the lower portion causes it to elongate more than the upper portion. This results in the upward bending of the stem. Either the auxin or its precursor moves downward under the influence of gravity.

In horizontally placed roots, the lower portion also contains more auxin than does the upper, but the root response is opposite to that of the stem. This seeming inconsistency may be explained by the fact that roots are much more sensitive to auxin than are stems (Fig. 29.2). The optimum concentration for elongation in stems inhibits the growth of roots, which experience maximum elongation at much lower concentrations of auxin. In most roots auxin probably exists in concentrations that are near the optimum for elongation. Therefore any appreciable addition of auxin becomes inhibitory. The downward movement of auxin (or its precursor) in a horizontal root then produces an inhibitory concentration in the lower portion and leaves a stimulatory concentration in the upper half. This causes the upper part to grow more than the lower part and produces the positive geotropic response. It is also possible that some other as yet unidentified hormone may affect geotropism in the root.

Although many processes occurring in plants contribute to **homeostasis,** or the steady state, tropisms are good examples. We first might draw an analogy with an inanimate system. A pail placed under an open tap soon fills with water, which then spills over its top. The quantity of water entering the pail during any period of time equals the amount leaving during that time. Thus, there is no net change within the pail. It now has reached a homeostatic condition; it remains full.

In spite of their name, homeostatic systems are not static in the sense that the image of a swimmer in mid-dive is preserved indefinitely in a photograph. Homeostatic systems are busy systems, with several events occurring simultaneously, each event tending to correct another whenever the second event tends to move the system away from a norm. The overflow from the pail equals the amount of water entering it from the tap.

In a living root which is placed in a horizontal position, positive geotropism corrects the situation by making the root bend into a vertical position. When the root grows straight downward again, the bending action ceases. Should the root be forced into a horizontal attitude again, the action of the auxin once more bends it until a vertical position is achieved. Roots growing through soil often meet impenetrable rocks, and once they have grown around them, their homeostatic mechanism sends them on a downward course again. Thus the positive geotropism of roots keeps them below the ground where they can perform their dual functions of anchorage and absorption.

Negative geotropism functions similarly in stems, but superimposed on this is their positive phototropism. The net result is that the stems bend in such a way that the leaves are kept as much as possible in the sunshine that they require. As the position or direction of the light changes, so does that of the stems and leaves.

## Apical Dominance

Under some conditions auxin promotes cell division; the auxin moving downward from reactivated terminal buds early in the year causes the vascular cambium to begin its divisions. On the other hand, auxin produced by a terminal bud inhibits the growth of lateral buds, whose apical meristems produce no auxin as long as they are dormant. This phenomenon called **apical dominance** prevents plants from becoming excessively bushy. Should a terminal bud be removed from a stem or should it be damaged by disease, one or more lateral buds on that stem will be released from inhibition and will resume growth and produce a branch (Fig. 29.5). When the terminal bud of this new branch becomes active, it produces its own auxin, and it exerts apical dominance over any lateral buds below it. This is another example of homeostasis, the function of a missing terminal bud being assumed by a lateral bud which becomes the terminal bud of the developing branch.

**29.5** Apical dominance. When this mock orange twig was pruned (arrow 1), the two lateral buds immediately below the cut surface were released from apical dominance; each developed into a branch. Later, when the branch at the right was pruned (arrow 2), its lateral buds produced new branches. The branch to the left also was pruned, but the cut stump does not show in this photograph.

Untrimmed, undamaged plants do have some branches, however. With increasing distance from a terminal bud, the concentration of its auxin diminishes. Eventually a point is reached where the concentration is so low that it exerts no dominance over the lateral buds, which then can develop into branches. The characteristic shapes of many species of trees and shrubs are affected by the amount of auxin produced by their terminal buds and the sensitivity of their lateral buds to it.

The buds ("eyes") of untreated Irish, or white, potatoes often sprout while they are in storage. Because the potatoes ordinarily are stored in darkness, the food they contain is the only energy source for any of their buds that might sprout; therefore any sprouting represents a loss in food value to the owner of the potatoes. Spraying storage potatoes with auxin retards or completely prevents sprouting and its concomitant food loss.

## Leaf Abscission

In autumn, the formation of a layer of weak, thin-walled cells called the **abscission zone** pre-

249

cedes the dropping of leaves. The base of the petiole, where this zone forms, receives auxin moving downward from the blade and from the terminal bud. As long as the auxin from the blade exceeds that from the terminal bud, as it does early in the growing season, the abscission zone does not form. Later in the year, as the blade ages, it produces less auxin and the abscission zone forms.

### Synthetic Auxins

After naturally occurring auxins were discovered, several additional auxins were synthesized, among them 2,4-D and 2,4,5-T (Fig. 29.1). When used at appropriate concentrations, 2,4-D kills broad-leaved weeds by disrupting their metabolism, but it does not affect the grasses, which include not only lawn grass, but the cereal grains and sugar cane. Therefore it has become a favorite weed killer for use on lawns and many farm fields. Because they decompose readily, 2,4-D and 2,4,5-T were believed to be good choices for the purpose, for they would not affect broad-leaved crops or ornamentals planted later in treated areas. Recently 2,4,5-T gained notoriety when it was used as a defoliating agent in southeastern Asia; mangrove and rubber trees sprayed several times show few signs of recovery, and the ecological damage is extensive. The assumption that these substances were harmless to humans is now brought into question, too.

Synthetic auxins find use as rooting agents. Many commercially valuable plants once were difficult to propagate by cuttings, because they do not form adventitious roots readily. Dipping the cuttings into a dilute solution of an auxin or allowing the cut ends to soak in it for several hours usually overcomes the difficulty and induces copious rooting.

Auxins also can be used to induce ovaries of unpollinated pistils to mature into fruits. No seeds form because pollination and fertilization did not occur. Seedless tomatoes, watermelons, and cucumbers have been produced this way. In Hawaii,

250

pineapple fields are sprayed with auxin to induce flowering at any desired time of year.

## Gibberellins

Discovery of the **gibberellins** preceded that of auxins only by a few years in the 1920's. The discovery was made in Japan, and because communications between scientists of distant countries were not of the best in those days and because World War II later disrupted them, it was only in the mid-1950's that Western scientists became aware of the existence of this new class of plant hormones. At first considered mere curiosities, gibberellins gradually commanded more and more interest when it was learned that they perform important functions in the growth and reproduction of flowering plants. At least nine gibberellins have been discovered; gibberellic acid ($GA_3$), illustrated in Fig. 29.1, has been used in many experiments. Unlike the auxins, gibberellins move either upward or downward in stems and roots.

Marked elongation of stems is one of the startling effects of gibberellin. Dwarf varieties of several plants exist because of their possession of genetic defects that keep them small. External applications of gibberellin to the stem tip of a dwarf or bush variety of bean convert it into the tall climbing vine of the pole variety.

Gibberellin has a noticeable effect on several biennial rosette plants. In their first year these plants produce a short stem that barely rises above the surface of the ground and a rosette of leaves also near the ground (Fig. 29.6a). The second year, after the plant has been exposed to a cold winter, the stem elongates markedly and produces flowers (Fig. 29.6b). In nature the exposure to cold temperatures seems to be a requisite for stem elongation and flowering, but applications of gibberellin can substitute for cold treatment. The gibberellin stimulates the cells of the apical meristem to divide and the resulting cells to elongate. Applications of gibberellin to the rosette plant cause stem elongation and flowering even though there has been no exposure to low temperatures (other conditions must be favorable for flowering, of course). Carrots, radishes, and other biennials can be made to flower in their first years.

A third important function of gibberellin is its role in germination. Produced by the embryo, it stimulates the production of several enzymes including amylase in the endosperm of corn (Chapter 28); thus it promotes the digestion of starch to the sugars that the embryo requires for germination. Gibberellin also stimulates production of auxin, which, in turn, promotes growth of the embryo.

## Cytokinins

The **cytokinins** (or simply kinins) comprise a third group of plant hormones, of which kinetin (Fig. 29.1) is perhaps the best known. Discovered in the mid-1950's, this group has not been so well studied as the auxins and gibberellins. Kinins are most abundant in young fruits and seeds and in plant tumors.

Kinins stimulate cell division, and in some cases they cause cell elongation. It is known that they stimulate protein synthesis, probably through some action on the genetic system of the cell. When applied to tissue cultures (tissues isolated from living organisms and grown under sterile conditions in test tubes), they cause the initiation of buds from previously undifferentiated tissue.

Kinins applied to plants are immobile; they do not move from the site of application. In some way not yet understood, they mobilize other substances and cause them to move toward the site of application.

The three groups of hormones discussed above do not function separately. Two or three are likely to be present simultaneously in a given tissue, and

(a)

**29.6** Effect of cold weather on rosette plants. (a) Great mullein plant in its first season of growth. As long as the weather is warm, the plant will continue to grow with its leaves arranged in a rosette near the ground. (b) Great mullein plant in its second season of growth. Having passed through a cold winter, the plant produced a tall stem with a cylindrical cluster of small flowers at the top. Without exposure to cold, the plant would have remained in its rosette form.

(b)

these interact with each other. In endosperm, gibberellin initiates the formation of auxin.

The absolute amount of a given hormone is not so important as the relative concentrations of the different hormones. With relatively low kinetin concentrations in combination with relatively high auxin concentrations, root formation is initiated; but buds are produced by tissues with high concentrations of kinetin and low concentrations of auxin.

In addition to the three groups discussed above, several other plant hormones are known: traumatic acid, which is produced by wounded tissue and promotes cell division and healing of the wound; dormin (abscisin II), which promotes dormancy of buds and abscission of leaves; thiamin (vitamin $B_1$), which roots require for normal growth; and perhaps florigen, the flowering hormone, whose existence has been postulated but not yet proved.

## Questions

1. Define the following terms:
hormone
homeostasis
tropism
apical dominance
2. In a general sense, what is the function of a hormone?
3. How does auxin function in elongation, tropisms, apical dominance, and abscission?
4. If you were to apply auxin to one side of a stem in the region of elongation, what result would you expect? If you applied it to one side of a tree trunk, would you expect the same result? Explain.
5. Name some common uses of synthetic auxin.
6. Explain why roots and stems have opposite geotropic responses.
7. How do tropisms benefit plants?
8. What effect would you expect if you sprayed gibberellin on a biennial plant in its first year?
9. What function do cytokinins have in plants?
10. How do hormones function in homeostasis? Give an example.

## Suggested Readings

Devlin, R. M. Plant Physiology, 2nd ed. New York: Van Nostrand Reinhold Company, 1969.
Galston, A. W. "Regulatory Systems in Higher Plants," American Scientist 55: 144–160 (1967).
———, and P. J. Davies. "Hormonal Regulation in Higher Plants," Science 163: 1288–1297 (1969).
Khan, A. A. "Cytokinins: Permissive Role in Seed Germination," Science 171: 853–859 (1971).
Letham, D. S. "Cytokinins and Their Relation to Other Phytohormones," BioScience 19: 309–316 (1969).
Ray, P. M. The Living Plant. New York: Holt, Rinehart and Winston, Inc., 1963.
Salisbury, F. B. "Plant Growth Substances," Scientific American 196(4): 125–134 (April 1957).
Stewart, F. C. Plants at Work. Reading, Mass.: Addison-Wesley Publishing Co., Inc., 1964.
Torrey, J. G. Development in Flowering Plants. New York: Macmillan Publishing Co., Inc., 1967.
van Overbeek, J. "The Control of Plant Growth," Scientific American 219(1): 75–81 (July 1968).

# The Human Body

# Part VII

Multicellular animals, like multicellular plants, are complex organisms composed of several kinds of tissues that form the organs (for example, stomach, heart, kidney) of the body. In all but the simplest animals, the organs are grouped into organ systems (for example, digestive, circulatory, and excretory systems), with all the members of one organ system contributing to the performance of one or several functions.

Because the reader probably has more interest in the functioning of his own body than in that of another animal, the next ten chapters consider the human body. The internal workings of other mammals are similar but not identical.

## Organ Systems

Several organ systems comprise the human body. Biologists differ somewhat in their classifications of these systems. This book recognizes ten organ systems:

1. The **digestive system,** which digests at least part of the food eaten, absorbs the digested portion, and eliminates what is not absorbed.

2. The **respiratory system,** which absorbs oxygen from the air and excretes carbon dioxide into it.

3. The **muscular system,** which moves the several parts of the body—not only our arms and legs, but the blood in our arteries and the food in our stomachs. It also contributes to the support of the body.

4. The **circulatory system,** which, in addition to transporting food, oxygen, minerals, hormones, waste products, and other compounds around the body, also protects it from disease by killing or inactivating microorganisms that invade it.

5. The **excretory system,** which rids the body of metabolic waste products.

# 30

# The Human Body—Its Organ Systems and Its Tissues

1 The human body consists of several organ systems, each of which performs one or more functions.

2 Several basic tissue types comprise the organs of the body.

6. The **endocrine system,** which regulates and integrates some body processes by the production and secretion of hormones.

7. The **nervous system,** which regulates and integrates some body processes by means of rapidly transmitted nervous impulses.

8. The **skeletal system,** which supports the body and forms some blood cells.

9. The **integumentary system,** which surrounds the body, provides some protection against invasion by many microorganisms, and prevents excessive loss of water.

10. The **reproductive system,** which produces the gametes in both sexes and in the female shelters and nourishes the offspring until it is capable of living a separate life.

It is not rare for a given organ to perform two functions associated with two different systems. The pancreas is part of the digestive system by virtue of its production and secretion of digestive enzymes, but it is also an endocrine gland secreting the hormone insulin. As a large mass of muscle tissue, the heart is part of the muscular system, but one could hardly deny that it is an essential constituent of the circulatory system.

## Tissues

The organs of the body are composed of varying combinations of several types of tissue. Again, biologists differ in the number of tissue types they recognize. We will consider six types (Fig. 30.1): epithelial, supporting and connective, muscular, blood, nervous, and reproductive tissues.

**Epithelium** covers the external surface of the body (it constitutes the outer part of the skin) and lines the internal body cavities as well as the surfaces of several internal organs and glands. The epithelium may consist of only one or of several layers of cells, but characteristically the cells are

closely packed with no spaces between them. Because it lines so many organs, epithelium must permit various substances to pass across it. For example, oxygen and carbon dioxide cross the epithelium of the lungs, and digested foods enter the tissues of the body by crossing the intestinal epithelium.

The **supporting and connective tissues** are bone, cartilage, and fibrous connective tissue. In each case the cells are separated from each other and are embedded in a matrix. In bone the matrix is a hard, rigid substance that provides most of the support of the human body and protection to organs such as the brain and spinal cord, which are almost completely encased in bone. Cartilage, sometimes called gristle, is firm and strong, but not so rigid as bone because its matrix is more flexible. The reader has encountered cartilage at the ends of a drumstick bone in his chicken dinner and, if he has carved the chicken himself, at the lower end of the breastbone. The reader's ears and the end of his nose are supported by cartilage and not bone. Fibrous connective tissues consist of both cells and long extracellular fibers formed from materials secreted by the cells. These tissues bind organs together. Ligaments bind bones to bones, and tendons bind muscles to bones. Skin is tough by virtue of its connective tissue, and so is leather, which is made from skin. Fat cells, which store fat, occur among fibrous connective tissue, which is called adipose tissue if these cells are numerous.

**Muscle cells** function by contracting, and their contractions move the several parts of the body. The skeletal muscles, each of which is attached to two (or sometimes three) different bones, move bones relative to each other. This results in the overt, voluntary movements of the body. Visceral muscles associated with some internal organs such as the stomach, intestines, and arteries keep these organs functioning involuntarily. The heart is composed of cardiac muscle, a special type of involuntary muscle.

**Blood cells** are of three main types: red blood cells, which transport oxygen throughout the body;

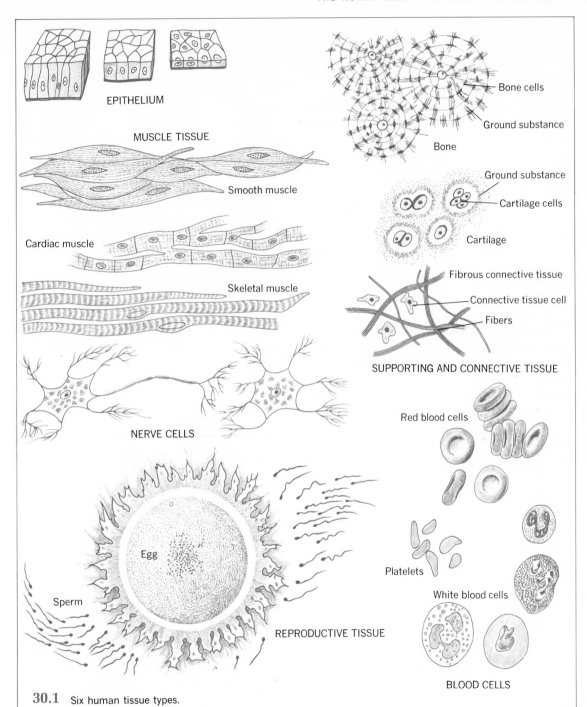

EPITHELIUM

MUSCLE TISSUE

Smooth muscle

Cardiac muscle

Skeletal muscle

NERVE CELLS

REPRODUCTIVE TISSUE

Egg

Sperm

Bone cells

Ground substance

Bone

Ground substance

Cartilage cells

Cartilage

Fibrous connective tissue

Connective tissue cell

Fibers

SUPPORTING AND CONNECTIVE TISSUE

Red blood cells

Platelets

White blood cells

BLOOD CELLS

**30.1** Six human tissue types.

257

white blood cells, of which there are several types that contribute in various ways to the defense of the body against disease; and platelets, which are important in the clotting of blood in wounds. Blood vessels reach essentially every part of the body and supply them with all of these services.

**Nervous tissue** possesses the peculiar property of being able to conduct rapid impulses. Although very fine, some neurons (nerve cells) extend from the spinal cord to the tips of fingers or toes; in a tall person some of these neurons may be more than 3 feet long. Like blood vessels, nervous tissue ramifies through essentially all parts of the body. Because of its ability to transmit impulses at the rate of many feet per second, nervous tissue coordinates the several parts of the body almost instantaneously. This is invaluable in emergencies when quick action is essential. Whether the stimulus be the sight of a car bearing down on one, a cry for help, or the odor of burning toast, nervous tissue relays messages to the muscles, which bring about action appropriate to the situation.

Unlike the five tissues mentioned, **reproductive tissues** are confined to only a small part of the body, where, by the process of meiosis (Chapter 41), haploid eggs and sperms are produced from diploid tissues. An egg is a relatively large spherical cell; only one egg matures at a time in the ovaries of a woman. The sperms, which are produced by the millions in the testes of a man, are small tadpole-shaped cells, each with a distinct head and a long tail. When sperms are in a liquid medium, their tails propel them by whiplike motions. If introduced into the female reproductive tract, sperms swim toward the egg, which may be fertilized by one of them.

These six tissue types in many different combinations form the ten organ systems mentioned earlier. Together they perform all of the processes essential for the well-being of the individual and for the reproduction of the species.

## Questions

**1.** List the organ systems of the human body. How does each system contribute to the functioning of the body?
**2.** List the types of tissues in the human body. How does each tissue contribute to the functioning of the body?

## Suggested Readings

Grollman, S. *The Human Body.* New York: Macmillan Publishing Co., Inc., 1964.

Every animal requires food as a source of energy. Only the smallest aquatic organisms and some parasites can absorb through their external surfaces all the food they need. Most multicellular animals actively seek food, which they then ingest into a digestive system especially modified for digesting and absorbing food.

## Passage of Food Through the Digestive Tract

The human digestive tract resembles a much-twisted tube open at both ends. It begins at the mouth, proceeds through the body, and opens at the anus (Fig. 31.1). Food that is swallowed is "outside" the body, for it is still in the lumen (hollow) of this open digestive tube. Small, soluble organic molecules, such as glucose, amino acids, fatty acids, and glycerol, can be absorbed into the body without further change; but larger molecules, such as starch, proteins, and fats first must be digested to glucose, amino acids, fatty acids, and glycerol (Chapter 8), which then can be absorbed. One of the functions of the digestive system is the secretion of digestive enzymes (Chapter 10) into the lumen where they are thoroughly mixed with the food. Table 31.1 summarizes information about some enzymes of the digestive system.

Food enters the digestive tract through the **mouth,** where the teeth break it mechanically by their biting and chewing action. This increases the surface area through which enzymes can act on the food, the smaller the particles the greater the amount of surface exposed to enzymes. Herein lies the benefit of the perennial parental admonition to children to "chew your food well."

The **salivary glands** of the mouth secrete saliva, which contains mucus and amylase. Mucus moistens and lubricates the food, and amylase is a starch-digesting enzyme. The tongue manipulates

# 31
# The Digestive System

1 **The functions of the digestive system include**
   a. **Mechanical breaking and chemical digestion of food.**
   b. **Absorption of digested foods, water, and minerals.**
   c. **Elimination of undigested foods.**

2 **Chemical digestion begins in the mouth and continues as the food passes through the esophagus, stomach, and small intestine, but most of the digestion occurs in the small intestine.**

3 **Absorption of digested food occurs mostly through the small intestine. Absorption of water and minerals occurs primarily through the large intestine.**

4 **Undigested food is eliminated through the anus.**

5 **Several large digestive glands—salivary glands, the liver, and the pancreas—secrete digestive enzymes and other substances into the digestive tract. These materials increase the rate of digestion.**

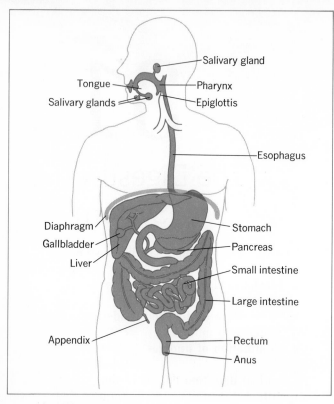

**31.1** Digestive system.

the food and initiates swallowing by pushing it into the pharynx.

The **pharynx,** a chamber in the neck, is part of both the digestive and respiratory systems. As part of the digestive system it connects the mouth and the esophagus; as part of the respiratory system it connects the nasal passages with the larynx and the trachea (Chapter 32). During swallowing, the larynx rises and meets the **epiglottis,** a flap of cartilaginous tissue that is bent over the opening of the larynx by the mass of food being swallowed. This temporarily stops breathing, but it also prevents the entrance of food into the respiratory system and possible choking or suffocation. Because the tongue presses against the roof of the mouth during swallowing, food can pass only into the esophagus.

The **esophagus** is a muscular tube that conducts food from the pharynx to the stomach. Contractions of the muscles in the wall of the esophagus push food downward to the stomach.

The **stomach** is a large muscular sac. Glandular cells located on the inner surface of the stomach wall secrete gastric juices that contain mucus,

Table 31.1

Enzymes of the Human Digestive System

| Organ in Which Enzyme Acts | Enzyme | Source | Reaction Catalyzed |
|---|---|---|---|
| Mouth | Amylase | Salivary glands | Starch ⟶ maltose |
| Stomach | Pepsin* | Stomach glands | Proteins ⟶ polypeptides |
| Small intestine | Lipase | Pancreas | Fats ⟶ glycerol and fatty acids |
| Small intestine | Amylase | Pancreas | Starch ⟶ maltose |
| Small intestine | Trypsin* | Pancreas | Proteins ⟶ polypeptides |
| Small intestine | Chymotrypsin* | Pancreas | Proteins ⟶ polypeptides |
| Small intestine | Carboxypeptidase | Pancreas | Polypeptides ⟶ amino acids |
| Small intestine | Aminopeptidase | Small intestine | Polypeptides ⟶ amino acids |
| Small intestine | Maltase | Small intestine | Maltose ⟶ glucose |
| Small intestine | Lactase | Small intestine | Lactose ⟶ glucose and galactose |
| Small intestine | Sucrase | Small intestine | Sucrose ⟶ glucose and fructose |

*Pepsin, trypsin, and chymotrypsin are secreted in inactive forms (pepsinogen, trypsinogen, and chymotrypsinogen, respectively) which are converted into the active forms only when they reach the lumen of the stomach or small intestine. This prevents the enzyme from digesting the cells in which they were formed. Mucus lining the inner surface of the stomach and intestinal walls protects them from digestion by the active enzymes.

Pepsin, trypsin, chymotrypsin, carboxypeptidase, and aminopeptidase are all proteases, but each acts at a different location in the protein molecule.

pepsin (a protein-digesting enzyme or protease), and hydrochloric acid. The hydrochloric acid increases the acidity of the stomach contents to a pH of 2 or lower. A low pH is optimal for the action of pepsin; it also kills many of the bacteria that enter on the food.

Water from body tissues enters the stomach and provides the medium in which digestion occurs; added to this is any water drunk during the meal. The action of the stomach muscles continues to break the food into smaller particles and to mix them well with the water and gastric juices. As digestion continues, the food acquires a soupy consistency, and then it passes from the stomach into the small intestine.

The **small intestine** resembles a long, hollow tube, about 9 feet in length; it fits within the abdomen because it is much coiled back and forth on itself. The first 10 inches of the small intestine comprise the duodenum, which receives additional water from body tissues and important secretions from two large glands, the liver and the pancreas.

The **liver** produces bile, an alkaline mixture of substances among which are the bile salts. Bile salts aid in the digestion of lipids by emulsifying them; that is, they physically break large masses of lipids into tiny droplets that are more readily digested by lipases (lipid-digesting enzymes) because of the increase in surface area, the smaller droplets having proportionately more surface area per volume (Chapter 20). Perhaps even more important is the fact that emulsification of digested or partly digested lipids allows them to be absorbed by the lymph system (see p. 262). No enzymes are known to be present in bile. The liver continuously produces bile, which it secretes into the duodenum when food is present there. At other times, the bile moves from the liver to the gall bladder, which stores it until food enters the duodenum.

The **pancreas** secretes into the duodenum a mixture of carbohydrate-, lipid-, and protein-digesting enzymes. The pancreatic fluid is alkaline and helps maintain a pH of 7 to 8 in the small intestine. The

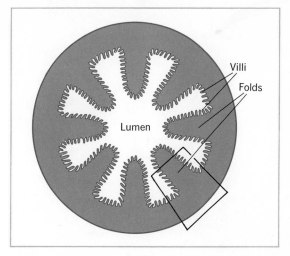

**31.2** Diagrammatic cross section of the small intestine, showing the inner folds of the wall and the villi on the folds. The photomicrograph in Fig. 31.3 corresponds to the area in the rectangle.

enzymes that act there function best at this pH.

The wall of the **small intestine** also secretes carbohydrate-, lipid-, and protein-digesting enzymes in alkaline solutions. The action of the muscles in the wall of the small intestine aids further in the mixing of food and enzymes and in moving them along the length of the intestine, where a great deal of digestible material—polysaccharides (such as starch), disaccharides, proteins, and some lipids (including some fats)—are digested to soluble molecules, simple sugars (such as glucose), amino acids, and fatty acids and glycerol. Nearly all the digested food is absorbed into the circulatory system through the walls of the small intestine, which is lined with epithelium. Absorption through the epithelium is not merely by diffusion, for many substances move against a concentration gradient. The epithelium therefore must be capable of active transport.

The inner surface of the small intestine wall is folded, and the folds bear millions of fingerlike projections called **villi** (Figs. 31.2 and 31.3). These structures increase many times the absorbing surface of the small intestine. The epithelial cells of

261

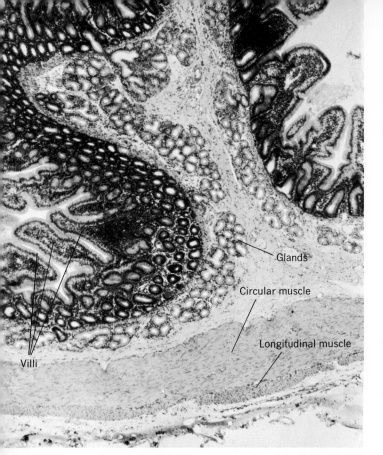

Villi

Glands

Circular muscle

Longitudinal muscle

**31.3** Cross section of the duodenum of a monkey, showing the villi which greatly increase the absorbing surface of the small intestine. The action of the circular and longitudinal muscles is responsible for moving food along the digestive tract and mixing the food well with digestive juices (see also Chapter 34). (Courtesy of Dr. Beth L. Wismar.)

the villi are equipped with many projections called **microvilli** (Fig. 31.4), which increase the absorbing surface even more. Microvilli are so small that under an ordinary student microscope they are invisible or appear as a fringe of very fine hairs. The folds, villi, and microvilli together increase the absorbing surface of the small intestine to about 185 square meters (a small house has less floor space on one level), or about 600 times what it would be without them.

Each villus contains a network of blood capillaries and a small lymph vessel called a lacteal (Chapter 34). Certain products of digestion—sugars, amino acids, glycerol, and some fatty acids—are absorbed into the blood flowing through the capillaries, but emulsified lipids (and some fat-soluble substances, such as vitamins A and D) enter the lacteals of the villi. These lipids

**31.4** Microvilli on the outer surface of a villus cell in the intestine of a rat. These microvilli increase still more the absorbing surface of the small intestine. (Dr. Keith R. Porter.)

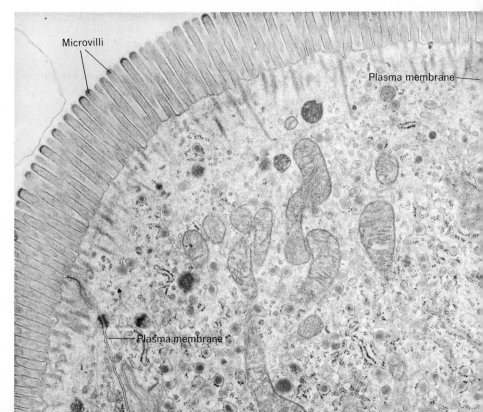

Microvilli

Plasma membrane

Plasma membrane

are carried at first by the lymph, but later the lymph flows into the bloodstream, which continues to distribute them to various parts of the body.

The blood carries food from the small intestine to the liver and then to other parts of the body. The liver converts glucose into the polysaccharide glycogen (animal starch) by dehydration synthesis and stores it. When the level of glucose in the blood falls, the liver digests some of the glycogen back to glucose, which the blood then distributes to other parts of the body where the glucose is utilized. Some glucose may be converted to fat and stored in adipose tissue under the skin. The liver detoxifies some toxic substances, and vitamins A, D, and $B_{12}$ are stored here. The bloodstream returns to the liver the bile salts that were used in the small intestine; these can be secreted again in bile.

The **large intestine** is only about 4 feet long, but its diameter is larger than that of the small intestine. Shortly beyond the junction of the small and large intestines there extends a short, finger-like projection, the **appendix.** This organ has no known function in human beings. If it becomes infected, appendicitis results.

Normally a large population of bacteria flourishes in the large intestine. These bacteria digest some materials—notably cellulose—not digestible by the enzymes produced by the human digestive tract. In so doing, the bacteria synthesize vitamins, especially some of the B vitamins and vitamin K, that are absorbed through the intestinal wall and then become available to the body. As in the small intestine, food moves along by the action of the muscles of the wall of the large intestine.

Absorption of water through the large intestine wall makes the intestinal contents drier and brings them to a semisolid consistency. Too slow a trip through the large intestine results in constipation because more water than usual is absorbed, and the material becomes very hard and dry. Diarrhea is caused by too rapid a movement of the intestinal contents along the large intestine, for then only a

small amount of water is absorbed by the body, and the intestinal contents remain very watery.

The last few inches of the large intestine comprise the **rectum.** Material reaching it is called **feces.** In a healthy individual feces consist mostly of undigested semisolid matter and bacteria. When the feces reach the rectum, one becomes aware of an urge to defecate. Feces are passed through the **anus,** the opening of the rectum.

# Vitamins

The ingredients of our meals include not only the food compounds—carbohydrates, fats, and proteins used as energy sources and structural molecules—but also vitamins and minerals. Some of the latter have been discussed in previous chapters and others will be introduced at appropriate places in some of the following chapters.

**Vitamins** are all organic compounds, but they belong to no one group of organic compounds. Figure 31.5 shows some of them and reveals their chemical diversity. Because they are needed in only minute quantities, it is unlikely that they would be used by the body as energy sources, for if oxidized, they could supply only an insignificant fraction of the body's energy needs. For this reason it was suspected that they might have an enzymatic function. This has proved to be true of many vitamins, some of which function as coenzymes or parts of coenzymes.

Vitamins in proper quantity are required to maintain a healthy body. An **avitaminosis** is a disease caused by the lack or insufficient quantity of a vitamin in the diet. Generally, supplementing the diet with the missing vitamin cures the disease, but if the diet has been neglected for a long time, permanent damage may have been done. A balanced diet contains all the vitamins an individual needs, and the taking of vitamin pills should not be necessary except under unusual circumstances.

263

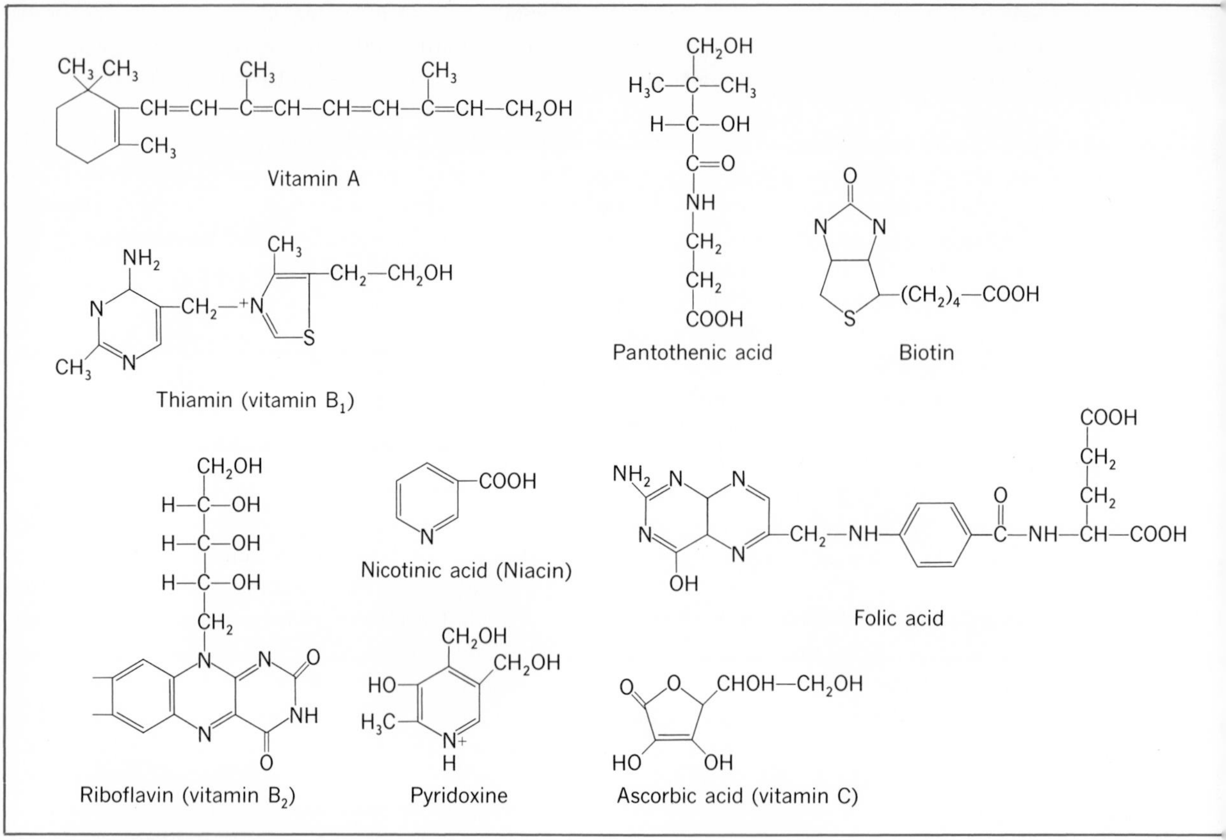

Vitamin A

Thiamin (vitamin B$_1$)

Pantothenic acid

Biotin

Riboflavin (vitamin B$_2$)

Nicotinic acid (Niacin)

Pyridoxine

Folic acid

Ascorbic acid (vitamin C)

Indeed, it is possible to take an overdose of at least some vitamins. A **hypervitaminosis** is a disease caused by the ingestion of excess quantities of a vitamin. Hypervitaminoses are known for vitamins A and D, and others may be discovered someday.

Humans generally receive their **vitamin A** in the form of carotene, which their bodies then digest to vitamin A:

$$C_{40}H_{56} + 2H_2O \longrightarrow 2C_{20}H_{29}OH$$

carotene                    vitamin A

The orange pigment carotene accompanies chlorophyll in plants and so is common in leaves. In some organs, such as carrot roots, it exists independently of chlorophyll. Vitamin A maintains mucous membranes, and the body utilizes it in the manufacture of rhodopsin and other pigments in the eye that are essential for normal vision. A deficiency of vitamin A in the diet results in night blindness and damage to mucous membranes, which become dry and horny. Foods rich in carotene include the green and yellow vegetables, butter, cream, eggs, and fish liver oils.

If somewhat more vitamin A than the body immediately needs is ingested, the excess is stored in the liver and can be utilized later if the diet becomes poor in the vitamin. If unusually large quantities of vitamin A are ingested, a hypervitaminosis results; the individual becomes irritable and suffers from headaches and nausea. Jaundice, falling of hair, and painful lumps in the arms and legs are other symptoms.

Once thought to be a single vitamin, the **vitamin B complex** includes at least a dozen vitamins with different functions. Some of them are thiamin,

**31.5** Structural formulas of some vitamins.

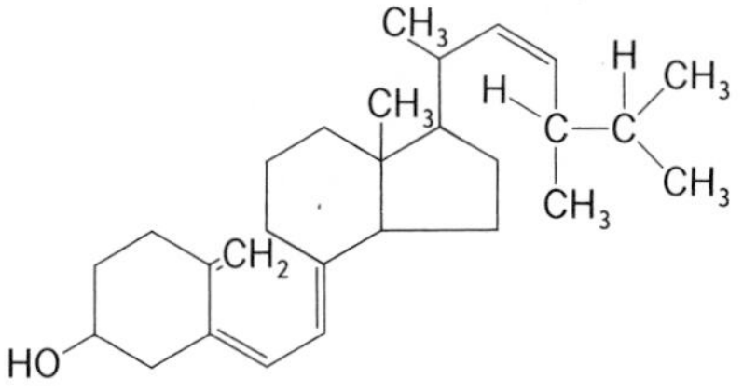

Vitamin D (calciferol)

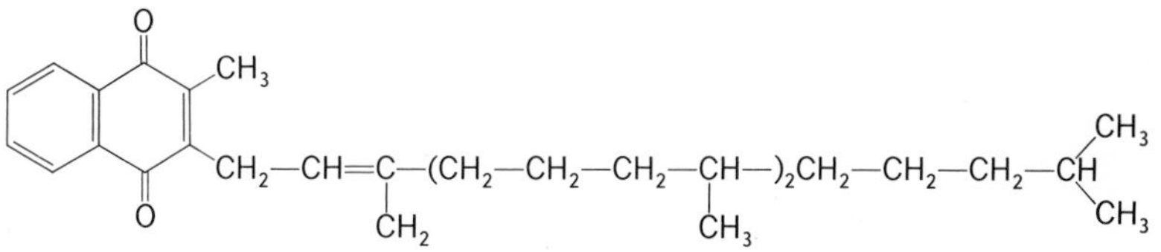

Vitamin K

riboflavin, nicotinic acid, pyridoxine, pantothenic acid, biotin, folic acid, and vitamin $B_{12}$.

**Thiamin (vitamin $B_1$)** is part of the enzyme co-carboxylase (thiamin pyrophosphate), which catalyzes decarboxylation reactions in respiration. Lack of thiamin produces the disease called beriberi, in which gastrointestinal and nervous disorders are followed by enlargement of the heart and edema. Sources of thiamin include pork, liver, whole grain cereals, green vegetables, and yeast. Thiamin exists primarily in the outer part of the cereal grain, and so polished rice and white flour are poor sources of this vitamin unless they have been enriched with it again. Rarely seen in America and Europe, beriberi is most common in the Orient wherever polished rice is the main item of diet.

**Riboflavin (vitamin $B_2$)** is part of the coenzyme

FAD (which acts as a hydrogen carrier in respiration). Its lack in the diet causes lesions around the mouth and tongue, a dermatitis near the nose and eyes, and indistinct vision. Sources include liver, kidney, heart, dairy products, green vegetables, and yeast.

**Nicotinic acid (niacin)** is part of the coenzymes NAD and NADP, which are hydrogen carriers in respiration and photosynthesis. Its lack in the diet produces pellagra, a disease that involves degeneration of the nervous system. It manifests itself in dermatitis, diarrhea, headaches, memory loss, and finally dementia (mental deterioration). Liver, kidney, heart, wheat germ, and yeast are good sources of nicotinic acid.

**Pyridoxine (vitamin $B_6$)** is part of a coenzyme that catalyses transaminations, the transfer of

265

amino groups ($-NH_2$) from one compound to another. As such, it is important in amino acid metabolism. A lack of pyridoxine in the diet causes skin disorders, insomnia, and irritability. Liver, meats, vegetables, and whole grain cereals are excellent sources.

**Pantothenic acid** is part of the coenzyme A molecule and so is essential in respiration and fatty acid metabolism. It is assumed to be essential for good human nutrition, but a deficiency of this vitamin in the human diet is rare, and the required daily intake is unknown. It is widely distributed in foods; liver, kidney, eggs, vegetables, and yeast are excellent sources.

**Biotin** (also called **vitamin H**) in the diet prevents a type of dermatitis in infants. It is found in liver, kidney, eggs, and yeast.

**Folic acid** is the name given to a group of closely related chemical compounds that perform a variety of metabolic functions. Lack of folic acid in the diet results in anemia. Liver, kidney, and yeast are good sources.

**Vitamin $B_{12}$ (cyanocobalamine)** is essential for normal blood formation, and its lack in the diet causes pernicious anemia. The vitamin $B_{12}$ molecule is a rather large, complex one; it consists in part of a porphyrin ring like that of chlorophyll (Fig. 11.1) and hemoglobin (Fig. 34.10), but a cobalt atom occupies the center of the ring. Liver, kidney, lean meat, and fish are excellent sources.

**Ascorbic acid (vitamin C)** is well known for its ability to prevent scurvy, a disease characterized by aching bones and joints, hemorrhaging in the mouth, swollen gums, loose teeth, and the failure of wounds to heal normally. Long before the vitamin itself was discovered, the connection between citrus fruits and the prevention of scurvy was known. British ships stocked limes to issue to their seamen, who soon were called "limeys." The metabolic role of ascorbic acid still is not known with certainty, but the vitamin is believed to function in maintenance of bones and connective tissue. The human body does not store ascorbic acid but excretes any excess in the urine. Good sources are citrus fruits (oranges, grapefruit, limes), tomatoes, and raw (unpasteurized) milk.

**Vitamin D (calciferol)** can be synthesized by the human body, but only if the skin is exposed to sunlight. For this reason, the human body is more likely to require this vitamin in the diet in winter than in summer. Vitamin D functions in the absorption of calcium in the intestine and also in the hardening of bones and teeth by the incorporation of calcium into them. Lack of vitamin D causes rickets, a disease characterized by soft bones; the bones bend, and the legs become bowed. If neglected for a long time, such distortions become permanent and no longer can be corrected by the addition of vitamin D to the diet. Excess of vitamin D in the diet (one or two hundred times that required) is toxic. It causes nausea, thirst, excessive urination, and the deposit of calcium in heart, lungs, and kidneys. Fish livers are excellent sources of vitamin D. (In the days before the vitamin pill and vitamin-enriched foods, concerned mothers dosed their unwilling children with a spoonful of cod liver oil every winter morning.) Liver, milk, and eggs contain some vitamin D. Exposure to sunshine decreases the need for this vitamin in the diet.

**Vitamin K** is essential for normal blood clotting, and a lack of it may result in hemorrhaging. It is found in green leafy vegetables, and it is produced by some of the bacteria living in the large intestine.

266

# Questions

1. Diagram the parts of the digestive system. How do they increase the internal surface area of the body? Specifically, how are the absorbing surfaces increased?

2. By what mechanical action does the digestive system increase the surface area of food exposed to enzymes? In what organs does most of this action occur?

3. What organs are the source of digestive enzymes? In what organs does most chemical digestion occur?

4. Where does most absorption of digested food occur? Where does absorption of water and minerals occur?

5. How do the following function in digestion?
salivary glands
liver
bile
gall bladder
pancreas
hydrochloric acid secreted by the stomach
villi
lacteal
bacteria in large intestine

6. List the vitamins. To how many of these can you assign a distinct biochemical role?

7. Follow a hamburger on a bun through the digestive system. What happens to it in each organ that it passes through?

## Suggested Readings

Bloom, W., and D. W. Fawcett. *A Textbook of Histology*, 9th ed. Philadelphia: W. B. Saunders Company, 1968.

Davenport, H. W. "Why the Stomach Does Not Digest Itself," *Scientific American* 226(1): 86–94 (January 1972).

Griffin, D. R., and A. Novick. *Animal Structure and Function*, 2nd ed. New York: Holt, Rinehart and Winston, Inc., 1970.

Vander, A. J., J. H. Sherman, and D. S. Luciano. *Human Physiology*. New York: McGraw-Hill Book Company, 1970.

267

# 32
# The Respiratory System

1 The respiratory system absorbs oxygen from the air and releases waste carbon dioxide to it.

2 Gaseous exchange occurs across the surfaces of the alveoli, the minute membranous sacs of the lungs. Their total surface area is enormous.

3 The breathing movements that force air into and out of the lungs are accomplished by the diaphragm and rib muscles.

The food digested and absorbed by the digestive system provides all the energy used by the body. However, only a small fraction of this energy can be released in the absence of oxygen, and absorbing oxygen is one of the functions of the respiratory system.

Without the oxygen supplied by the respiratory system, the human body would die quickly. Although the outer surface of the body is exposed to the oxygen of the air, the skin absorbs essentially none of this gas. The character of the skin is such that it is relatively impermeable to gases. Even if it were permeable, its total surface area probably would not permit the absorption of sufficient oxygen to support an active human being, and in most climates much of the skin is covered by several layers of clothing for at least part of the year. One of the important features of the respiratory system is that it exposes to air a surface area within the chest equal to about 50 times the external surface area of the body. It is across this internal surface that oxygen diffuses into our bodies and waste carbon dioxide diffuses out.

An aside about the word *respiration,* which has several meanings. In Chapter 12 and elsewhere in this book, respiration refers to the chemical process by which foods are oxidized, with the release of metabolically useful energy. The word also can be used to indicate breathing movements, "artificial respiration" being a method of assisting these breathing movements. A third meaning of the word is the physical exchange of oxygen and carbon dioxide between the air in the lungs and the blood in the circulatory system and between the blood and the body tissues. To avoid confusion, in this book *respiration* will refer only to chemical oxidation of food, and the terms *breathing movements* and *exchange of gases* will convey the other meanings. Nonetheless, the organ system involved in the exchange of gases is called the respiratory system. Thus the respiratory system does not have a monopoly on respiration. All the systems of the body have living cells that respire, but only the

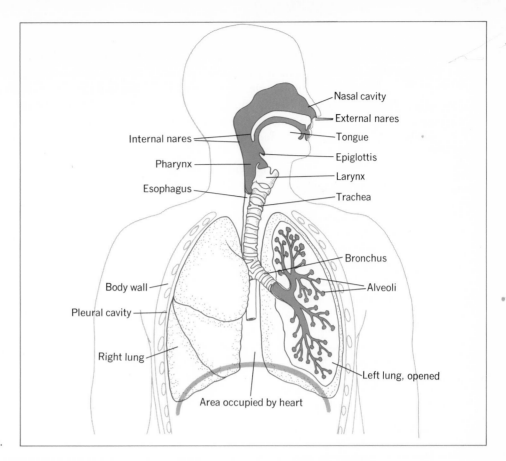

**32.1** Respiratory system.

respiratory system exchanges any great quantity of gases with the environment.

## Anatomy of the Respiratory System

The parts of the respiratory system include the nose with its nasal passages, the pharynx and larynx in the neck, and the trachea and lungs in the chest (Fig. 32.1). Closely associated with the functioning of the lungs are the rib muscles and the diaphragm, muscles that are responsible for breathing movements.

The air we breathe is a mixture of gases (Table 32.1). If we exclude water vapor, the amount of which varies greatly in different samples, most of the air—about 79 per cent—is nitrogen, an inert gas of no metabolic value to human beings or to any other organisms except the nitrogen fixers. Oxygen is the next most common gas, accounting for almost 21 per cent of the air. All the other gases

Table 32.1
Differences in Inhaled and Exhaled Air (exclusive of water vapor)

| | Percentage | |
| Gas | Inhaled Air | Exhaled Air |
| --- | --- | --- |
| Nitrogen | 79 | 79 |
| Oxygen | 21 | 17 |
| Carbon dioxide | 0.03 | 4 |
| Rare gases | trace | trace |

together comprise only a fraction of 1 per cent. Among them is carbon dioxide, which measures 0.03 per cent in most air samples; this figure is slightly higher in cities, where combustion of fuels produces carbon dioxide, and in forests at night, when the carbon dioxide produced by respiration is not utilized in photosynthesis.

Inhaled air enters the two **external nares,** or **nostrils,** of the nose and passes through the **nasal cavity.** The nostrils are equipped with short hairs that prevent the entrance of many small solid particles, most of them laden with a variety of bacteria. Even without bacteria, the particles themselves can be potentially dangerous. Workers who spend long hours inhaling small particles—miners breathing coal dust, for example—are subject to lung diseases. The presence of hairs in the nose reduces (but does not eliminate) these dangers.

The nasal cavity actually is two chambers separated by a wall. These open through a pair of internal nares into the **pharynx,** a chamber in the neck. Because the pharynx is shared by the digestive and respiratory system, a person can breathe through his mouth as well as his nose. A disadvantage of using the mouth is that the inhaled air is not filtered by the nasal hairs, nor is it then warmed by traveling through the nasal passages.

Except when one is swallowing, the pharynx communicates with another chamber in the neck, the **larynx,** popularly called the voice box or Adam's apple. During the process of swallowing food, the opening between the pharynx and larynx closes and allows free passage of the food to the esophagus (Chapter 31). This closing is accomplished by the rising of the larynx to meet the epiglottis, a process sometimes externally visible as a bobbing of the Adam's apple. This ordinarily prevents the entrance of food into the larynx, but an unexpected sneeze or hiccup coinciding with swallowing may allow the inhalation of solid food,

**32.2** Terminal bronchioles. (a) Section of a bronchiole with several alveoli. (b) Surrounding the alveoli are blood capillaries. Oxygen and carbon dioxide are exchanged between the alveoli and the capillaries.

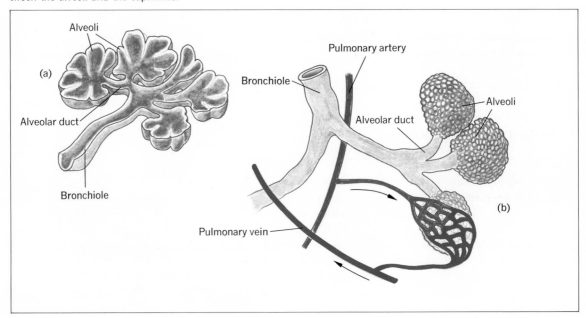

an event that at least is accompanied by some sputtering or mild choking; if the breathing passage becomes blocked, suffocation and death may result if the passage is not soon cleared.

The larynx is not merely a passageway for air. It contains the **vocal cords,** a pair of fibrous bands stretched across the larynx. Air passing through here sets the vocal cords vibrating, and this produces sound—and, hence, voice and speech. The vocal cords are attached to the larynx by muscles that can change the tension on the vocal cords, and thus change the pitch of the sound. Tense vocal cords produce high notes, and relaxed vocal cords produce low notes.

Descending from the larynx is a short tube, the **trachea** (popularly called the windpipe), which descends into the thoracic cavity (chest cavity). Here it divides into two smaller branches, the right and left **bronchi.** Each bronchus divides again, forming smaller and smaller tubes called **bronchioles;** and the finest tubes, called **alveolar ducts,** terminate in microscopic pouches called **alveoli** (Fig. 32.2). Each bronchus with all its bronchioles, alveolar ducts, and alveoli is called a **lung.** The lungs have no muscles. They are elastic, however, and can respond passively to the action of rib muscles and the diaphragm.

Surrounding each lung and firmly attached to it is a membranous covering, the **pleura.** Each pleura extends to the thoracic wall and the diaphragm; it also is firmly attached to them (Fig. 32.3a). The pulmonary pleura is the part that surrounds the lung, and the parietal pleura lines the thoracic wall and diaphragm. The two layers of pleura are so close to each other that only a minute space, the pleural cavity, separates them; a watery fluid, the **intrapleural fluid,** fills this space. This thin layer of fluid plays an important role in breathing.

Many of the epithelial cells that line the walls of most of the respiratory passages—nasal cavity, trachea, bronchi, and bronchioles—secrete mucus, and many of the bacteria and other small particles that escape entrapment by the nasal hairs become

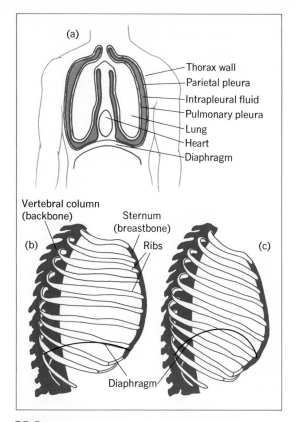

**32.3** Breathing movements. The inextensibility and incompressibility of the intrapleural fluid (a) causes the lungs to expand and contract with the movements of the diaphragm and ribs. During an inspiration the diaphragm contracts and flattens and the ribs rise (b) because of contraction of rib muscles (not shown); this increases the size of the chest cavity and forces air into lungs. During an expiration the diaphragm relaxes and rises and the ribs lower (c) because of relaxation of the rib muscles; this decreases the size of the chest cavity and forces air out of the lungs.

caught in this mucus. The cells also are equipped with cilia; these are microscopic, hairlike projections that beat upward. This action moves the mucus upward to where it can be swallowed to the stomach, where most of the bacteria are destroyed, or it can be discharged by coughs and sneezes, thus ridding the body of potentially dangerous material. That it is not always successful is attested to by the lung cancer and emphysema

271

of persistent cigarette smokers. Nicotine paralyzes the cilia and decreases their efficiency.

Although the alveoli of the lungs are small, they are so numerous—several hundred million—that their total surface area in an adult man is about 100 square meters, approximately 50 times his external body surface of about 2 square meters. It is across these 100 square meters of delicate alveolar surface that we absorb all our oxygen.

A network of tiny blood vessels called capillaries surrounds each alveolus. Their thin walls, like the alveolar walls, are only one cell thick. Thus, the blood passing through the capillaries is separated from the air by only two cells (Fig. 32.4). Some of the oxygen in the alveoli diffuses across these two cell layers into the bloodstream, which then carries it throughout the body. At the same time, waste carbon dioxide diffuses from the blood into the alveolar air. Because the movement of both gases across the alveolar and the capillary walls is solely by diffusion and not by active transport, it depends in part on the difference in concentration of the gases on either side of the membranes. A higher concentration of oxygen in the alveolar air than in the blood causes oxygen

to diffuse from the lungs into the blood, and a higher concentration of carbon dioxide in the blood than in the alveolar air causes this gas to diffuse from the blood into the lungs. At the same time, some water from the blood enters the alveoli. Chapter 34 discusses in more detail the function of the circulatory system in the exchange of gases. Exhaled air contains less oxygen and more carbon dioxide than inhaled air, and it always is saturated with water vapor.

# Breathing Movements

Diffusion of oxygen from the external atmosphere into the nose and through the several parts of the respiratory system to the alveoli would be too slow to supply a large, active organism with enough oxygen for its needs. Neither could diffusion rid the body of waste carbon dioxide rapidly enough. The forcing of air into and out of the lungs by breathing movements hastens the exchange of gases with the environment.

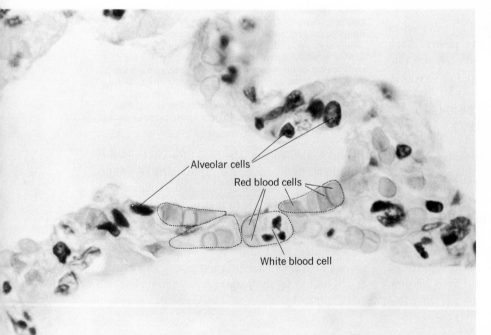

**32.4** A small section of an alveolar wall. Some of the capillaries have been outlined to make them more readily visible. (Courtesy of Dr. Beth L. Wismar.)

Alveolar cells

Red blood cells

White blood cell

Ordinarily, breathing movements are involuntary; they proceed without one's taking thought about them. They continue when one is asleep or unconscious. Despite this, it is possible to exercise voluntary control over breathing. You can hold your breath, for example, or breathe more deeply or more shallowly than usual. But one soon tires of this, and involuntary control takes over. Anyone with enough will power to hold his breath until he loses consciousness then resumes breathing as the involuntary mechanism takes over.

One large muscle, the **diaphragm,** lies slightly above the waistline and separates the thoracic cavity from the abdominal cavity. Only a few structures pierce this muscle—the esophagus and some of the major blood vessels. All of these make a tight fit, and the separation of thoracic and abdominal cavities is complete. The diaphragm is not flat, but dome-shaped, bulging upward into the thoracic cavity. When it contracts, the diaphragm flattens somewhat (Fig. 32.3). Acting in concert with the diaphragm are the **rib muscles,** which are attached to the ribs. The 12 pairs of ribs encircle the thorax, and they protect the lungs and heart contained therein. At the same time that the diaphragm contracts, the rib muscles contract also, moving the ribs slightly upward and forward. As the thorax expands, the parietal pleura, which is firmly attached to the thorax wall and diaphragm, moves outward and downward with them. Because water is not appreciably extensible or compressible the intrapleural fluid does not expand. It moves with the parietal pleura and brings with it the pulmonary pleura, which is firmly attached to the lung. This causes the elastic lungs to expand and lowers the air pressure in them. Because the air in the lungs is in direct communication with the external air, the difference in pressure cannot be maintained, and air moves in through the nose and proceeds down to the lungs. This constitutes an inspiration or inhalation of air.

Reversing the steps of an inspiration produces an expiration. When the diaphragm relaxes, it assumes its maximum curvature and pushes upward. Simultaneously, the rib muscles relax, allowing the ribs to move downward and backward. Both of these movements push against the parietal pleura. Because the intrapleural fluid is not compressible, the force is transmitted to the pulmonary pleura and the lungs. The lungs decrease in size, and the air pressure within them rises above that of the external air. This forces air out of the lungs in an expiration or exhalation of air.

The relaxation of rib muscles and diaphragm ordinarily is sufficient to bring about an expiration. Persons having difficulty in breathing may actively use other thoracic muscles and certain abdominal muscles to force air out in an expiration. Contraction of these muscles decreases the size of the thorax and thus the size of the lungs.

Breathing movements thus are controlled immediately by muscles and not by the lungs themselves. The muscles are stimulated by nervous impulses from the brain, which in turn is influenced by nerves that convey impulses to the brain, informing it whether an inspiration or expiration has just occurred. This system of brain, nerves, lungs, and muscles constitutes one of the many negative feedback mechanisms that maintain homeostasis in the human body and other organisms as well.

The concept of homeostasis has been introduced elsewhere in this book, but the body's control of its breathing provides a good opportunity to discuss negative feedback systems in homeostasis. A **negative feedback mechanism** has at least two components, each exercising such control over the other that certain conditions remain essentially the same or vary only within narrow limits. Compare this with the positive feedback mechanism (Chapter 15), which produces ever-changing conditions.

We might digress to discuss a negative feedback mechanism familiar to most persons: a thermostat and a furnace connected by an electric circuit. When working properly these are expected to keep a building warm in winter. We set the thermostat, which is merely a switch in the circuit, at some comfortable temperature. If the room temperature

273

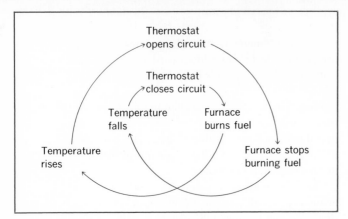

**32.5** An example of a negative feedback system outside of a living thing. See text for details. Notice that each component in the system is mentioned twice, because for each complete turn of the cycle it functions in two different ways.

falls below that indicated on the thermostat, the switch closes the circuit. This causes the furnace to start working. As the furnace warms the house, the temperature rises. When it surpasses the temperature that we set on the thermostat, the heat opens the switch and breaks the electric circuit. This turns off the furnace, and the house cools until the cooling of the thermostat closes the switch and starts the process all over again. If all parts of the system are in good working order, the temperature within the house does not vary more than a degree or two. Notice that neither the furnace alone nor the thermostat alone could accomplish this, but that by turning each other "on" and "off" (Fig. 32.5), they maintain a constant temperature while the parts remain in working order and fuel for the furnace and electrical energy to run its motor are in supply.

This example is sufficient to describe a negative feedback system, but the perceptive reader will see that when the temperature rises in summer, the furnace and thermostat no longer can maintain a comfortable temperature. When pressed beyond its limits, a negative feedback system can fail; in this particular example the furnace is not intended to cool. To maintain a comfortable temperature in the building throughout the year, we can add an-

other negative feedback system—an air conditioner with its thermostat. The living organism contains many negative feedback mechanisms that work in concert, but it will be convenient to consider some of them separately.

We might return now to the negative feedback mechanism that controls breathing. The **respiratory center** of the brain is connected to the respiratory muscles (diaphragm and rib muscles) by motor neurons that transmit impulses in one direction—from the respiratory center to the muscles. A set of sensory neurons conducts impulses from the lungs to the respiratory center. At one moment the respiratory center may be sending a stimulatory impulse via the motor neurons to the respiratory muscles, thus causing them to contract (Fig. 32.6). This inflates the lungs in an inspiration. The expansion of the lungs initiates impulses in the sensory neurons that extend from the lungs to the brain. These impulses inhibit the breathing center, which then ceases to send its stimulatory impulses to the respiratory muscles. No longer stimulated, these muscles relax, and the lungs deflate in an expiration. The deflated lungs cease stimulating the sensory neurons to send inhibitory impulses to the respiratory center of the brain. No longer inhibited, the respiratory center once again sends out stimulatory impulses to the respiratory muscles, and the process repeats.

This set of events maintains regular breathing. The muscles alone could not do this, neither could the lungs nor the brain; but the entire feedback mechanism with its various parts turning each other "on" and "off" does maintain regular breathing movements under ordinary conditions. Disease, drugs, or damage to nerves or muscles may disrupt the mechanism.

Most of the time, breathing movements supply oxygen as rapidly as the body needs it and remove carbon dioxide as fast as it forms. Occasionally, however, carbon dioxide may begin to accumulate in the blood or the oxygen concentration may fall—two changes that often occur simultaneously. Either of these conditions increases the rate and-

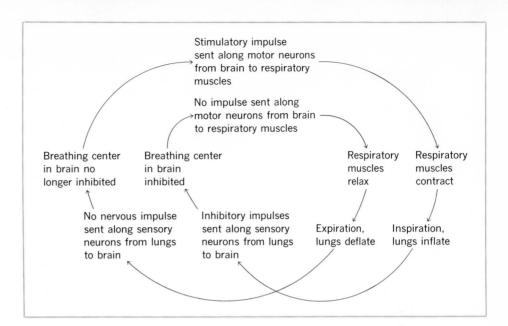

**32.6** A negative feedback mechanism controls normal breathing. See text for details. Each component is indicated twice because each functions in two different ways at different places in the cycle.

Diagram labels:
- Stimulatory impulse sent along motor neurons from brain to respiratory muscles
- No impulse sent along motor neurons from brain to respiratory muscles
- Breathing center in brain no longer inhibited
- Breathing center in brain inhibited
- Respiratory muscles relax
- Respiratory muscles contract
- No nervous impulse sent along sensory neurons from lungs to brain
- Inhibitory impulses sent along sensory neurons from lungs to brain
- Expiration, lungs deflate
- Inspiration, lungs inflate

/or the depth of breathing, but the body seems to be more sensitive to the carbon dioxide concentration in the blood than the oxygen concentration. The change in breathing accelerates the removal of carbon dioxide from the blood and brings oxygen to it more rapidly.

Carbon dioxide dissolved in the blood reacts with water and forms carbonic acid, which ionizes to bicarbonate and hydrogen ions:

$$CO_2 + H_2O \rightleftharpoons H_2CO_3 \rightleftharpoons HCO_3^- + H^+$$

As more carbon dioxide enters the blood, the hydrogen ion concentration rises. It is believed that the high concentration of hydrogen ions rather than the dissolved carbon dioxide gas stimulates breathing. Futhermore, there is some evidence that the change in hydrogen ion concentration in the blood itself does not directly influence breathing, but rather it is the resulting change in hydrogen ion concentration in the fluid that surrounds the brain.

Neurons with endings in blood vessels (specifically, the aorta and the common carotid arteries) are sensitive to oxygen concentration. These neurons monitor the blood continuously, and when the oxygen concentration begins to fall, they stimulate the respiratory center.

During rest, the lungs of an adult usually hold about 3 liters of air after an inspiration and about 2.5 liters after an expiration, the difference of 0.5 liters representing the amount of air inspired and expired with each set of breathing movements. During violent exercise, greater volumes of air are inspired and expired—perhaps 1, 2, or 3 liters, depending on the size of the person and his body's immediate need for oxygen. The maximum amount of air that the respiratory system (including all passageways) of an adult man can hold during a forced inspiration is about 6 liters.

## Questions

1. Define the following terms:
   pharynx
   larynx
   vocal cords
   pleura
   alveolus
   diaphragm
   negative feedback mechanism

**2.** How does the structure of the respiratory system increase the surface area through which the body exchanges gases with the environment?

**3.** What gases does the respiratory system exchange with the environment, and in what direction does each gas move?

**4.** Follow some air from the time it enters the nostrils until it leaves through them. Through what organs does the air pass? How does its composition change? Where does the change occur?

**5.** Distinguish between the several uses of the term *respiration*.

**6.** What role does diffusion play in the exchange of gases?

**7.** In the case of a large active animal such as a human being, diffusion alone is inadequate to supply the body with sufficient oxygen. How is the difference made up?

**8.** Why do you not ordinarily swallow food into the trachea?

**9.** How does the respiratory system rid itself of small airborne particles?

**10.** How does the muscular system contribute to breathing movements? How does the nervous system contribute to breathing movements?

**11.** Describe the negative feedback mechanism that permits you to breathe without conscious effort.

**12.** Why would it be impossible to commit suicide by holding your breath until you die?

## Suggested Readings

Bloom, W., and D. W. Fawcett. *A Textbook of Histology*, 9th ed. Philadelphia: W. B. Saunders Company 1968.

Comroe, J. H., Jr. "The Lung," *Scientific American* 214(2): 56–68 (February 1966).

Fenn, W. O. "The Mechanism of Breathing," *Scientific American* 202(6): 138–148 (June 1960).

Griffin, D. R., and A. Novick. *Animal Structure and Function*, 2nd ed. New York: Holt, Rinehart and Winston, Inc., 1970.

Langley, L. L. *Homeostasis*. New York: Reinhold Publishing Corporation, 1965.

Larimer, J. *An Introduction of Animal Physiology*. Dubuque, Iowa: William C. Brown Company, Publishers, 1968.

Vander, A. J., J. H. Sherman, and D. S. Luciano. *Human Physiology*. New York: McGraw-Hill Book Company, 1970.

Muscles move the body. Walking to the store, riding a bicycle, operating a typewriter, moving of food through the digestive system, breathing, and beating of the heart all are accomplished by muscles. Large multicellular animals cannot depend on sufficient food and oxygen to come to them and so must bring these commodities to themselves. Some muscles enable the individual actively to seek food and to force air into and out of the lungs. Other muscles help to move substances around the body once they are obtained. Muscles comprise almost half of the human body weight.

Muscles function by virtue of the ability of their cells, often called muscle fibers, to contract. When the individual cells contract, the muscle contracts. Contractility of protoplasm is a property shared by many kinds of cells, but few possess it to the degree that muscle cells do. Three types of muscles keep the several parts of the body moving: skleletal (striated) muscle, visceral (smooth) muscle, and cardiac muscle. Most of the muscular system of the body consists of skeletal muscle; this is what constitutes steak, hamburger, breast of chicken and many other meats. Table 33.1 summarizes some of the characteristics of muscles.

## Skeletal Muscle

Skeletal muscles (Fig. 33.1) receive their name from the fact that most of them are connected to bones (not directly, but by tendons, a type of connective tissue); some skeletal muscles connect with skin. For the most part, skeletal muscles are under control of the will and so are called voluntary muscles. They are responsible for the overt movements of the body, such as moving of the arms and legs. A few skeletal muscles do function automatically without our taking thought; notable among them are the rib muscles and the diaphragm, which keep us breathing whether we are awake or asleep.

# 33

# The Muscular System

1 **The function of the muscles is the movement of parts of the body.**

2 **Skeletal muscles, which are under voluntary control, are responsible for overt movements of the body, including breathing. They also help to support the body.**

3 **Visceral muscles, which are not under voluntary control, are responsible for movements of most internal organs.**

4 **Cardiac muscle, or heart muscle, keeps the heart beating.**

5 **Muscles move body parts by contraction; when muscles relax they perform no work.**

| | Skeletal Muscle | Visceral Muscle | Cardiac Muscle | Table 33.1 Characteristics of Muscle Cells |
|---|---|---|---|---|
| | Striated | Smooth | Striated | |
| | Voluntary | Involuntary | Involuntary | |
| | Contracts rapidly | Contracts slowly | Contracts at inter-mediate speed | |
| | Connected mostly to bones, but also to skin, used in locomo-tion and other overt motions of arms, legs, face, and trunk | Serve several internal organs—digestive tract, blood vessels, urinary bladder | Heart muscle, keeps heart beating | |

Individual skeletal muscle cells are long and slender. Under an ordinary student microscope, these multinucleate cells show distinct transverse bands (Fig. 33.2), and for this reason they are called striated muscle. These cells lie parallel to each other in spindle-shaped bundles, several such bundles forming a muscle. Connective tissue surrounds each muscle and each of its bundles of cells. Penetrating the muscle are the blood vessels and nerves that serve them. Blood brings to the muscles food and oxygen, both of which are in great demand when the muscle is actively contracting. Blood likewise removes carbon dioxide and lactic acid, products of muscle activity. Nerve endings reach each skeletal muscle cell. When stimulated by a nerve impulse, the muscle cell contracts and remains contracted as long as it is stimulated and as long as energy is available to it. When the stimulation ceases or the energy supply is exhausted, the muscle cell relaxes. The response of skeletal muscle is very rapid, 0.01 to 0.1 second being sufficient for a contraction and a relaxation. This speed, coupled with the rapidity of transmission of impulses along neurons, allows for essentially immediate response in emergencies.

A muscle cell of either skeletal, visceral, or cardiac muscle contracts only when it receives a stimulus of sufficient intensity, and then it contracts fully. Stimuli of higher intensity do not elicit greater contraction. Stimuli below a certain intensity cause no contraction at all. Muscle cells do not contract partially. This all-or-none principle applies only to the individual cells and not to entire muscles. Depending on the number of its cells in the contracted state, a muscle may contract only slightly or to its maximum degree of contraction. In this way, just the appropriate amount of energy is expended in picking up a pencil or hitting a baseball with a bat.

All the muscle cells served by a single neuron are stimulated simultaneously and contract simultaneously. One neuron and all the muscle cells it serves comprise a **motor unit.** A motor unit may contain as few as only four or five muscle cells or it may contain hundreds.

Paler than the transverse striations of skeletal muscle cells and much more difficult to see under an ordinary student microscope are the **myofibrils**—long, fine fibrils that lie within the cell parallel to its long axis. The higher magnifications of the electron microscope not only reveal the internal structure of the myofibrils, but suggest the mechanism of muscle contraction.

Each myofibril consists of a linear series of repeating units called **sarcomeres.** Dark, narrow, transverse bands (Z lines) mark the ends of the sarcomeres (Figs. 33.3 and 33.4), and a dark, wide, transverse band (A band) occupies the center of each sarcomere. In a relaxed cell a smaller, lighter band (H zone) can be seen within the A band. Between the A bands are the lighter I bands (a Z line lies in the center of an I band). The sarcomeres of one myofibril coincide with the sarcomeres of adjacent myofibrils, and so do their bands. This

278

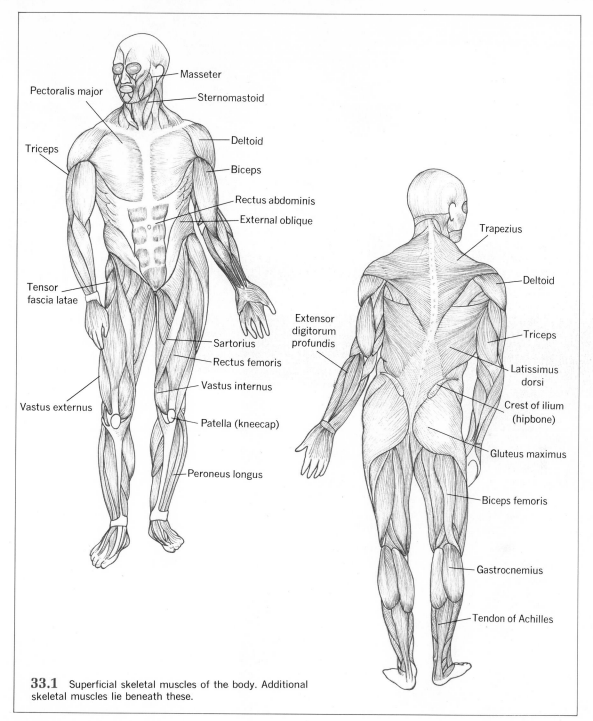

**33.1** Superficial skeletal muscles of the body. Additional skeletal muscles lie beneath these.

279

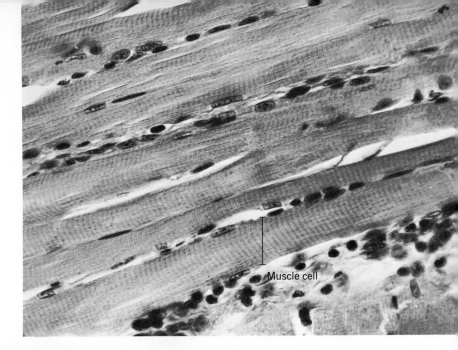

**33.2** Skeletal muscle of opossum tongue. Each cell or fiber is very much elongated. Note the striations that give this muscle its other name of striated muscle. (Courtesy of Dr. Beth L. Wismar.)

Muscle cell

**33.3** Electron micrograph of skeletal muscle of rabbit. Myosin filaments are darker than the actin filaments. Many small bridges can be seen between the actin and myosin filaments. (Dr. H. E. Huxley.)

H zone

A band

I band

Z line

Actin filament

Myosin filament

is what gives striated muscle cells their striated appearance.

Close inspection of very high magnifications shows many extremely fine proteinaceous filaments lying parallel to the length of the myofibril. Some of them, composed of the protein **myosin,** run the length of the A band. The myosin filaments are thick, and they are partially responsible for the darkness of the A bands. The remaining filaments are composed of the protein **actin.** They extend from the Z line into the A bands; they are thinner than the myosin filaments and form the light I bands and contribute to the darkness of the A bands as well.

## Contraction of Skeletal Muscle

The contraction and relaxation of a sarcomere depends on the sliding of actin and myosin filaments back and forth along each other. In a relaxed sarcomere, the actin filaments extending from opposite ends of the sarcomere do not meet; their failure to meet is responsible for the light H zone in the middle of the dark A band. When the actin filaments from opposite ends of the sarcomere move over the myosin filaments, they approach each other, thus narrowing the H zone and also the I bands; both of them may disappear. What is more important, the movement of the filaments causes the sarcomere to contract. This process repeated in all the sarcomeres of the myofibril shortens the myofibril, and the contracting of the myofibrils contracts the cells. It is this process that permits a muscle to support an object many times its own weight. (If you hang from a rope by one hand, the muscles in your arm support the weight of your entire body.)

When a sarcomere relaxes, the actin filaments move back to their original position, and the sarcomere is restored to its original length. This relaxes the myofibrils and the cell. The H zone and I band reappear and widen.

The exact mechanism of the movement of actin filaments along the myosin filaments is not under-

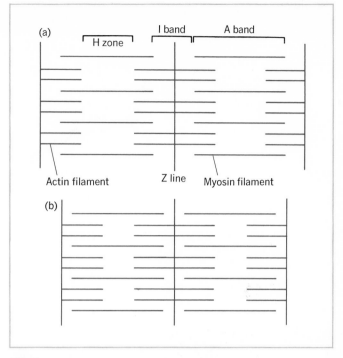

**33.4** Relaxed sarcomere (a) and contracted sarcomere (b).

stood completely, but it is believed to involve the repeated forming and breaking of bridges between the actin and myosin filaments. With the forming and breaking of each bridge, the actin filaments are pulled a short way along the myosin filaments—possibly in a rachetlike fashion.

Whatever the method of contraction, it requires great quantities of ATP generated by respiration. In the cytoplasm between the myofibrils lie the mitochondria, sites of aerobic respiration, which produces more ATP per quantity of glucose oxidized than does the glycolysis of glucose to lactic acid (Chapter 12). The latter occurs in muscle cell cytoplasm but not in the mitochondria. However, if the muscle is active for a long time, the circulatory system no longer is able to supply it with sufficient oxygen to oxidize glucose completely. When this happens, the cells rely on lactic acid fermentation, a type of respiration that generates a small amount of ATP as glucose is converted to lactic acid; in this process one glucose molecule

281

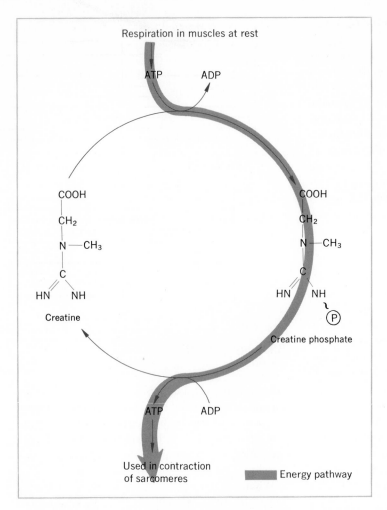

Respiration in muscles at rest

ATP      ADP

COOH                    COOH

CH₂                     CH₂

N—CH₃                   N—CH₃

C                       C

HN   NH                 HN   NH

Creatine                         P

                        Creatine phosphate

ATP      ADP

Used in contraction
of sarcomeres            ▬▬▬ Energy pathway

**33.5**  Creatine cycle by which muscles store energy.

282

forms two lactic acid molecules. Although lactic acid fermentation is much less efficient in terms of energy made available per unit of fuel used, at least it releases some energy when the lack of oxygen prevents aerobic respiration from proceeding maximally. Later, when the muscle relaxes and more oxygen becomes available, some of the lactic acid is oxidized with the generation of more ATP, and some is transported to the liver, which converts it back to glucose and then to glycogen, which is stored.

Muscles have still another method of supplying energy quickly to the sarcomeres. ATP never is stored as such in muscle cells, but the cells contain creatine, a compound capable of reacting with ATP. The resulting product is **creatine phosphate,** which contains a high-energy phosphate bond (Fig. 33.5). Creatine phosphate forms from ATP produced by respiration in resting muscle cells. The cells store a limited amount of creatine phosphate until renewed muscular activity requires ATP. Then creatine phosphate transfers its phosphate group to ADP, forming the needed ATP. Later, when the muscle relaxes, the creatine is used in the formation of more creatine phosphate again.

A muscle cell that has exhausted all its sources

of ATP no longer can contract and is said to be fatigued. Until its respiration can produce more ATP, it is unresponsive to stimulation by neurons.

In addition to supplying the energy necessary for movement, respiration of muscle cells produces most of the heat that keeps the body warm. On a brisk winter day, a little exercise brings welcome warmth. Shivering, which consists of rhythmic muscular tremors, rapidly generates heat that warms the body in cold weather.

Skeletal muscles have two (occasionally more) distinct ends attached by tendons to two (or more) different bones. Because muscles function only by contracting, a given skeletal muscle can move a given bone only by pulling it. Relaxation of the

**33.6** A pair of antagonistic muscles. Contraction of the biceps bends the arm at the elbow, and contraction of the triceps straightens the arm.

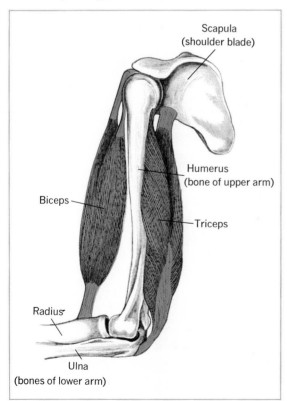

Scapula
(shoulder blade)

Humerus
(bone of upper arm)

Biceps

Triceps

Radius

Ulna
(bones of lower arm)

muscle does not push the bone back to its original position. Rather, another muscle attached in a somewhat different way to the same bone pulls it back. Most skeletal muscles occur in pairs, each performing opposite or antagonistic functions that "undo" what the other muscle does.

The biceps and triceps muscles of the upper arms constitute an antagonistic pair (Fig. 33.6). If we consider an arm bent at the elbow, then the biceps occupies the inner, smaller angle, and the triceps extends around the outside of the elbow. The upper end of the biceps is attached at the shoulder, and the lower end is attached to a bone in the lower arm just below the elbow. Now if the biceps muscle contracts, the lower arm will be drawn up against the upper arm. The triceps is also attached at the shoulder and to a bone in lower arm, but because it is on the outside of the elbow, its contraction causes the arm to straighten at the elbow. Most of the overt movements of the body are controlled by pairs of antagonistic muscles bending and straightening joints or raising and lowering structures at appropriate times.

Sphincter muscles are circular muscles; the long axes of their cells are arranged parallel to the circumference of the muscle. When they contract they close an opening. The anal sphincter, which is partially under voluntary control, closes the opening of the rectum. Unlike most muscles, it has no antagonistic muscle.

Few movements involve only a single muscle. So simple an action as walking across a room requires the coordinated contraction and relaxation of many muscles. This coordination is provided by the nervous system.

In addition to moving the parts of the body, the muscles contribute to its support. The bony skeleton provides a framework that alone would collapse. Simultaneous contraction of antagonistic skeletal muscles supports the skeleton, with the degree of contraction of the several muscles determining the position of the bones and the posture of the body. Not only do the resurrected skeletons of Halloween fiction move mysteriously without

283

muscles, they defy the law of gravity by managing to remain erect.

# Visceral Muscle

The cells of visceral muscles are long and tapered, each with a single nucleus (Fig. 33.7). Rather than forming spindle-shaped muscles, the cells are arranged in flat sheets that surround the organs they serve. They are called smooth muscles, because they lack obvious transverse striations and appear to be smooth under an ordinary optical microscope. They also are called involuntary muscles, because ordinarily they are not under control of the will, but function automatically when their action is needed.

Visceral muscles control the movements of many internal organs, including those of the digestive tract, arteries, and urinary bladder. Here, too, there exist antagonistic muscles. Within the walls of the esophagus, stomach, and small and large intestines, for instance, are two layers of involuntary muscle. In a cross section they appear as two concentric rings (Fig. 31.3). The muscle cells in the inner layer of muscle, called **circular muscle,** lie with their long axes encircling the organ; when they contract they constrict the organ, decreasing its width but increasing its length. The cells in the outer layer, called **longitudinal muscle,** lie with their long axes parallel to the long axis of the organ. When they contract, they cause the organ to become wider and shorter.

From the time that swallowing pushes food into the esophagus until the unabsorbed portion leaves the body in defecation, **peristalsis** moves the food along the digestive tract. Peristalsis is the forward movement of a ring of contraction along an organ such as the esophagus, stomach, or intestine, and it is followed by a wave of expansion that restores

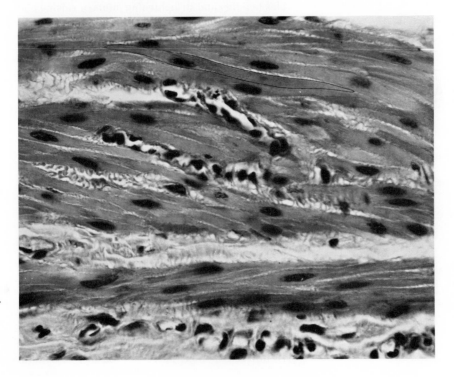

**33.7** Smooth muscle of monkey uterus. Smooth muscle cells are shorter than those of skeletal muscle. Each cell is spindle-shaped and has one nucleus. One cell is outlined. (Courtesy of Dr. Beth L. Wismar.)

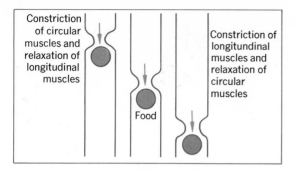

**33.8** Peristalsis in esophagus.

the original diameter (Fig. 33.8). Swallowed food moves down the esophagus not so much by being pulled by gravity as by being pushed by the advancing ring of constriction. (The movement of an earthworm along a surface might serve as a model of peristalsis in the esophagus. A wave of contraction clearly proceeds from one end of the earthworm to the other.)

Superimposed on peristalsis are other muscular actions. Food usually resides for a few hours in the stomach, the walls of which are thick and muscular. Here the muscles subject the food to the violent churning action that breaks it into small particles and mixes them well with the gastric juices. Another mixing action, called **rhythmic segmentation,** happens in the small intestine. Here, at one moment, several rings of constriction alternate with areas of expansion. At the next moment the expanded areas constrict and the constricted areas expand (Fig. 33.9). This action serves further to mix food well with the digestive juices.

**33.9** Rhythmic segmentation in the small intestine.

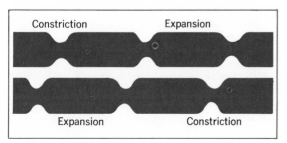

Other internal organs served by involuntary muscle include the urinary bladder, some arteries and a few veins, and parts of the reproductive system. Strong contractions of the smooth muscles of the uterine wall expel the baby during childbirth.

Several sphincters are smooth muscle. The sphincter between the esophagus and stomach ordinarily is in a state of contraction, but it relaxes, permitting food to enter the stomach, and then constricts again. Rarely does it allow food to return back to the esophagus, but this does happen during vomiting. Another sphincter muscle between the stomach and the small intestine keeps food in the stomach until the time comes for it to move into the small intestine. The pupil of the eye constricts and dilates because it consists of two antagonistic smooth muscles. The contraction of a sphincter muscle constricts the pupil. The dilator muscle, like the sphincter, is ring-shaped, but the muscle cells are arranged radially, and when they contract, the pupil dilates.

Contraction of smooth muscle is less well understood than contraction of skeletal muscle. Special stains show fine longitudinal myofibrils that probably function like those of skeletal muscle cells. They do contain actin and myosin, but they do not seem to have the same ordered structure that the myofibrils of skeletal muscles do. Contraction of smooth muscle is slower than that of skeletal muscle, and a contraction may last for several seconds. Not all smooth muscle cells are connected to neurons. A smooth muscle cell stimulated to contract by a nervous impulse may spread the stimulation to neighboring muscle cells.

# Cardiac Muscle 285

The heart is one mass of cardiac muscle that acts as a pump, keeping blood flowing throughout the

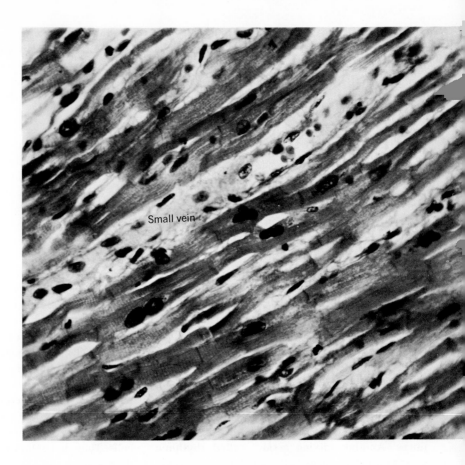

Small vein

**33.10** Human cardiac muscle. The cells are striated but much shorter than those of skeletal muscle. The Y-shaped cells give the tissue the appearance of a network. (Courtesy of Dr. Beth L. Wismar.)

body. Unlike other involuntary muscle, it appears striated, but the cells usually have one nucleus each (Fig. 33.10). The cells are cylindrical, with flat ends that fit against the ends of adjacent cells. Some of the cells are branched, or Y-shaped, and these connect with three cells instead of just two. This intercommunicating network may assist in coordination of all parts of the heart. Unlike other striated muscle, cardiac muscle does not require nervous stimulation to contract, but it is richly supplied with nerve endings that control the rate of beating. The heart usually continues its rhythm of contraction and relaxation without any appreciable interruptions for an entire lifetime. Any but short interruptions may result in death.

286

## Questions

1. Define the following terms:
   motor unit
   sarcomere
   rhythmic segmentation
   peristalsis
   circular muscle
   longitudinal muscle
   sphincter
2. Distinguish between skeletal, visceral, and cardiac muscles. Which of these are voluntary? Which are involuntary? Which are smooth muscles? Which are striated muscles?

**3.** What kinds of muscles accomplish the following actions?

walking

heartbeat

breathing

typing

mixing of food in the small intestine

moving food from small intestine to large intestine

**4.** Explain the advantage of having muscles function in antagonistic pairs. Describe the working of one antagonistic pair of skeletal muscles and one antagonistic pair of visceral muscles.

**5.** How do muscles assist in the support of the body?

**6.** Describe what happens when a muscle cell contracts and when it relaxes. What is meant by the all-or-none principle of muscular contraction?

**7.** What are the several sources of energy available to muscle cells? From a biochemical standpoint what is meant by a fatigued muscle?

**8.** What causes shivering? Of what value to the body is shivering?

**9.** In what ways does cardiac muscle differ from striated muscle? From visceral muscle?

## Suggested Readings

Bloom, W., and D. W. Fawcett. *A Textbook of Histology,* 9th ed. Philadelphia: W. B. Saunders Company, 1968.

Griffin, D. R., and A. Novick. *Animal Structure and Function,* 2nd ed. New York: Holt, Rinehart and Winston, Inc., 1970.

Hayashi, T. "How Cells Move," *Scientific American* 205(3): 184–204 (September 1961).

Hoyle, G. "How Is Muscle Turned On and Off?" *Scientific American* 222(4): 84–94 (April 1970).

Huxley, H. E. "The Contraction of Muscle," *Scientific American* 199(5): 66–82 (November 1958).

———. "The Mechanism of Muscular Contraction," *Scientific American* 207(6): 18–27 (December 1962).

Vander, A. J., J. H. Sherman, and D. S. Luciano. *Human Physiology.* New York: McGraw-Hill Book Company, 1970.

# 34

# The Circulatory System

1 The circulatory system transports food, oxygen, wastes, and other substances throughout the body.

2 Exchange of materials between the blood and the body tissues occurs exclusively in the capillaries.

3 Oxygen is transported primarily by the hemoglobin molecules in the red blood cells.

4 Carbon dioxide is transported by both red blood cells and plasma.

5 White blood cells provide resistance to contagious diseases by ingesting microorganisms that have invaded the body.

6 Plasma cells, which are a type of white blood cell, produce gamma globulins, antibodies that confer immunity to some diseases.

An animal that absorbs its food in one part of its body and its oxygen in another and uses them both in still other places requires a mechanism of distributing these substances throughout the body as they are needed. Wastes also must be carried to the organs that dispose of them. These functions are carried on by the circulatory system, which not only distributes substances within the body, but also plays an important role in immunity and protection from contagious diseases.

The circulatory system consists of a system of channels that transport two fluids, blood and lymph, throughout the body. For convenience we may separate the circulatory system into the blood vascular system, which carries blood, and the lymphatic system, which carries lymph. The thymus and the spleen are associated organs.

## Blood Vascular System

The blood vascular system is a closed one, consisting of the heart, arteries, capillaries, and veins, with the blood confined to these structures and circulating through them. The **heart,** which is composed of cardiac muscle tissue (Chapter 33), pumps blood through itself into the blood vessels called **arteries.** The arteries (Fig. 34.1) branch into successively smaller and smaller arteries, the smallest of which are the **arterioles.** These finally branch into **capillaries,** the finest blood vessels of all. The capillaries proceeding from each of the arterioles branch several times and unite with each other forming a network. Each capillary net leads into a **venule,** or small vein. Venules join into larger **veins,** and these join into still larger veins. The largest of them return blood to the heart (Fig. 34.2).

The walls of arteries and veins consist of three layers: an inner endothelium, a middle layer of smooth muscle, and an outer layer of connective tissue. The smooth muscles control to some extent the amount of blood that flows through different

**34.1**   The major arteries of the body.

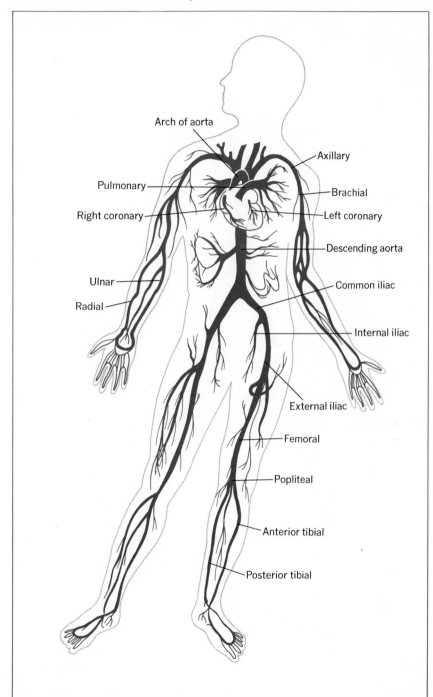

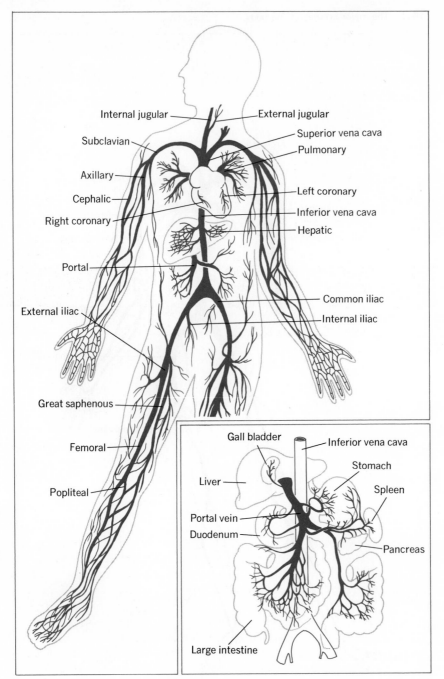

Internal jugular

External jugular

Subclavian

Superior vena cava

Pulmonary

Axillary

Cephalic

Left coronary

Inferior vena cava

Right coronary

Hepatic

Portal

Common iliac

External iliac

Internal iliac

Great saphenous

Femoral

Popliteal

Gall bladder

Inferior vena cava

Liver

Stomach

Spleen

Portal vein

Duodenum

Pancreas

Large intestine

**34.2** The major veins of the body.
Insert: hepatic portal system.

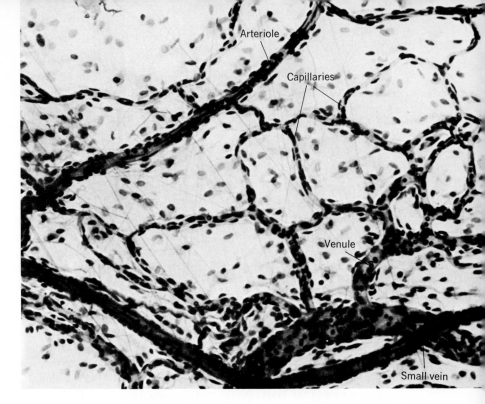

Arteriole

Capillaries

Venule

Small vein

**34.3** A capillary bed. Exchange of food, oxygen, minerals, and wastes between the blood and body tissues occurs across the surfaces of the capillaries. Though small, they are numerous and provide a great deal of surface area. (Courtesy of Dr. Beth L. Wismar.)

vessels. This is especially true of the arteries, which are better supplied with muscles than the veins. The capillaries consist of endothelial tissue alone, the walls being only one cell thick.

Arteries always conduct blood away from the heart, and the veins always return blood to the heart. However, it is the capillary nets that accomplish some of the most important work of the circulatory system—that of picking up and delivering materials. Substances in the blood and substances in body tissues are exchanged only across the capillary endothelium, and not across the walls of arteries and veins. Capillary nets serve nearly all the living tissues of the body, their concentration in the tissue depending on the local need for exchange of materials. Muscles, which are called upon frequently to move the body, and the kidneys, which must remove waste products constantly, require great quantities of food and oxygen and are well supplied with capillaries. The lens of the eye, a very inactive tissue, has none.

Because these minute capillaries are so numerous, they present a great deal of surface area that

contacts other tissues; more than 6,000 square meters (about 3,000 times the total external surface area of the body) allow for a great deal of exchange between the blood and the tissues (Fig. 34.3). The thinness of the capillary wall, only one cell thick, also facilitates exchange.

Muscles in the arteries and arterioles leading to capillaries control the amount of blood reaching the capillaries. A precapillary sphincter of smooth muscle surrounds the entrance to each capillary. By opening and closing, the sphincter regulates the movement of blood from an arteriole into a capillary. The more the sphincter remains open, the more blood enters and the more the capillary dilates. The more the sphincter remains closed, the less blood enters and the more the capillary constricts. In the tissues with the greatest immediate need for blood, precapillary sphincters are open, allowing more blood to move through the capillaries. When the need for blood declines, the precapillary sphincters close, and so the capillaries transport less blood. When one takes a brisk walk before dinner, his active skeletal muscles receive more blood than do the tissues of the empty stomach and small intestine. When he rests after dinner, the stomach and small intestine receive more blood, the skeletal muscles less.

## Heart

The human **heart,** approximately centrally located in the chest, is a tough muscular organ divided into four chambers (Fig. 34.4). The two on the right side communicate with each other, as do the two on the left, but there is no opening between the two sides. (This is not true before birth. Fetal circulation is discussed in Chapter 40.) The two upper chambers of the heart are called atria, the two lower chambers are ventricles.

Blood from any body tissue other than the lungs returns to the heart through either of two veins: **superior vena cava** and **inferior vena cava.** The former drains blood from the arms, head, and the upper part of the body; the latter drains the lower part of the body and the legs. Having come from living and therefore respiring tissues, the blood in these veins, called spent blood, has a low oxygen content and a high carbon dioxide content. It passes from the venae cavae into the **right atrium** of the heart. A contraction of the atrium forces blood into the lower chamber, the **right ventricle.**

The **atrioventricular (AV) valve** between the two chambers prevents backward flow of blood from the ventricle to the atrium (Fig. 34.5). The valve consists of three flaps of tissue that together form a more or less funnel-shaped arrangement, the narrow end extending into the ventricle. The pressure of blood in the atrium forces the valve open, but when pressure develops in the ventricle, the pressure pushes the flaps against each other, effectively closing the opening.

Because of the presence of the AV valve, a contraction of the right ventricle can force blood out only through the **pulmonary trunk,** the only artery departing from the right ventricle. Another valve, the **pulmonary valve,** between the right ventricle and the pulmonary trunk prevents backward flow of blood. Shortly after leaving the right ventricle, the pulmonary trunk branches into the two **pul-**

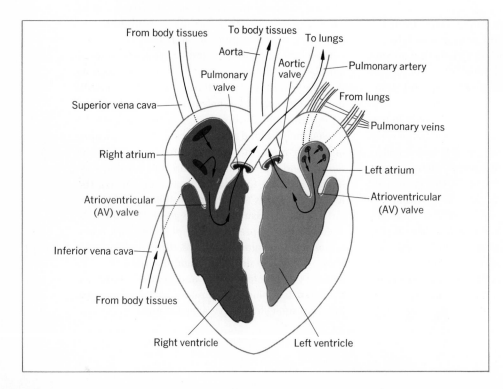

**34.4** Circulation within the heart. The arrows indicate the flow of blood. Actually all four valves are not open simultaneously (see Fig. 34.5).

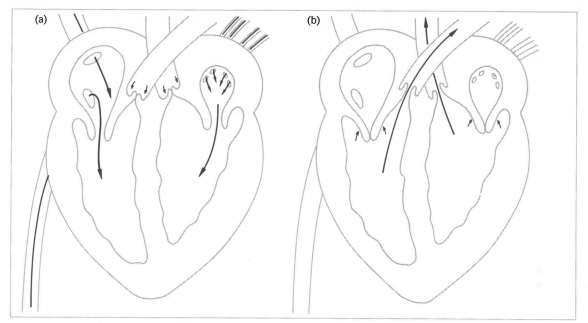

**34.5** Beating of the heart. (a) Contraction of the atria. The AV valves are open, permitting blood to flow from the atria to the ventricles. The pulmonary and aortic valves prevent backward flow (small arrows) into the ventricles from the pulmonary artery and the aorta. (b) Contraction of the ventricles. The pulmonary and aortic valves are open, permitting blood to flow from the ventricles into the pulmonary artery and the aorta. The AV valves prevent backward flow (small arrows) from the ventricles to the atria.

**monary arteries,** each of which serves a different lung. In the capillaries of the lungs, spent blood releases the carbon dioxide it received from the body tissues and receives a fresh supply of oxygen.

After passing through the lungs, oxygenated blood returns to the heart by way of the **pulmonary veins,** which enter the **left atrium.** Contraction of this chamber forces blood into the **left ventricle.** Another AV valve between the left atrium and the left ventricle prevents backward flow into the left atrium. Contraction of the left ventricle forces blood into the **aorta,** the largest artery of the body. The **aortic valve** prevents backward flow into the left ventricle.

The branches of the aorta carry oxygenated blood to all parts of the body except the lungs. In the brain, a muscle, a gland, or some other organ, the oxygenated blood becomes spent blood as it releases its oxygen and accepts carbon dioxide from the tissue. The veins that drain capillaries carry spent blood to either the superior vena cava or the inferior vena cava, both of which return blood to the right atrium of the heart.

The heart may be considered as two pumps, the right side pumping spent blood from body tissues to the lungs, the left side pumping oxygenated blood from the lungs to body tissues. In one complete circuit around the body, blood passes through the right side of the heart, the lungs, the left side of the heart, and one (rarely two) other organ of the body and back to the right side of the heart again. All the arteries except pulmonary arteries carry oxygenated blood, and all the veins except pulmonary veins carry spent blood.

293

With a few exceptions, oxygenated blood passes through only one capillary net in only one organ during its trip from the left ventricle back to the right atrium. For example, blood traveling from the heart to a muscle in the left leg returns immediately to the heart and does not pass through the pancreas or any other organ on its return. Neither does blood going from heart to pancreas serve a bone or muscle on its way back to the heart. One exception is the passage of blood from the stomach, small intestine, and some other abdominal organs to the liver before it returns to the heart. Capillaries in the stomach wall and in the villi of the small intestine receive food which the **hepatic portal system** (Fig. 34.2) carries to the liver where the food is released from a second set of capillaries. From the liver, blood returns to the heart by way of the inferior vena cava.

When the blood sugar level begins to fall, the liver digests some of its glycogen to glucose. Blood passing through the liver receives the glucose, which it carries throughout the body (after having passed through the heart). When the blood reaches other tissues, the glucose passes through the capillary walls into these tissues.

Every tissue contributes its waste products to the blood as it passes through the capillaries. The two main metabolic wastes are carbon dioxide and ammonia; the latter is a waste product of the metabolism of amino acids and other nitrogen-containing compounds. We have already seen that much of the carbon dioxide is delivered to the lungs, which then dispose of it (Chapter 32). Some of the carbon dioxide together with the ammonia reaches the liver, which converts them to urea. The overall reaction is: *Compound found in urine due to protein metabolism*

$$CO_2 + 2NH_3 \longrightarrow H_2N-\overset{\text{O}}{\underset{\|}{C}}-NH_2 + H_2O$$

More detail is given in Fig. 35.1. The urea passes into the blood, which carries most of it to the kidneys, which then absorb it and dispose of it (Chapter 35). Some urea also leaves the body through the skin in perspiration.

Blood also transports hormones from the glands that produce them to the sites at which they exert their effect. Heat, a product of respiration, especially in muscle cells, is carried to the skin, from which it dissipates into the environment.

Being an active muscle, the heart requires a steady supply of oxygen, more than can be supplied by the blood that the heart pumps through its four chambers. The walls of the chambers do not absorb nutrients, and in any event they provide only a fraction of the surface that would be required for adequate absorption. Instead, the heart is served like any other organ, by its own arteries, the coronary arteries, which branch directly from the aorta. Their capillaries are embedded within the thick heart muscle.

The walls of arteries sometimes become thickened on the inside by the accumulation of fatty deposits. Their lumens become smaller and smaller, permitting less and less blood to pass through them. This condition is known as hardening of the arteries, or arteriosclerosis. The presence of the fatty material seems to encourage the formation of blood clots that block the arteries completely. If this occurs in the coronary arteries, it prevents blood from reaching the cardiac muscle, and a heart attack, or coronary thrombosis, results. There is some evidence that a high cholesterol content in the diet leads to thickening of artery walls and thus predisposes one to heart attacks, but a final decision on this awaits more research. Because the body manufactures cholesterol from

**34.6** During one heartbeat, an atrial systole (contraction) lasts one eighth of a second, and it is followed by a ventricular systole lasting three eighths of a second. An atrial diastole (relaxation) lasts seven eighths of a second, and a ventricular diastole lasts five eighths of a second.

| | 8/10 second | | | | | | | |
|---|---|---|---|---|---|---|---|---|
| Atrium | S | D | D | D | D | D | D | D |
| Ventricle | D | S | S | S | D | D | D | D |

S=systole  D=diastole

saturated fatty acids, diets rich in these compounds are suspected of contributing to heart attacks, too.

The beating of the heart pumps blood through the body, each beat requiring approximately eight tenths of a second (= 75 beats per minute), although this varies with age, exercise, state of health, and emotion. Unlike other striated muscles, the heart requires no nervous stimulation to contract. A heart continues to beat if all nerve connections to it are severed. However, the nerves serving the heart do control the rate at which it beats.

A single beat of the heart consists of a contraction of the atria followed by a contraction of the ventricles and a period of relaxation of all the chambers. Contractions are called **systoles,** and relaxations are called **diastoles.** An atrial systole lasts one eighth of a second, and it permits blood to flow from it into the ventricles. It is followed by an atrial diastole lasting seven eighths of a second that allows blood from the venae cavae and pulmonary veins to enter the atria. The beginning of the atrial diastole coincides with the beginning of a ventricular systole, which lasts three eighths of a second and forces blood into pulmonary arteries and the aorta. The ventricular diastole lasts through the atrial systole of the next beat. For half of the beat (half a second) all the chambers of the heart relax (Fig. 34.6). It is this alternate contracting of atria and ventricles that pumps blood throughout the body. The valves determine the direction of flow.

Although isolated heart muscle can contract and relax rhythmically without benefit of nervous connections, in the intact body a negative feedback mechanism involving both the heart and the nervous system controls the rate of heartbeat (Fig. 34.7). As blood fills the right atrium, the stretching of the muscle there initiates nervous impulses that are carried to the medulla, a portion of the brain that controls heartbeat. From here stimulatory nervous impulses return to the heart and increase the rate of its beat. This forces blood into the aorta

and causes it to stretch. Another set of nerves with endings in the aorta transmit impulses to the medulla, which in turn sends inhibitory nervous impulses to the heart and slows its beat. As blood re-enters the right atrium, the cycle begins again. The negative feedback mechanism maintains a steady rate of beating if nothing interferes. Exercise, for example, which increases requirement for food and oxygen in body tissues and increases the need for disposal of waste products, also increases the rate of heartbeat and so accomplishes the more rapid moving of these substances to their various destinations. Fear, love, rage, and other emotional experiences may quicken the heartbeat also.

The action of the heart forces the blood into the arteries under pressure. Because of the branching of the arteries, their total cross-sectional area increases with distance from the heart. This causes the rate of blood flow to decrease with distance from the heart, the blood moving most slowly in the capillaries. This allows more time in which food, oxygen, wastes, and other substances can be exchanged with body tissues. The rate of movement of blood increases as it returns to the heart through the veins, because the total cross-sectional area of the veins decreases as the heart is approached.

The friction of blood flowing through the vessels causes its pressure to fall. Blood pressure falls gradually in the arteries, then more rapidly in the arterioles and capillaries as the total surface area increases. The pressure is very low in the veins, and when blood reaches the heart, it may have no pressure at all. Contraction and relaxation of adjacent skeletal muscles assist the movement of blood in the veins. This has the effect of repeatedly squeezing the veins and forcing the blood to move. Valves in the veins prevent blood from running backward (Fig. 34.8). This is especially important in the legs, where the trip is upward, and gravity might otherwise cause the blood to drop downward. Standing perfectly still for a long time can result in the failure of blood to return to the heart from the lower part of the body, for the skeletal

295

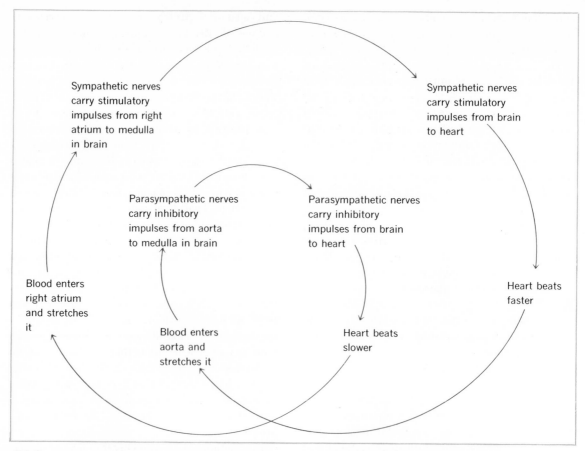

**34.7** Negative feedback mechanism controlling the rate of heartbeat. (See Chapter 37 for more details about the sympathetic and parasympathetic nervous systems.)

muscles of the legs do very little work. It can result in the embarrassment of a soldier's fainting while standing at attention because his circulatory system no longer brings his brain enough oxygen.

## Blood

Blood consists of a liquid portion called plasma and three kinds of cells: red blood cells, white blood cells, and platelets. From 50 to 60 per cent of whole blood is plasma, the cells comprising the remaining portion.

The **plasma** is mostly water (about 90 per cent)

296

in which are various substances in solution or suspension. Among these are proteins, glucose, amino acids, fats, inorganic ions (especially sodium, potassium, magnesium, and calcium), and waste materials such as carbon dioxide, ammonia, and urea. The proteins include prothrombin and fibrinogen, both of which function in the clotting of blood; gamma globulins, which contribute to immunity to diseases; and a variety of enzymes and hormones, each with its own particular function. Serum is plasma from which fibrinogen has been removed; it does not clot and sometimes can be observed as the pale straw-colored liquid that

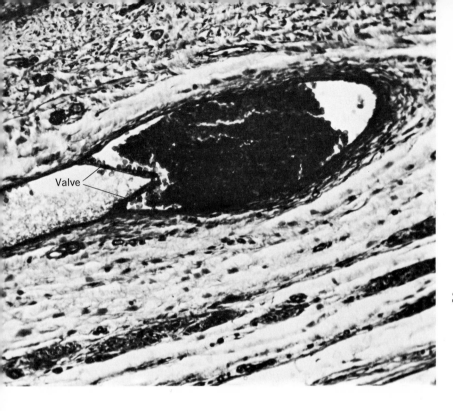

Valve

**34.8** Vein with valve. This valve permits blood to flow from left to right in this photograph but not in the reverse direction. The dark material beyond the valve is an accumulation of blood cells. (Courtesy of Dr. Beth L. Wismar.)

**34.9** Human blood cells. Red blood cells are most numerous. They lack nuclei. Their clear centers and "donut" appearance are due to the fact that these flattened cells are thicker around the edge and very thin near the center. White blood cells are larger than red blood cells; the stain used in this preparation makes their nuclei dark. Platelets appear as small fragments. (Courtesy of Dr. Beth L. Wismar.)

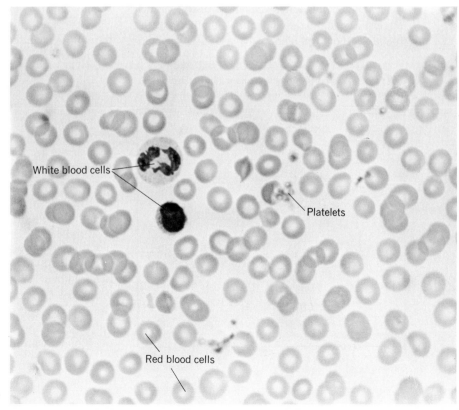

White blood cells

Platelets

Red blood cells

oozes out from a scab freshly forming over a wound.

The **red blood cells** (also called red corpuscles, erythrocytes, or rubicytes, Fig. 34.9), which transport oxygen, are the most numerous of the blood cells; they number about 5 million per cubic millimeter of blood. They give blood its red color because of their red pigment **hemoglobin,** which makes up 95 per cent of the dry matter of the cell. Hemoglobin (Fig. 34.10) resembles chlorophyll of plants in consisting of a porphyrin ring bound to a protein. The side chains on the ring are different, the protein is different, and an iron atom rather than magnesium occupies the center of the ring. It is the iron in the hemoglobin molecule that accepts oxygen in the lungs and releases it to the body tissues. Where the oxygen concentration is relatively high, as in the lungs, hemoglobin combines with oxygen. Where the oxygen concentration is relatively low, as in most body tissues, the oxygen separates from the hemoglobin and diffuses into the tissues.

The hemoglobin carries about 60 times as much oxygen as could be dissolved in the plasma of the blood. Without red blood cells and their hemoglobin, the blood could not bring to the tissues of the human body all the oxygen it requires. Hemoglobin not only has an affinity for oxygen, but has an even greater one for carbon monoxide (CO). When a person inhales this poisonous gas, the carbon monoxide molecules combine with hemoglobin molecules making them incapable of transporting oxygen. If enough hemoglobin molecules are so occupied, the individual dies for lack of sufficient oxygen in his tissues. Carbon monoxide is a colorless, odorless gas that cannot be detected by the human senses. It is formed when organic compounds are burned with insufficient oxygen to oxidize completely the carbon in the burning material. For these reasons, we are cautioned never to run the engine of an automobile in a closed garage, for the internal combustion engine always produces carbon monoxide. Many cars, with their engines idling in the huge traffic jams that are becoming typical of rush hours in large cities, may raise the carbon monoxide concentration of the atmosphere above the streets to the danger level. Among the array of noxious substances inhaled by cigarette smokers is carbon monoxide, for the burning tobacco in the cigarette never is oxidized completely.

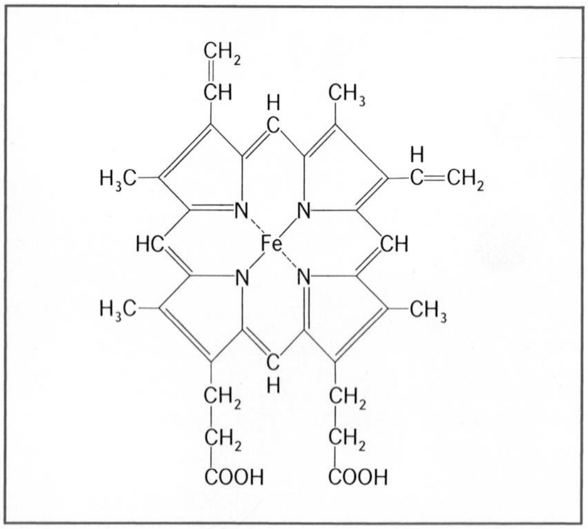

The blood transports waste carbon dioxide to the lungs in several forms. About 25 per cent of the carbon dioxide combines chemically with hemoglobin in the capillaries in body tissues and so is transported by red blood cells; in the lungs it separates from the hemoglobin and is exhaled. About 8 per cent of the carbon dioxide is transported as a dissolved gas in both the red blood cells and the plasma. The remaining 67 per cent ionizes to the bicarbonate and hydrogen ions, which are highly soluble in the water of both red blood cells and plasma.

Red blood cells of humans are formed in the bone marrow of long bones, ribs, and the skull. When first formed, they have nuclei, mitochondria, Golgi bodies, and other parts typical of other cells, but they lose these parts as they mature and enter the bloodstream. A mature red blood cell is little

**34.10** Heme portion of the hemoglobin molecule. The globin portion, not shown here, is a protein.

298

more than a sac of cytoplasm, an important constituent of which is its hemoglobin molecules.

When viewed from one direction, red blood cells appear perfectly round with a diameter of about 7.7 microns, but when seen on edge, they show two concave sides. This peculiar shape provides additional surface area through which oxygen can diffuse. It also ensures that all the oxygen held in the cell is near the surface and so can be exchanged quickly. Because they lack nuclei, red blood cells cannot reproduce themselves, nor can they live indefinitely. They live for an average of four months and are replaced by new cells formed in bone marrow. In a healthy person the rate of production of red blood cells equals that of their destruction.

Pernicious anemia is a disease caused by the absorption of insufficient vitamin $B_{12}$ by the digestive tract. The number of red blood cells becomes abnormally low. The condition ordinarily can be cured by injections of vitamin $B_{12}$.

Sickle-cell anemia is a genetic disease in which the protein portion of the hemoglobin molecule is defective. Only one amino acid (glutamic acid) is missing and replaced by another (valine), but this is sufficient to cause the red blood cells to assume a sickle shape. The disease is a serious one, and most persons afflicted with it die young. More will be said about the genetics of this disease in Chapter 42.

**White blood cells** (also called white corpuscles or leukocytes, Fig. 34.9) are larger than red blood cells but are less numerous—about 5,000 to 9,000 per cubic millimeter. Several types exist; they may be irregular in shape and are capable of ameboid movement. These cells protect the body against bacteria that have invaded the blood; they engulf the bacteria and digest them in ameboid fashion, a process called **phagocytosis.** Because of their ameboid qualities, they can move between the cells of capillary walls and enter the body tissues where they also phagocytize bacteria that may be present there. The pus formed in a wound consists of white blood cells, bacteria, and dead tissue cells.

Some white blood cells become converted into plasma cells, which manufacture antibodies (see p. 302). The life span of the different kinds of white blood cells varies from less than 12 hours to several years. New cells form in bone marrow, lymph nodes, spleen, and the thymus gland. Leukemia is a fatal, malignant disease in which the number of white blood cells rises dramatically.

**Platelets** (also called thrombocytes, Fig. 34.9) arise in bone marrow by the fragmentation of bits of cytoplasm of larger cells that are confined to the marrow. Platelets are smaller than red blood cells and less numerous. There are about 250,000 of them per cubic millimeter, but this figure is variable. They live no longer than 10 or 12 days.

Platelets function in the clotting of blood and repairing of damaged blood vessels. One definitely wants clotting of blood to happen when it is needed, but just as certainly one does not want it to happen when it should not. Even a small wound would continue to lose blood indefinitely if a clot did not form and close it off. On the other hand, if a clot formed within an intact blood vessel, the circulation would cease in that vessel, and it would no longer perform its functions. The several steps in the clotting mechanism ensure that clotting ordinarily occurs only when it is required to stop the flow of blood. In addition to platelets, several plasma proteins, among them prothrombin and fibrinogen, are necessary for clotting. The manufacture of prothrombin in the liver requires the presence of vitamin K. Doctors sometimes prescribe vitamin K prior to surgery to reduce the chances of hemorrhaging.

When a blood vessel is cut, platelets that come in contact with the damaged tissue adhere to the broken vessels, break up, and release a phospholipid to the plasma. This phospholipid enters into reactions that produce the protein thromboplastin. Thromboplastin is an enzyme that catalyzes the conversion of prothrombin to thrombin. Calcium ions must be present for the reaction to occur:

299

$$\text{Prothrombin} \xrightarrow{\text{thromboplastin, } Ca^{++}} \text{thrombin}$$

Thrombin itself is another enzyme, and it catalyzes the conversion of the soluble fibrinogen to the insoluble fibrin:

$$\text{Fibrinogen} \xrightarrow{\text{thrombin}} \text{fibrin}$$

Fibrin consists of long, fibrous protein molecules that form a meshwork across the opening in the blood vessel. The clot slows the flow of blood and eventually stops it if the amount of damage is not too great. Blood cells usually become trapped in the meshwork and help to plug the spaces within it.

Clots usually do not form in intact vessels where the platelets also remain intact, but the case of blood clots in heart attacks has been mentioned earlier. Presumably the thickening of the walls of the arteries presents a rough surface on which the

**34.11** Major lymph vessels of the body.

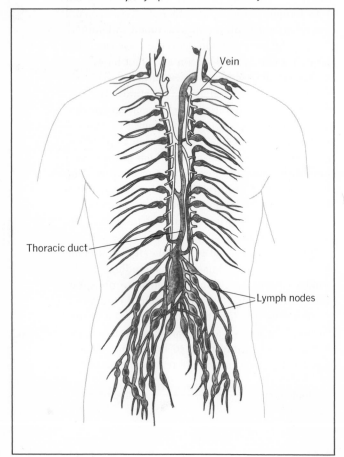

platelets break up and initiate the clotting reactions. Strokes result from clots forming in blood vessels in the brain. Circulating in the blood is heparin, an anticoagulant that normally prevents clotting in intact blood vessels. It does this by interfering with the conversion of fibrinogen to fibrin. Another anticoagulant is fibrinolysin, an enzyme that catalyzes the decomposition of fibrin.

Clotting fails in persons affected with hemophilia, a hereditary blood disease. Hemophiliacs sometimes are in danger of bleeding to death from relatively minor wounds and often suffer painfully from internal hemorrhaging. Hemophilia afflicted several of Queen Victoria's descendants; one of them was her great-grandson, the son of Czar Nicholas II and heir to the Russian throne. His mother's desperate faith in the healing powers of the infamous monk Rasputin contributed to the downfall of the Romanoff family and affected the subsequent history of their country.

## Lymphatic System

The pressure of blood passing through the capillaries forces some of the fluid from the blood and into the body tissues. Much of it returns to the blood because of the differences in osmotic pressure between blood and tissue fluids. The rest enters the other part of the circulatory system, the lymphatic system. The liquid, which resembles blood but contains no red blood cells and so is colorless, is called **lymph.**

The lymphatic system (Fig. 34.11) consists of capillaries and **lymph vessels** that correspond to the capillaries and veins of the blood vascular system. There is nothing that corresponds to the heart or arteries. The lymph capillaries do not form a net like the blood capillaries with which they often are intertwined, but they resemble microscopic fingers that connect with the lymph vessels. The smaller lymph vessels unite, forming larger ones. Like veins, they depend on adjacent muscles to keep their fluid moving and have valves that direct its flow. The largest lymph vessels conduct lymph into veins in the shoulder, thereby returning the lymph to the blood.

Lymph capillaries in the villi of the small intestine (Fig. 34.12) are called **lacteals.** They absorb lipids and fat-soluble substances which become somewhat diluted before they reach the blood.

Along the lymph vessels are enlargements called **lymph nodes.** They consist of a network of connective tissue that filters out bacteria and other small particles such as dust and soot that may have gained entrance to the body. Lymph nodes also manufacture white blood cells, some of which phagocytize the trapped bacteria. Numerous small lymph nodes are scattered throughout the body, but they are most numerous under the arms and in the neck and groin. Two large lymph nodes that some persons still can see in their mirrors are their tonsils, which lie on either side of the back of the mouth. Occasionally they trap bacteria faster than they can destroy them, and then these organs become infected. The problem customarily is solved by tonsilectomy—an operation so common that many persons no longer have their tonsils.

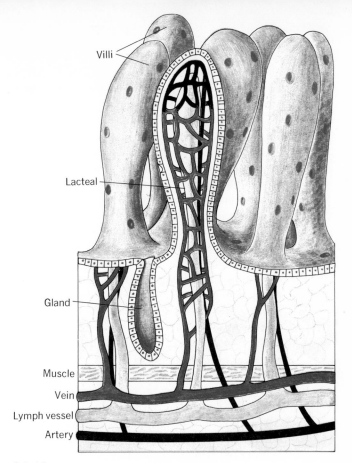

**34.12**  Lacteal and blood capillary in a villus.

## Thymus Gland and Spleen

The thymus gland and the spleen are two large organs not immediately part of the circulatory system. But their functions of producing or destroying blood cells definitely associates them with the circulatory system.

The **thymus gland** (Fig. 34.13), once believed to be an endocrine gland, has no known endocrine function. In young persons it does manufacture white blood cells, and for this reason it is considered in this chapter. Located in the chest, the thymus gland is relatively large in children, but it begins to atrophy after puberty and seems to have

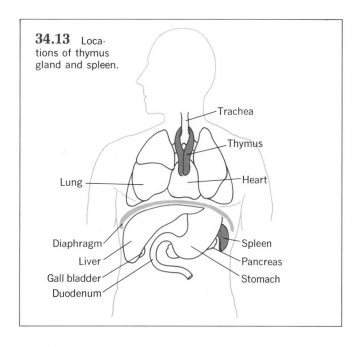

**34.13**  Locations of thymus gland and spleen.

no function in adults. The white blood cells it produces can be converted to plasma cells, which function in the development of immunity to some diseases (see p. 303). Experimental animals from which the thymus glands have been removed surgically immediately after birth develop no immunity to diseases to which they may be exposed later. Animals with intact thymus glands do develop immunity after exposure to many kinds of disease-causing organisms.

The **spleen** (Fig. 34.13), which lies in the abdomen somewhat to the left of the stomach, destroys dead blood cells. Iron from the hemoglobin molecule returns to the bone marrow, which uses it in the manufacture of more hemoglobin. The porphyrin rings are transported to the liver where they become the bile pigments and are excreted in the feces, to which they lend their color. The spleen also holds a reserve supply of blood that the body can call upon in case of emergency. Before and shortly after birth it manufactures red blood cells, but does not seem to do so in the adult. All the functions of the spleen are performed by other organs as well, and therefore the spleen can be removed with relative safety.

## Immunity

As unpleasant as they may be, attacks of the so-called children's diseases have the happy property of conferring immunity to these diseases once they have been suffered. If the pathogenic bacteria or viruses should penetrate the skin or mucous membranes and enter the blood, and should they also multiply faster than the white blood cells can remove them, then immunity presents another line of defense.

302

### Antigen–Antibody Reaction

The microorganisms bring with them their proteins—proteins foreign to the host's body. In response to the presence of foreign proteins, called **antigens,** the circulatory system produces other proteins, called **antibodies.** A reaction between the antibodies and the antigens renders the microorganisms ineffective in causing disease. The nature of the reaction varies with the nature of the antigen. In some cases the reaction causes the microorganisms to come together in clumps, thus making it possible for the phagocytes to destroy several of them at one time. If the microorganisms release a toxin, the antigen neutralizes the toxin and renders it harmless.

The shape of an antibody molecule corresponds to that of its antigen molecule, somewhat like the shapes of enzyme molecules and their substrate molecules (Chapter 10). Because of this geometrical "lock-and-key" fit, antibodies are specific for their antigens just as enzymes are specific for certain substrates and certain chemical reactions. An antibody that protects against typhoid fever has no effect against any but the typhoid fever bacteria.

The first time a given species of microorganism invades the body, the production of antibodies may not happen fast enough to prevent illness. However, after recovery, the patient very likely will be immune for the rest of his life. This is why many diseases are typical of childhood. One is likely to be exposed to them sometime during childhood and so become immune to them thereafter. Measles and chicken pox are examples. In some cases the original illness may be so mild that it may not be recognized as such. Some children have had mild cases of polio no more severe than a bad cold; nonetheless they become immune to polio thereafter.

The mechanism of antibody formation in the body still is not completely understood. Of the several theories proposed from time to time, the one described here currently enjoys popularity among researchers in this field. The theory assumes that early in embryonic development a wide variety of **lymphocytes** (a type of white blood cell) develop in the thymus gland. Each type of lymphocyte cell has the ability to synthesize one type of antibody, and the total lymphocyte complement of the body includes cells capable of forming antibodies specific for many proteins foreign to the body. When a specific antigen enters the blood, some of its molecules come in contact with the lymphocytes. Those lymphocytes capable of manufacturing antibodies specific for that antigen are converted into **plasma cells,** which begin to multiply, and soon large numbers of this particular type of plasma cell begin to appear. These cells produce their particular antibodies, which are released into the bloodstream; these antibodies are the proteins called **gamma globulins.** Once one has recovered from an illness, the concentration of the specific antibodies declines in the blood, but later infections initiate more rapid multiplication of the appropriate plasma cells and production of their antibody—production rapid enough to confer immediate immunity.

There is some question whether there possibly can be as many kinds of lymphocytes (or plasma cells) in the body as there are foreign proteins; the potential number is astronomical. Undoubtedly the theory will undergo modifications in the future. It seems certain, however, that plasma cells do manufacture antibodies.

A few persons suffer from the disease called agammaglobulinemia (which, translated into everyday English, means that they have no gamma globulins in their blood), and these persons are very susceptible to many contageous diseases. Their lack of gamma globulins stems from their lack of plasma cells.

Immunity to some diseases can be acquired artificially. **Vaccinations** are injections of living or dead pathogenic organisms in quantities sufficient to initiate antibody formation but not enough to cause illness (although one may not feel perfectly well for a few hours or days after receiving certain vaccinations). Living organisms usually are attenuated by treatment with chemicals or heat before injection to reduce the chances of their causing disease. This is most important when the disease in question is a dangerous one or if there is no known cure for it; smallpox, cholera, and rabies are examples. Vaccinations usually do not confer lifelong immunity, but most of them can prevent the disease for several years.

Another method of conferring immunity artificially is by the injection of serum from animals that previously were injected with the appropriate microorganisms (or occasionally serum from human patients who have recovered from the disease if such are available). The animals (or human patients) produce their own specific antibodies, and these are present in the serum, which is called **antiserum** or immune serum. Antiserum injections provide only relatively short-term immunity, but they are used when it is suspected that a susceptible person may have been exposed recently to a disease. Antisera against tetanus, diphtheria, German measles, rabies, and some other diseases are used under these circumstances. One disadvantage of using antisera is that the patient may be allergic to proteins other than the antibodies in the serum and may die from shock. Horses often are used to produce antisera, and many persons are allergic to their serum proteins. The possibility of contracting hepatitis exists when human antiserum is used, for many apparently healthy persons carry the hepatitis virus in their blood.

## Allergy

Allergy (hypersensitivity) is a modification of the antigen–antibody reaction, bringing discomfort rather than relief. Again, the body forms antibodies in response to the presence of antigens in the body; but rather than circulating in the blood, the antibodies become fixed to tissue cells. The first exposure to the antigen elicits the formation of specific antibodies but causes no adverse effect. Upon later exposure (though not necessarily the second exposure) the cells to which the antibodies are fixed release histamine, a substance that is damaging to several types of tissue, especially the smooth muscles of blood vessels. The symptoms of allergies are many and varied, depending on the

303

person and the type of antigen—rash, itching, watery eyes, running nose, sneezing, local swellings, difficulty in breathing, and others. More serious are shock and even death. Administration of antihistamine can relieve some of the milder discomforts.

A wide variety of proteins and some other compounds combined with proteins cause allergies. Common sources are pollen grains (producing the misnamed hay fever), nearly any part of the poison ivy plant, the tuberculosis organism, and a great many foods. Some allergies, like hay fever, produce an immediate response to an exposure to the antigen if the person has already been sensitized to it. Others, like the poison ivy allergy, require several hours or days before the unpleasant symptoms become apparent.

The use of enzymes in laundry detergents has given rise to concern, for being proteins foreign to the human body, these enzymes can cause allergies. Asthmalike allergies have been reported among the factory workers manufacturing enzyme detergents. Yet to be determined is how much difficulty enzyme detergents might cause among housewives, who ordinarily would receive less exposure than factory workers.

## Blood Types

When blood transfusions first were employed by physicians, they met with haphazard success. Some patients benefited from them, but others died shortly after receiving the transfusions. Research revealed that humans have different blood types and that some combinations of blood are compatible for transfusions and others are not.

### A-B-O System

Blood typing is done on the basis of several substances in the blood. One system of blood typing is the A-B-O system. It is based on the presence or absence of two polysaccharides, A and B. These are called **agglutinogens** and are present on the surface of the red blood cells of some persons. Persons with blood type O have neither of these agglutinogens; those with blood type A have agglutinogen A; those with blood type B have agglutinogen B; and those with blood type AB have both.

**Agglutinins** are plasma substances that react with agglutinogens in much the same fashion that antibodies react with antigens. In this case they cause clumping of red blood cells. Agglutinin a causes clumping of red blood cells with agglutinogen A, and agglutinin b causes clumping of red blood cells with agglutinogen B. O type blood contains both agglutinins a and b; A type blood contains agglutinin b; B type blood contains agglutinin a; and no agglutinins are present in AB type blood. This is summarized in Table 34.1, which shows that no one has agglutinins in his plasma specific for the agglutinogens in his red blood cells. Therefore clumping of his own blood cells does not occur. He can receive blood identical with his own safely. (Care must be taken that other factors, such as the Rh factor described on p. 305 are compatible also.)

Difficulties arise when a person receives a transfusion with incompatible blood. If a patient of blood type B receives type A blood, agglutinins a in the recipient's serum cause clumping of the donor's red blood cells which contain agglutinogen A; death of the recipient may result. If the donor

**304**

Table 34.1

Agglutinogens and Agglutinins in the A-B-O Blood Type System

| Blood Type | Agglutinogens in Red Cells | Agglutinins in Serum |
|---|---|---|
| O | none | ab |
| A | A | b |
| B | B | a |
| AB | AB | none |

**Table 34.2**
Compatibilities and Incompatibilities in the A-B-O Blood Type System

— compatible, no agglutination.
+ incompatible, agglutination.

| Blood group of donor | Agglutinogens of donor | O | A | B | AB | Blood group of recipient |
|---|---|---|---|---|---|---|
| | | ab | b | a | none | Agglutinins of recipient |
| O | none | − | − | − | − | |
| A | A | + | − | + | − | |
| B | B | + | + | − | − | |
| AB | AB | + | + | + | − | |

is of blood type B, this does not happen. For this reason doctors must be sure that only compatible types are transfused. Table 34.2 shows all possible combinations of donors and recipients and the results to be expected from each. Note that the reaction between the recipient's agglutinins and the donor's agglutinogens is what is important and not the recipient's agglutinogens and the donor's agglutinins. The dilution of the donor's agglutinins in the recipient's blood renders them ineffective. For this reason persons of blood type O are called universal donors, for they can give blood to persons of all four blood types (assuming no incompatibility because of other factors). Persons with AB blood are universal recipients, for they can receive from persons of all four blood types (again assuming no other incompatibilities). If only plasma rather than whole blood with its red blood cells is transfused, the problem of clumping does not arise.

### Rh System

Another blood typing system is the Rh system. Persons who are Rh positive have the Rh antigen in their blood, and persons who are Rh negative lack it. If an Rh negative person receives a single transfusion of Rh positive blood, ordinarily he experiences no difficulty as a result of this, but his body does form antibodies specific for the antigen. Should he have a second transfusion with Rh positive blood, his antibodies cause clumping and destruction of red blood cells and perhaps death.

Rh negative women with Rh positive husbands may have Rh positive babies. Although the blood-streams of the pregnant mother and her fetus do not mix, occasionally there is some leakage of blood across the placenta. During the first pregnancy with an Rh positive child, his antigens may pass into her blood, and then she forms Rh antibodies. Usually this does not affect that child, but any subsequent Rh positive child will react if he receives her Rh antibodies. Such babies, commonly called Rh babies, suffer from a disease called erythroblastosis fetalis. Their red blood cells are destroyed, and death often comes shortly before or after birth. Usually massive transfusions are necessary to save the child.

Any Rh negative children of an Rh negative mother run no risk from any Rh antibodies the mother may have formed, for they have no Rh antigens.

There are other blood types in addition to the A-B-O and Rh types, but these two are the only ones that seem to be of great medical concern.

## Questions

1. Define the following terms:

| | |
|---|---|
| lymph | antigen |
| diastole | antibody |
| systole | fibrin |
| valve | thrombin |
| AV valve | allergy |
| plasma | hemophilia |
| serum | sickle-cell anemia |
| hemoglobin | gamma globulin |
| phagocytosis | |

**2.** What are the main channels of blood flow in the body?

**3.** In what way does the circulatory system increase the internal surface area of the body?

**4.** What is the actual site of exchange of substances between the blood and body tissues?

**5.** Trace a drop of blood on one complete trip around the body from the time it enters the right atrium until it enters the right atrium again. Assume that it passes through muscle tissue on this trip. Through how many sets of capillaries does the blood pass on this trip? What happens in each set of capillaries?

**6.** Describe what happens when the heart beats.

**7.** In what way does the concentration of capillary nets in a tissue reflect the activity of that tissue?

**8.** Do most of the body's arteries or veins carry oxygenated blood? What arteries and veins are exceptions to this generality?

**9.** Describe the negative feedback mechanism that keeps the heart beating.

**10.** How do red blood cells, white blood cells, and platelets differ in structure and function?

**11.** Describe the clotting mechanism of blood.

**12.** Describe two ways in which the circulatory system protects the body from disease-causing organisms.

**13.** What is the function of the lymphatic system?

**14.** What is the relationship between blood and lymph?

**15.** Trace the path of lymph from the time it leaves the bloodstream until it enters it again.

**16.** What relationship do the thymus gland and the spleen have to the circulatory system?

**17.** How does vaccination confer immunity?

**18.** Which of the following represent compatible transfusions?

| Donor | Recipient |
|-------|-----------|
| Type A | Type O |
| Type A | Type A |
| Type A | Type B |
| Type A | Type AB |

# Suggested Readings

Adolph, E. "The Heart's Pacemaker," *Scientific American* 216(3): 32–37 (March 1967).

Bloom, W., and D. W. Fawcett. *A Textbook of Histology,* 9th ed. Philadelphia: W. B. Saunders Company, 1968.

Edelman, G. H. "The Structure and Function of Antibodies," *Scientific American* 223(2): 34–42 (August 1970).

Griffin, D. R., and A. Novick. *Animal Structure and Function,* 2nd ed. New York: Holt, Rinehart and Winston, Inc., 1970.

Laki, K. "The Clotting of Fibrinogen," *Scientific American* 206(3): 60–66 (March 1962).

Langley, L. L. *Homeostasis.* New York: Reinhold Publishing Corporation, 1965.

Larimer, J. *An Introduction to Animal Physiology.* Dubuque, Iowa: William C. Brown Company, Publishers, 1968.

Mayerson, H. S. "The Lymphatic System," *Scientific American* 208(6): 80–90 (June 1963).

Perutz, M. F. "The Hemoglobin Molecule," *Scientific American* 211(5): 64–76 (November 1964).

Ponder, E. "The Red Blood Cell," *Scientific American* 196(1): 95–102 (January 1957).

Spain, D. M. "Atherosclerosis," *Scientific American* 215(2): 48–56 (August 1966).

Vander, A. J., J. H. Sherman, and D. S. Luciano. *Human Physiology.* New York: McGraw-Hill Book Company, 1970.

Wiggers, C. J. "The Heart," *Scientific American* 196(5): 74–87 (May 1957).

Wood, J. E. "The Venous System," *Scientific American* 218(1): 86–96 (January 1968).

Zucker, M. B. "Blood Platelets," *Scientific American* 204(2): 58–64 (February 1961).

Zweifach, B. W. "The Microcirculation of the Blood," *Scientific American* 200(1): 54–60 (January 1959).

As part of its metabolism, every living cell forms waste products. If permitted to accumulate in the cell or even in its immediate vicinity, these wastes reach toxic concentrations and eventually kill the cell. The disposal problem is small for single-celled organisms; the wastes merely diffuse into the environment. In multicellular organisms of any complexity, however, the environment of any cell consists of other cells, and disposal of wastes merely by diffusion would result in the poisoning of one cell by another. The need for a special excretory system thus arises.

An aside is in order to explain the difference between excretion and elimination. **Excretion** is the removal from the body of waste products of metabolism; the substances excreted have been part of the living tissues in one form or another. The urea of urine, for example, forms from carbon dioxide and ammonia, which, in turn, derive from carbohydrates, fats, proteins, and other organic compounds that once were integral parts of the living cells. **Elimination** (or defecation) is the removal from the body of substances that pass unused through the digestive tract. They are the undigested substances that leave the body as feces. Because the digestive tract topologically is outside the body (Chapter 31), none of this material has been part of the body tissues. (One exception is the bile pigments, which once were part of the hemoglobin molecules in the red blood cells. They are excreted by the liver and gall bladder into the small intestine and there join material to be eliminated.)

Most wastes are excreted as urine by the kidneys, but considerable amounts of carbon dioxide are released through the lungs (Chapter 32). The sweat glands in the skin (Chapter 39) release perspiration, which is a dilute solution of urea, salts, and several other wastes in water. It is therefore not incorrect to consider the lungs and the skin as part of the excretory system, but for convenience we will consider the excretory system in a narrower sense as consisting of only the kidneys and their associated structures.

# 35
# The Excretory System

1 The excretory system rids the body of wastes, especially urea.

2 By excreting materials present in excess and conserving those in insufficient supply, the excretory system also helps to maintain homeostasis within the body.

## Formation of Urea

$$R-\underset{\underset{NH_2}{|}}{\overset{\overset{H}{|}}{C}}-COOH + H_2O \xrightarrow{\text{NAD NADH}_2} R-\underset{\underset{O}{\|}}{C}-COOH + HN_3$$

The two main metabolic wastes of the body are carbon dioxide and ammonia. Oxidation of any organic compounds produces the former. The metabolism of nitrogenous substances such as amino acids produces the latter. The amino groups $(-NH_2)$ removed from amino acids by deamination reactions form ammonia. One type of deamination, an oxidative deamination, can be summarized:

The bloodstream accepts the carbon dioxide and ammonia released from metabolizing body cells and carries them to the liver, where they are converted to urea, which is less toxic than ammonia and can be more safely handled by the body. The formation of urea requires energy which is sup-

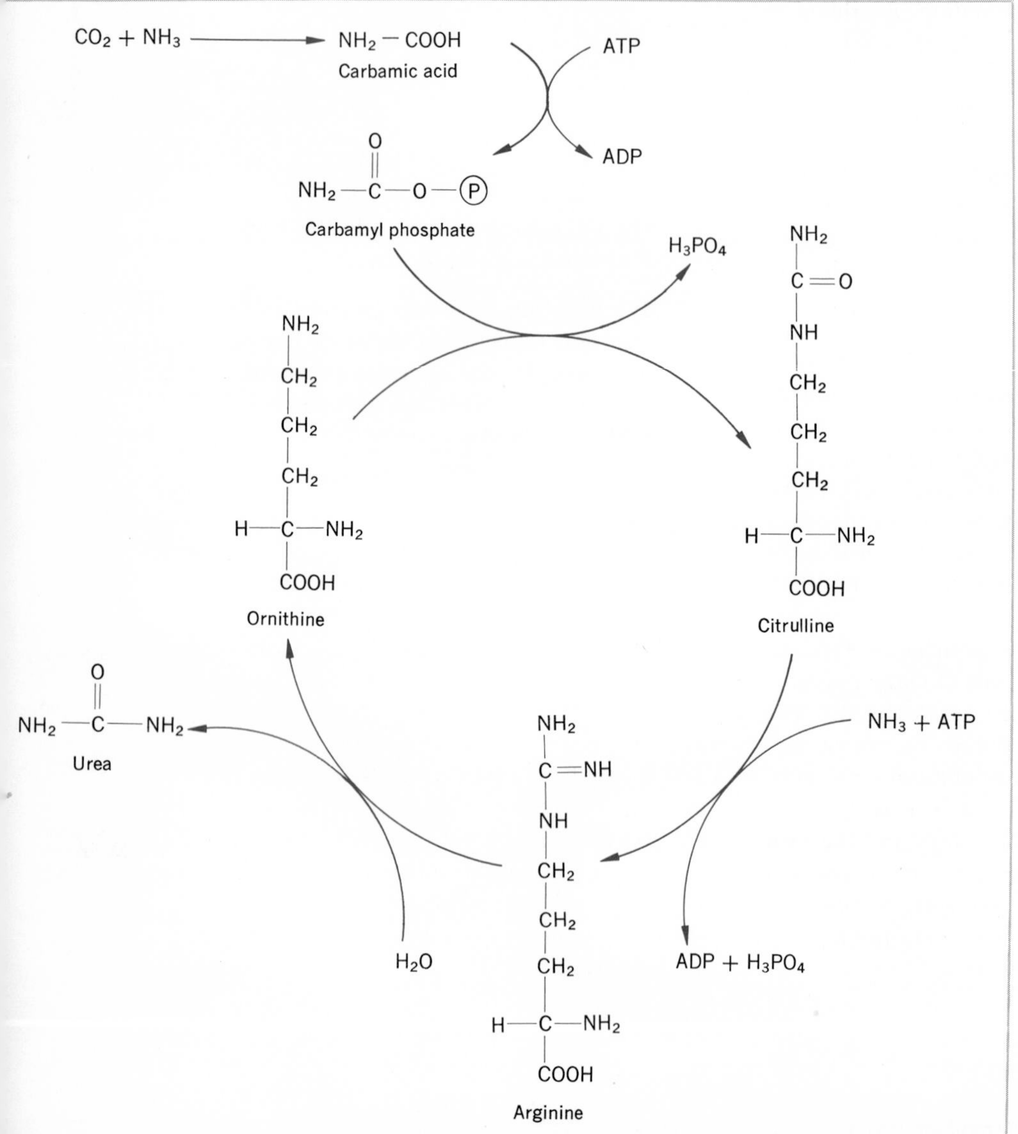

**35.1** Ornithine cycle by which the liver forms urea from carbon dioxide and ammonia.

plied by ATP. The overall reaction may be summarized:

$$CO_2 + 2NH_3 \xrightarrow[\text{ATP ADP}]{} NH_2-\overset{\displaystyle O}{\underset{\displaystyle \text{urea}}{C}}-NH_2 + H_2O$$

The complete reaction is a cyclic one; its individual steps are shown in Fig. 35.1. Once formed, the urea is transported by the circulatory system from the liver to the kidneys, which remove it from the blood and excrete it.

## Anatomy and Functioning of the Excretory System

The excretory system consists of two kidneys, two ureters, the urinary bladder, and the urethra (Fig. 35.2). The kidneys are two bean-shaped organs lying near the back of the abdominal cavity about the level of the waistline. The ureters lead from the kidneys to the urinary bladder, which lies low in the abdominal cavity. The urethra extends from the urinary bladder to the exterior of the body.

Most of the important work of the excretory system occurs in the kidneys. Each kidney is composed of a million or more similar units called **nephrons,** each of which functions in the same manner. Each nephron is a hollow tubule folded in a somewhat complex manner. The tubule functions as more than a mere conduit, for here the work of extracting wastes from the blood is done.

One end of each nephron (Fig. 35.3) is modified into a spherical structure called **Bowman's capsule.** This has the form of a cup with a hollow wall, the lumen of the wall being continuous with that of the rest of the tubule, which curves back and forth on itself. The renal artery, which departs from the aorta and enters the kidney, branches profusely. One arteriole enters the center of each Bowman's capsule and there forms a capillary bed

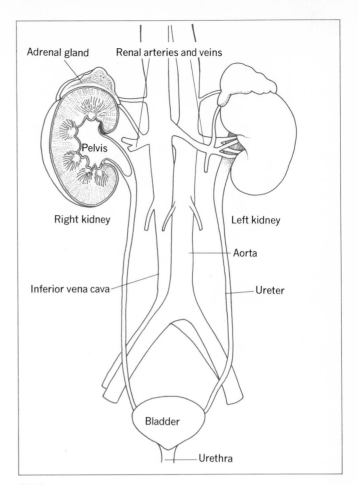

**35.2** Excretory system. The right kidney is shown in section.

called a **glomerulus.** Instead of returning directly to a venule and then to the heart by way of the veins as blood ordinarily does when it leaves a capillary bed, the blood leaving the glomerulus collects in another arteriole that leads to another capillary network—the peritubular capillaries—laced among the curves of the tubule. This second capillary net leads into a venule that empties blood into veins leading to the renal vein. Thus the blood traveling through the kidney passes through two capillary nets before returning to the heart. Different things happen in each set of capillaries.

Because the arteriole leading into the glomerulus has a wider diameter than the arteriole leading

309

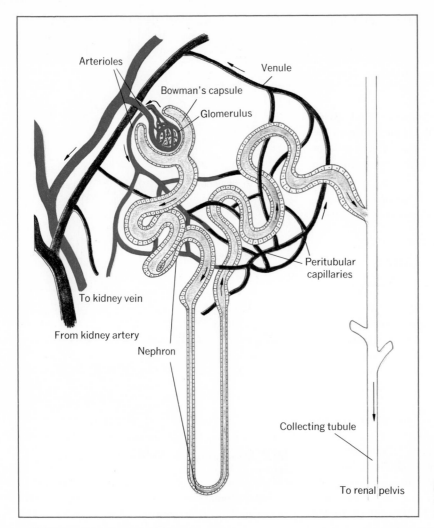

Arterioles

Bowman's capsule

Glomerulus

Venule

To kidney vein

From kidney artery

Nephron

Peritubular capillaries

Collecting tubule

To renal pelvis

**35.3** Detail of a nephron showing relationships with capillary networks.

310

away from it, blood within the glomerulus is under pressure. This forces some of its water out through the capillary walls and into the lumen of Bowman's capsule. Dissolved in the water are the waste urea that was formed in the liver and other soluble substances, including glucose, amino acids, salts, and other compounds. Proteins of large molecular size and the blood cells remain within the blood in the capillaries. The glomerulus therefore acts as a force filter, and its filtrate resembles plasma except for the lack of proteins.

Glomerular filtrate contains many substances that the body can use. The peritubular capillaries retrieve these as the glomerular filtrate passes along the rest of the tubule. The glucose, amino acids, salts, and water are reabsorbed here, but most of the urea is not. For each 100 liters of water absorbed by the glomerulus, 99 are reabsorbed by the second capillary net. Without the work done by this second set of capillaries, we would have to drink about 100 times more water each day than we do. The filtration and reabsorption of different substances is under hormonal and nervous control.

By the time the glomerular filtrate completes its trip through the tubule, its urea and salt content have become higher than that of blood, and normally nothing but a trace of the glucose and amino acids remain; the liquid then is called **urine.** The compositions of plasma, glomerular filtrate, and urine are summarized in Table 35.1.

Several nephric tubules empty into one collecting tubule, and all the collecting tubules of a kidney empty into the **ureter,** which departs from the concave side of the kidney. The process of urine formation is a continuous one, occurring day and night, and the ureter continuously receives small amounts of urine. Peristalsis of the smooth muscles of the ureters keep the urine moving into the **urinary bladder.** The oblique angle at which the ureters traverse the bladder wall acts as a valve; the presence of urine in the bladder tends to keep the ureter closed except while urine is entering and prevents backward flow.

The bladder stores the urine until it is full. The bladder has some capacity for stretching, but when about 250 cubic centimeters (about a half pint) of urine accumulates, one begins to experience an uncomfortable feeling. Emptying of the bladder is under voluntary control except in the very young, who require diapers until they acquire control. Contraction of a sphincter between the bladder and the urethra prevents passage of urine; relaxation of the muscle allows the urine to move

through the **urethra** and hence out of the body. The urethra traverses the length of the penis in men and hence is longer in men than in women.

### The Kidneys and Homeostasis

The kidneys do not function only as concentrators of metabolic wastes. Another of their functions is maintaining a homeostatic condition with regard to the concentration of several substances in the body—water, glucose, and minerals, for example. In turn this helps to regulate the total volume of blood and other body fluids and also their osmotic pressure and pH. The amount of each substance excreted by the kidneys depends on how much of it is filtered from the blood in the glomerular capillaries into the kidney tubules, how much is reabsorbed by the tubule walls and returned to the blood in the peritubular capillaries, and how much is secreted from the peritubular capillaries into the tubules. The amounts in each case differ for each substance and are greatly influenced by its concentration in the blood or in the glomerular filtrate.

Low blood pressure throughout the body (resulting perhaps from loss of blood through a wound or loss of water and salts in diarrhea) is reflected in a lowering of the blood pressure in the glomeruli. This decreases the amount of water and salts filtered into the kidney tubules and helps to conserve them. High blood pressure, on the other hand, results in an increase in the volume of the glomerular filtrate and the excretion of more water and salts from the body. The glomerular filtrate is not the final urine, however, and it is subject to change by the activities of the tubular cells.

When the concentration of some substances in the blood begins to fall below normal, the cells of the kidney tubules increase the rate at which these substances are reabsorbed. The reabsorption process usually is an energy-requiring one, for the substances must travel back against a concentration gradient. If the concentration of a substance in the blood rises above normal, the tubule cells reabsorb

Table 35.1

A Comparison of the Compositions of Plasma, Glomerular Filtrate, and Urine

| | Grams per 100 cc of Fluid | | |
|---|---|---|---|
| Components | Plasma | Glomerular Filtrate | Urine |
| Urea | 0.030 | 0.030 | 2.0 |
| Glucose | 0.100 | 0.100 | trace |
| Amino acids | 0.050 | 0.050 | trace |
| Inorganic salts | 0.720 | 0.720 | 1.50 |
| Proteins | 8.000 | 0.000 | 0.000 |

311

less of this substance and increase the amount lost in the urine. In either case, the kidney tubules tend to maintain a constant concentration of the substance in the blood. In normal persons the concentrations of glucose, water, and many minerals are maintained in this way.

Persons with diabetes mellitus have more than the normal 0.1 per cent of glucose in their blood. Their kidneys reabsorb only some of this glucose (a normal person reabsorbs virtually all the glucose in the glomerular filtrate), and the rest is excreted in the urine. Analyzing urine for its sugar content is one way of testing for diabetes mellitus.

It should be mentioned that decreasing the rate of loss of a given substance by increasing its rate of reabsorption cannot alone maintain a homeostatic condition indefinitely, for it does not replace substances lost by perspiration, diarrhea, or hemorrhaging. Eventually these must be replaced by eating or drinking. (Hunger and thirst are other parts of the total homeostatic system of the body.)

Only a few substances, potassium among them, are actively excreted from the peritubular capillaries into the kidney tubules. In this case, a high plasma concentration of the substance increases the rate of secretion and a low plasma concentration decreases it. Again, the living cells of the kidney tubules actively transport the substance against a concentration gradient, this time from the blood to the lumen of the kidney tubules.

Because tubular reabsorption and tubular secretion of many substances is against a concentration gradient, the tubule cells must expend energy which they obtain from respiration-generated ATP. In terms of its rate of respiration, kidney tissue is one of the most active in the body. Although other organs contribute to homeostasis, the kidney is an excellent example of why living things must constantly expend energy just to maintain a steady state—not necessarily to grow or to reproduce, but just to stay alive.

Failure on the part of diseased kidneys to remove wastes and regulate the concentrations of several substances in the body can result in death if the condition is not corrected. However, the human body does not require all of its kidney tissue. A single kidney is adequate if it functions properly. For this reason one kidney can be removed surgically with relatively little danger.

312

## Questions

**1.** What major waste product of metabolism does the excretory system remove from the body? What is the source of this waste in the body?

**2.** How does the circulatory system contribute to homeostasis?

**3.** Distinguish between excretion and elimination.

**4.** Describe the structure and function of a nephron.

**5.** Trace the path of urine from the time it forms in the kidneys until it leaves the body.

**6.** Why can chemical analysis of urine tell the physician something about his patient's state of health?

## Suggested Readings

Bloom, W., and D. W. Fawcett. *A Textbook of Histology,* 9th ed. Philadelphia: W. B. Saunders Company, 1968.

Griffin, D. R., and A. Novick. *Animal Structure and Function,* 2nd ed. New York: Holt, Rinehart and Winston, Inc., 1970.

Langley, L. L. *Homeostasis.* New York: Reinhold Publishing Corporation, 1965.

Larimer, J. *An Introduction to Animal Physiology.* Dubuque, Iowa: Willaim C. Brown Company, Publishers, 1968.

Merrill, J. P. "The Artificial Kidney," *Scientific American* 205(1): 56–64 (July 1961).

Smith, H. W. "The Kidney," *Scientific American* 188(1): 40–48 (January 1953).

Vander, A. J., J. H. Sherman, and D. S. Luciano. *Human Physiology.* New York: McGraw-Hill Book Company, 1970.

The problem of being multicellular, which has introduced the last five chapters, is solved by the division of labor within the body. The human body, like that of most multicellular organisms, possesses several different organ systems that perform different functions. Such a situation creates a new problem: that of coordinating the many functions occurring within the organism and maintaining a homeostatic condition. This function is assumed by hormones in plants (Chapter 29). In humans and other animals two organ systems take over this function. The endocrine system, which produces hormones, is discussed in this chapter. The nervous system will be considered in the following chapter.

# Hormones

The endocrine system functions by means of its hormones, chemicals that act as messengers. A **hormone** is a substance produced in one part of the body and transported to another, where it exerts its effect. The endocrine glands, which synthesize the hormones, are ductless glands. Unlike some of the digestive glands, the liver, for instance, they have no ducts through which to empty their products. But like many organs, they are served by capillaries, and they secrete their hormones directly into the blood passing through the capillaries. Because the kidneys excrete hormones and because other tissues destroy hormones, the endocrine glands must continue to produce and secrete their hormones if the concentrations of these substances are to remain relatively constant within the body.

The target organs (the organs affected by the hormones) are many and varied; but each hormone produces only one specific effect or one specific set of effects. Together they affect essentially every part of the body and its processes directly or indirectly. In turn, the endocrine glands are affected

# 36
# The Endocrine System

1 The endocrine system coordinates many processes occurring within the body.

2 Endocrine control is by chemical messengers called hormones.

3 Most hormones function as part of negative feedback mechanisms that maintain homeostasis in the body.

4 Hormones usually are carried throughout the body by the blood.

313

**Table 36.1**

A Summary of Endocrine Glands and Their Hormones

| Gland | Hormone | Chemical Nature | Function |
|-------|---------|-----------------|----------|
| Thyroid | Thyroxin | Amino acid | Increases rate of respiration |
| | Calcitonin | Polypeptide | Controls calcium metabolism |
| Parathyroids | Parathormone | Polypeptide | Controls calcium metabolism |
| Pancreas | Insulin | Protein | Stimulates formation of glycogen from glucose |
| | Glucagon | Polypeptide | Stimulates digestion of glycogen to glucose |
| Adrenal medulla | Adrenalin and noradrenalin | Modified amino acid | Associated with "fight-or-flight" reaction |
| Adrenal cortex | Glucocorticoids | Steroids | Regulates carbohydrate metabolism |
| | Mineralocorticoids | Steroids | Regulates salt and water relationships |
| | Sex hormones | Steroids | Controls development of secondary sexual characteristics |
| Posterior pituitary | Oxytocin | Polypeptide | Stimulates contraction of smooth muscles and ejection of milk from mammary glands |
| | Antidiuretic hormone (ADH, Vasopressin) | Polypeptide | Controls reabsorption of water by kidney tubules |
| Anterior pituitary | Growth hormone | Protein | Stimulates growth |
| | Thyrotropin | Protein | Regulates thyroxin production by thyroid gland |
| | Adrenocorticotropin | Protein | Regulates production of cortical hormones by adrenal cortex |
| | Follicle-stimulating hormone | Protein | Stimulates development of follicles in the ovaries of women and seminiferous tubules in the testes of men |
| | Luteinizing hormone | Protein | Stimulates secretion of sex hormones from ovaries and testes, induces ovulation |
| | Prolactin | Protein | Stimulates secretion of milk in mammary glands |
| Ovaries | Estrogens | Steroids | Controls development of female secondary sexual characteristics, initiates growth of uterine lining during part of menstrual cycle |
| | Progesterone | Steroid | Maintains uterine lining during part of menstrual cycle and during pregnancy |
| Testes | Testosterone | Steroid | Controls development of male secondary sexual characteristics |

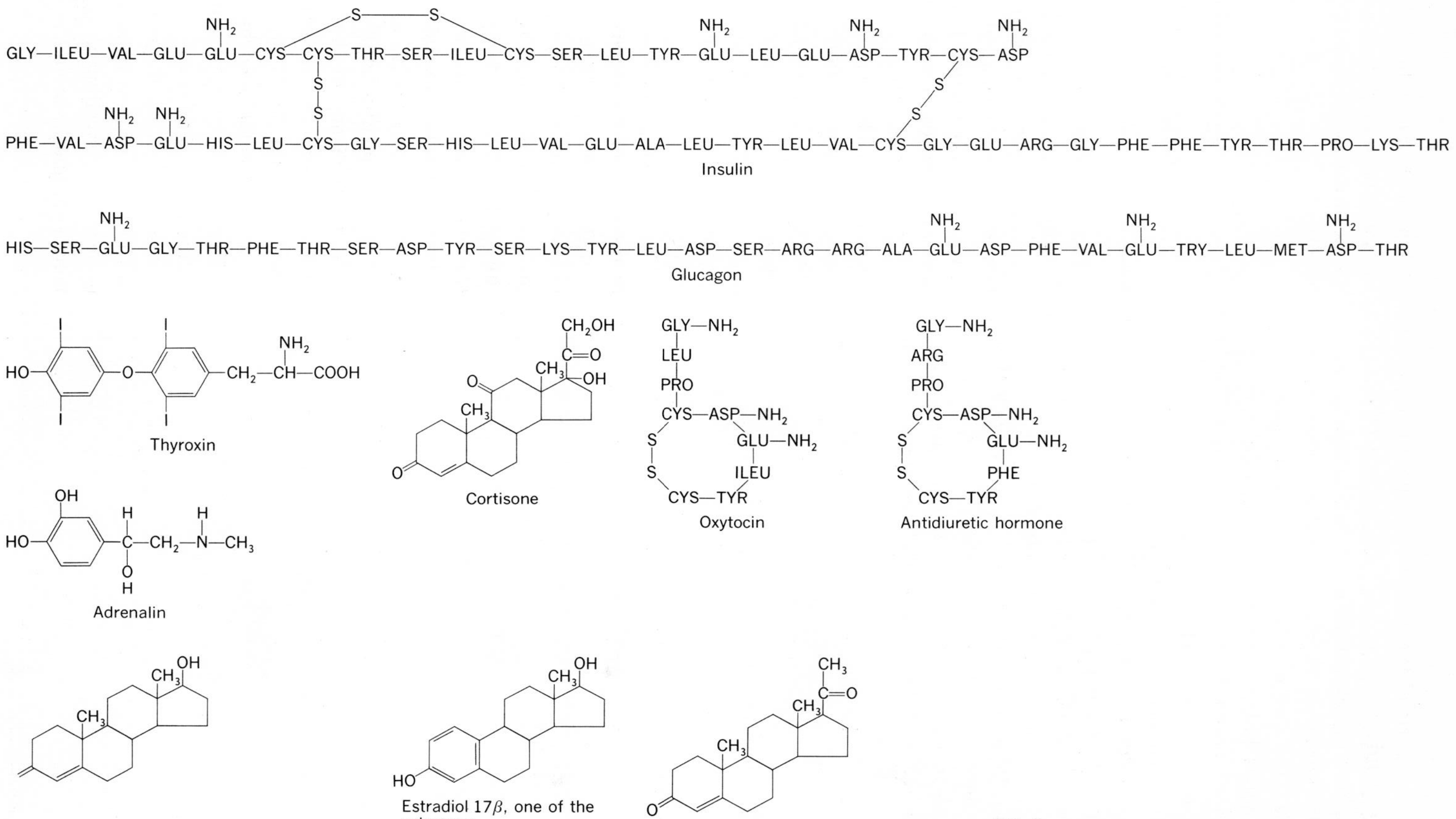

**36.1** Molecular structures of some hormones.

Insulin

```
              NH₂                                    NH₂
               |                                      |
GLY—ILEU—VAL—GLU—GLU—CYS—CYS—THR—SER—ILEU—CYS—SER—LEU—TYR—GLU—LEU—GLU—ASP—TYR—CYS—ASP
                          |                                                    |
                          S                                                    S
                          |                                                    |
                          S                                                    S
         NH₂ NH₂          |                                                    |
          |   |           S                                                    S
PHE—VAL—ASP—GLU—HIS—LEU—CYS—GLY—SER—HIS—LEU—VAL—GLU—ALA—LEU—TYR—LEU—VAL—CYS—GLY—GLU—ARG—GLY—PHE—PHE—TYR—THR—PRO—LYS—THR
```

Glucagon

```
 NH₂
  |                    NH₂                           NH₂
HIS—SER—GLU—GLY—THR—PHE—THR—SER—ASP—TYR—SER—LYS—TYR—LEU—ASP—SER—ARG—ARG—ALA—GLU—ASP—PHE—VAL—GLU—TRY—LEU—MET—ASP—THR
```

Oxytocin

```
GLY—NH₂
  |
LEU
  |
PRO
  |
CYS—ASP—NH₂
  |      |
  S    GLU—NH₂
  |      |
  S     ILEU
  |      |
CYS—TYR
```

Oxytocin

Antidiuretic hormone

```
GLY—NH₂
  |
ARG
  |
PRO
  |
CYS—ASP—NH₂
  |      |
  S    GLU—NH₂
  |      |
  S     PHE
  |      |
CYS—TYR
```

Antidiuretic hormone

Thyroxin

Adrenalin

Cortisone

Estradiol 17β, one of the estrogens

Progesterone

by other parts of the body, including other endocrine glands. Many negative feedback mechanisms have been discovered within the endocrine system.

The hormones themselves are a varied group. Unlike the enzymes, all of which are proteins, hormones fall into no single group, but most of them are proteins, polypeptides, or steroids (Table 36.1 and Fig. 36.1). The steroid molecules consist of four connected rings of carbon atoms with varying side chains; steroids are fat-soluble compounds. Like the hormones of plants, hormones in animals are effective in relatively low concentrations. Relatively little is known about the specific actions of hormones, but it is believed that they may change the permeability of plasma membranes to various substances, modify enzyme activity within cells, or control the action of genes. More is known about the results of these reactions than about the nature of the reactions themselves.

**36.2** Endocrine system.

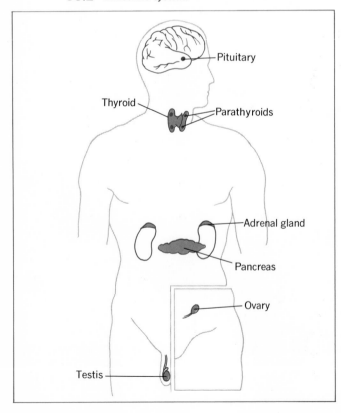

Generally, the endocrine system acts slowly in comparison with the nervous system. Minutes, hours, or even days may be required for noticeable effects.

# Endocrine Glands

The endocrine glands include the thyroid gland, parathyroid glands, pancreas, adrenal glands, pituitary gland, and the ovaries and testes (Fig. 36.2). In addition, many small glands in the stomach and intestinal walls secrete hormones that affect the flow of digestive enzymes from digestive glands. Some endocrine glands have nonendocrine functions as well. The reader will remember the pancreas as a digestive gland producing several digestive enzymes (Chapter 31). The ovaries and testes are reproductive organs producing gametes as well as sex hormones. Within these multifunctional organs, different cells perform the different functions.

The **thyroid gland** lies in the throat below and to the front of the larynx. Appearing almost like two glands, it is a bilobed structure, the two lobes connected by only a narrow band of tissue. It produces the hormone **thyroxin,** which influences the general rate of metabolism, a high concentration increasing the rate.

Persons with an excess of thyroxin usually are very active physically, and they frequently are nervous. Their thin bodies never seem to lack energy. A deficiency of thyroxin is likely to result in fatigue and overweight. A deficiency in the growing child leads to cretinism, a form of dwarfism accompanied by mental retardation.

Thyroxin is an amino acid that contains four iodine atoms, and it cannot be formed unless the diet contains a small quantity of iodine. Without it, the thyroid gland enlarges into a goiter, which appears as a swelling in the throat. Once goiters were common in inland places such as the Great

Lakes area where people were unlikely to eat much iodine-rich seafood. Now considered disfigurements, goiters once were regarded as marks of beauty. The slight swelling in the throat of Leonardo da Vinci's Mona Lisa has been interpreted as a small goiter. With the coming of iodized salt to the market, goiters quickly disappeared from the Great Lakes region. Rapid transportation also makes possible the enjoyment of fresh seafood no matter how far inland one lives.

**Calcitonin** is a second hormone produced by the thyroid gland. An increase in the calcium content of the blood stimulates the thyroid gland to release calcitonin. The bones of the body contain a great deal of calcium, which they obtain from the blood and which they may also return to it. Calcitonin acts by inhibiting the release of calcium from the bones, and this lowers the blood calcium level. As the calcium blood level falls, the thyroid gland no longer is stimulated to produce calcitonin. If an increase in the blood calcium level follows, this immediately stimulates the production of more calcitonin. This is a negative feedback mechanism that functions in concert with another, for although calcitonin can lower a high calcium level, it does not raise a low one. A hormone of the parathyroid glands assumes this function.

Four **parathyroid glands** are embedded in the back of the thyroid gland, two associated with either lobe. They secrete **parathormone,** which raises a low calcium level of the blood. A rise in the calcium level depresses the production of parathormone by the parathyroid glands. This tends to decrease the calcium concentration, and the parathyroid glands are stimulated to synthesize more parathormone again.

Figure 36.3 summarizes the double feedback

**36.3** Feedback mechanism by which calcitonin from the thyroid gland and parathormone from the parathyroid glands control the calcium level of the blood.

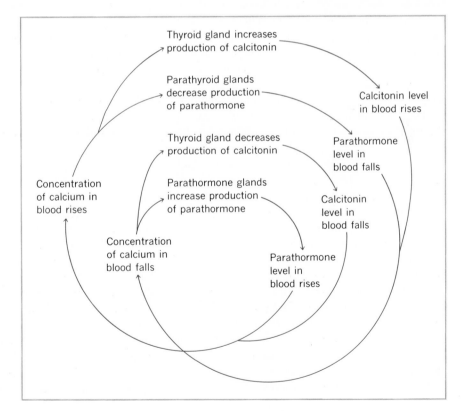

mechanism involving both calcitonin and parathormone. In a healthy body functioning properly, the two hormones maintain a constant calcium blood level, calcitonin preventing a rise and parathormone preventing a drop.

Parathormone has three major targets in the body: the bones, the kidneys, and the intestine. It causes calcium to be released from the bones, inhibits calcium excretion by the kidneys, and increases the amount of calcium absorbed by the intestine.

If the parathyroid glands enlarge beyond their normal size or if tumors begin to grow on them, they secrete abnormally high amounts of parathormone. Bones become soft and weak as calcium is drawn from them. Nerves and muscles, which receive more than their normal share of calcium, become slow to respond, and the individual becomes unresponsive to stimuli. A deficiency of the hormone lowers the calcium level in the blood. Calcium may be deposited in bones in greater than normal quantities, and the joints may become calcified. Muscles and nerves suffer from lack of calcium, and convulsive twitching can result in death.

The **pancreas,** which produces several digestive enzymes, also synthesizes **insulin,** a hormone that controls the conversion of glucose to glycogen in the liver. Glucose absorbed by capillaries in the small intestine is transported to the liver where it is converted to glycogen, which is stored here. Not all of it follows this course, however, for enough remains in the blood to keep the glucose level at its normal 0.1 per cent.

Whenever the glucose content of the blood begins to fall below 0.1 per cent, the liver digests some of the glycogen to glucose, which it then secretes into the bloodstream. If the glucose content of the blood rises above 0.1 per cent, glucose is converted to glycogen in the liver until the normal level is reached. It is the latter step that insulin controls. Insulin also increases the rate at which glucose enters cells by altering the permeability of the plasma membrane to glucose.

A deficiency of insulin produces diabetes mellitus (commonly called sugar diabetes or merely diabetes). The liver converts glycogen to glucose, and the glucose level in the blood rises above 0.1 per cent. The body uses only some of this glucose, and some is excreted in the urine. When the glycogen supply is exhausted, the body begins to digest fat and proteins and the patient loses weight. Periodic injections of insulin obtained from animal sources bring relief to diabetics. Insulin cannot be administered orally, for it is a protein, and the protein-digesting enzymes in the stomach and small intestine destroy it before it can be absorbed into the bloodstream. Some successful oral medications for diabetes are drugs that stimulate the pancreas to increase its insulin production.

**Glucagon,** another hormone secreted by the pancreas, raises the concentration of glucose in the body by digestion of glycogen. In a normal person, a balance between insulin and glucagon maintains a normal carbohydrate metabolism.

The **adrenal glands** (also called suprarenal glands) lie like small caps atop the kidneys. Functionally, each adrenal gland is two glands—the inner part, or medulla, and the outer part, or cortex.

The **adrenal medulla** produces **adrenalin** (also called epinephrine). In times of stress (anger, fear, and so on) the hormone is released into the bloodstream. This brings more blood to tissues by increasing the rate of heartbeat; blood pressure rises; blood sugar content rises temporarily. Blood from the smooth muscles is directed to where it is most likely to be needed, the skeletal muscles. These changes—sometimes called the "fight-or-flight" reactions—help one to survive emergencies. High adrenalin levels enable persons to perform "superhuman" feats, such as lifting heavy weights that they otherwise could not raise. That the adrenal medulla can be removed without danger probably is due to the fact that part of the nervous system performs these same functions, and adrenalin merely reinforces them. In fact, the adrenal medulla is modified nervous tissue.

**Noradrenalin** (also called norepinephrine),

which is produced by the adrenal medulla, has a structure similar to that of adrenalin and performs some of the same functions as adrenalin.

The **adrenal cortex** produces several hormones, all of which are steroids. They fall into three groups: glucocorticoids, mineralocorticoids, and sex hormones. The **glucocorticoids,** which include cortisone, regulate protein and carbohydrate metabolism; they stimulate the conversion of proteins to carbohydrates. This favors high glycogen content in body tissues and helps to maintain the glucose level of the blood. Administration of cortisone to persons suffering from rheumatism or arthritis may alleviate their symptoms, but if given in large quantities cortisone may produce several undesirable side effects, including high blood pressure and brittle bones. The **mineralocorticoids** regulate the salt and water relationships of the body by stimulating reabsorption of sodium by the kidney tubules. The **sex hormones** of the adrenal cortex are similar to those produced by the sex organs, and they control production of secondary sexual characteristics.

Removal of the adrenal cortex causes Addison's disease, which is characterized by bronzing of the skin, muscular weakness, low blood pressure, and digestive disturbances. The concentrations of sodium, chlorine, and sugar in the blood fall below normal, and the concentration of potassium increases. Glycogen is lost from tissues, and proteins are not easily converted to carbohydrates.

The **pituitary gland** (also called the hypophysis) is another gland of two parts—a posterior lobe and an anterior lobe. Because of its effect on many other parts of the body, especially some of the other endocrine glands, it has been called the master gland. It occupies a position below the brain.

The **posterior lobe of the pituitary** secretes two polypeptide hormones, oxytocin and antidiuretic hormone (ADH, also called vasopressin). **Oxytocin** stimulates smooth muscles to contract and is important in childbirth when the smooth muscles of the uterus expel the fetus. After birth, oxytocin stimulates ejection of milk by the mother in response to stimulation by the sucking infant. **ADH** regulates blood pressure by constriction of arterioles; it affects the water content of the body by regulating reabsorption of water by the kidney tubules; thereby it also affects blood pressure. An excess of ADH raises the blood pressure above normal by causing the reabsorption of abnormal quantities of water. A deficiency of this hormone causes the excretion of excessive amounts of urine, a condition called diabetes insipidus. Increased urination following the drinking of alcohol is due to the inhibition of ADH release from the posterior pituitary by alcohol.

Oxytocin and ADH are not produced in the posterior pituitary but rather in the **hypothalamus,** a portion of the brain adjacent to the pituitary gland. Nerve cells extending from the hypothalamus to the posterior pituitary produce these two hormones. The hormones travel the length of the nerve cells to the capillaries of the posterior pituitary, where they are released into the blood (Fig. 36.4).

The **anterior lobe of the pituitary** produces a variety of hormones. One of them, the growth hormone (also called somatotropic hormone), regulates the growth of the long bones during childhood. A deficiency of this hormone in children retards their growth in height; if the deficiency is severe enough they mature into midgets. An excess in childhood causes gigantism. If an excess develops only in adulthood, the bones of the hands, feet, and head grow, making these parts disproportionately large; the condition is called acromegaly.

The relationships of the anterior pituitary with some other endocrine glands are negative feedback mechanisms, the hormones produced by the anterior pituitary stimulating the hormone output of the other glands and their hormones depressing the productions of the anterior pituitary. Among the anterior pituitary hormones are thyrotropin, adrenocorticotropin, follicle-stimulating hormone, luteinizing hormone, and prolactin.

**Thyrotropin** (or thyroid-stimulating hormone,

319

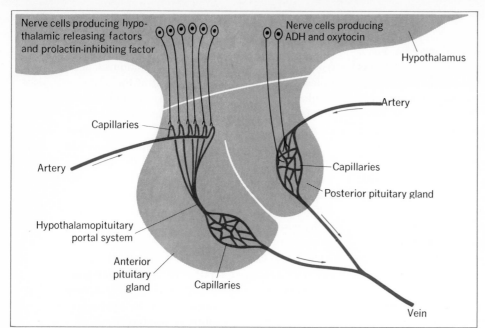

**36.4** Relationship between the hypothalamus and the pituitary gland. At least five releasing factors and prolactin-inhibiting factor are secreted by nerve cells into the hypothalamo-pituitary portal system, which conducts these factors to the anterior pituitary. Other nerve cells produce ADH and oxytocin, which they release into the capillaries in the posterior pituitary. The several hypothalamus substances are produced by different nerve cells. Each nerve cell in this drawing actually represents a group of cells.

TSH) produced by the anterior pituitary stimulates the thyroid gland to produce its thyroxin. The thyroxin inhibits thyrotropin production by the anterior pituitary. Inhibited, the anterior pituitary produces much less thyrotropin. This reduces the stimulation of the thyroid gland, which then produces less thyroxin. As the amount of thyroxin reaching the anterior pituitary decreases, its inhibition is released, and it increases its production of thyrotropin again (Fig. 36.5). As long as the two glands function normally, this mechanism keeps the level of thyroxin nearly constant. This is another example of homeostasis.

**Adrenocorticotropin** (usually better known as ACTH) stimulates the adrenal cortex. The relationship between the anterior pituitary and the adrenal cortex is another feedback mechanism, ACTH stimulating the production of adrenal cortical hormones which, in turn, inhibit the production of ACTH.

Two hormones called **gonadotropins** affect the gonads, or sex organs. Both sexes produce both hormones. **Follicle-stimulating hormone** (FSH), one of the gonadotropins, gets it name from the fact that it stimulates production of the ovarian follicles in women. In men it stimulates development of seminiferous tubules in the testes; it is these tubules that produce sperms (Chapter 40).

In women, **luteinizing hormone** (LH), another gonadotropin, controls ovulation, the formation of the corpus luteum, and the release of female sex hormones from the ovary. FSH and LH are involved in the control of the menstrual cycle, which is another feedback mechanism and is discussed in Chapter 40. In men, LH controls production of male sex hormones in the testes. The relationship between the anterior pituitary and the ovaries or testes is still another feedback mechanism. LH stimulates the production of sex hormones, which depress the LH production of the anterior pituitary.

**Prolactin** stimulates production of milk after pregnancy and controls the production of the sex hormones estrogen and progesterone by the ovary.

In addition to the negative feedback systems existing between the anterior pituitary and the target organs of its several hormones, another negative feedback system exists between the anterior

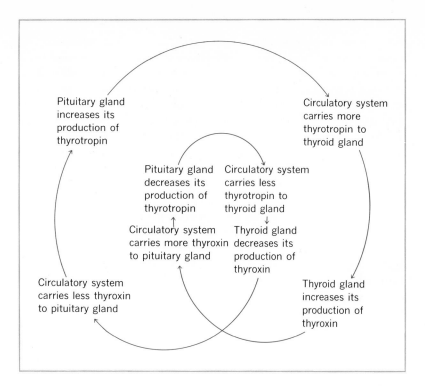

**36.5** Feedback mechanism by which the pituitary gland controls production of thyroxin by the thyroid gland.

pituitary, the target organs, and the hypothalamus. Nerve cells in the hypothalamus produce at least five substances that stimulate the anterior pituitary to release five of its hormones. These substances, called hypothalamic releasing factors, are growth-hormone-releasing factor, TSH-releasing factor, ACTH-releasing factor, FSH-releasing factor, and LH-releasing factor.

Hypothalamic nerve cells produce these factors and secrete them into the **hypothalamopituitary portal system.** Recall that from the time it leaves the heart until it returns, blood usually passes through only one set of capillaries. We have seen two exceptions—in the kidneys and in the hepatic portal system. Here between the hypothalamus and the anterior pituitary is another (Fig. 36.4). Nerve cells produce the releasing factors and secrete them into capillaries in the hypothalamus. The capillaries lead into the very short hypothalamopituitary portal system, which enters the ante-

rior pituitary where it branches into a second capillary net. Thus the releasing factors need not pass through the heart on their way to the anterior pituitary but arrive there rather directly. Each releasing factor stimulates the anterior pituitary to release its specific hormones. For example, TSH-releasing factor stimulates the release of thyrotropin. Thyrotropin, in turn, stimulates the thyroid gland to secrete thyroxin. The elevated thyroxin level in the blood inhibits the production of TSH-releasing factor in the hypothalamus, and this lowers the release of TSH from the anterior pituitary. The resulting lowered production of thyroxin by the thyroid gland decreases the inhibition on the hypothalamus nerve cells that produce TSH-releasing factor.

Similar negative feedback systems exist between ACTH, FSH, and LH and their corresponding releasing factors. The hypothalamus produces no releasing factor for prolactin; rather it secretes

321

prolactin-inhibiting factor, which inhibits release of prolactin from the anterior pituitary.

As part of the brain, the hypothalamus is part of the nervous system, but it could be considered part of the endocrine system as well. Not only do its several releasing factors and prolactin-inhibiting factor intimately affect the endocrine system, but these factors really are hormones. They are substances produced in one organ (hypothalamus) and transported via the circulatory system to another organ (anterior pituitary) where they exert their effects.

The **sex organs,** which are considered in more detail in Chapter 40, produce the sex hormones, which are steroid molecules. The main male sex hormone produced by the testes is **testosterone. Estrogens** and **progesterone** are formed in the ovaries. They affect not only the feedback mechanisms described earlier but also the secondary sexual characteristics. These are the physical traits typical of the two sexes, but not the sexual organs themselves. Examples are the beards of men and their low voices, and the curvaceous figures of women and their higher voices.

Sex hormones also influence sex drive. Breeders of animals castrate some of their male animals by removing their testes. Sometimes this is done because the animal has undesirable traits that the breeder does not want passed on to the next generation. Sometimes it is done to reduce the sex drive and keep the animal from wasting his energies in amorous pursuits; the animals also become more docile and easier to handle. When castrated young, bulls and roosters become steers and capons, respectively; they produce more and better meat than their fertile brothers.

In addition to the relatively large, well-defined glands mentioned, the linings of the stomach and small intestine assume some endocrine function. It would be wasteful and, what is more important, harmful for the digestive juices to be secreted continuously into the digestive tract. Rather, they reach their site of action only when they are needed, and this is due to the action of hormones

like gastrin, enterogastrone, cholecystokinin, and secretin.

The mechanical stretching of the stomach when food enters it causes some cells of the stomach wall to secrete **gastrin** into the bloodstream. Gastrin stimulates the flow of hydrochloric acid from the stomach glands and the flow of bicarbonate from the pancreas.

Fats entering the small intestine stimulate the intestine wall to secrete **enterogastrone,** which has an action on the acid glands of the stomach opposite to that of gastrin. It depresses the flow of hydrochloric acid. This protects the body from the ulcers that may form as a result of the exposure of the stomach or intestinal linings to too much hydrochloric acid. Doctors often prescribe diets rich in fats to their ulcer patients.

Fats, partially digested proteins, and dilute acids stimulate the small intestine to produce **cholecystokinin.** The gall bladder releases its bile when this hormone reaches it.

Entering foodstuffs also stimulate the small intestine to produce **secretin,** a hormone that stimulates the flow of pancreatic juice from the pancreas.

# Questions

1. In a general sense, what is the function of the endocrine system?

2. Describe in more detail the functions of a few endocrine glands and their hormones.

3. Describe a few negative feedback mechanisms involving hormones.

4. Give examples of pairs of hormones that have opposite effects.

5. Consider the endocrine glands one by one. What result would you expect if each one was removed from the body? How many of these results would be serious?

6. Hormones, enzymes, and vitamins all function at very low concentrations. In what ways do these substances resemble each other? How do they differ?

# Suggested Readings

Bloom, W., and D. W. Fawcett. *A Textbook of Histology,* 9th ed. Philadelphia: W. B. Saunders Company, 1968.

Clegg, P. C., and A. G. Clegg. *Hormones, Cells and Organisms.* Stanford, Calif.: Stanford University Press, 1968.

Davidson, E. "Hormones and Genes," *Scientific American* 212(6): 36–45 (June 1965).

Fieser, L. F. "Steroids," *Scientific American* 192(1): 52–60 (January 1955).

Gillie, R. B. "Endemic Goiter," *Scientific American* 224(6): 92–101 (January 1971).

Gorbman, A., and H. A. Bern. *A Textbook of Comparative Endocrinology.* New York: John Wiley & Sons, Inc., 1962.

Griffin, D. R., and N. Novick. *Animal Structure and Function,* 2nd ed. New York: Holt, Rinehart and Winston, Inc., 1970.

Guillemin, R., and R. Burgus. "The Hormones of the Hypothalamus," *Scientific American* 227(5): 24–33 (November 1972).

Langley, L. L. *Homeostasis.* New York: Reinhold Publishing Corporation, 1965.

Larimer, J. *An Introduction to Animal Physiology.* Dubuque, Iowa: William C. Brown Company, Publishers, 1968.

Levine, R., and M. S. Goldstein. "The Action of Insulin," *Scientific American* 198(5): 99–106 (May 1958).

Li, C. H. "The Pituitary," *Scientific American* 183(4): 18–22 (October 1950).

———. "The ACTH Molecule," *Scientific American* 209(1): 46–53 (July 1963).

Pike, J. E. "Prostaglandins," *Scientific American* 225(5): 84–92 (November 1971).

Rasmussen, H. "The Parathyroid Hormone," *Scientific American* 204(4): 56–63 (April 1961).

———, and M. M. Pechet. "Calcitonin," *Scientific American* 223(4): 42–50 (October 1970).

Thompson, E. O. P. "The Insulin Molecule," *Scientific American* 192(5): 36–41 (May 1955).

Turner, C. D. *General Endocrinology,* 4th ed. Philadelphia: W. B. Saunders Company, 1966.

Vander, A. J., J. H. Sherman, and D. S. Luciano. *Human Physiology.* New York: McGraw-Hill Book Company, 1970.

Wilkins, L. "The Thyroid Gland," *Scientific American* 200(3): 119–129 (March 1960).

Zuckerman, Sir S. "Hormones." *Scientific American* 196(3): 76–87 (March 1957).

# 37

# The Nervous System

1  **The function of the nervous system is the rapid coordination of events occurring within the body.**

2  **The somatic nervous system controls voluntary muscles, and the autonomic nervous system controls involuntary muscles and glands.**

3  **Nervous impulses, which travel along neurons, are self-propagating electrochemical impulses.**

4  **At synapses, the spaces between neurons, the impulses are carried by chemical transmitters.**

The endocrine system, described in the preceding chapter, regulates the internal environment of the body primarily by correcting any unfavorable changes that may have occurred within it. The nervous system shares this function, but in addition it also enables the organism to respond to its environment in such a way that many potentially harmful internal changes are avoided completely. The sight of food or danger sends along the nerves electrochemical messages that result in muscular activity appropriate to the situation; the organism takes the food or runs away and thus avoids starvation or being injured by an enemy. Together with the endocrine system, the nervous system coordinates the activities of the body. Unlike the endocrine system, it transmits its messages rapidly, less than a second being required for some responses. The basic units of the nervous system are neurons, specialized cells that transmit electrochemical messages.

## Overview of the Nervous System

The nervous system (Fig. 37.1 and Table 37.1) is divided into two main anatomical parts: the **cen-**

Central nervous system
  Brain
    Forebrain
      Cerebrum
      Thalamus
      Hypothalamus
    Midbrain
    Hindbrain
      Cerebellum
      Pons
      Medulla
  Spinal cord
Peripheral nervous system
  Cranial nerves
  Spinal nerves

**Table 37.1**
Major Parts of the Nervous System

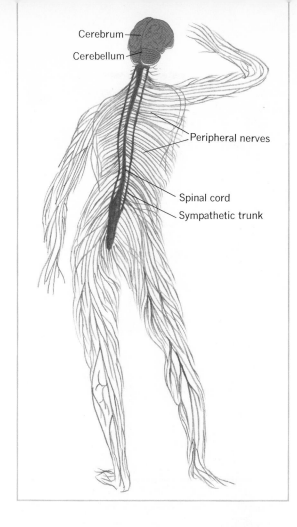

**37.1**  Nervous system.

tral nervous system, which consists of the brain and spinal cord, and the peripheral nervous system, which consists of the cranial and spinal nerves. The nerves lead from the brain and spinal cord to other parts of the body.

The brain occupies a protected position within the skull. Its function of sorting out the many nervous impulses that reach it and relaying them to the proper parts of the body has led to its being compared with a switchboard. The analogy fails, however, when one realizes that no switchboard possesses the almost unbelievable complexity of the human brain (Figs. 37.2 and 37.3). There are more than 9 billion neurons in the cerebral cortex of the brain, and some of them have as many as 200,000 endings which communicate with other neurons.

The spinal cord is a nearly cylindrical mass of nervous tissue that extends from the brain down the length of the trunk, where it tapers to a fine ending. It runs through openings in the vertebrae that compose the spinal column and so, like the brain, is protected by bone.

Forty-three pairs of peripheral nerves extend from the central nervous system. Twelve of the

**37.2**  A small portion of the cerebral cortex of the human brain. Note the complexity of arrangement of neurons. (Courtesy of Dr. Beth L. Wismar.)

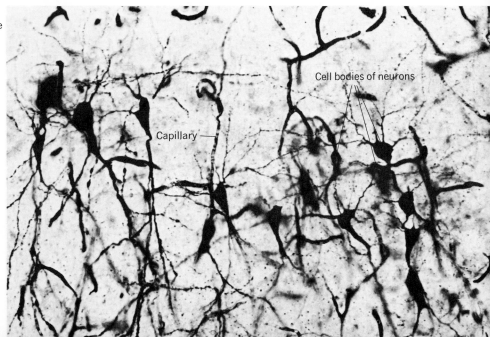

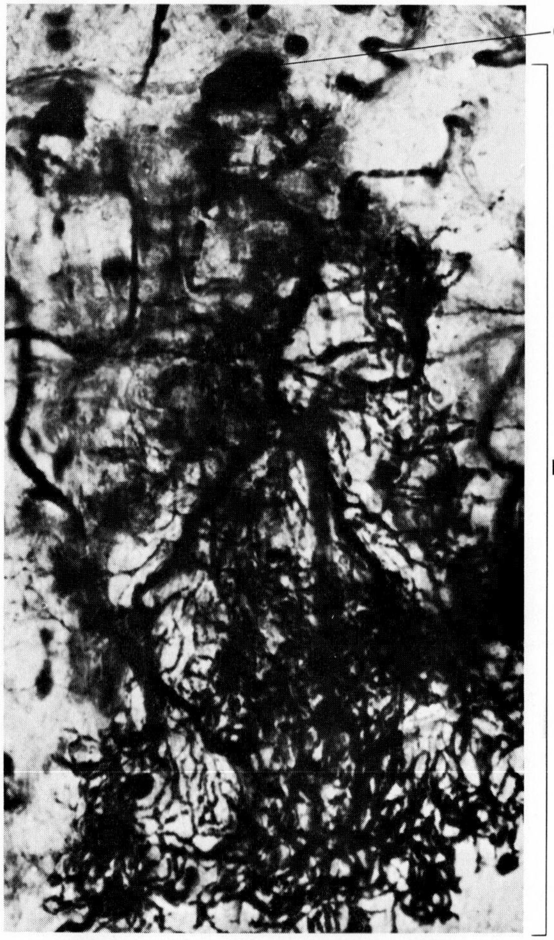

**37.3** A neuron from the cerebellum of the human brain. The dendrite is so profusely branched that it is difficult to see clearly. It does give some idea of the number of other neurons with which it might communicate. (Courtesy of Dr. Beth L. Wismar.)

pairs are cranial nerves that proceed from the brain to various parts of the head and some parts of the trunk. The 31 pairs of spinal nerves extend from the spinal cord to many parts of the trunk and limbs. Both cranial and spinal nerves branch profusely and reach into all organs of the body. Each nerve is a bundle of neurons held together by connective tissue and served by small blood vessels.

## Neurons

The anatomy of a **neuron** (Fig. 37.4) differs from that of other cells of the body. Each neuron consists of a **cell body** with several elongated processes called **dendrites** and **axons.** The cell body is a more or less compact mass of cytoplasm containing the nucleus of the cell. Extending from the cell body are one or more dendrites and a single axon. These cytoplasmic processes may be very short or may reach the length of several feet; some neurons, for example, reach from the spinal cord to the toes. Each dendrite or axon branches several times, enabling it to communicate with several other neurons. Nervous impulses ordinarily travel along neurons in one direction only—from dendrite to cell body to axon. Where the axon of one neuron meets the dendrite of another neuron is a small gap called a **synapse.** The impulse travels across the synapse and then proceeds along the dendrite of the second neuron. When it reaches the axon of the second neuron, it must cross another synapse to reach a third neuron, and so on. Not all axons synapse with dendrites; some terminate on the cell body of another neuron.

The processes of neurons may be surrounded by one, both, or neither of two sheaths—the gray neurilemma and the white myelin sheath. The **neurilemma** is formed of living cells (called Schwann cells) that are wrapped around the axons of neurons in the peripheral nervous system, but not those of the central nervous system. Axons with a neurilemma can regenerate missing parts if they are damaged, but those lacking a neurilemma have no power of regeneration. This is one reason why damage to the brain or spinal cord can be so serious, for ordinarily it results in permanent damage, whereas there is some hope for repair of damage to peripheral nerves. Neurons are incap-

326

able of division, and those that die are not replaced.

The **myelin** sheath is the plasma membrane of a Schwann cell and is wrapped around the neuron process several times in "jelly roll" fashion. Like other cell membranes, the myelin sheath is rich in fatty substances, and this gives it its white color. Some but not all neurons in both the central nervous system and the peripheral nervous system are myelinated. These cells transmit impulses about 10 to 20 times more rapidly than unmyelinated cells. In man, myelinated neurons may transmit impulses at a rate of about 300 feet per second. In general, motor neurons serving skeletal muscles are the most heavily myelinated, and those serving the viscera are moderately myelinated. Sensory neurons usually have little myelin or none. Multiple sclerosis, a disease of uncertain origin, but possibly caused by viruses, is characterized by degeneration of the myelin sheaths of neurons in the central nervous system with resultant severe impairment of transmission of impulses.

Although the neuron is the basic unit of the nervous system, neurons do not function singly. Most impulses in the living body travel along at least three neurons, and often more. A **sensory neuron,** an **association** or **connector neuron,** and a **motor neuron** form a **reflex arc** (Fig. 37.5). The dendrite endings of sensory neurons lie in the skin, in special sense organs such as the ear, and in

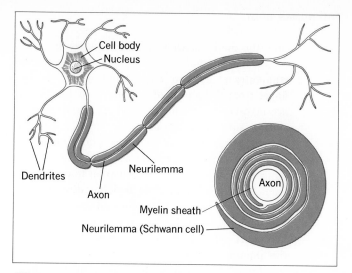

**37.4**  A neuron with both neurilemma and myelin sheath. A cross section of the axon is shown at right.

many internal organs. The axons of motor neurons terminate in muscles or glands. In one example of a reflex arc, a sensory neuron extends along a spinal nerve and reaches the spinal cord, where its axon synapses with the dendrite of an association neuron. The axon of the association neuron synapses with the dendrite of a motor neuron that proceeds from the spinal cord along the spinal nerve to a muscle.

When a stimulus impinges on a dendrite of the sensory neuron and initiates an impulse—perhaps a finger is burned by a match—the impulse travels along that neuron, the association neuron, and

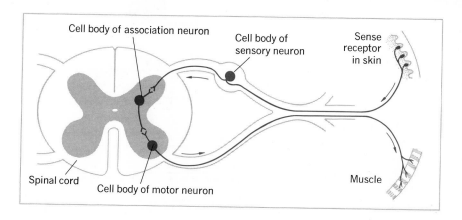

**37.5**  A reflex arc consisting of three cells: sensory neuron, connector neuron, and motor neuron. The connector neuron lies completely within the spinal cord.

then the motor neuron leading back to the arm. As it leaves the motor neuron, the nervous impulse stimulates a muscle to contract and so removes the finger from the source of irritation. Such a reflex arc involves no conscious thinking about the resulting act, for the brain is not involved. This allows for rapid action; the person does not have to pause to make a decision about how great the danger might be to his finger, whether to withdraw from the danger, and if so in what direction to move. Admittedly the time required to make these decisions might be less than a second, but even this short delay might make a great difference in the amount of injury received.

Although the brain is not involved in this reflex arc, additional association neurons do carry impulses up the spinal cord to the brain, which receives conscious sensation of the burning of the finger. This sets in motion some conscious thinking that the person voluntarily acts upon. He checks to see if the match is still burning, and, if so, he stamps it out to prevent its setting the house on fire. He hurries to cool the burned finger in cold water, or, if the injury seems to require it, he seeks medical aid. All these actions involve many association neurons in the brain and spinal cord and many motor neurons and muscles.

Reflex arcs keep involuntary muscles (smooth muscles of the viscera and cardiac muscles of the heart) and glands working without our having conscious knowledge of what is going on. Even if one should take thought about his heartbeat or some uncomfortable activity in his intestine, he cannot will them to change.

## Transmission of Nerve Impulses

The transmission of an impulse along a neuron is an electrochemical phenomenon and not a pure electric current, as once was thought. In a neuron at rest—that is, one that is not transmitting an impulse—there is an unequal distribution of positive and negative ions inside and outside the cell. The charge is positive outside the plasma membrane and negative inside the membrane (Fig. 37.6). Such a resting neuron is said to be **polarized.**

A stimulus applied to a neuron initiates a nervous impulse by increasing the permeability of the plasma membrane to ions at the point of application of the stimulus. With the sudden change in the nature of the membrane, the membrane re-

**37.6** Transmission of a nerve impulse along a portion of a neuron. (a) Resting neuron with positive charges on the outer surface of the membrane and negative charges on the inner surface. (b–e) Electrochemical impulse moves along the neuron as outer surface becomes negatively charged and inner becomes positively charged. (f) Neuron returns to the resting condition.

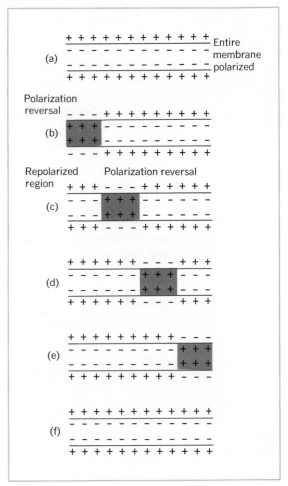

verses its polarization, and the interior of the cell becomes more positively charged than the outside. The polarization reversal of one small segment of a neuron causes the polarization of the adjacent area to become reversed as well.

When this happens, the part that was stimulated first becomes repolarized and returns to its original condition. The process repeats in the adjacent region—polarization reversal at this point causes polarization reversal of the region beyond it and the second segment becomes repolarized itself. In this way the impulse propagates itself along the length of the neuron. During polarization reversal, a neuron cannot transmit a second impulse, but it becomes receptive again when it is repolarized.

The analogy between the propagation of a nervous impulse and the burning of a fuse often has been made. The heat from the burning of one portion of the fuse heats the adjacent part enough to start it burning, too, and the spark travels the length of the fuse. If a fuse is set afire somewhere in its middle, sparks will proceed in either direction. As a fuse ordinarily is arranged, however, only one of the two sparks accomplishes what it was intended to do. The spark reaching the dynamite sets off the desired explosion; the one proceeding to the free end of the fuse merely sputters out. So, too, if the middle of a neuron is stimulated, two impulses proceed in either direction, one going to the endings of the axon, the other to the endings of the dendrite (or dendrites). However, only the impulse reaching the axon ending is propagated across a synapse to the next neuron (or to a muscle or gland); the impulse reaching the dendrite ending dies out there, the dendrite not having the required cellular machinery to send the impulse across a synapse. The analogy between the propagation of a nervous impulse and the burning of a fuse breaks down in one important aspect. A burning fuse consumes itself, whereas a neuron quickly becomes repolarized and is ready for work again. Repolarization requires only 0.001 to 0.005 second, the exact length of time varying with the type of neuron and its condition. Several hundred to a

thousand impulses can be transmitted by a neuron in a second.

The rate of conduction of an impulse depends on the neuron and its condition rather than the strength of the stimulus that initiated the impulse. Like muscle cells, neurons operate on an "all-or-none" principle—either an impulse is initiated or it is not. If a stimulus is of sufficient intensity to initiate an impulse, the impulse travels at a rate characteristic of that neuron and not of the impulse. Drugs and temperature may change the state of the neuron and so alter its ability to conduct. Surgeons use anesthetics for their ability to impair temporarily the transmission of impulses and so dull pain. Low temperatures applied to nerves have similar effects.

If a stimulus is very weak, then no impulse is initiated. If it is very strong, it may stimulate many neurons, and it may stimulate them repeatedly, causing them to transmit a series of impulses, the frequency of the impulses being limited by the ability of the neurons to repolarize themselves. The brain perceives the strength of the stimulus by the frequency of the impulses it receives.

The sensation that the brain perceives depends on the part of the brain receiving the impulses and not on the nature of the impulses, for all nervous impulses are alike. Light striking nerve endings in the retina of the eye stimulates the optic nerves, and these carry their impulses to only one portion of the brain. Pressure applied to the eyes may stimulate the same neurons in the retina and the impulses carried to the brain give the sensation of light ("seeing stars," for instance) even though no light was present at the time.

Because one neuron does not touch another, there is no contact of membranes between them, and the electrochemical impulse cannot pass directly from one cell to another. The synapse separating two neurons may be only 20 nanometers wide—a space so small as to be invisible under an optical microscope, but large enough to block the electrochemical impulse. The endings of axons contain cytoplasmic vesicles filled with compounds

329

called **transmitter substances.** One of these compounds produced by axons in the sympathetic nervous system (see page 334) is noradrenalin (which is also a hormone produced by the adrenal glands). Axons of neurons in the parasympathetic nervous system (see page 334) and the axons of motor neurons in the somatic nervous system release acetylcholine (Fig. 37.7). An electrochemical impulse reaching the end of an axon stimulates the vesicles to release their transmitter into the synapse. The substance diffuses across the synapse to the dendrite of the adjacent neuron, where it initiates a new electrochemical impulse that continues its way along the neuron to the axon, where the next synapse is bridged in a similar manner. Acetylcholine diffusing across the space between a motor neuron axon and a skeletal muscle cell

(neuromuscular junction) stimulates the muscle cell to contract. Heart muscle, on the other hand, is induced to contract by the noradrenalin secreted by sympathetic nerves and is inhibited by the acetylcholine released from parasympathetic nerves. The reader might refer again to Fig. 34.7, which illustrates the negative feedback mechanism by which these two opposing parts of the autonomic nervous system control the beating of the heart.

Enzymes destroy the transmitters shortly after they are released into synapses and neuromuscular junctions. Acetylcholinesterase, for example, digests acetylcholine to acetic acid and choline; these compounds diffuse back to the axon, where they are used again in the synthesis of acetylcholine which awaits the arrival of another electrochemi-

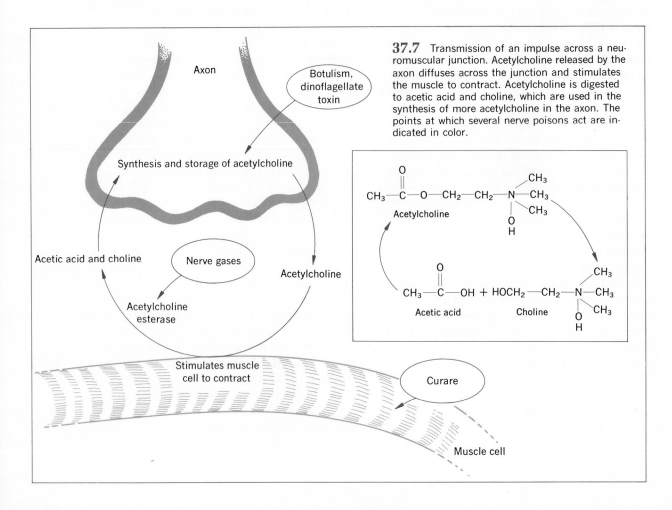

**37.7** Transmission of an impulse across a neuromuscular junction. Acetylcholine released by the axon diffuses across the junction and stimulates the muscle to contract. Acetylcholine is digested to acetic acid and choline, which are used in the synthesis of more acetylcholine in the axon. The points at which several nerve poisons act are indicated in color.

cal impulse. If this did not happen, the transmitters would continue to initiate impulses in dendrites indefinitely and cause muscles to contract indefinitely.

A variety of nerve poisons disrupt the different events occurring in synapses and neuromuscular junctions. Botulism is a most unpleasant form of food poisoning caused by eating certain improperly canned foods, especially beans and meats. If the cans are not properly sterilized, the botulism bacterium may grow there and escrete its toxin, one of the most potent poisons known; ingesting only a small morsel of contaminated food can be lethal. The toxin hinders either the synthesis or release of acetylcholine from axons and so paralyzes muscles. Death usually is by asphyxiation because the respiratory muscles no longer function. The famous red tides that occasionally plague Florida, California, and some other warm-water coasts consist of an abundance of microscopic marine organisms called dinoflagellates. Their toxin is similar to the botulin toxin in effect and potency.

Curare renders muscles less sensitive to acetylcholine and so reduces their ability to respond to nervous impulses. It has been used by South American Indians, who obtain it from plants and dip the tips of their hunting arrows into it. If the arrow pierces an animal, the poison paralyzes it and makes it easier for the hunters to catch their prey.

Atropine and belladonna, two similar drugs, also prevent acetylcholine from acting; belladonna is used to dilate the pupil of the eye during eye examinations.

Some nerve gases unite with the enzyme acetylcholinesterase and prevent the digestion of acetylcholine to acetic acid and choline. The acetylcholine remaining in synapses and neuromuscular junctions continally stimulates neurons and muscles, and the victim suffers uncontrollable muscle spasms and sometimes death. This is the method of action of some of the organophosphate insecticides, such as parathion, malathion, and diazinon, which are chemically related to the nerve gases developed in Germany during World War II. The effect of these insecticides is greatest on insects, but they kill other animals as well. However, unlike some other pesticides, they do not persist long in the environment.

Strychnine, which enjoys some popularity as a poison for rodents, kills by increasing the irritability of nerves, and the animal dies in uncontrollable convulsions.

The meeting of the axons of two neurons at a synapse with the dendrite of a third neuron may result in either summation or inhibition of their impulses. Summation is caused by the release of the same transmitter from the two axons. If the amount released by each axon is insufficient to stimulate the dendrite, the total amount released by the two axons may be sufficient to do the job. Repeated release of the transmitter by a single axon over a short period of time also produces enough to transmit the impulse across the synapse.

The nature of the mechanism of inhibition is not well understood, but it is known that one impulse can inhibit another arriving simultaneously at the same synapse. The sight and sound of an interesting television program in the living room can make us completely unaware of the sounds and odors of dinner preparation in the kitchen. The sudden awareness of an emergency can make all these stimuli seem to fade away even though they actually do continue to impinge on our nerve endings.

Repeated use of a neural pathway produces facilitation; later impulses travel more readily over the same pathway. For this reason, practice makes a person more skillful at a sport or craft, and repeated recitation commits information to memory.

# Anatomy of the Nervous System

The **brain** (Fig. 37.8) is a large mass of nervous tissue, its delicate parts very precisely arranged

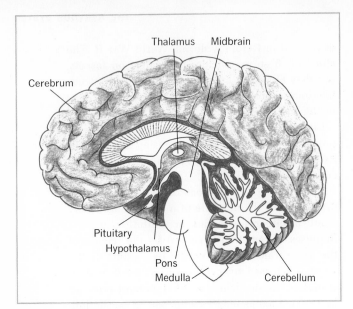

**37.8** Section of a human brain.

within the skull which shelters it. It is divided into three parts: forebrain, midbrain, and hindbrain. The **forebrain,** which is the largest of the three, contains, among other parts, the cerebrum, thalamus, and hypothalamus. The small **midbrain** connects the forebrain to the **hindbrain,** which consists of the cerebellum, pons, and medulla oblongata.

The **cerebrum** comprises most of the brain and is divided into two wrinkled hemispheres that superficially resemble the half of the meat of a walnut. The outer part of the cerebrum, called the cerebral cortex, is gray and contains mostly the cell bodies of neurons; the inner part is white and consists mostly of axons and dendrites, the white coloring being due largely to the white myelin sheaths surrounding the axons. The cerebrum is concerned with consciousness; the posterior part perceives conscious sensations and the anterior part controls voluntary actions. The functions of the cerebrum also include thinking and reasoning, and it is here that memory is stored. Other parts of the brain have to do with unlearned, automatic behavior.

The **thalamus** lies between the cerebrum and the

midbrain. It plays an important role in monitoring impulses arriving from the midbrain and proceeding down to the midbrain. It allows some sensory impulses to continue on to the cerebrum and suppresses others; similarly it regulates motor impulses traveling down from the cerebrum.

The **hypothalamus,** the floor of the thalamus, controls many visceral functions; it regulates body temperature, thirst, hunger, blood pressure, and emotions. It also contains special neurons that form vasopressin and oxytocin, hormones that are conducted to the posterior pituitary gland from which they are released to other parts of the body. Other neurons, produce the hypothalamic releasing factors that stimulate the anterior pituitary to release several of its hormones (Chapter 36).

The **midbrain** connects the thalamus to the hindbrain. It relays messages within the brain, and it contains the optic reflex centers which control the dilation of the pupil and the focusing of the eye.

The **cerebellum** lies low in the back of the head. It coordinates complex muscular movements. The **pons** (which means "bridge") contains fibers that connect the two halves of the cerebellum; it coordinates muscle movements on the two sides of the body. The **medulla oblongata,** which extends into the neck and connects with the spinal cord, controls involuntary actions such as digestion, circulation, and breathing movements.

The **spinal cord** runs through the openings in the vertebrae of the vertebral column ("backbone"). In cross section (Fig. 37.9), the spinal cord has a central butterfly-shaped gray area, which contains cell bodies of neurons, and a surrounding white matter, which contains axons and dendrites ascending to and descending from the brain. Three layers of connective tissue called **meninges** sheathe the spinal cord and brain. Meningitis is an infection of these sheaths.

Each of the **spinal nerves** branches into a **dorsal root** and a **ventral root** as it enters the spinal cord. Sensory neurons enter by way of the dorsal root, and motor neurons leave by way of the ventral

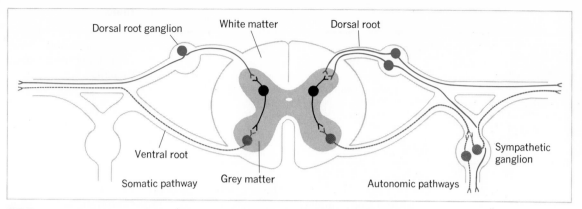

**37.9** Some nervous pathways through the spinal cord. For clarity somatic neurons are shown on the left, autonomic neurons on the right. Actually both types are found on both sides of the body. All cell bodies are either in ganglia or the gray matter of the spinal cord. Colored solid line: sensory neurons. Black line: association neurons. Colored dashed line: motor neurons.

root. If the dorsal root of a nerve should be severed or damaged, the sensory impulses traveling along that nerve would fail to reach the spinal cord and brain; therefore the brain would receive no sensations from the tissues served by that nerve. Impulses traveling through motor neurons in the ventral root would remain undamaged, however, and the brain would still have control over the action of the muscles served by that nerve. If, on the other hand, the ventral root were to be severed or damaged, motor impulses would fail to reach the muscles, and they would become paralyzed. Poliomyelitis (infantile paralysis) is an infection of the ventral roots of nerves and not of muscles, although the result is the paralysis and atrophy of muscles.

**Spinal ganglia** (dorsal root ganglia) are enlarged portions of the dorsal roots of the spinal nerves. They contain the cell bodies of sensory neurons, the axons of which extend into the gray matter of the spinal cord. Association neurons lie completely within the gray matter of the spinal cord, where they synapse with the dendrites of motor neurons. The cell bodies of motor neurons lie within the gray matter and their axons extend through the white matter and then into the ventral root. These three neurons are the minimum required for a reflex arc.

The motor portion of the peripheral nervous system is divided into two parts on the basis of function: the **somatic nervous system** and the **autonomic nervous system.** The former consists of the motor neurons that control the skeletal (voluntary) muscles, and the latter consists of motor neurons that control the smooth (involuntary) muscles of the viscera and the cardiac (involuntary) muscle of the heart.

At least four neurons comprise a typical pathway involving the autonomic nervous system, for there are two motor neurons within the pathway (Fig. 37.9). A sensory neuron enters the spinal cord by way of the dorsal root, and the association neuron lies completely within the gray matter of the spinal cord. It synapses with the first motor neuron, which leaves by way of the ventral root and then enters another ganglion called a sympathetic ganglion. Here it synapses with the second motor neuron, which is the one that serves a muscle or gland.

The autonomic nervous system is divided further into the **sympathetic nervous system** and the **parasympathetic nervous system,** which affect the

333

| Organ Served | Sympathetic System | Parasympathetic System |
|---|---|---|
| Heart | Strengthens and accelerates beat | Weakens and slows beat |
| Digestive tract | Slows peristalsis, decreases activity | Speeds peristalsis, increases activity |
| Urinary bladder | Relaxes bladder | Constricts bladder |
| Muscles in bronchi | Dilates passages | Constricts passages |
| Muscles of iris | Dilates pupil | Constricts pupil |
| Muscles of hair | Causes erection of hair | Causes hair to lie flat |
| Sweat glands | Increases secretion | Decreases secretion |

**Table 37.2**

A Comparison of the Sympathetic and Parasympathetic Nervous Systems

viscera in opposite ways. The effects of the sympathetic nervous system prepare the body quickly when a stress situation arises. The parasympathetic nervous system controls the more routine, day-by-day functions of the viscera.

Sympathetic neurons lie in the spinal nerves between shoulder and waist, and their ganglia lie near the spinal cord (Fig. 37.1). Parasympathetic neurons lie in a few of the cranial nerves and in some spinal nerves in the hip region; their ganglia are near the organs they serve. The two systems have opposite effects on the organs they serve, one stimulating an organ to action, the other inhibiting it (Table 37.2). When danger threatens, the sympathetic nervous system rapidly brings about the "fight-or-flight" reactions mentioned in the preceding chapter. Individual sympathetic nerves affect specific organs. One of these is the adrenal medulla, which is stimulated to release adrenalin into the circulatory system. The adrenalin carried by the blood affects most of the same organs affected by the sympathetic nervous system and so reinforces it. Together they prepare a person almost immediately for an emergency at hand.

---

334 | ## The Senses and Sense Organs

---

The traditional list of five senses has been enlarged to include at least 15 senses, and the future may reveal even more. Sensory nerve endings throughout the body receive stimuli, but not all of them are sensitive to the same stimuli—at least not at the intensities at which these stimuli ordinarily present themselves.

Five types of nerve endings, sensitive to **touch, pressure, pain, heat,** and **cold** are present in the skin. These are somewhat unevenly distributed over the body. Touch receptors, for instance, are common on the finger tips, which are used often for feeling objects, and rare on the back, which is not so used. Muscles and tendons contain the endings of stretch receptors, sensory nerves that transmit to the brain information about the state of contraction or relaxation of a muscle or tendon. At least five sensations reach us from our viscera—**thirst, hunger, nausea,** and the **urges to urinate and defecate.** Four special sense organs—tongue, nose, eye, and ear—contain nerve endings that record **taste, odor, vision, sound,** and **equilibrium;** only these last five senses will be discussed in any more detail here.

### The Sense of Taste

Along the sides of the papillae that give the tongue its rough surface are the **taste buds.** Each contains a few nerve endings sensitive to four basic tastes: sweet, sour, bitter, and salty. Taste buds at the forward tip of the tongue are most sensitive to sweet substances, those at the sides to sour materials. Salty foods are best perceived by taste buds at both the sides and front of the tongue, and the back of the tongue is most receptive to bitter flavors.

Everyone can recall experiencing more than these few tastes; broiled steak, creamed asparagus, or chocolate cookies certainly do not fall into any one of these categories. The reason for this is that these foods affect several kinds of taste buds simultaneously and stimulate them to different degrees; many different combinations of the four flavors produce many different taste sensations. In addition to this, much of what we call taste is odor. Volatile substances in the food enter the nasal passages, where their odors enhance the flavor of the food. Probably everyone has had the experience of his favorite foods sadly lacking in flavor when he suffers from a cold. Actually, he is tasting as much of the food as he ordinarily does, but the nasal passages blocked by mucus receive no odors.

Taste buds are rather sensitive; they may detect quinine diluted to one part per million in water. To be detected, however, a substance must be dissolved in water. Because the mouth ordinarily is wet, this condition usually is met.

## The Sense of Smell

Odors are detected in the nasal passages, where the endings of the olfactory nerves reside. As with the sense of taste, substances must be dissolved before the dendrites are stimulated, and again, the nasal passages ordinarily are moist. The sense of smell is at least a thousand times more sensitive than the sense of taste; one can perceive substances at concentrations of less than one part per billion in the air and sometimes even one part per 30 billion. The sense of smell tends to become fatigued, however, after continuous exposure to an odor. When a person enters a gas station or a dry cleaning store, he immediately becomes aware of the odor peculiar to that establishment. If he works there all day, however, he soon fails to notice the odor. Return to fresh air makes the nose sensitive to the odor again.

Research indicates that there may be seven basic odors, which, in different combinations, produce many different smell sensations. The seven suggested odors are camphorlike, etherlike, floral, pepperminty, pungent, and putrid. The evidence for this is not conclusive, however.

## Vision

The human **eye** (Fig. 37.10) consists of two chambers forming a nearly spherical eyeball. A clear crystalline **lens** separates the two chambers, and two clear fluids, the aqueous humor and vitreous humor, fill the chambers before and behind the lens.

The outermost coat of the eyeball, the **sclera,** consists of connective tissue that surrounds the larger, posterior chamber and continues to the front of the eye as the **cornea,** which is clear. Next to the sclera lies the **choroid,** richly supplied with blood; its dark color prevents light from being reflected within the eye and so blurring the image formed on the retina. Toward the front of the eye the choroid coat is continuous with the **iris,** the opening of which is the **pupil.** In humans the color

**37.10** The human eye in section.

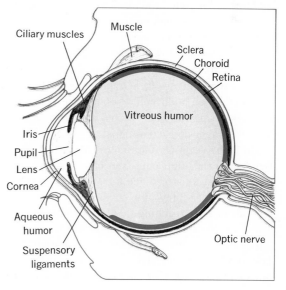

335

of the iris usually ranges from brown to blue, green, and gray. The pupil appears to be black because little or no light is reflected back out of the eye. The iris, which is under control of the autonomic nervous system, opens and closes, regulating the amount of light passing through it. In bright light, the iris closes down, constricting the pupil; in dim light, it opens wider, dilating the pupil. Next to the choroid coat is the retina, which contains neurons sensitive to light. The **retina** transmits its images to the brain by way of the **optic nerve.** In the brain the impulses produce a conscious impression of the object viewed. The retina does not continue to the front of the eye.

Both the cornea and the lens focus light on the retina, but only the lens adjusts for distant or near vision. **Suspensory ligaments** hold the lens in place and change its shape as the eye focuses on distant or nearby objects. For distant vision the lens becomes flattened, and for near vision it becomes more nearly spherical.

Eyeballs too long for the lens to focus a clear image of distant objects on the retina cause near-sightedness; the plane of focus lies within the eye slightly to the front of the retina, and distant objects appear indistinct. An abnormally short eyeball brings the retina too close to the lens, and the plane of focus lies behind the retina when the eye attempts to focus on nearby objects. This condition is farsightedness; nearby objects appear indistinct. Farsightedness tends to plague us in middle or advanced age, when the lens of the eye becomes less pliable and is unable to accommodate its shape for near vision. Properly ground glasses correct both conditions.

The light-sensitive neurons of the retina are of two types; rods and cones, so named because of their shapes. **Rods** provide only black and white vision because they are sensitive only to the amount of light reaching them, but the cones are sensitive to colors. **Cones** function primarily during daylight hours or wherever there is bright illumination; in dim light the rods take over. At sunset one can see colors fade slowly even though there

may still be enough light to see objects fairly distinctly.

The light-sensitive pigment in rods is **rhodopsin,** a protein (opsonin) combined with retinene. Retinene is a derivative of vitamin A, usually present in foods as carotene. Lack of vitamin A in the diet results in night blindness (impaired vision in dim light). Three kinds of cones exist with different pigments sensitive to blue, green, or red light. The ability of human beings to see an almost endless variety of colors results from different degrees of stimulation of the three kinds of cones. Cone pigments are similar to rhodopsin, but their proteins are different.

Human vision encompasses the so-called rainbow colors: violet, blue, green, yellow, orange, and red. Their wavelengths range from about 400 nanometers to about 700 nanometers (Fig. 6.1). Shorter wavelengths, such as ultraviolet, and longer wavelengths, such as infrared (heat), are not visible to the human eye, but some animals can see into these parts of the electromagnetic spectrum.

### Hearing

The **ear** is receptive to the vibrations we call sound waves. It is divided into three parts: the outer ear, middle ear, and inner ear (Fig. 37.11). The nerve endings sensitive to sound lie in the inner ear, the functions of the other parts merely being the transmission of vibrations from the atmosphere to the inner ear.

The **outer ear** consists of the pinna, auditory canal, and eardrum. The **pinna,** which extends outward from the head, catches sound waves and channels them into the **auditory canal,** which extends into the skull. Cupping one's hand behind the pinna aids in hearing because it reflects more sound waves toward the auditory canal. The inner end of the canal is covered by a membrane, the **eardrum,** which is set to vibrating when sound waves reach it.

The **middle ear** is a small chamber with the

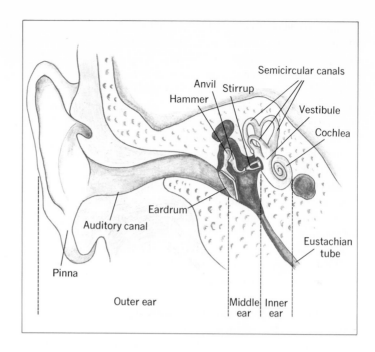

**37.11**  The human ear in section.

eardrum separating it from the outer ear, and an-other membrane, the oval window, which sepa-rates it from the inner ear. Extending across the middle ear from eardrum to oval window are three small bones called the **hammer, anvil,** and **stirrup** because of their shapes. Vibrations of the eardrum set the hammer to vibrating, the hammer moves the anvil, the anvil moves the stirrup, and the stirrup starts the **oval window** vibrating.

The **inner ear** contains three fluid-filled canals wound into a snail-shaped **cochlea.** Vibration of the oval window sets the fluid moving.

The central canal of the cochlea contains den-drites of sensory neurons. When these are stimu-lated by the fluid, they transmit their impulses along the **auditory nerve** to the brain. The fre-quency of the sound waves impinging on the ear-drum is reproduced in the vibrations of the ear-drum, hammer, anvil, stirrup, oval window, and fluid of the inner ear. Different sensory neurons are sensitive to different frequencies of vibrations and are stimulated only by these. Impulses from different neurons give the conscious sensation of sound of different pitches.

## Equilibrium

The inner ear also contains the organs of equi-librium: two sacs (**utricle** and **saccule**) and three semicircular canals. Each sac has an area of sensi-tive hair cells on which rests a small stone of calcium carbonate and protein. When the head changes position, these stones either fall away from the hair cells or rest more firmly on them, depending on the change of position. Nerve end-ings associated with the hair cells transmit to the brain information regarding the amount of pres-sure on these cells. Because different positions of the head affect the four sacs of the two ears differ-ently, the brain can become aware of its exact position even if darkness, blindness, or closed eyes preclude the use of vision to determine one's posi-tion. These sacs provide only information about static equilibrium.

337

The three **semicircular canals,** which are organs of dynamic equilibrium, stand perpendicular to each other in each inner ear. The fluid-filled canals contain sensitive hair cells connected to nerve endings of the auditory nerve. Movement of the head in any given direction moves the canals, for they are embedded in the skull. The fluid within them, however, tends to remain in place. As a result, the hairs bend as they are dragged in the fluid. This stimulates dendrites of neurons in the auditory nerves. Because hair cells in the six semicircular canals are stimulated somewhat differently by any movement, the brain receives precise information about that movement.

The organs of equilibrium in the ear transmit to the brain information about the position and movement of the head only. The brain integrates this with additional information coming from neurons with endings in muscles, joints, skin, and eyes. Together, these give an awareness of the position and direction of movement of all parts of the body.

# Questions

1. Define the following terms:
   nerve
   neuron
   dendrite
   cell body
   axon
   synapse
   neuromuscular junction
   neurilemma
   myelin sheath
   transmitter substance
   ganglion
   meninges
2. In a general sense, what is the function of the nervous system? How does this compare with the function of the endocrine system? Contrast the ways these two systems carry messages in the body.
3. What is the advantage of having two such systems in the body?

4. What are the main components of the central nervous system? Of the peripheral nervous system?
5. Differentiate between sensory neuron, association neuron, and motor neuron.
6. Describe a reflex arc.
7. Why are impulses not transmitted from motor neurons to association neurons to sensory neurons?
8. Describe the electrochemical transmission of a nerve impulse.
9. Describe the transmission of an impulse across a synapse or a neuromuscular junction.
10. Explain how different nerve poisons can have different effects on the transmission of an impulse.
11. What are the parts of the brain and what are their functions?
12. Contrast the functions of the somatic and autonomic nervous systems.
13. Contrast the functions of the sympathetic and parasympathetic nervous systems.
14. Name the five traditional senses. How many other senses beside these have been discovered? What are they?
15. Neurons operate on the all-or-none principle. How then does the brain perceive the intensity of the stimulus impinging on a burned finger? How does the brain distinguish between light, sound, taste, and the other senses?
16. Diagram the eye and label its parts. How does each of its parts contribute to its function of vision?
17. Diagram the ear and label its parts. How does each of its parts contribute to its functions of hearing and equilibrium?
18. Contrast the functions of rods and cones.

# Suggested Readings

Baker, P. F. "The Nerve Axon," *Scientific American* 214(3): 74–82 (March 1966).

Bloom, W., and D. W. Fawcett. *A Textbook of Histology,* 9th ed. Philadelphia: W. B. Saunders Company, 1968.

Clayton, R. K. *Light and Living Matter,* Vol. 2: New York: McGraw-Hill Book Company, 1971.

Eccles, Sir J. "The Synapse," *Scientific American* 212(1): 56–66 (January 1965).

Griffin, D. R., and A. Novick. *Animal Structure and Function,* 2nd ed. New York: Holt, Rinehart and Winston, Inc., 1970.

Heimer, L. "Pathways in the Brain," *Scientific American* 225(1): 48–60 (July 1971).

Hubel, D. H. "The Visual Cortex of the Brain," *Scientific American* 209(5): 54–62 (November 1963).

Katz, B. "How Cells Communicate," *Scientific American* 205(3): 209–220 (September 1961).

Larimer, J. *An Introduction to Animal Physiology.* Dubuque, Iowa: William C. Brown Company, Publishers, 1968.

MacNichol, E. F., Jr. "Three-Pigment Color Vision," *Scientific American* 211(6): 48–65 (December 1964).

Marrazzi, A. S. "Messengers of the Nervous System," *Scientific American* 196(2): 86–94 (February 1957).

McCashland, B. W. *Animal Coordinating Mechanisms.* Dubuque, Iowa: William C. Brown Company, Publishers, 1968.

Miller, W. H., F. Ratcliff, and H. K. Hartline. "How Cells Receive Stimuli," *Scientific American* 205(3): 222–238 (September 1961).

Neisser, U. "The Processes of Vision," *Scientific American* 219(3): 204–214 (September 1968).

Rushton, W. A. H. "Visual Pigments in Man," *Scientific American* 207(5): 120–132 (November 1962).

Snider, R. S. "The Cerebellum," *Scientific American* 199(2): 84–90 (August 1958).

Sperry, R. W. "The Great Cerebral Commissure," *Scientific American* 210(1): 42–52 (January 1964).

Stent, G. S. "Cellular Communication," *Scientific American* 227(3): 42–51 (September 1972).

Vander, A. J., J. H. Sherman, and D. S. Luciano. *Human Physiology.* New York: McGraw-Hill Book Company, 1970.

von Békésy, G. "The Ear," *Scientific American* 197(2): 66–78 (August 1957).

Wald, G. "Molecular Basis of Visual Excitation," *Science* 162: 230–239 (1968).

Young, R. W. "Visual Cells," *Scientific American* 223(4): 80–91 (October 1970).

# 38

# The Skeletal System

**1 A skeleton of bone and cartilage contributes to the support of the body.**

The organs and tissues described in previous chapters are relatively soft and flexible. In large masses they cannot support their own weight or maintain their shape. For this reason, all large animals require some system of support. In the vertebrates (mammals, birds, reptiles, amphibians, and fish), an internal skeleton of bone or cartilage, together with the muscular system (Chapter 33), provides the support for other tissues. The moving of the bones by the skeletal muscles provides locomotion for the body. The skeleton gives protection to some organs of the body, and the marrow of some bones manufactures red and white blood cells (Chapter 34).

## Anatomy of the Skeletal System

A total of 206 bones comprises the bony skeleton of the human body (Fig. 38.1). Some bones, such as those of the arms and legs, are fairly distinct, separate bones held together by ligaments. Others, such as most of the bones of the skull, are fused together and appear at first glance to be a single bone. Closer inspection shows that these bones are fused along wavy edges that fit each other exactly and preclude the possibility of the bones slipping in relation to each other. The fusion of these bones into a single, rigid skull protects the brain from mechanical damage. The only openings in the skull are those required for the entrance of several blood vessels and nerves and the spinal cord.

Twenty-six vertebrae comprise the vertebral column or "backbone." Each vertebra has projections that permit it to articulate with adjacent vertebrae. Most of the vertebrae can be moved slightly with reference to their neighbors, and this allows a small amount of bending forward and backward. Each vertebra has an arch that surrounds an opening. The openings of all the vertebrae form a continuous canal that houses and protects the spinal cord. Spinal nerves leave the spinal cord through small spaces between the vertebrae.

Twelve pairs of ribs form a cage around the chest, where they protect the heart and lungs. Each rib articulates with a vertebra at the back, and small pieces of cartilage connect most of the ribs to the sternum ("breastbone") at the front of the chest. The two lower pairs of ribs, the so-called floating ribs, do not meet the sternum and have no connections in the front.

Abdominal organs are not nearly so well protected as those of the chest, but the bones of the hip do provide some protection from the rear.

## Bone and Cartilage

**Bone** is a hard, rigid substance composed of protein and inorganic materials. The former is chiefly a fibrous protein called collagen; the latter are chiefly calcium and phosphates and some magnesium and carbonates. The proportion of organic to inorganic material varies with age. Children have more organic than inorganic matter in their bones; as a result their bones break easily, but they repair easily also. As the child becomes older, the bones accumulate more inorganic material, and they become stronger. In old age the quantity of organic material decreases, leaving the bones very brittle; again they break easily, but they do not mend so readily as in childhood.

Most bones consist of compact bone, spongy bone, and bone marrow. These three tissues can be seen in a cut bone, with the marrow occupying the center, and spongy bone lying between the marrow and the outer hard bone. Rib bones have a great deal of spongy bone, but the long bones of the arms and legs have little spongy bone except at their ends.

Despite its hard, "dead" appearance, **compact bone** contains the living cells that formed it. The

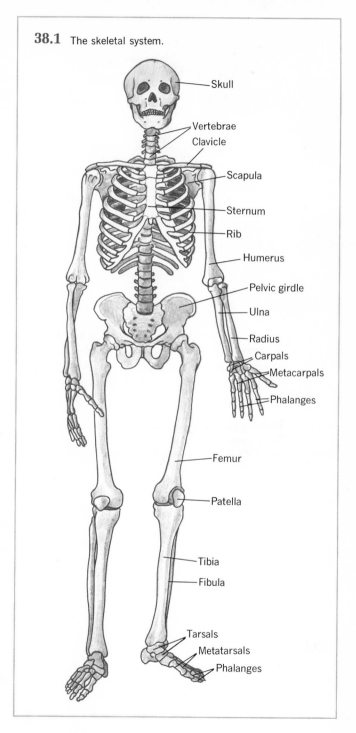

**38.1** The skeletal system.

Skull

Vertebrae
Clavicle

Scapula

Sternum

Rib

Humerus

Pelvic girdle

Ulna

Radius

Carpals

Metacarpals

Phalanges

Femur

Patella

Tibia

Fibula

Tarsals

Metatarsals

Phalanges

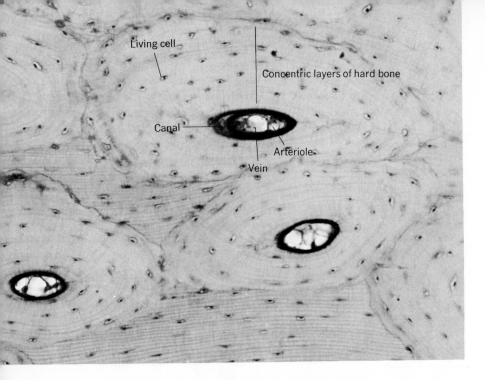

Living cell

Concentric layers of hard bone

Canal

Arteriole

Vein

**38.2** A section of human bone. Bone consists of a hard ground substance laid down in concentric layers around small canals. Blood vessels occupy the canals and serve the living cells in the bone. Very fine canals can be seen radiating from each large canal; it is through these that the cells receive nourishment. (Courtesy of Dr. Beth L. Wismar.)

cells are nearly isolated from each other, trapped in concentric layers of compact bone that they themselves secreted as a product of their metabolism (Fig. 38.2). The cells do, however, have long, thin, processes by which they maintain contact with each other. Blood vessels ramifying through compact bone supply the cells with food and oxygen and remove their wastes. Nerves also extend into bone.

Old bone tissue dies, and living cells form new

**38.3** A section of cartilage from a human larynx. Living cells are embedded in the firm but not rigid substance they secrete. (Courtesy of Dr. Beth L. Wismar.)

342

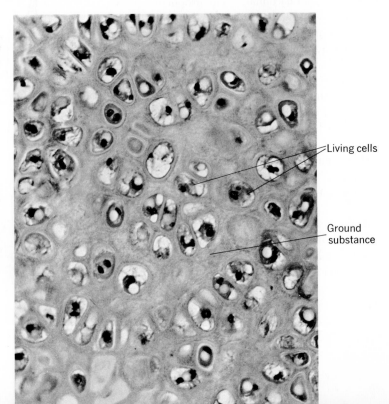

Living cells

Ground substance

bone that replaces it. The replacement is not quite exact, however, for as an individual grows, his weight and his proportions change. This changes the stress borne by bones, and this requires modification in their structure. The new bone forms to fit the new requirements. Because bone is a living tissue, broken bones are capable of mending.

The matrix of bones represents a store of calcium and phosphate. If the blood levels of these substances fall, the bone releases them to the blood. If their blood levels rise, the bone accepts some of the excess (the kidneys excrete some, too) and use it in the formation of new bone matrix.

Spongy bone is spongy in appearance only—not in texture. It consists of spicules of tissue resembling compact bone mixed with bone marrow. The arrangement of these two components of spongy bone is such that it affords maximum strength while burdening the body with only minimum weight.

Bone marrow is of two types: red and yellow. Red blood cells form in red marrow, which is characteristic of the bones of young children. Inactive, yellow, fatty marrow replaces about half of the red marrow in adults. Vertebrae, ribs, sternum, and the long bones of arms and legs contain mostly red marrow in adults.

Cartilage, or gristle, is firm and strong, but not so rigid as bone. Here, too, the cells are separated from each other by a substance they secrete (Fig. 38.3). As in bone, the protein collagen comprises an important part of the extracellular organic material. In the adult human being, cartilage is found mostly at the ends of bones, where they articulate with other bones. Cartilage also supports the pinna of the ear and part of the nose; bending these parts reveals the flexibility of cartilage. In the human embryo, the skeleton first forms as cartilage; bone later replaces much of it.

# Joints

A joint is the articulation of two bones. Cartilage protects the ends of the bones, and those joints that permit free movement are filled with fluid that acts as a lubricant and allows smooth movement.

Joints may be grouped into four types: ball-and-socket, hinge, slightly movable, and immovable (Fig. 38.4). The shoulder and hip joints are ball-and-socket joints, the rounded head of one bone fitting into the rounded concavity of the other; these joints permit the arms and legs to rotate. Hinge joints allow movement in only one plane; examples are the joints of fingers and toes. The slightly movable joints between the vertebrae and the immovable joints of the skull were mentioned earlier.

At the time of birth, small spaces between the bones of the infant's skull make it more flexible

**38.4** Four kinds of joints.

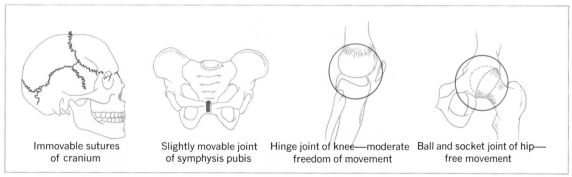

Immovable sutures of cranium

Slightly movable joint of symphysis pubis

Hinge joint of knee—moderate freedom of movement

Ball and socket joint of hip— free movement

343

and so permit the head to change its shape as the baby passes through the narrow birth canal. At the same time, the mother's pubic symphysis, a slightly movable joint, provides a little more room for the baby's head. In the months after birth, the bones of the child's skull grow together, forming a tight fit.

## Questions

1. What are the functions of the skeletal system?
2. How does bone differ from cartilage?
3. What is the difference between hard bone, spongy bone, and bone marrow?
4. Name four types of joints and describe the kinds of movements they allow.

5. Consider the joint at your shoulder and those in your fingers. If these two types of joints were interchanged (that is, if you had a hinge joint at your shoulder and ball-and-socket joints in your fingers), how would this affect your life?

## Suggested Readings

Bloom, W., and D. W. Fawcett. *A Textbook of Histology,* 9th ed. Philadelphia: W. B. Saunders Company, 1968.

Griffin, D. R., and A. Novick. *Animal Structure and Function,* 2nd ed. New York: Holt, Rinehart and Winston, Inc., 1970.

Gross, J. "Collagen," *Scientific American* 204(5): 120–130 (May 1961).

The skin, or integument, is more than a sack that holds the body together and prevents parts from slipping out. Its structure varies somewhat in the several parts of the body, but sections of most samples of skin have a very busy appearance (Fig. 39.1). Although the outermost cells are dead, the rest of the skin is living tissue provided with sweat glands that remove some wastes and contribute to maintenance of constant body temperature, nerve endings that convey to the central nervous system information about the environment, and blood capillaries that exchange compounds with the skin and release heat conveyed from other body parts. Except for its sweat glands, the skin is waterproof and prevents evaporation of body fluids. Most bacteria cannot enter the body through undamaged skin (though some gain entrance through the agency of biting insects), and the harmful ultraviolet rays in sunlight are screened out by skin, especially heavily pigmented skin. In other words, the skin, like other body parts, is involved in homeostasis.

## Anatomy and Functions of the Skin

The skin consists of an epidermis and dermis, the epidermis to the outside and resting on the dermis. The **epidermis** is a very thin layer consisting of epithelial cells and it lacks both nerves and capillaries. The innermost epidermal cells divide frequently, forming new ones that force the older cells outward. The outermost cells produce keratin, a horny protein similar to that of hair and nails. Repeated friction and wear causes formation of calluses, which are thickenings of the keratinized layer of the epidermis. Walking barefoot gradually develops tough calluses on the bottoms of the feet and protects them from the discomfort and dangers of gravel, rough rocks,and sun-heated

# 39

# The Integumentary System

1 The skin is multifunctional, providing temperature regulation; some sensory perception; and protection against mechanical injury, water loss, and invasion of the body by bacteria.

345

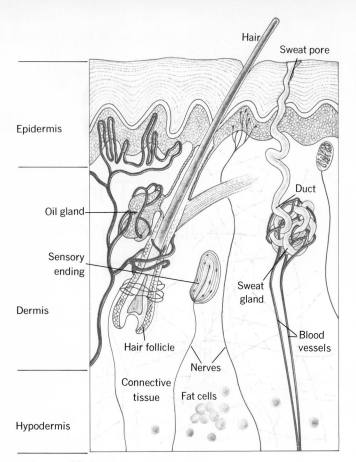

Hair
Sweat pore

Epidermis

Oil gland

Sensory
ending

Dermis

Hair follicle

Connective
tissue

Nerves

Fat cells

Hypodermis

Duct

Sweat
gland

Blood
vessels

**39.1** Section through human skin.

that extend across the epidermis and open at the surface of the skin. The portion of a sweat gland deep within the dermis is much coiled among blood capillaries from which it absorbs water, sodium chloride, and urea, which are the major constituents of perspiration.

When perspiration reaches the openings of the sweat glands, its water usually evaporates, leaving on the surface of the skin a deposit of sodium chloride and urea and giving the skin a salty taste if it is not cleaned occasionally. Evaporation of water requires energy (Chapter 6) and the body provides at least part of it in the form of heat and so is cooled in hot weather.

Under the control of the autonomic nervous system the capillaries associated with sweat glands dilate when the body temperature begins to rise above normal (about 37°C) either because of physical exercise, hot weather, or illness. The increased flow of blood to the sweat glands brings to them heat that is dissipated in perspiration, and this contributes to cooling of the body. Lowering of body temperature constricts the capillaries and so conserves heat.

Unless perspiration is copious, it ordinarily goes unnoticed, for it evaporates before it can accumulate on the skin. In humid weather, however, the rate of evaporation declines, and perspiration adheres to the skin and fails to cool the body. On such days a breeze is welcome, for it blows moist air away from the immediate vicinity of the body and replaces it with somewhat drier air. An artificial breeze generated by a fan does the job just as well—unless the person operates a simple hand-held fan himself and so produces additional heat within his body by increasing his activity.

Perspiration is nearly odorless, but it does provide a good growth medium for bacteria, and body warmth affords an excellent temperature for many bacteria and encourages their rapid metabolism. It is their by-products that are offensive to the nose. Possibly these odors had some significance or value to early man or his ancestors. Now we take care to remove them by frequent bathing and

pavements. Hands used in hard labor become similarly callused.

The outer surface of the epidermis is not smooth but is thrown into grooves and ridges that form various patterns on different parts of the body. The fingers bear patterns that are different in each person, and for this reason fingerprints can be used for identification purposes.

Gradually epidermal cells age, die, and slough off in minute, dry, almost imperceptible flakes. On the scalp they sometimes appear in the more embarrassing form of dandruff, and, in the case of sunburn, they peel off in large patches.

The **dermis,** which is much thicker than the epidermis, consists of connective tissue supplied with nerves, blood vessels, sweat glands, and oil glands. The **sweat glands** are long narrow tubes

then replace them with a variety of perfumes and aftershave lotions.

Female mammals have mammary glands that are modified sweat glands, but they secrete milk rather than perspiration.

Beneath the dermis is the **hypodermis,** a connective tissue that connects the dermis with the underlying muscle. Fat accumulates in the hypodermis and serves not only as a food reserve, but as a protective padding that minimizes mechanical injuries. It also insulates the body by slowing heat loss. The hypodermis is rich in collagen. Leather, which is made by tanning hides, owes its toughness to collagen.

## Skin Color

The color of the skin depends on pigment present in the epidermal cells, blood flowing through the dermis, and the fat content of the hypodermis.

**Melanin,** a brown pigment, gives various shades of color ranging from the pale "white" of Caucasians through tan and dark brown to almost black; the more concentrated the pigment, the darker the color. Melanin absorbs ultraviolet radiations, and in general it is more concentrated in the skins of persons of tropical ancestry and only sparsely represented in the Arctic races, who are exposed to less intense sunlight.

Tanning of the skin is a result of exposure to sunlight, and it increases the protection against the ultraviolet portion of sunlight. Ultraviolet light striking the skin causes both the darkening of the melanin already present and the synthesis of additional melanin. Repeated exposure to sunlight produces darker skin, and the darker the epidermis becomes, the more ultraviolet radiation it can absorb and the greater protection it affords the underlying tissues. In spite of this, persons who work outdoors in sunshine for many years contract skin cancer more often than those who work indoors.

The ability to tan is modified by heredity; southern Europeans, for instance, tend to tan readily. Northern Europeans, on the other hand, are more likely to become red with sunburns; for these people especially, the prolonged exposure of untanned skin to sunlight during the first beautiful days of summer is unwise, for it almost certainly leads to painful sunburn.

**Carotene,** an orange pigment, is especially abundant in the skins of Chinese, Japanese, and other eastern Asian peoples. It also is responsible for the reddish color of the American Indians, who are of Asian descent.

Blood flowing through capillaries in the dermis lends color to skins that have little melanin or carotene. The pale pinkish color of Caucasian skin depends partly on its small melanin and carotene content and partly on the blood in the dermis. In acute embarrassment, the dilation of capillaries in the skin of the face brings on an even redder blush.

Albinos, who possess neither melanin nor carotene, usually have obviously pink skin because of the blood showing through the translucent epidermis. Albinism is an inherited condition and may appear in persons of any race.

Fat in the hypodermis lends a yellow color to the skin.

## Outgrowths of the Skin

Fingernails, toenails, and hair are dead outgrowths of the skin. The nail grows from a root embedded in the skin. Here the live cells divide frequently, but some of the cells they produce become highly keratinized and die. They are pushed forward, becoming the visible nail. Because nails are dead and contain no nerves, cutting them is a painless procedure.

Human beings, like other mammals, are hairy animals; but unlike most mammals, they have only small, sparse, often unpigmented hairs over much

347

of the body, although some men can boast of a fairly thick covering of dark body hair. The scalp usually bears an abundance of long hair; but some newborn babies lack it temporarily, and advancing age deprives others of what they once had. Scalp hair is pigmented and relatively coarse; its colors range from pale yellow through many shades of red and brown to black, and finally it may become gray with age. With the onset of puberty, coarse, fairly long hairs appear in the armpits and pubic region. At this time facial hair begins to grow on young men, and their body hair may become pigmented, coarse, and longer than it was during childhood.

A hair consists of a long shaft and a root which develop from a hair follicle deeply embedded in the dermis. Like a nail, the hair shaft contains dead, keratinized cells, and growth of a hair depends on the activity of living cells in the follicle. Cutting hairs or plucking them does not stop growth as long as the follicles remain intact.

Associated with each hair is an oil gland that secretes oil and keeps the hair in good condition. Some oil on the skin is desirable, for it prevents chapping in winter and gives the skin a smooth, young appearance. Very frequent washings deplete the oil supply of the skin and result in "detergent hands." Conversely, infrequent bathing permits the accumulation of a great deal of oil on the skin and may contribute to the growth of bacteria.

Attached to each hair is a muscle, which, when it contracts, causes the hair to stand erect and produces bumpy gooseflesh. This happens as a result of exposure to cold temperatures. In animals more hairy than humans, erection of hairs is a method of heat conservation, the many air pockets among the hairs providing a layer of insulation near the skin.

# Questions

1. What are the functions of the skin? Draw a cross section of the skin and label its parts. How do these parts contribute to its functions?
2. Describe the role of the skin in maintaining body temperature.
3. What is the difference between epidermis, dermis, and hypodermis?
4. What substances contribute to skin color?

# Suggested Readings

Bloom, W., and D. W. Fawcett. *A Textbook of Histology*, 9th ed. Philadelphia: W. B. Saunders Company, 1968.

Daniels, F., Jr., J. C. van der Leun, and B. E. Johnson. "Sunburn," *Scientific American* 219(1): 38–46 (July 1968).

Gross, J. "Collagen," *Scientific American* 204(5): 120–130 (May 1961).

Langley, L. L. *Homeostasis*. New York: Reinhold Publishing Corporation, 1965.

Lerner, A. B. "Hormones and Skin Color," *Scientific American* 205(1): 98–108 (July 1961).

Loomis, W. F. "Skin-pigment Regulation of Vitamin-D Biosynthesis in Man," *Science* 157: 501–506 (1967).

Montagna, W. "The Skin," *Scientific American* 212(2): 56–66 (February 1965).

The entire human body functions as a complex of negative feedback systems that maintain a homeostatic condition over many years. The previous chapters showed briefly that the systems of the body contribute to maintaining this state, although only a few of the individual feedback systems were described in any detail.

The maintenance of homeostasis requires the constant expenditure of energy, and this is provided by food. As long as food remains in adequate supply and as long as the mechanisms of the body are not disrupted by disease or injury, they maintain the homeostatic condition for many years. But even under the best of conditions, old age brings with it a wearing out of the machinery of the body, the homeostatic systems fail, and death inevitably results.

Our only hope for immortality on earth—if indeed we have such—is through our children. The birth of each child brings into the world a new body with a fresh set of homeostatic systems potentially capable of living out his traditional three score years and ten, during which time he, too, has the opportunity of reproducing himself.

We see, then, that nature provides homeostasis on a level higher than that of the individual. By the process of reproduction, an individual replaces himself and so perpetuates his species. In the natural course of events, disease, starvation, or predation reduce the numbers of individuals of a species whenever they become too numerous, the birth rate and the death rate becoming opposing parts of the homeostatic system.

# Asexual and Sexual Reproduction

Reproduction among plants and animals is of two types, asexual and sexual. **Asexual reproduction,** which involves only one parent, produces offspring genetically identical with the parent, but

# 40

# The Reproductive System

1 The male reproductive system produces sperms with which eggs are fertilized.

2 The female reproductive system produces eggs; it also shelters and nourishes the offspring during its nine months of development.

3 Birth control methods interfere with the meeting of eggs and sperms or with the normal development of the embryo or fetus.

if the two organisms live in different environments they may not resemble each other very much.

Cutting of a worm or a geranium into two individuals that survive is asexual reproduction. So, too, is mitotic cell division in unicellular organisms. Asexual reproduction does not occur in humans, but the theoretical possibility of obtaining a complete human being from a single cell removed from the body and carefully nurtured is an interesting idea today and may become a reality in the future.

In higher animals **sexual reproduction** requires two parents, more or less obviously differentiated into male and female. Even if the two sexes are similar externally, their internal reproductive apparatus is different. Each parent contributes one gamete or sex cell to each offspring. The gametes of the female are called **eggs** and those of the male **sperms.** These are produced in reproductive organs, the **ovaries** of females and the **testes** (testis, singular) of males.

Briefly, sexual reproduction is the production of haploid gametes by both parents and the fertilization of the haploid egg by a haploid sperm forming a diploid zygote from which the new individual grows. In virtually all mammals, including humans, fertilization and early development of the young offspring occur within the body of the female parent. The several parts of the reproductive systems of both male and female contribute to these events.

# The Male Reproductive System

The male reproductive system (Fig. 40.1) consists of a pair of testes, a pair of sperm ducts that lead from the testes into the urethra and penis, and several accessory glands—a pair of seminal vesicles, the prostate gland, and a pair of bulbourethral glands.

The testes form within the abdominal cavity during embryological development of a male baby, but at about the time of birth they descend into

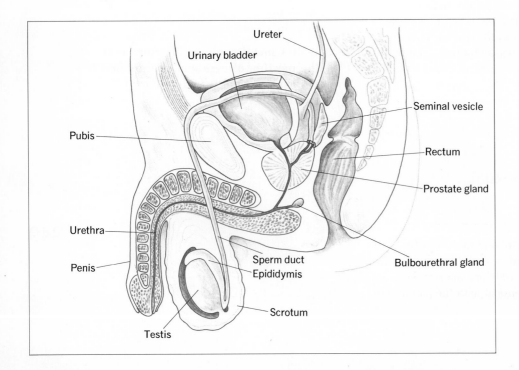

**40.1** Male reproductive system.

the **scrotum,** a pouch formed from the lower part of the abdominal wall. At first the abdominal cavity and that of the scrotum are continuous, but connective tissue forms between them. Lifting of heavy weights sometimes results in a hernia, a rupture of this tissue. Large hernias permit a portion of the intestine to drop into the cavity of the scrotum. Such a situation should be corrected by surgery, for gangrene can result if blood vessels serving the intestine become squeezed closed in such close quarters. The position of the scrotum outside the abdomen is responsible for the slightly lower temperature of the testes that is required for the production of fertile sperms. Men with undescended testes are sterile.

At puberty, the anterior pituitary produces **follicle-stimulating hormone** (FSH) and **luteinizing hormone** (LH) and secretes them under the influence of the appropriate releasing factors from the hypothalamus (Chapter 36). Within each testis are about a thousand coiled **seminiferous tubules** the maturation of which depends on FSH. Occupying the spaces between the tubules are the **interstitial cells** which secrete the male hormone **testosterone** and thus make the testes endocrine glands as well as reproductive organs. Testosterone, which is formed under the influence of LH from the anterior pituitary, also is essential for the production of sperms. Within the seminiferous tubules sperms form by meiosis (Chapter 41) from diploid cells called primary spermatocytes (Fig. 40.2). Each primary spermatocyte divides into two secondary spermatocytes, and these divide into four spermatids, each of which matures into a sperm. As primary spermatocytes are used up in forming sperms, they are replenished by mitotic divisions of other diploid cells (spermatogonia) in the seminiferous tubules.

A sperm has a streamlined shape with a head that contains a nucleus surrounded by only a very thin layer of cytoplasm, a narrow middle piece with many mitochondria, and a long, slender tail. When first formed, the sperms are nonmotile, but later they swim actively by movements of their

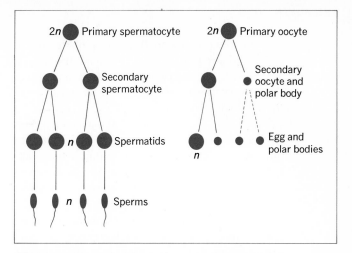

**40.2** Formation of sperms and eggs by meiosis. A primary spermatocyte produces four spermatids, which mature into sperms. A primary oocyte forms one egg and two or three polar bodies. In some animals the first polar body does not divide. $2n$ = diploid, $n$ = haploid.

tails; the energy they use is provided by ATP produced by respiration in the mitochondria of the middle piece.

Continuous with the seminiferous tubules are the tubules of the **epididymis** on the upper part of each testis. The pressure of later-formed sperms forces older ones into the epididymis, where they are stored temporarily before entering the sperm ducts. In their passage through the sperm ducts, sperms receive fructose in an alkaline secretion from the **seminal vesicles.** The fructose serves as a food supply for the sperms, and the alkalinity neutralizes the acidity they will encounter if they reach the female reproductive system. The **prostate gland** adds more alkaline secretions, and the **bulbourethral glands** contribute a lubricating substance. Seminal fluid, the products of all these accessory glands, makes the sperms motile. The fluid and the sperms together are called **semen.**

Under sexual excitement the arteries leading into the **penis** dilate and the veins leading away from it constrict. As a result, it becomes filled with blood, the pressure of which causes the penis to become firm and erect. In this condition it can be inserted into the vagina of the female partner. At the climax of sexual excitement, or orgasm, the prostate gland and smooth muscles of the sperm ducts and urethra contract, forcibly discharging semen into the upper part of the vagina. From there the sperms swim through the uterus and into the oviducts, where fertilization can occur if an egg is present. In fertile males at least 100 million sperms are discharged in a single ejaculation, but only one fertilizes an egg. Those that do not fuse with an egg may remain alive in the female reproductive tract for several days—possibly as many as six—after which they die.

Frozen sperms remain alive for months or even years; when thawed, they can be used in artificial insemination. Animal breeders take advantage of this. Frozen semen from valuable bulls, for instance, can be shipped anywhere in the world while the future father remains at home, never to meet the mother of his future offspring; he also may continue to father offspring long after his death. As vasectomy (see p. 360) becomes more and more popular as a birth control method, men so sterilized choose to have a supply of their semen frozen before submitting to the operation so that they may have additional children if they should so desire later.

Sperm production begins with puberty and may continue through the rest of a man's life, although vigor and desire for sexual intercourse may decline with age.

## The Female Reproductive System

The female reproductive system consists of a pair of ovaries, a pair of oviducts (Fallopian tubes), a uterus, and a vagina (Fig. 40.3). The vagina receives the penis of the male in sexual intercourse, or coitus. The ovaries lie within the abdomen and do not descend as do the testes of the male. They produce eggs, and the oviducts conduct the eggs from the ovaries to the uterus where any fertilized egg develops into an embryo and then a fetus. Extending from the uterus into the vagina is a short neck called the cervix. In the human female the reproductive and excretory systems are completely separate, the urethra opening slightly forward of the vagina. Homologous with the penis, but much smaller, is the clitoris; its stimulation may lead to sexual arousal. Two folds of tissue, the labia, surround the clitoris and the openings of the vagina and urethra.

Eggs form within the **ovary** by the process of meiosis (Fig. 40.2). Within the ovary is a large but limited supply of diploid primary oocytes. A primary oocyte divides into two cells of different sizes, a large secondary oocyte and a small polar body. The secondary oocyte then divides unequally, forming a large, fertile egg and another small polar body. In some animals the first polar body divides into two polar bodies, but this division does not occur in human females. One primary oocyte thus produces only one fertile egg which can be fertilized by a sperm. The infertile polar bodies disintegrate. A spherical follicle of cells closely surrounds the egg.

Unlike the case in men, all of a woman's eggs are formed within the ovaries before she is born. Later, during her childbearing years, one egg matures about every 28 days, only one ovary functioning at a time. Although the total supply of eggs in the two ovaries runs into the hundreds of thousands, only a few hundred (13 times the number of childbearing years—about 30 or 35) ever are released from the ovaries. The others merely disintegrate.

### Menstrual Cycle

During the 28 days that are the average time required for the completion of the menstrual cycle (Fig. 40.4), a complex series of hormonally controlled events occurs. FSH and LH produced by the anterior pituitary and progesterone produced by the ovaries function in negative feedback systems that affect each other's production and

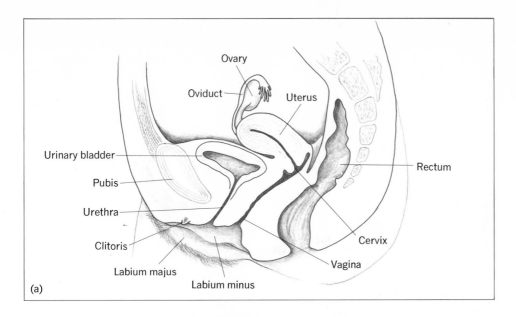

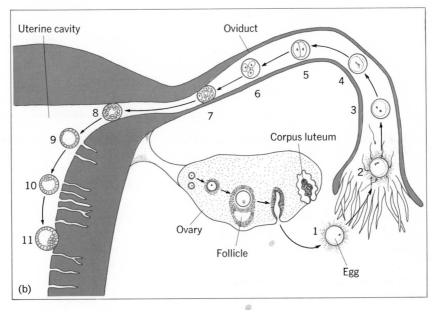

**40.3** (a) Female reproductive system. (b). Numbers 1 to 11 indicate the passage of the egg and developing embryo along the oviduct to the uterus.

effects. Within the 28 days of the cycle their mutual interactions result in the preparation of a thick, capillary-rich uterine lining that receives and nourishes any embryo that might result from sexual intercourse and the destruction of that lining should no embryo be formed. A somewhat more detailed description follows.

**FSH** from the anterior pituitary stimulates ovarian follicles to produce **estrogens,** which have two effects. They cause the inner lining of the uterine

353

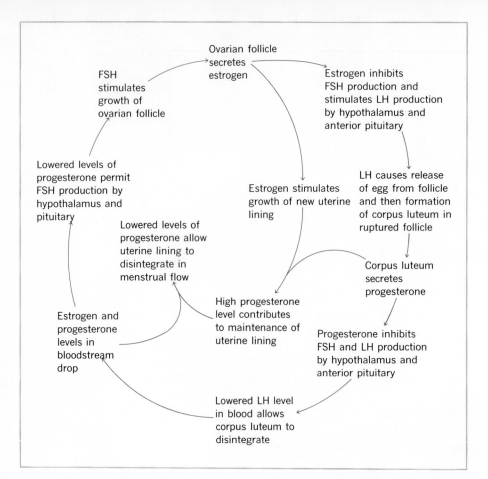

FSH stimulates growth of ovarian follicle

Ovarian follicle secretes estrogen

Estrogen inhibits FSH production and stimulates LH production by hypothalamus and anterior pituitary

Lowered levels of progesterone permit FSH production by hypothalamus and pituitary

Estrogen stimulates growth of new uterine lining

LH causes release of egg from follicle and then formation of corpus luteum in ruptured follicle

Lowered levels of progesterone allow uterine lining to disintegrate in menstrual flow

Corpus luteum secretes progesterone

Estrogen and progesterone levels in bloodstream drop

High progesterone level contributes to maintenance of uterine lining

Progesterone inhibits FSH and LH production by hypothalamus and anterior pituitary

Lowered LH level in blood allows corpus luteum to disintegrate

**40.4** Human menstrual cycle, somewhat simplified.

wall to begin to thicken appreciably and to become well supplied with many blood capillaries through which an embryo—should one be produced—can be fed and relieved of its wastes. At the same time, estrogens depress the FSH production by the hypothalamus and anterior pituitary, and this permits these parts to produce and secrete LH.

LH causes one **follicle** to break open and release a mature, fertile egg into the abdominal cavity. Each **oviduct** opens near an ovary. The beating action of hairlike cilia in the oviduct and peristalsis of muscles in the oviduct wall create a current that sweeps the egg into the oviduct and on toward the uterus. After the follicle ruptures and releases its egg, the remaining cells of the follicle divide, and the space left by the egg becomes filled with a yellow mass of cells called the **corpus luteum.**

It produces another female sex hormone, **progesterone,** which stimulates the further growth of the uterine lining. Progesterone also inhibits the production and secretion of both FSH and LH by the hypothalamus and anterior pituitary. As the LH level in the blood falls, it no longer maintains the corpus luteum, which then ceases its production of progesterone. As the progesterone level in the blood falls, the thick uterine lining is maintained no longer, and it disintegrates in the menstrual flow. No longer inhibited by progesterone, the hypothalamus and anterior pituitary once again produce and secrete FSH and begin the cycle again. As long as no egg is fertilized, menstrual cycles continue from puberty in the early teens to menopause in the mid- or late forties, when a woman becomes infertile.

# Pregnancy

If, when an egg travels down the oviduct, sperms are present, one of them may fertilize it. Only the sperm nucleus enters the egg and fuses with the egg nucleus. The cytoplasmic portion of the sperm remains outside the egg. The resulting **zygote** continues down the oviduct to the uterus, undergoing its first few cell divisions along the way. Within eight to ten days it becomes implanted in the uterine lining, which at this part of the menstrual cycle is thick and prepared to receive it. The embryo becomes attached to the wall and remains there during the ensuing nine months of pregnancy.

When fertilization and implantation have occurred, the corpus luteum persists for some time, its progesterone helping to maintain the uterine lining and so continuing the pregnancy. Later in pregnancy the placenta assumes the function of progesterone production.

Pregnancy interrupts the menstrual cycle. No new eggs are released, and so there is no opportunity for another fertilization until some time after birth when the cycle begins again.

Shortly after fertilization the zygote divides into a two-celled **embryo,** these two cells divide into a four-celled embryo, and so on. Early in its existence, the young embryo is a spherical mass of similar cells, but even before it reaches the uterus, some differentiation of tissues occurs within it. At the age of eight weeks, all the systems of the body are represented and many of them are functional. No longer called an embryo, but a **fetus,** it is recognizable as a miniature human being with arms, legs, fingers, toes, and facial features easily distinguishable.

Not all the cells descending from the original zygote become part of the embryo proper and later

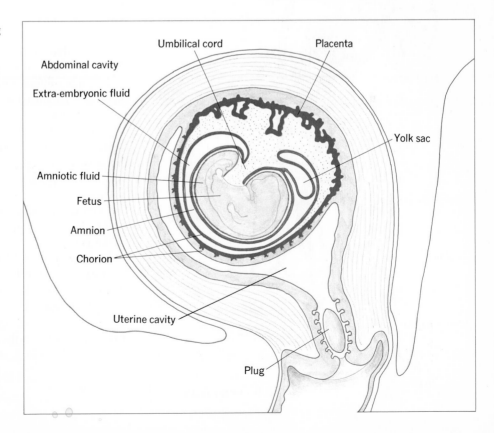

**40.5** Fetus and surrounding membranes.

Abdominal cavity

Extra-embryonic fluid

Umbilical cord

Placenta

Yolk sac

Amniotic fluid

Fetus

Amnion

Chorion

Uterine cavity

Plug

the fetus. Some of them form two multicellular membranes, the **chorion** and the **amnion,** which surround the new individual, and some form the **umbilical cord.** Between the chorion and amnion and between the amnion and the fetus are fluid-filled cavities, the fluids providing a protective cushion against mechanical shocks if the mother should fall or receive a severe blow (Fig. 40.5).

The umbilical cord, which extends from the infant's abdomen to the mother's uterine lining, contains an umbilical vein that brings to the fetus all its food, oxygen, and minerals and two umbili-cal arteries that remove its wastes. Where the cord joins the uterine wall, branched villi grow from the chorion deep into the uterine lining and form a great deal of surface area through which materials are exchanged between mother and child. In this area, composed of both maternal and fetal tissue called the **placenta,** the villi contain blood capillaries from the umbilical arteries and veins. The mother's capillaries in the placenta release small pools of blood that bathe the villi and provide the most intimate arrangement of the two blood supplies short of mixing. The membranes separating the two blood supplies generally allow only small molecules like oxygen, carbon dioxide, digested foods, and urea to pass between mother and child, but occasionally larger molecules do cross the membranes. Erythroblastosis fetalis, the disease of Rh babies (Chapter 34), would not occur were it not for the passage of proteinaceous antigens and antibodies across the membranes of the placenta.

As pregnancy advances, the several body systems of the fetus come to resemble more and more those of the adult body. However, because of the fetus' peculiar situation, some parts do not have the opportunity to function, and others work in a way that differs from their future rules after birth. For instance, life in a small pool of amniotic fluid provides the lungs with no air for breathing. The lungs of the fetus remain collapsed until the birth events initiate breathing movements. Because the fetal lungs do not absorb oxygen or rid the body of carbon dioxide, there is no need for circulation of much blood in the pulmonary arteries and veins. (Before progressing further, the reader might like to refresh his memory on the subject of the general plan of circulation of blood in the adult, described in Chapter 34.) Instead, a small opening called the **foramen ovale** (oval window) exists in the wall between the right and left atria of the fetal heart. It permits some of the blood to move directly from the right atrium to the left atrium and then to the aorta and, in so doing, to bypass the lungs (Fig. 40.6). Most of the rest of the

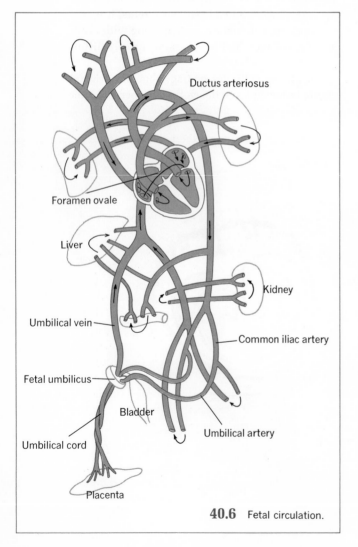

Ductus arteriosus

Foramen ovale

Liver

Kidney

Umbilical vein

Common iliac artery

Fetal umbilicus

Bladder

Umbilical artery

Umbilical cord

Placenta

**40.6** Fetal circulation.

blood that reaches the right ventricle and leaves it through the pulmonary artery is immediately shunted to the aorta by a short vessel, the **ductus arteriosus.** Circulation in the umbilical vessels assumes the functions of gaseous exchange in the placenta.

After nine months in the uterus, the fetus reaches a stage of development that permits it to assume some degree of independence of its mother. The chorion and amnion break; and **oxytocin,** one of the posterior pituitary hormones, causes contractions of the smooth muscles of the uterus which force the fetus out through the vagina. This process, called labor, usually lasts for several hours.

Normally the child is born head first. After his entire body has emerged, he remains physically attached to the uterus by way of the umbilical cord through which he still obtains his oxygen and disposes of his carbon dioxide. When the doctor or midwife cuts and ties the umbilical cord, there is no place for the baby's carbon dioxide to escape to; its accumulation in his blood soon starts breathing movements of the respiratory muscles, and the infant begins breathing on his own. The new baby's cries following the traditional slap on his bottom assures those present that breathing movements have been initiated. With the inflation of the lungs, blood moves into the pulmonary arteries and then into the pulmonary veins and back to the heart. The foramen ovale and the ductus arteriosus close, and adult circulation is established. Blue babies are infants in which the foramen ovale or ductus arteriosus fails to close. The mixing of unoxgenated blood with oxygenated blood gives the baby's skin a slightly bluish color. The condition may be corrected surgically.

The doctor or midwife ties the severed umbilical cord near a newborn baby's abdomen to prevent infection or excessive bleeding. The scar remains as the **navel.** Shortly after the birth of the baby the placenta detaches itself from the uterus. It and the chorion and amnion then are expelled as the **afterbirth.**

During pregnancy the mammary glands within the mother's breasts begin to enlarge in preparation for forming milk with which the infant will be fed. Both the nervous and endocrine systems control the production of milk. Suckling by the infant sends nervous impulses to the hypothalamus and pituitary glands. Prolactin from the anterior pituitary causes secretion of milk into the mammary glands, and oxytocin from the posterior pituitary stimulates ejection of milk from the breasts.

During the first three months of development, when most of the differentiation of tissues is being completed, the baby is most sensitive to the effects of drugs, viruses, and radiations.

Thalidomide is a sedative, now infamous for its deforming effects on children whose mothers took the drug early in pregnancy. When taken at the time of limb formation, thalidomide prevents the proper development of the limbs. Many of the affected children were born without arms but with fingers growing from the shoulders. LSD and other drugs that enjoy an illicit popularity among young people today are suspected of producing abnormalities in developing embryos.

If a mother contracts German measles (rubella) during the first three months of pregnancy, her child runs the risk of being born with heart disease, deafness, cataracts, or mental retardation. A significant number of such children die early in childhood. Because one attack of German measles usually confers lifelong immunity, young girls were often encouraged to associate with friends suffering from the disease in the hope that they, too, would contract the disease and so protect themselves from the risk of getting it during a pregnancy later in life. Today a German measles vaccine is available, and in some communities schoolchildren are vaccinated to avoid the possibility of their infecting their mothers, who still might be susceptible to the disease and who might be pregnant.

Exposure to X-rays, gamma rays, and other radiations of short wavelengths produces a wide

357

variety of fetal abnormalities. Abdominal X-ray examination early in pregnancy once was popular, but doctors now avoid this practice unless it is absolutely necessary. Not only can deformities result, but cancers also may develop in the children.

### Multiple Births

Twins arise in two different ways. Occasionally a woman's ovaries may release two eggs simultaneously. If both eggs are fertilized by different sperms, the two zygotes develop into separate fetuses. The twins are called fraternal (even though one or both may be girls), and they bear the same relationship to each other as do any two children born of the same parents but at separate births. They may be of the same sex or may be brother and sister.

Identical twins result from the mitotic division of a zygote and the separation of the two daughter cells, each then acting like a zygote and developing into a fetus. Such twins share an identical heredity for they come from the same egg and the same sperm. Identical twins are always of the same sex, and sometimes it is difficult for any but their closest friends and relatives to tell them apart. However, different environmental effects, such as disease, malnutrition, or accident, may change the appearance of one and not the other.

Siamese twins result from incomplete splitting of a two-celled embryo; the two individuals remain attached to each other by some parts of their bodies. If they are almost completely separate and do not share between them a single essential organ, surgery often can complete the separation and permit them to follow normal lives. When the connection is impossible to sever safely, Siamese twins must live their entire lives together. Extremely close unions, such as a single body with two heads, are not viable and die shortly after birth.

The members of a set of triplets, quadruplets, quintuplets, and larger sets may all be identical, or some may be identical and others fraternal.

## Birth Control

Despite the fact that the official policies of governments and religions throughout most of history have included the encouragement of high birth rates, many individuals, recognizing that large families are expensive and often have poorer health than small ones, have sought to control their fertility by practicing birth control. The social, religious, and legal attitudes toward the many methods of contraception have varied from general acceptance to almost complete rejection. As more governments recognize the present problem of overpopulation, birth control gradually is attaining ever more widespread acceptance.

For a live baby to be born, a sperm and an egg must meet and fuse, the resulting zygote must travel down the oviduct to the uterus where it must become implanted in the thick ovary wall, and here it must develop for approximately nine months (babies born a month or two prematurely often survive). Anything that interferes with this succession of events prevents the birth of a baby. Birth control methods are many, and they employ a variety of techniques and materials: coitus interruptus, the rhythm method, condoms, diaphragms, intrauterine devices, foams and jellies, the "Pill," sterilization, and abortion. All methods have their advantages and disadvantages, and each couple desiring to practice birth control must choose the method or methods they find most suitable.

Coitus interruptus and the rhythm method are inexpensive; they require no chemicals or equipment other than a calendar for the rhythm method. **Coitus interruptus** is an ancient method of birth control practiced by men, the man withdrawing the penis from the vagina before orgasm; the ejaculated sperm thus are not deposited in the female reproductive tract and do not fertilize an egg. It is not a very effective method for it requires self-control on the part of the man. Withdrawal may be too late, or sperms may be released from the penis before orgasm or even before coitus.

The **rhythm method,** the only method of birth control acceptable to the Roman Catholic Church, depends on abstinence from sexual intercourse during those days of the menstrual cycle in which fertilization could occur. Because the unfertilized egg remains alive for about one day and sperms can survive up to six days, there are only a few days in each cycle when intercourse might result in conception. But few women can determine with certainty the exact day of ovulation, and for practical purposes the "unsafe" period usually is considered to be eight days long, beginning with the eleventh day after the first day of the last period if the woman has a normal 28-day cycle. For women with shorter or longer cycles, the "unsafe" period occurs earlier or later, respectively, in the cycle. For women with irregular cycles the knowledge that ovulation occurs 14 days before the onset of the next menstrual period is useless. This method has the disadvantage of requiring a considerable amount of self-control on the part of both partners. Its failure rate is higher than that of any other method commonly used.

The condom, the diaphragm, and the intrauterine device (I.U.D.) are mechanical devices. The **condom** is a rubber or plastic sheath placed over the penis during sexual intercourse. It holds the ejaculated sperms and prevents fertilization. Use of condoms is popular, probably because they require no special fittings or prescriptions and can be purchased freely in drug stores; they also provide some protection against venereal disease. The **diaphragm** is a rubber cap with a metal rim; it fits over the cervix and so covers the opening to the uterus. Because the fit is not perfectly tight, the diaphragm must be used with spermicidal (sperm-killing) creams or jellies. Diaphragms are not so easily obtained as condoms; each woman must be fitted by her physician. The **I.U.D.** is a plastic or metal object of any of several shapes: coil, ring, bow, or other shape. A physician inserts the I.U.D. into the uterus, where it remains indefinitely. Its method of action is not known, but it is believed to interfere with implantation of the young embryo in the uterus. Some women use the I.U.D. successfully, but in others it may perforate the uterine wall and cause infection, and in others it may be expelled spontaneously and so offer no more contraceptive protection. I.U.D.'s used by women who can retain them have lower failure rates than condoms, which have lower failure rates than diaphragms.

Two general types of chemical protection are available. Those inserted into the vagina before intercourse come in the forms of **creams, jellies, foams,** and **suppositories** containing some spermicidal compound; used alone they provide some protection, but ordinarily they are used with the diaphragm. **Oral contraceptives,** commonly known as the "Pill" even though there are several types, are pills containing the female hormones, estrogens and progesterone. One pill is taken each day for 21 of the 28 days of the menstrual cycle. (In women with long, short, or irregular cycles, use of the pill brings the cycle to 28 days.) The pills are believed to prevent either ovulation or implantation of the embryo in the uterus. Its failure rate is very low—lower than that of the I.U.D. The pills must be prescribed by a physician, and the prescription must be changed if uncomfortable or dangerous side effects appear. By middle-class American standards, the pills are not expensive, but the cost is beyond the means of most of the world's poor women.

**Sterilization** usually renders the individual permanently incapable of having children, although in some cases the operation may be reversed. In women it usually is performed by cutting and sealing the cut ends of the oviducts, thus keeping

359

the eggs from reaching the uterus. Once a serious abdominal operation, it now can be achieved by the use of a laparoscope, a small device equipped with operating tools, which is inserted into a tiny incision in the abdomen. In men, severing of the sperm ducts is called a **vasectomy.** Sperms remain in the testes and die there and then are resorbed by the tissues. The sperm ducts lie near the skin of the scrotum, and only small incisions are necessary to cut them. A doctor may perform the operation during an office call, although occasional patients may require a short hospital stay. These operations in no way affect hormonal balance, for the ovaries and testes remain in the body. Sexual desire and sexual performance may be affected favorably or adversely, the change probably being brought about by the individual's anticipations rather than by the operation itself.

Sterilization should be considered a permanent operation and should not be performed on those persons who wish to have additional children. Attempts may be made to repair cut oviducts or sperm ducts, but there is no guarantee of success. Perhaps half of the attempts fail. One advantage of sterilization is that, once it is performed, it requires no equipment, no calculations, and no abstinence. Formerly seriously frowned upon, if not illegal, sterilization slowly is gaining in popularity among couples who have had all the children they desire. In India, where the population problem looms exceedingly large, the government actively encourages vasectomies and even pays men small sums of money for having the operation.

**Abortion** is the removal of the embryo or fetus from the uterus—usually before the sixth or seventh month of pregnancy, after which time it can survive if given proper care. Spontaneous abortions (miscarriages) occur because of illness, hormonal imbalance, or accident. Induced abortions are illegal in most countries unless the life of the mother would be endangered by continuing the pregnancy. In spite of this, in most societies some women always have resorted to illegal abortions to terminate unwanted pregnancies. In attempts to limit their growing populations, several countries have made nontherapeutic abortions legal—among them Japan, Sweden, England, and Hungary. In 1973 the Supreme Court of the United States legalized abortions performed during the first three months of pregnancy.

Although legal in some places, abortion still raises a difficult moral problem, for many persons equate it with murder. Others see in the birth of a new baby into an overpopulated world the death of another whom he deprives of food or poisons by the pollution he will generate in his lifetime. The question is not an easy one to resolve.

---

# Questions

1. What is asexual reproduction? Give some examples. Contrast it with sexual reproduction.

2. Beginning with a primary oocyte in an ovary, describe the formation of an egg and follow it until it is fertilized by a sperm. How do the parts of the female reproductive system contribute to the production of a ripe egg? What contribution does the endocrine system make?

3. Beginning with a primary spermatocyte in a testis, describe the formation of a sperm and follow it until it fertilizes an egg. How do the parts of the male reproductive system contribute to the production of a mature sperm? What contribution does the endocrine system make?

4. In humans, how many sperms can be produced from one primary spermatocyte? How many eggs can be produced from one primary oocyte?

5. Follow an embryo from the time of fertilization until it becomes a fetus.

6. Describe how a fetus receives food, oxygen and minerals, and how it disposes of carbon dioxide and other wastes. How does this differ from the way it will be done after the fetus is born?

7. Contrast fetal circulation with adult circulation. Why is fetal circulation more suitable before birth than after?

8. Describe two ways in which twins can originate.

9. List several methods of birth control and describe how each one interferes with the birth of a live baby.

## Suggested Readings

Allen, R. D. "The Moment of Fertilization," *Scientific American* 201(1): 124–134 (July 1959).

Barry, J. M. "The Synthesis of Milk," *Scientific American* 197(4): 121–128 (October 1957).

Bloom, W., and D. W. Fawcett. *A Textbook of Histology,* 9th ed. Philadelphia: W. B. Saunders Company, 1968.

Csapo, A. "Progesterone," *Scientific American* 198(4): 40–46 (April 1958).

Fieser, L. F. "Steroids," *Scientific American* 192(1): 55–60 (January 1955).

Griffin, D. R., and A. Novick. *Animal Structure and Function,* 2nd ed. New York: Holt, Rinehart and Winston, Inc., 1970.

Hardin, G. *Birth Control.* New York: Pegasus, 1970.

Peel, J., and M. Potts. *Textbook of Contraceptive Practice.* New York: Cambridge University Press, 1969.

Smith, C. A. "The First Breath," *Scientific American* 209(4): 27–35 (October 1963).

Spratt, N. T., Jr. *Introduction to Cell Differentiation.* New York: Reinhold Publishing Corporation, 1964.

Vander, A. J., J. H. Sherman, and D. S. Luciano. *Human Physiology.* New York: McGraw-Hill Book Company, 1970.

# Genetics

Part VIII

Two events happen in life cycles of all sexually reproducing species: fertilization and meiosis. Because the zygote resulting from fertilization receives chromosomes from two gametes, the number of chromosomes per cell would double with each succeeding generation if there were not some mechanism that halved the number in each generation. The mechanism that accomplishes this is a special type of nuclear division called meiosis. In different species meiosis occurs in different parts of the life cycle, but in most animals and in flowering plants it is associated with the formation of gametes in the reproductive organs (Chapters 27 and 40). As a result, somatic (nonreproductive) tissues are diploid in these organisms, and only the gametes (and a few associated cells in plants) are haploid.

Each species has its characteristic number of chromosomes, and these are present in pairs in diploid cells. In humans, for instance, somatic cells have 46 chromosomes in 23 pairs. The egg contributes one set of 23 chromosomes to a zygote, and the sperm contributes the other set of 23.

The two members of a pair are called **homologous chromosomes;** they carry genes that affect the same traits. For example, if one of the maternal chromosomes carries at one particular locus (site) a gene that affects eye color, the homologous paternal chromosome carries at its corresponding locus a gene that affects eye color. The two genes are not necessarily identical, however; one might be a gene for blue eye color and the other a gene for brown eye color.

**Meiosis** differs from mitosis (which the reader might want to review in Chapter 19) in involving two divisions of a cell (or sometimes only its nucleus) while the chromosomes duplicate themselves only once. As a result, four cells form, each with half as many chromosomes as the parent cell. Not only is the chromosome number halved, but each of the four daughter cells receives only one member of each pair of chromosomes. Each daughter cell then has one haploid set, but as we shall see in the more detailed description

# 41
# Meiosis

1 By halving the chromosome number in eggs and sperms, meiosis keeps the chromosome number constant from generation to generation in sexually reproducing organisms.

2 Meiosis contributes to genetic diversity in sexually reproducing organisms.

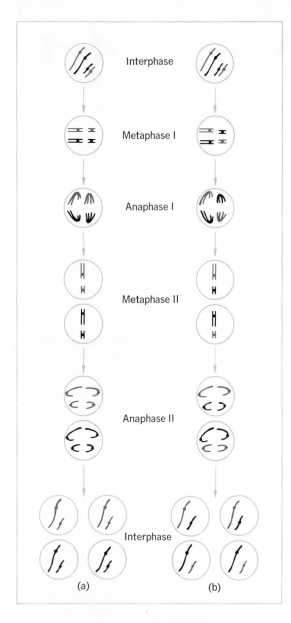

Interphase

Metaphase I

Anaphase I

Metaphase II

Anaphase II

Interphase

(a)          (b)

**41.1** A cell with two pairs of chromosomes can undergo meiosis in two ways (a and b). Four different arrangements of chromosomes in the gametes are possible. Diagrammatic, only chromosomes are indicated. The two colors indicate maternal and paternal chromosomes.

mother cell or megaspore mother cell in flowering plants, primary spermatocyte or primary oocyte in animals) has the full diploid complement of chromosomes. Preceding the first division, the chromosomes are duplicated, but the double nature is not obvious under an ordinary student microscope. During the prophase of the first division (prophase I), something occurs that does not happen in the prophase of mitosis—the homologous chromosomes pair physically with each other. This forms a tetrad of chromatids, with the two sister chromatids formed from each chromosome still attached to each other at the centromere (Fig. 41.1a). The nuclear membrane and the nucleoli disappear, and spindle fibers form.

At metaphase I the tetrads line up along the equator of the cell; compare this with metaphase of mitosis in which only the two chromatids formed from a single chromosome act as a unit.

During anaphase I, homologous chromosomes separate from each other and proceed to opposite sides of the cell. At this time the identical chromatids have not yet separated from each other. Therefore, unlike the daughter cells of mitosis, the products of the first division of meiosis are not identical with each other. For each pair of homologous chromosomes, each daughter cell receives only the maternal chromosome or the paternal chromosome; in mitosis each daughter cell receives both.

Following the first division, the cells enter an interphase during which there is no duplication of chromosomes, but the chromosomes are still double from the duplication that occurred in the preceding interphase. The cells then proceed into prophase II, and at metaphase II the chromosomes line up along the equators of the two cells. It is only during anaphase II that the identical chro-

below, the maternal and paternal chromosomes do not remain together as a set; the chromosomes in any egg or sperm usually are derived in part from the individual's maternal chromosomes and in part from the paternal chromosomes.

A cell about to undergo meiosis (microspore

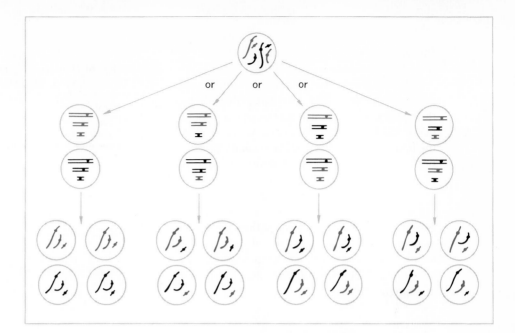

**41.2** A cell with three pairs of chromosomes can undergo meiosis in four ways. Eight different arrangements of chromosomes in the gametes are possible. Diagrammatic. The two colors indicate maternal and paternal chromosomes.

matids separate from each other. At the end of the second division, four cells have formed, each with half the number of chromosomes possessed by the original parent cell. Depending on the original cell, these four haploid cells may be four microspores, four megaspores (of which three disintegrate), four sperms, or an egg and three polar bodies. Notice that each haploid cell contains one member of each pair of chromosomes.

The importance of meiosis does not lie only in the reduction of chromosome number; this division also produces cells that are different. The significance of this second feature will become clearer after a consideration of genetics presented in the following chapter, but some suggestion of it is possible here. Figure 41.1a shows only one possible distribution of chromosomes during the meiosis of a cell with two pairs of chromosomes. In this illustration, the two maternal chromosomes move in the same direction at anaphase I and the two paternal chromosomes move in the opposite direction. Because the arrangement of one pair of chromosomes along the equator at metaphase I in no way influences the arrangement of another pair, it is equally likely that the maternal chromosome

of one pair and the paternal chromosome of another pair might move toward one pole of the cell during anaphase I while their homologues move to the other. This possibility is illustrated in Fig. 41.1b. Notice that the resulting cells differ from those in Fig. 41.1a. In an organism with two pairs of chromosomes in the diploid cells, these two alternate sets of events are equally likely, and four different types of gametes usually form in equal or nearly equal numbers. If, for example, 5 million primary spermatocytes with four chromosomes as the diploid number produce 20 million sperms, each of the four kinds of haploid cells shown in Fig. 41.1 would be represented by about 5 million sperms.

An organism with three pairs of chromosomes produces eight different kinds of gametes in equal proportions (Fig. 41.2). Organisms with more chromosomes produce more genetic combinations in their gametes, each additional pair of chromosomes doubling the number of possible combinations in the gametes (Table 41.1). The formula for determining this number is $2^n$, where $n$ equals the numbers of pairs of chromosomes in the diploid cells. Any human, with 23 pairs of chromosomes, can produce any of $2^{23}$, or approximately 8.4 million, different kinds of gametes.

The probability that a pair of parents would produce a second child genetically identical with their first child is one in $2^{46}$. (The probabilities are one in $2^{23}$ that the second sperm is identical with the first and one in $2^{23}$ that the second egg is identical with the first. The probability that that particular sperm and that particular egg will unite is equal to the one probability multiplied by the other, or $2^{23} \times 2^{23}$, which equals $2^{46}$.) This number is about 70 trillion, vastly more than the number of persons that inhabit the world today. It is no wonder that no two persons except identical twins have the same heredity. Each human being indeed is a unique individual.[1]

## Questions

1. Diagram the several stages of meiosis. What effect does meiosis have on the chromosome number of cells?

---

[1]Actually the numbers $2^{23}$ and $2^{46}$ are too small, for another phenomenon, called crossing over, increases greatly the number of possible gametes that any individual can produce. During prophase I some chromatids may break in one or more places, and the broken fragments may be exchanged between chromatids of homologous chromosomes. In this way a given chromosome in a gamete may contain some maternal and some paternal material. The numbers and locations of the breaks vary from one cell to another, and they are sufficiently frequent that the number of possible genetic combinations in the gametes of one human being becomes much greater than $2^{23}$, and the number of different possible combinations from which each of us receives only one genetic endowment becomes astronomically large. But this only serves to emphasize the uniqueness of the individual.

**Table 41.1**
Relation of Number of Chromosomes and Number of Different Kinds of Gametes

| Number of Pairs of Chromosomes in Diploid Cells | Number of Different Kinds of Gametes Possible |
| --- | --- |
| 1 | 2 |
| 2 | 4 |
| 3 | 8 |
| 4 | 16 |
| 5 | 32 |
| $n$ | $2^n$ |

2. Review mitosis. Contrast the events occurring in mitosis and meiosis. Contrast the end results of these two processes.

3. Why would you expect both meiosis and fertilization to occur in the life cycle of a sexually reproducing species but not in a life cycle of a species reproducing only asexually?

4. Explain how the chromosome number influences the number of kinds of gametes that an individual can produce.

5. What kinds of cells are the products of meiosis in flowering plants? What kinds in human beings?

6. If an animal has 20 as the diploid number, how many chromosomes would be present in a primary spermatocyte? In a sperm?

## Suggested Readings

Bonner, D. M., and S. E. Mills. *Heredity*, 2nd ed. Englewood Cliffs, N.J.: Prentice-Hall, Inc., 1964.

McLeish, J., and B. Snoad. *Looking at Chromosomes*. London: Macmillan & Company, Ltd., 1959.

Swanson, C. P. *The Cell*, 3rd ed. Englewood Cliffs, N.J.: Prentice-Hall, Inc., 1969.

The history of modern genetics dates back to the middle of the nineteenth century, when an Austrian monk, Gregor Mendel, performed the first experiments to give a clear-cut understanding of some of the patterns of inheritance. His experiments still serve as excellent illustrations of some of the simpler genetic patterns that bear his name—Mendelian genetics.

Mendel worked with garden peas. Unlike most flowering plants, garden peas ordinarily self-pollinate and produce lines of plants that breed true for many generations; in any line, the same characteristics appear in every generation. By using such plants for his experiments, Mendel could feel that he was dealing with plants of genetic purity. He selected several stocks of true-breeding plants with contrasting characters, such as tall plants and dwarf plants, yellow seeds and green seeds, round seeds and wrinkled seeds.

## Monohybrid Crosses

Mendel's first crosses were those that involved only one pair of contrasting characters; today we call this a **monohybrid cross.** In one such experiment he crossed true-breeding tall plants with true-breeding dwarf plants by transferring pollen from a plant of one type to a plant of the other type, and then removing the stamens from the pollinated flowers to prevent self-pollination. (In these and many other experiments it does not matter which plant is used as the male parent and which as the female parent. Among the exceptions to this general rule are the cases of sex-linked traits, which are discussed later in this chapter.) When the plants of this parental generation ($P$) set seed, he planted the seeds and found that all the plants of the first generation of offspring ($F_1$, for first filial generation) were tall. When he crossed two $F_1$ plants with each other, three fourths of the

# 42
# Genetics

1 Genes are the physical factors that we inherit from our parents. Diploid cells contain genes in pairs.

2 Where contrasting inherited traits occur, one member of a pair of genes often is dominant and the other is recessive.

3 Because most genes are located on chromosomes, they segregate from each other at meiosis. Each gamete receives one gene from each pair of genes.

4 Pairs of genes located on different pairs of chromosomes segregate independently of each other at meiosis.

plants in the second generation ($F_2$, for second filial generation) were tall and one fourth were dwarf:

$$
\begin{array}{ll}
P & \text{tall} \times \text{dwarf} \\
F_1 & \text{all tall} \\
F_2 & \tfrac{3}{4} \text{ tall and } \tfrac{1}{4} \text{ dwarf}
\end{array}
$$

In other sets of crosses with other pairs of characters, he obtained similar ratios:

$$
\begin{array}{ll}
P & \text{yellow seeds} \times \text{green seeds} \\
F_1 & \text{all yellow} \\
F_2 & \tfrac{3}{4} \text{ yellow and } \tfrac{1}{4} \text{ green} \\
& \\
P & \text{round seeds} \times \text{wrinkled seeds} \\
F_1 & \text{all round} \\
F_2 & \tfrac{3}{4} \text{ round and } \tfrac{1}{4} \text{ wrinkled}
\end{array}
$$

Other traits that he studied and that produced the same ratios were the colors of the flowers, the positions of the flowers on the stems, the color of the pods, and the shapes of the pods.

From the data that he gathered, Mendel concluded that some material factors, which we now call **genes,** were inherited from one generation to the next. He concluded, furthermore, that the plants possessed the genes in pairs; that one gene of each pair was **dominant** to its partner, which is the **recessive** gene; and that each parent contributes only one gene to each offspring. Today we call these alternate forms of a gene **alleles.** Dominant genes are those which find expression regardless of the presence or absence of their recessive alleles. A recessive gene finds expression only in the absence of its dominant allele. In the plants Mendel used, tall is dominant to dwarf, yellow seed color is dominant to green seed color, and roundness of the seeds is dominant to the wrinkled condition.

The concepts of pairs of genes and of dominance and recessiveness can explain the "skipping" of generations by the recessive dwarf, green, and wrinkled characters. Being the only physical link between the parental and $F_2$ generations, the $F_1$ generation must have possessed genes for the recessive characters even though the $F_1$ plants did

not display them. Mendel assumed then that members of the $F_1$ generation possessed two genes, the dominant because the plants displayed the dominant characteristic, and the recessive because they passed that gene from one generation to another. If the $F_1$ contained two genes, it was likely that plants of other generations also possessed genes in pairs.

Mendel also assumed that because the parental plants used in each experiment came from long lines of true-breeding plants, the two genes of each plant would be the same. That is, the tall parent in the first cross described earlier had two genes for tall, and the dwarf parent had two genes for dwarf. By convention the dominant usually is indicated by a capital letter and the corresponding recessive gene is indicated by the same lower-case letter. Thus the tall parent had two genes, $T$ and $T$, and the dwarf parent had two genes, $t$ and $t$.

Mendel had no knowledge of chromosomes or of meiosis, but we know now that his discoveries coincide well with the events that occur during meiosis. The genes are small units located on the chromosomes, perhaps a thousand or several thousand to a chromosome, the pairs of genes being found on homologous chromosomes. If both genes of a pair are identical (such as $T$ and $T$ or $t$ and $t$), the individual is **homozygous** for that pair of genes; if the genes are different ($T$ and $t$, for instance), the individual is **heterozygous.** The total of an organism's genes is its **genotype** ($TT$, $Tt$, or $tt$, for instance); the appearance of an organism (tall or dwarf, for instance) is called its **phenotype.** Two plants, one with the genotype $TT$ and the other with $Tt$, both have the same phenotype (tall); the former is homozygous, however, and the latter is heterozygous. Dwarf plants, with the genotype $tt$, are all homozygous.

The genes duplicate whenever the chromosomes duplicate, and separate whenever the chromosomes separate. As a result, in an individual heterozygous for a pair of genes, half of the gametes carry the dominant gene and half carry the recessive gene (Fig. 42.1). Of course, in a homozygous

370

**42.1** Segregation of the two members of a pair of genes during meiosis. Not all phases are shown here. Much cellular detail is omitted for we are interested primarily in the chromosomes and genes.

The four cells formed here might represent four sperms or an egg and three polar bodies in animals. In plants they might represent four microspores (and hence sperms which are genetically identical to the microspores from which they come) or a fertile megaspore (and hence an egg) and three sterile megaspores.

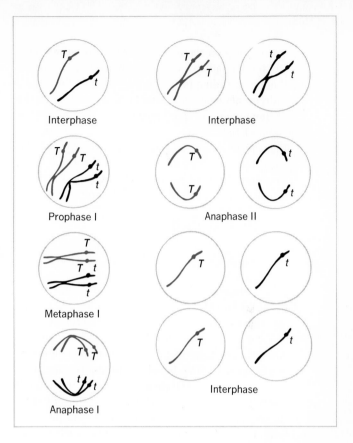

Interphase    Interphase

Prophase I    Anaphase II

Metaphase I

Anaphase I    Interphase

individual, all the gametes carry the same gene, either dominant or recessive, according to the genotype of the individual. Because of this, the tall plant of the parental generation in Mendel's experiment could contribute only gametes with the

**42.2** A monohybrid cross. The $F_1$ individuals all show the dominant phenotype. In the $F_2$, three-fourths of the individuals show the dominant phenotype and one-fourth show the recessive phenotype. Rectangles represent the genotypes of individuals; circles represent the genes borne by gametes.

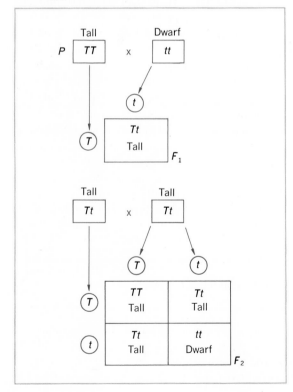

gene T to its offspring, and the dwarf plant of the same generation could contribute only gametes with the gene t. Therefore, each of their offspring had both genes T and t; but because T is dominant to t, all the $F_1$ individuals were tall. This is shown diagrammatically in Fig. 42.2, where genotypes are enclosed in rectangles and gametes are indicated by circles.

The $F_1$ individuals likewise could contribute only one gene to each offspring, but being heterozygous each $F_1$ individual could contribute two different kinds of gametes, those with T and those with t, the two kinds being formed in equal quantity. Figure 42.2 shows that the four possible combinations of genes from the two heterozygous individuals produce a ratio of three tall to one dwarf in the $F_2$ generation. Of the three tall individuals, one is homozygous and two are heterozygous.

371

Garden peas set seed after self-pollination, and so Mendel had an excellent opportunity to test his theory according to which one third of the tall $F_2$ individuals would be homozygous and two thirds of them would be heterozygous. He self-pollinated some of these plants and found that one third of them produced all tall offspring, which was to be expected of *TT* plants, and two thirds of them produced tall and dwarf offspring in a 3:1 ratio, which was to be expected of *Tt* plants:

It is impossible to tell merely from appearance whether an individual with a dominant phenotype is homozygous or heterozygous. Mendel's test by self-fertilization is useful only for those species that can produce offspring this way. Species with separate male and female individuals cannot be self-fertilized, and those with bisexual individuals may be self-sterile. For this reason, the organism to be tested usually is crossed with another individual with the recessive phenotype; such a cross is called a **test cross.** If a plant breeder wishes to know whether a tall corn plant is homozygous or heterozygous, he crosses it with a dwarf corn plant, whose genotype is known because the dwarf condition is a recessive one. All the gametes formed by the dwarf plant contain the gene *t*. The ratio of phenotypes of the offspring, therefore, depends on the gametes produced by the tall parent. If the tall parent is homozygous, all its gametes carry the gene *T*, and so all the offspring will have the genotype *Tt* and will be tall (Fig. 42.3a). If the tall plant is heterozygous, half of its gametes carry the gene *T* and half will carry *t*. Therefore about half of the offspring will have the genotype *Tt* and will be tall, and about half will have the genotype *tt* and will be dwarf (Fig. 42.3b). Therefore, if the results of a test cross give all offspring with the dominant phenotype, the breeder infers that the parent tested is homozygous. If half of the off-

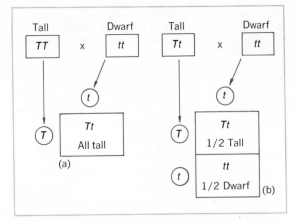

**42.3** Two test crosses. (a) A homozygous tall plant produces all tall offspring when crossed with a dwarf plant. (b) A heterozygous tall plant produces tall and dwarf offspring in a 1:1 ratio when crossed with a dwarf plant.

spring have the dominant phenotype and half have the recessive phenotype, then he infers that the parent is heterozygous.

We should mention that the fertilization of any egg by any sperm is a chance process analogous to tossing coins. If we intend to toss a coin ten times, our best guess is that it will land heads up five times and tails up five times. Yet six heads and four tails or six tails and four heads would not surprise us. So, too, the genetic ratios obtained in actual crosses often deviate slightly from the expected ratios. If the cross tall × dwarf produces 48 tall offspring and 52 dwarf offspring, this is accepted as being virtually a 1:1 ratio. Geneticists employ statistical procedures to determine if an actual ratio is compatible with the expected ratio.

## Dihybrid Crosses

After determining the patterns of inheritance for monohybrid crosses, Mendel proceeded to **dihybrid crosses,** crosses involving two pairs of genes. Using plants he knew to be homozygous for all the genes involved, he crossed plants raised from round yellow seeds with those raised from wrinkled green seeds. Knowing from previous experiments that round is dominant to wrinkled, and yellow is dominant to green, Mendel well might have anticipated that all the $F_1$ individuals would be round yellow, which they were. But the $F_2$ showed a new ratio: nine sixteenths round yellow, three sixteenths round green, three sixteenths wrinkled yellow, and one sixteenth wrinkled green, a ratio of 9:3:3:1. From these results Mendel concluded that the inheritance of one pair of characters is independent of the inheritance of another pair.

This has a somewhat familiar ring for those who have studied meiosis, for the distribution of one pair of chromosomes at anaphase I is independent of the distribution of another pair of chromosomes in the same division. If the two pairs of genes are located on two different pairs of chromosomes, Mendel's dihybrid ratio of 9:3:3:1 is easily understood. Let us review meiosis once more, this time following two pairs of chromosomes and two pairs of genes. Following the steps in Fig. 42.4, we find that the genes duplicate and segregate as do the chromosomes. The segregation of genes in the first division of meiosis depends on the arrangement of the chromosomes at metaphase I. Either both chromosomes bearing the dominant genes are oriented toward one pole of the cell and both chromosomes bearing the recessive genes are oriented toward the other pole (left side of illustration), or the chromosome with a dominant gene of one pair and the chromosome with a recessive gene of the other pair are oriented toward one pole and the partners of these are oriented toward the other pole (right side of illustration). These two arrangements are equally probable, and their results are different. In this way an individual heterozygous for two pairs of genes can produce four different kinds of gametes, and these in equal proportions: one fourth with both dominant genes ($RY$), one fourth with one dominant gene and the recessive gene of the other pair ($Ry$), one fourth with the

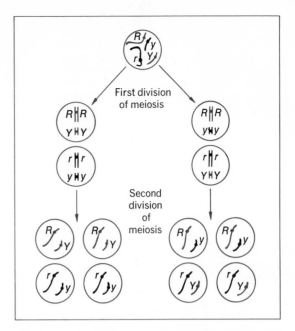

**42.4** Independent assortment of two pairs of genes on two pairs of chromosomes. Four types of gametes that differ genetically are produced in equal numbers. Only interphases are shown, and much cellular detail is omitted. Highly diagrammatic.

other dominant gene and the other recessive gene ($rY$), and one fourth with both recessive genes ($ry$). (It should be emphasized that these ratios happen only if the two pairs of genes are on different pairs of chromosomes. If they were on the same pair, the ratios would be different.)

Figure 42.5 follows the genes through two generations of one of Mendel's dihybrid crosses. All the $F_1$ individuals are heterozygous for both pairs of genes, and each one can produce four different kinds of gametes. When these gametes combine, sixteen different combinations are possible; among them are nine genotypes (some occur more than once) and four phenotypes. Notice that two of these phenotypes—round green and wrinkled yellow—are different from those of the parents. Herein lies one of the important features of sexual reproduction. Fertilization does more than produce new individuals, and meiosis does more than

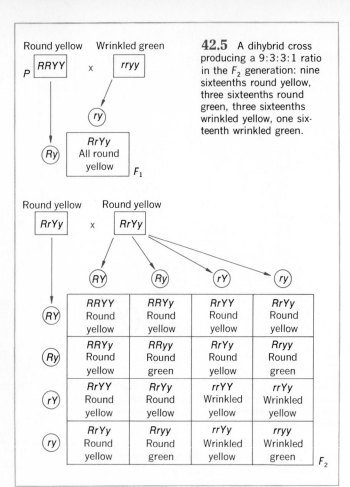

**42.5** A dihybrid cross producing a 9:3:3:1 ratio in the $F_2$ generation: nine sixteenths round yellow, three sixteenths round green, three sixteenths wrinkled yellow, one sixteenth wrinkled green.

reduce the number of chromosomes and thus counteract the doubling effect of fertilization. Together they permit the formation of new individuals with new genotypes and new phenotypes. Occasionally some of these individuals are better suited to the environment than the parent generation and so bring about an improvement in the species.

Of course, only two pairs of genes provide for little variety, but as the numbers of genes and chromosomes increases, so does the variety. In a **trihybrid cross,** the $F_2$ generation has eight phenotypes, of which six differ from those of the parental generation. Most organisms have more than three pairs of chromosomes; some have a few dozen, each with many genes, and the number of

374

possible variations becomes enormous. As the preceding chapter suggested, the number of genetic combinations possible in the human species far exceeds the number of persons on earth.

## Some Traits Inherited by Human Beings

Unfortunately Mendel's works received little recognition during his lifetime. Published in the mid-1860's, his works lay forgotten until the beginning of the twentieth century, when they were rediscovered and confirmed. Geneticists began searching for examples of dominance and recessiveness in other organisms and found many of them. In human beings, the Rh-positive blood type is dominant to the Rh-negative blood type. Persons who are Rh-negative are homozygous for the recessive *rh* gene. Should an Rh-negative woman marry an Rh-positive man, at least some of her children are likely to be Rh-positive (all the children if he is homozygous, about half if he is heterozygous). Because of an antigen–antibody reaction (Chapter 34), the second, third, and any subsequent Rh-positive children born of such parents run a high risk of suffering from erythroblastosis fetalis. Blood tests that require only a few seconds of time can tell prospective parents their blood types and allow them to decide whether they wish to have children or not. The tests also forewarn physicians of the likelihood of erythroblastosis fetalis in the babies and permit preparation for any treatment required.

The A-B-O system also displays dominance and recessiveness, but here three genes are involved: *A*, *B*, and *O*. (These three different genes exist in the human population. Any one person has only two of these genes because he receives only one from each parent.) *A* is dominant to *O*, *B* is dominant to *O*, *O* is recessive to both *A* and *B*, and neither *A* nor *B* is dominant to the other. Table

| Blood Types (Phenotypes) | Possible Genotypes |
|---|---|
| O | OO |
| A | AA, AO |
| B | BB, BO |
| AB | AB |

**Table 42.1** Phenotypes and Corresponding Genotypes in the A-B-O System

42.1 summarizes the phenotypes and genotypes in this system.

A knowledge of the inheritance patterns of these blood types has made it possible sometimes to determine if two babies were exchanged accidentally in the hospital and to settle some paternity suits. For instance, given a baby with type A blood, a baby with type B blood, one pair of parents with blood types A and O, and another pair of parents with blood types A and B, we must conclude that the parents and babies match up as follows:

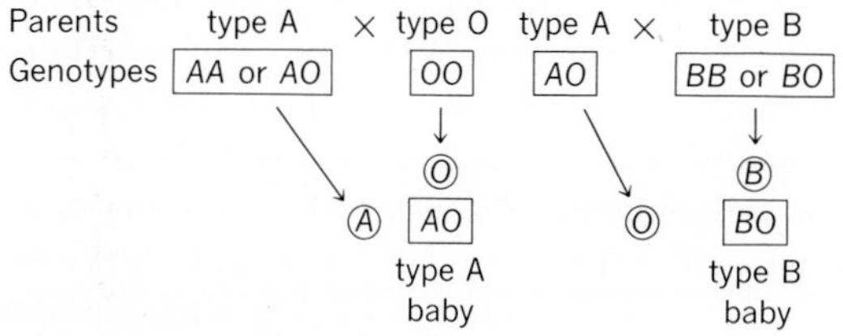

Parents          type A    ×  type O    type A   ×    type B
Genotypes    [AA or AO]    [OO]        [AO]        [BB or BO]
                                            ↓                          ↓
                                          Ⓞ                        Ⓑ
                      Ⓐ  [AO]          Ⓞ  [BO]
                          type A                    type B
                           baby                      baby

Although the type A baby could have been the child of either set of parents, the type B baby must have had a parent with at least one gene *B*, and only the type B parent could provide that.

Sometimes the A-B-O system does not provide enough information to make a decision. For instance, the following children cannot definitely be assigned to either group of parents:

Parents          type A    ×   type A      type A    × type O
Genotypes    [AA or AO]    [AA or AO]   [AA or AO]   [OO]
Children                 type O and type A

If the Rh factor is determined for all six individuals, there is a possibility (but no guarantee) that

useful information might be obtained. The following shows one possible example:

Parents   A Rh+ × A Rh+      A Rh− × O Rh−
                       ↓                        ↓
Children         O Rh+              A Rh−

The Rh-positive child could not be the offspring of two Rh-negative parents.

If a man with AB type blood (genotype *AB*) is accused by a woman with blood type A of fathering her child with type O blood, he easily can claim not to be the father, for he lacks the gene O which the child must have received from both parents. If the man's blood type had been either A, B, or O, he might have been the father. Blood types cannot prove paternity, but they disprove it in some cases and indicate the possibility of paternity in other cases.

Since the rediscovery of Mendel's work, new genetic ratios have been discovered. Some genes have only **incomplete dominance**, the heterozygous individuals having an intermediate phenotype. In shorthorn cattle, for instance, the gene for red coat and the gene for white coat show incomplete dominance, and heterozygous animals are roan (Fig. 42.6).

Sickle-cell anemia, an inherited disease that afflicts many persons of African descent, is due to an incompletely dominant gene (Fig. 42.7). The hemoglobin of persons homozygous for the gene is abnormal, and the red blood cells assume a shape that roughly approaches that of a sickle (Fig. 42.8). The cells clump and interfere with the flow of blood in small vessels; the heart, brain, spleen, and kidneys often are damaged. The disease is a grave one, and persons afflicted with it usually die young. Heterozygous individuals have sickle-cell trait; they have some normal hemoglobin and some abnormal hemoglobin, but they appear to be normal under most conditions. They possess an advantage over other persons; they have great resistance to malaria. In areas where malaria is prevalent they are likely to live longer than either

375

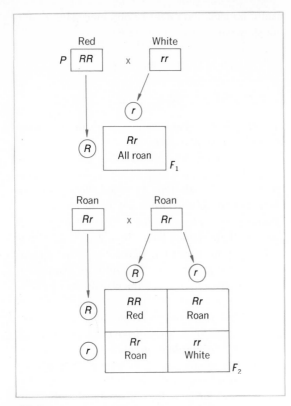

42.6 A monohybrid cross with incomplete dominance in cattle. The $F_2$ individuals show a 1:2:1 ratio: one fourth red, one half roan, one fourth white.

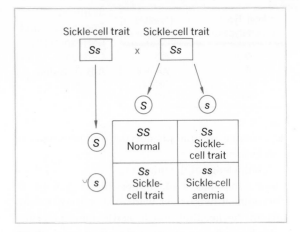

42.7 Inheritance of sickle-cell anemia. Heterozygous persons generally appear to be normal but they carry the sickle-cell gene. Two heterozygotes could expect about one-fourth of their children to be normal, about one-half to have sickle-cell trait, and about one-fourth to have sickle-cell anemia.

persons homozygous for the normal gene or those homozygous for the sickle-cell gene.

Some traits are affected by more than one pair of genes; among them are height, weight, skin color, and intelligence—characters that do not fall into distinct groups. Because environmental factors alter the expression of these characters, it is difficult to determine exactly how many genes affect these traits. It has been estimated that at least four or five pairs of genes influence human skin color.

We might consider a simplified example by assuming that skin color of blacks and whites is determined by only two pairs of genes, $A$ and $a$ and $B$ and $b$. Let us assume that genes $A$ and $B$ contribute equally to blackness of skin, and genes $a$ and $b$ contribute equally to whiteness. Then persons with genotypes $AABB$ and $aabb$ would be extremely dark and extremely light, respectively. Persons of mixed racial origin would have some other combination of genes and have some intermediate shade of skin color. Figure 42.9 shows the origin of several intensities of skin color possible among persons with ancestors of different races. If two persons, one homozygous for both pairs of genes for black and the other homozygous for both pairs of genes for white, were to marry, all their children would be heterozygous for both pairs of genes and would have an intermediate color of skin. If one of these children were to marry a person who also is heterozygous for both genes, their children could range in color from black through several intermediate shades to white. Notice that although most of the children would be expected to have intermediate skin color, some would be darker than their parents, and some would be lighter.

With four or five pairs of genes actually influencing skin color, more shades of color are possible. Environmental factors, such as exposure to

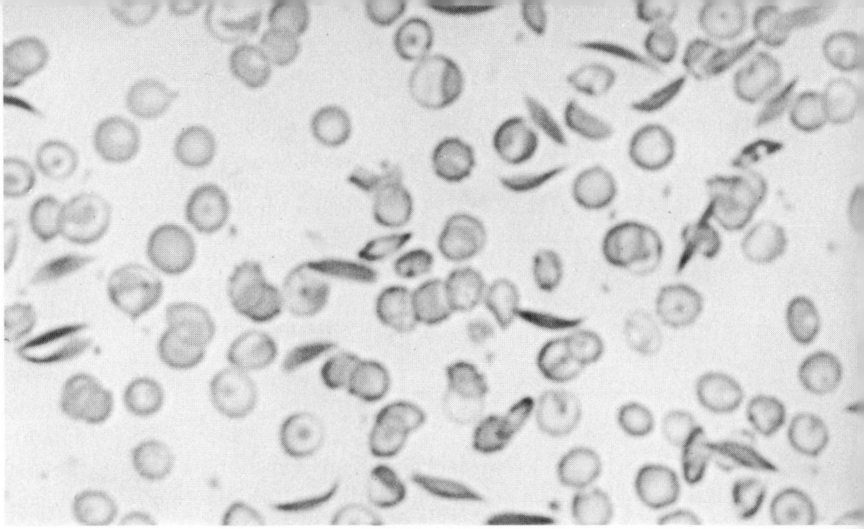

**42.8** Blood of a person with sickle-cell anemia. Some of the red blood cells have an abnormal sickle shape. These cells block small blood vessels and cause painful swelling. More serious is damage to vital organs such as the brain, heart, or kidneys. (Courtesy Carolina Biological Supply Company.)

| | AB | Ab | aB | ab |
|---|---|---|---|---|
| AB | AABB Black | AABb Medium dark | AaBB Medium dark | AaBb Intermediate |
| Ab | AABb Medium dark | AAbb Intermediate | AaBb Intermediate | Aabb Medium light |
| aB | AaBB Medium dark | AaBb Intermediate | aaBB Intermediate | aaBb Medium light |
| ab | AaBb Intermediate | Aabb Medium light | aaBb Medium light | aabb White |

Grandchildren

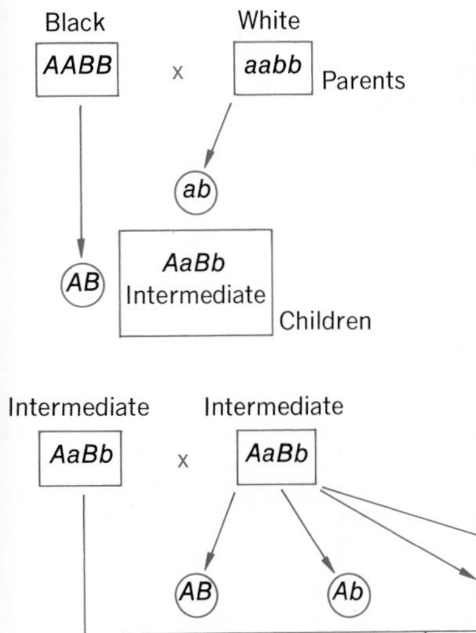

**42.9** Inheritance of skin color, somewhat simplified. If only two pairs of genes influence skin color, then five shades of color are possible. Actually the situation is somewhat more complex, for several more pairs of genes as well as environmental factors (exposure to sunlight, for example) influence skin color and produce almost a continuum of shades from black to white.

sunlight, further increase the possibilities. Almost a continuum of skin colors exist.

Similar relationships exist for other inherited traits influenced by several pairs of genes. Parents may have children taller or shorter than they are, and two persons of average intelligence may produce a genius or an extraordinarily stupid child. Although extremes like this are possible, the characteristics of most of the children tend to approach those of the parents. Where the parents are dissimilar, most of the children tend to have characteristics intermediate between those of the two parents.

## Sex Determination and Sex Linkage

The sex of an individual is determined not by a single pair or a few pairs of genes but by a pair of chromosomes called the **sex chromosomes.** In human females the twenty-third pair of chromosomes is a pair of sex chromosomes called X chromosomes. In human males the twenty-third pair is not a matched pair, for it consists of one X chromosome and another smaller and dissimilar chromosome called the Y chromosome. A girl inherits one X chromosome from each parent; a boy

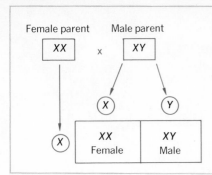

**42.10** A zygote receiving two *X* chromosomes matures into a female. A zygote receiving an *X* chromosome from the mother and a *Y* chromosome from the father matures into a male.

inherits an X chromosome from his mother and a Y chromosome from his father. All the mother's eggs contain X chromosomes; half of the father's sperms contain an X chromosome, and half contain a Y chromosome. Therefore, the sex of a child depends on which type of sperm fertilizes an egg (Fig. 42.10).

Most genes located on the X chromosome have no partner on the Y chromosome. They exhibit a peculiar pattern of inheritance and are said to be **sex-linked.** Therefore even a recessive gene on the X chromosome in a male finds expression for there can be no dominant to mask it. Red-green color-blindness, which is more common among men than women, is a common sex-linked trait. A red-green colorblind male can pass the gene only to his daughters, for only his daughters receive an X chromosome from him; his sons receive his Y chromosome, which does not carry the gene. His daughters will pass it to about half of their sons and half of their daughters. Of the grandchildren that do receive the gene, all the boys will be color-blind (for they have no homologous dominant gene), but the girls will have normal vision—unless they also inherit a recessive gene from their father. This type of colorblindness is inherited by a male from his mother and not from his father. For these reasons a typical pattern of inheritance of a sex-linked recessive trait is from a man through his unaffected daughters to his grandsons (Fig. 42.11).

Hemophilia, or bleeder's disease (Chapter 34), is another sex-linked recessive disease. Queen Victoria of Britain was heterozygous for the disease and did not suffer from it; but one of her sons had it, and several of her grandsons and great-grandsons received it through those of her daughters who were heterozygous. The present British royal family is free of the gene. Queen Elizabeth II descends from Queen Victoria's oldest son, who became Edward VII; he was not a bleeder and therefore could not transmit it to his descendants. Although Prince Philip descends from Victoria's daughter Princess Alice, who was heterozygous for the gene (one of her sons had hemophilia), Philip himself does not possess it. Less fortunate were other royal families who acquired the gene by marriage—among them the Russian and Spanish royal families.

**42.11** Typical inheritance of a sex-linked trait. A dash is used to indicate the *Y* chromosome, which bears neither the gene *N* (normal) nor *n* (red-green color blindness). In this example only males are color-blind. A color-blind female could arise if both parents have the gene *n* (for example, *Nn* × *n*−).

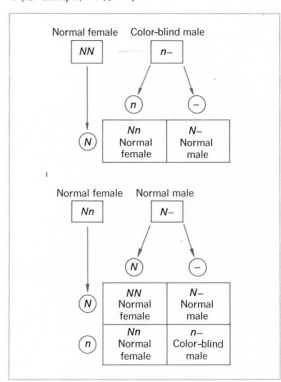

**42.12** Chromatid break. These are chromosomes from a human cell in prophase. The two chromatids of each chromosome are visible. Arrow indicates a break in a chromatid. Having no centromere, the small fragment will not move with the rest of its chromatid during mitosis and eventually will become lost. This chromatid break is believed to have been caused by drugs. (Courtesy of Dr. Frank A. Walker.)

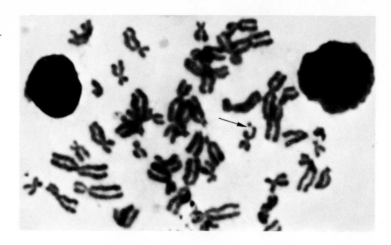

## Chromosomal Abnormalities

Occasionally, an error occurs during mitosis or meiosis, and one or more chromosomes move in the wrong direction. As a result, some cells have abnormal chromosome numbers, either lacking some chromosomes or possessing extra ones. Humans with only one X chromosome and neither another X or Y chromosome suffer from a condition known as Turner's syndrome; these persons are underdeveloped females. A corresponding condition in males with an extra X chromosome (XXY) is called Klinefelter's syndrome. Down's syndrome results from the presence of an extra chromosome of pair 21; this condition of mental and physical retardation is also called Mongolian idiocy because of the somewhat oriental appearance of the eyes.

**Polyploidy,** the possession of extra complete sets of chromosomes, is rare in animals and well may be lethal in many animal species. It is fairly common in plants, however, and some tetraploid varieties are in demand as garden plants for they are larger and produce more showy blossoms.

Breaking of a chromosome (Fig. 42.12) occasionally happens, and the chromosomal fragment lacking a centromere does not move correctly when the nucleus divides and usually it becomes lost. This may not be serious for most body cells, but it could result in the formation of abnormal gametes and might have a deleterious effect on offspring. Exposure to X-rays is known to cause chromosome breakage. Medical X-rays cause some slight damage, but this is small compared with the benefit received; the X-rays are administered in such a fashion that those parts of the body not requiring treatment or examination are shielded. At the present there is concern about prolonged exposure of persons to X-radiation from other high-voltage electronic equipment such as some color television sets. Recent evidence indicates that LSD and other drugs may be responsible for chromosome breakage.

## Inbreeding

Most societies have strong taboos against incest, and with reason, for the mating of two closely related individuals is more likely to produce defective offspring than marriages between unrelated persons. Many inherited defects are recessive, and most persons carry recessive genes for at least a

379

(a)

(b)

**42.13** The effects of inbreeding. (a) This mature corn plant is the result of several generations of inbreeding. The legs of the men in the background give some idea of the size of the plant. (b) These are hybrid corn plants, the result of several generations of outbreeding. Compare the height of these plants with that of the bus.

few such defects. Close relatives are more likely to carry the same recessive genes than are two unrelated persons, for relatives have a common ancestor from whom they both could have inherited the gene. Although the defect thus far may be unknown in the family, if two cousins who marry are heterozygous for it, about one fourth of their children would be expected to suffer from it.

Repeated inbreeding increases the percentage of individuals homozygous for a given gene, and it also increases the number of pairs of genes for which individuals are homozygous. Because of this, inbreeding frequently weakens a population (Fig. 42.13). The severity of the results depends on the genetic quality of the original stock. Homozygosity for desirable genes actually would improve a population, and theoretically the original breeding stock might have no undesirable genes. Then even brother–sister matings would have no ill effects on the children, but few such highly endowed individuals are likely to exist.

Plant and animal breeders do use inbreeding to improve their stocks, but for every superior plant or animal produced there may be many inferior ones. The latter can be destroyed or otherwise removed from the breeding population, leaving only the superior individuals. The improvement, of course, is improvement from the point of view of the breeder. Under the care and protection of humans they thrive, but if released to the wild many of these individuals might not survive, and if they did, unsupervised breeding soon would destroy the purity of the artificially produced varieties.

## Cytoplasmic Inheritance

Not all genes are located on the chromosomes, for mitochondria and chloroplasts also contain genes. Because only the sperm nucleus fertilizes an egg, a zygote receives all its mitochondria, and, if it is a green plant, all its chloroplasts, from the egg. This is called **cytoplasmic or maternal inheritance.** Its pattern of inheritance differs from nuclear inheritance in that genes are passed from a female parent to all offspring and never from male parent to offspring.

## Questions

1. Define the following terms:
   dominance
   recessiveness
   allele
   incomplete dominance
   homozygous
   heterozygous
   genotype
   phenotype
   test cross
   sex linkage
   polyploidy
2. Distinguish between segregation and independent assortment.
3. Review meiosis. How is it responsible for segregation and independent assortment?
4. What are X and Y chromosomes?
5. Explain why approximately equal numbers of boy and girl babies are born.
   In guinea pigs black coat color (*B*) is dominant to white (*b*). Use this information to answer the next questions.
6. If a homozygous black guinea pig mates with a homozygous white guinea pig, what will be the phenotype of the $F_1$ generation? Of the $F_2$ generation? Be sure to indicate ratios.
7. A black guinea pig mates with a white one. If all the offspring are black, what is the genotype of the black parent? If half of the offspring had been black and half white, what would you conclude about the genotype of the black parent? Could all the offspring have been white? Why or why not?
8. What are the genotypes of all the individuals in the following family tree?

white × black
black × white
black × black
white

381

**9.** In guinea pigs short hair is dominant to long hair. If a homozygous black short-haired guinea pig mates with a white long-haired guinea pig, what will be the phenotype of the $F_1$ generation? Of the $F_2$ generation? Be sure to indicate ratios.

**10.** Indicate the genotypes of the parents in the following crosses.

    a. black short hair × white long hair
        all black short hair

    b. black short hair × white long hair
        $\frac{1}{2}$ black short hair
        $\frac{1}{2}$ black long hair

    c. black short hair × white long hair
        $\frac{1}{2}$ black short hair
        $\frac{1}{2}$ white short hair

    d. black short hair × white long hair
        $\frac{1}{4}$ black short hair
        $\frac{1}{4}$ black long hair
        $\frac{1}{4}$ white short hair
        $\frac{1}{4}$ white long hair

**11.** Mr. and Mrs. Smith both have blood type A. Mr. and Mrs. Jones have types O and AB, respectively. Each couple has one child. One of the children named Mary has blood type A. The other child named John has blood type O. What is Mary's last name?

**12.** If Mary and John should marry, what blood types could their children have?

**13.** Which woman would be concerned about having an Rh baby: an Rh-positive woman with a Rh-negative husband, or an Rh-negative woman with an Rh-positive husband?

In cats, the gene for yellow coat color is incompletely dominant to the gene for black coat color, heterozygotes having a color called tortoise-shell. These genes are sex-linked. Use this information to solve the following problems.

**14.** A female tortoise-shell cat mates with a black tomcat. What colors would you expect to find among the female kittens? Among the male kittens?

**15.** Why would you not expect a tortoise-shell male cat from any cross?

**16.** A female yellow cat has a litter of yellow and tortoise-shell kittens. How much do you know about the cats in her neighborhood?

## Suggested Readings

Beadle, M., and G. Beadle *The Language of Life.* Garden City, N.Y.: Doubleday & Company, Inc., 1966.

Bonner, D. M., and S. E. Mills. *Heredity,* 2nd ed. Englewood Cliffs, N.J.: Prentice-Hall, Inc., 1964.

Gray, G. W. "Sickle-cell Anemia," *Scientific American* 185(2): 56–59 (August 1951).

Herskowitz, I. H. *Genetics.* Boston: Little, Brown and Company, 1965.

Levine, R. P. *Genetics,* 2nd ed. New York: Holt, Rinehart and Winston, Inc., 1968.

McKusick, V. A. "The Royal Hemophilia," *Scientific American* 213(2): 88–95 (August 1965).

Peters, J. A., ed. *Classic Papers in Genetics.* Englewood Cliffs, N.J.: Prentice-Hall, Inc., 1959.

Snyder, L. H., and P. R. David. *The Principles of Heredity,* 5th ed. Boston: D. C. Heath & Company, 1957.

Genes, the genetic units that determine many of our physical and mental characteristics, were known during the first quarter of the twentieth century only to be very small particles located on the chromosomes. Because the chromosomes consist primarily of nucleoprotein (nucleic acids in combination with protein), it was assumed that the genes also must be nucleoprotein. Pressed to be more specific, many geneticists predicted that the protein portion one day would be revealed as the genetic material, for the theoretical number of different combinations of amino acids in proteins is astronomical, and the great variety among living things probably could best be explained by a great deal of variation in the genetic material. The nucleic acids seemed less likely as candidates for this role. However, experimental work performed in the 1940's suggested that the hereditary material consists primarily of nucleic acid, and subsequent work confirmed this. By the middle of the 1950's the general structure of several nucleic acids had been determined with some degree of certainty.

## Nucleic Acids

The nucleic acids consist of repeating units called **nucleotides,** and each nucleotide consists of three units: a phosphate group, a sugar, and a nitrogen base. The sugar is a 5-carbon sugar, either **ribose** ($C_5H_{10}O_5$) or **deoxyribose** ($C_5H_{10}O_4$); the latter is so called because it has one less oxygen atom than ribose. The nitrogen base may be either a purine or a pyrimidine; the **purine** is either **adenine** or **guanine,** and the **pyrimidine** is either **cytosine, thymine,** or **uracil.** Figure 43.1 illustrates the molecular structures of these substances.

Two types of nucleic acid have been discovered in cells: **deoxyribonucleic acid (DNA),** which is the genetic material, and the **ribonucleic acids (RNA),** which function in the expression of the genes. Each type of nucleic acid consists of only four

# 43

# Nucleic Acids and Gene Action

1 The genetic material of virtually all living things consists of DNA; it codes for the synthesis of enzyme (protein) molecules.

2 The genetic code consists of triplets of nucleotides in the DNA molecules. Each triplet codes for one amino acid.

3 The genetic code is transferred to mRNA molecules, which carry it to the ribosomes where proteins are synthesized.

4 Two factors known to stimulate genes to produce mRNA are the presence of the substrates of the reactions controlled by the genes and the presence of specific hormones.

383

## PHOSPHORIC ACID

OH
|
O=P—OH
|
OH

### SUGARS

Ribose

Deoxyribose

### PURINES

Adenine

Guanine

### PYRIMIDINES

Cytosine

Thymine

Uracil

**43.1**  Components of nucleotides.

### Table 43.1

### Comparison of DNA and RNA

| DNA Components | RNA Components |
| --- | --- |
| Phosphate | Phosphate |
| Deoxyribose | Ribose |
| Adenine | Adenine |
| Guanine | Guanine |
| Cytosine | Cytosine |
| Thymine | Uracil |

DNA nucleotides:

| | |
| --- | --- |
| Phosphate—deoxyribose—adenine | (P—D—A)* |
| Phosphate—deoxyribose—guanine | (P—D—G) |
| Phosphate—deoxyribose—cytosine | (P—D—C) |
| Phosphate—deoxyribose—thymine | (P—D—T) |

RNA nucleotides:

| | |
| --- | --- |
| Phosphate—ribose—adenine | (P—R—A) |
| Phosphate—ribose—guanine | (P—R—G) |
| Phosphate—ribose—cytosine | (P—R—C) |
| Phosphate—ribose—uracil | (P—R—U) |

*These abbreviations derive from the initials of the components of the nucleotides. Note that in this context P stands for phosphate and is not the symbol for the element phosphorus.

kinds of nucleotides. A DNA nucleotide is formed from a phosphate, deoxyribose, and either adenine, guanine, cytosine, or thymine. A RNA nucleotide is formed from a phosphate, ribose, and either adenine, guanine, cytosine, or uracil. Table 43.1 summarizes this information.

## DNA

Nucleotides unite chemically with each other, the phosphate of one nucleotide reacting with the sugar of another nucleotide (Fig. 43.2). The linking of a few hundred nucleotides forms a long chain consisting of alternating sugars and phosphate groups with a nitrogen base extending from each

384

sugar. RNA consists of a single such chain, but DNA is a double-stranded molecule formed of two long chains united by their nitrogen bases, adenine always being paired with thymine and cytosine being paired with guanine (Fig. 43.3). The bonds holding the two strands together are weaker than other bonds in the DNA molecule. For this reason,

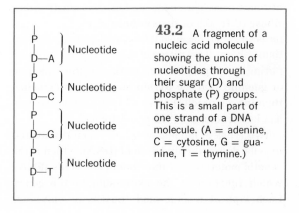

**43.2**  A fragment of a nucleic acid molecule showing the unions of nucleotides through their sugar (D) and phosphate (P) groups. This is a small part of one strand of a DNA molecule. (A = adenine, C = cytosine, G = guanine, T = thymine.)

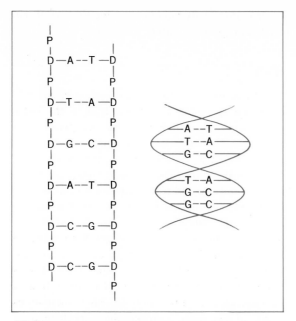

**43.3** At the left is a portion of a DNA molecule, showing the two strands and their union through their nitrogen bases. Complementary nitrogen bases are joined by hydrogen bonds that are weak individually but strong collectively. The flat, ladderlike arrangement shown on the left actually is twisted into a double helix like the one on the right.

the two strands can separate when conditions are right. This happens when new DNA is synthesized prior to a nuclear division.

Preceding each mitotic division of a nucleus and preceding the first division of meiosis, all the chromosomes of a cell are duplicated, and this means that the genetic material must duplicate as well. When this happens, the two strands of the DNA molecules unwind and separate from each other (Fig. 43.4a); each strand then acts as a template for the synthesis of a complementary strand. From a pool of nucleotides in the nucleus, separate nucleotides move toward complementary nucleotides on the template strand (Fig. 43.4b). When the new nucleotides unite with each other and with the old strands, two new DNA molecules are formed. Each new molecule consists of one of the old strands and one newly formed strand (Fig. 43.4c). The two new molecules are identical with the original molecule, and they can duplicate themselves again before the next nuclear division. In this way the genetic material passes unchanged from one cell generation to another, and, through the gametes, from parents to children.

The many chemical reactions of which a cell is capable do not all proceed at once; if they did, the results would be chaotic. Nearly every reaction, the reader will recall, is controlled immediately by its specific enzyme (Chapter 10), with most cellular reactions not proceeding at any perceptible rate in the absence of their specific enzymes. Anything that controls the synthesis of enzymes might then control the activities of the cell. Enzymes are proteins. They consist of long chains of amino acids,

**43.4** Duplication of a DNA molecule. (a) The two strands separate. (b) Complementary nucleotides (color) approach the strands. (c) Two new DNA molecules; the new portions in color. For simplicity, the molecules are depicted as flat rather than in helices.

(a)

(b)

(c)

**43.5** Formulation of mRNA on a template of DNA.
(a) Two strands of the DNA molecule separate. (b) One
DNA strand acts as a template for the assembly of RNA
nucleotides. (c) A portion of an mRNA molecule.

the sequence of the amino acids in the molecule
determining the properties of the protein and so
its ability to function as an enzyme. Anything
controlling the activities of the cell must be able
to determine the specific sequence of amino acids
in the enzyme molecules as they are synthesized.
The controlling agent, then, must carry some kind
of code to indicate the incorporation of each of
the amino acids into the growing protein molecule.
DNA molecules carry this code in the sequence of
their nucleotides.

Although the nucleotides of DNA differ from
each other only by their nitrogen bases, and although there are only four nitrogen bases in DNA,
if these are considered in groups, a sufficient
amount of variation becomes evident to account
amply for the coding of 20 amino acids. The **genetic code** now is known to consist of groups of
three adjacent nucleotides called triplets, each
triplet coding for one amino acid. A few minutes
of calculation reveal that with four different kinds
of nucleotides, not only 20 but 64 triplets are possible. This provides a certain amount of redundancy,

some amino acids being coded for by two, three,
four, and even six different triplets.

## RNA

DNA remains in the chromosomes in the nucleus, but proteins are synthesized in the cytoplasm. Obviously there must be an intermediary
that carries the code from the nucleus to the cytoplasm, and it is in this capacity that RNA functions. Three forms of RNA exist: messenger RNA,
ribosomal RNA, and transfer RNA. It is messenger
RNA that actually carries a copy of the genetic
code to the cytoplasm.

When a molecule of **messenger RNA (mRNA)**
forms, a double-stranded molecule of DNA un-
winds, and one of the strands functions as a template for the formation of mRNA rather than a
complementary strand of DNA. RNA nucleotides
move to positions opposite the complementary
nucleotides of the template strand of DNA (Fig.
43.5). Cytosine approaches guanine, guanine approaches cytosine, and adenine moves into position opposite thymine. RNA nucleotides have no
thymine, and those containing uracil move toward
the adenine of the DNA template. The newly assembled RNA nucleotides combine with each
other, and the resulting single strand of mRNA
moves away from the DNA. It carries the same
genetic code as the DNA template from which it
was formed, albeit in complementary form. The
genetic code, summarized in Table 43.2, usually is
given in terms of mRNA triplets rather than DNA
triplets. For example, the triplet AAU (adenine-
adenine-uracil) codes for asparagine, and so does
AAC (adenine-adenine-cytosine).

## RNA and Protein Synthesis

The mRNA molecules move across the nuclear
membrane and into the cytoplasm. Here they approach the ribosomes which consist of ribosomal
RNA and protein; and one or more ribosomes pass
along the length of each mRNA molecule. Only one

Table 43.2

The Genetic Code*

*U = uracil, C = cytosine,
A = adenine, G = guanine.

| | | | |
|---|---|---|---|
| UUU phenylalanine | UCU serine | UAU tyrosine | UGU cysteine |
| UUC phenylalanine | UCC serine | UAC tyrosine | UGC cysteine |
| UUA leucine | UCA serine | UAA terminator | UGA terminator |
| UUG leucine | UCG serine | UAG terminator | UGG tryptophane |
| | | | |
| CUU leucine | CCU proline | CAU histidine | CGU arginine |
| CUC leucine | CCC proline | CAC histidine | CGC arginine |
| CUA leucine | CCA proline | CAA glutamine | CGA arginine |
| CUG leucine | CCG proline | CAG glutamine | CGG arginine |
| | | | |
| AUU isoleucine | ACU threonine | AAU asparagine | AGU serine |
| AUC isoleucine | ACC threonine | AAC asparagine | AGC serine |
| AUA isoleucine | ACA threonine | AAA lysine | AGA arginine |
| AUG methionine | ACG threonine | AAG lysine | AGG arginine |
| | | | |
| GUU valine | GCU alanine | GAU aspartic acid | GGU glycine |
| GUC valine | GCC alanine | GAC aspartic acid | GGC glycine |
| GUA valine | GCA alanine | GAA glutamic acid | GGA glycine |
| GUG valine | GCG alanine | GAG glutamic acid | GGG glycine |

triplet is in place on a ribosome at a time, and the amino acid corresponding to the triplet in place is brought into position. The positioning is done through the agency of transfer RNA (tRNA) molecules. The single strand of a tRNA molecule bends back on itself and winds around itself (Fig. 43.6). An amino acid is attached at the "open" end of the molecule, and at the opposite end a triplet of nitrogen bases is exposed. This triplet is the complement of the mRNA triplet coding for that amino acid. For instance, AAU is one of the mRNA codes for asparagine; a tRNA with the complementary triplet, UUA, carries asparagine. As the tRNA molecules position themselves opposite the mRNA triplets, they bring amino acids into close proximity to each other, and as the amino acids combine with each other, a polypeptide chain begins to form (Fig. 43.7). The tRNA molecules are released and can unite with more molecules of the appropriate free amino acids if they are available in the cell.

**43.6** Two transfer RNA molecules with their amino acids. The coiled portions of the molecules consist of nucleotides, but only the triplets complementary to triplets in mRNA are indicated here.

Three of the mRNA triplets, UGA, UAG, and UAA, do not code for any amino acid but are called terminators; they serve as "punctuation marks" in the code. A given mRNA molecule may code for several different polypeptides, and the

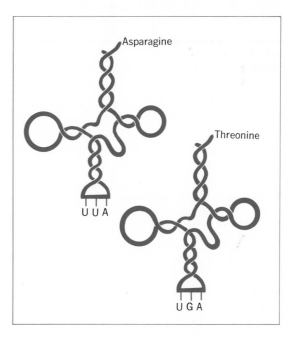

387

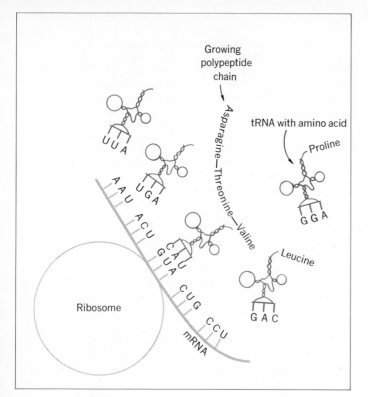

Growing
polypeptide
chain

tRNA with amino acid

Proline

Asparagine—Threonine—Valine

Leucine

G G A

G A C

U U A

A A U A C U

U G A

C A U

G U A

C U G

C C U

mRNA

Ribosome

**43.7** A small portion of an mRNA molecule moving across a ribosome. The growing polypeptide chain consists of only three amino acids, and two more are being brought into position by tRNA molecules. The tRNA molecules that brought asparagine and threonine into position are now free.

terminator triplets indicate the end of the code for one polypeptide and the beginning of the code for the next.

The events leading from DNA to a chemical reaction in the cell are summarized in Fig. 43.8.

**43.8** DNA control of reactions occurring within a cell. DNA acts as a template for the synthesis of mRNA, and mRNA controls the synthesis of a particular enzyme. The enzyme, in turn, catalyzes a specific chemical reaction.

$$\xrightarrow{\text{DNA}} \text{mRNA} \downarrow$$

$$\text{reactants} \xrightarrow{\text{enzyme}} \text{products}$$

### Effects of Reactants on Gene Action

At any moment only a few of the many genes in a cell actively form mRNA. To put it another way, genes can be turned "on" and "off." One factor known to turn a gene on is the presence of the substrate of the chemical reaction controlled by that gene. For example, if the reaction in Fig. 43.8 is the digestion of the sugar lactose

$$\text{Lactose} \xrightarrow{\text{lactase}} \text{glucose} + \text{galactose}$$

it can proceed only in the presence of the enzyme lactase. In certain bacteria, in the absence of lactose, the lactase gene is turned off, but the presence of lactose turns it on. Then the DNA in the gene synthesizes the mRNA carrying the code for the synthesis of lactase. When this enzyme forms, the digestion of lactose proceeds. When the supply of lactose dwindles, the lactase gene is turned off again.

Often the products of a reaction are used in a second reaction, or, if they are waste products, they diffuse away from the cell. But if they accumulate, they may turn the gene off.

## Hormones and Genes

In earlier chapters the reader learned that the processes occurring in living organisms are controlled by the environment (Chapter 6), by hormones (Chapters 29 and 36), by the nervous system in animals (Chapter 37), and by DNA (this chapter). These are not contradictory statements, but each is a part of the total picture of integration of internal processes.

Except for the results of an occasional abnormal mitosis, every diploid cell of a plant or animal has exactly the same set of chromosomes and therefore the same genes and the same DNA. Yet a nerve cell in an animal does not resemble a blood cell, and a xylem cell in a plant differs from a palisade parenchyma cell. Cells with the same genetic potential come to be so different because only certain genes function in each cell.

There now exists strong evidence that some hormones function by turning particular genes on at appropriate times. Thus a gene functions not only when a suitable substrate is present, but when the proper time comes in the organism's life cycle. The proper time for a given gene to function may arrive at a certain stage in the embryology or later development of the individual or at a certain time of the year. In a small embryo consisting of only two or four cells, all the cells may be very much alike, but as more cells form, they differentiate and assume different appearances and functions; some become liver cells, muscle cells, or kidney cells; cambium cells, pith cells, or phloem cells.

The endosperm of cereal grains (Chapter 28) is rich in starch which supplies the young seedling with much of its food before its leaves expand and synthesize their own food. Before the insoluble starch can be used, however, it must be digested to soluble sugars that can diffuse into the embryo. In a dormant seed no amylase (starch-digesting enzyme) is present, and the starch remains unchanged within the endosperm. When given adequate water and kept at temperatures suitable for germination, the cereal embryo secretes the hormone gibberellin, which diffuses into the endosperm. Here it stimulates the synthesis of mRNA from DNA. At least some of the mRNA carries the code for the synthesis of amylase. After amylase forms, it catalyzes the digestion of starch, and the resulting sugars diffuse into the embryo and provide it with the energy it requires for germination (Fig. 43.9). Another enzyme coded for by the mRNA produced in the endosperm is a protease that catalyzes the digestion of proteins present in the aleurone layer of the endosperm. One of the products of the digestion is the amino acid tryptophane, which travels to the embryo where it is converted to IAA.

$$\underset{\text{Tryptophane}}{\text{CH}_2\text{—}\overset{\overset{\text{H}}{|}}{\underset{\underset{\text{NH}_2}{|}}{\text{C}}}\text{—COOH}} \longrightarrow \underset{\text{Indole acetic acid (IAA)}}{\text{CH}_2\text{—COOH}}$$

The IAA stimulates the elongation of cells in the coleoptile of the embryo (Chapter 29).

Ecdysone, the molting hormone of insects, has been shown to have physical and chemical effects on chromosomes. In most organisms, the chromosomes are so small that little detail can be observed, but cells in the salivary glands of fruit flies have banded chromosomes so large that the bands

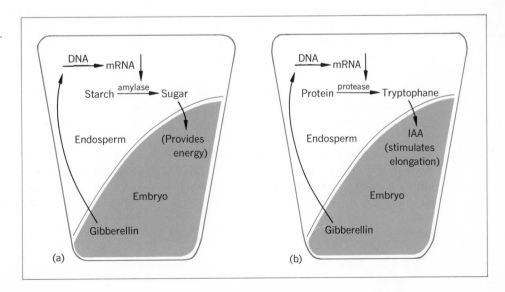

**43.9** Two examples of gibberellin control of chemical reactions in cereal grains.

**43.10** Cyclic AMP (cAMP).

are readily visible under a microscope. The locations of many of their genes have been determined, and each gene always is associated with a particular band on a particular chromosome. When ecdysone is applied to salivary gland cells, certain bands swell. These swellings, called puffs, are rich in RNA, evidence that the ecdysone stimulated genes, and only certain genes, to produce RNA.

Recent research indicates that at least some hormones do not enter the cells they affect. An enzyme, **adenyl cyclase,** is located in the cell membranes of many animal cells. The presence of a specific hormone at the cell surface stimulates adenyl cyclase to catalyze the conversion of ATP to a form of AMP called **cyclic adenosine monophosphate (cyclic AMP** or cAMP), the structure of which is shown in Fig. 43.10. Cyclic AMP then activates a certain enzyme within the cell; it does this by converting an inactive form of the enzyme to an active form. Because of its role within the cell, cyclic AMP has been called the second messenger, hormones being first messengers.

Any one hormone activates only certain cells this way because only certain cells possess receptor sites on their surfaces specific for that hormone. For example, thyrotropin produced by the pituitary gland circulates in the blood and reaches all the organs of the body, but it stimulates only

the thyroid gland to produce thyroxin because only thyroid gland cells have receptor sites specific for thyrotropin. Therefore thyrotropin causes the production of cyclic AMP only in these cells and causes only these cells to produce their particular secretions. Other hormones have their particular target organs (some have more than one, adrenalin, for example), the cells of which have receptor sites specific for those hormones.

We have not yet fully explored the functions of cyclic AMP. It has been found to regulate cell activity at the level of the genes as well as at the level of enzymes. It functions in the release of transmitter substances at synapses. Cancerous cells seem to have low supplies of cyclic AMP. This fact may lead to new areas of exploration in the search for the prevention and cure of cancer.

When answering the question, "What turns the genes and enzymes on?" with the answer "hormones," we merely invite another question, "What causes the production of the hormones?" The answer in some cases at least seems to be certain environmental conditions. Throughout both the plant and animal kingdoms we can find many examples of the effects of external conditions on a variety of biological processes. Reproduction of many species coincides with certain seasons of the year, young animals often being born as the abundance of summer presents itself. Animals with short gestation periods often mate early in the year, and those with long gestation periods mate in summer or fall, their offspring being born in the spring or early summer of the following year. In the Northern Hemisphere some birds migrate north in the spring and south in autumn; this not only relieves the population pressure in the south during the breeding season, it also takes the migrants to areas where day length is longer and allows for more foraging time. The dropping of leaves from deciduous trees in autumn shortly before the arrival of cold, dry winter weather prevents excessive loss of water by transpiration. Superficial examination of these activities seems to reveal a certain anticipation of good or bad weather ahead, but of course an animal has no way of knowing how far in the future its offspring will be born, nor could it anticipate the weather at that time. When a bird in Florida departs its winter home, it cannot know what conditions will be like when it arrives in Canada. Neither does a deciduous tree understand that the oncoming winter could dehydrate it if it were to retain its leaves.

The mechanism of "anticipation" in these examples is the response organisms make to the day length at some appropriate time before the coming season. In many cases the response is through hormones, and in animals, the nervous system as well. As the days lengthen late in winter and early spring, the pituitary glands of birds secrete hormones which cause the gonads to enlarge and become functional. But if the birds are blinded first, their gonads remain small. This suggests that the light affects the retina and that nervous stimuli traveling to the brain exert an effect more or less directly on the pituitary, causing it to secrete its gonad-stimulating hormones.

In flowering plants, the exposure of the leaves to day lengths appropriate for flowering causes some of the stem tips to cease producing new leaf primordia and to produce flower buds instead. After exposure to only a few short days, the RNA content and then the protein content rise dramatically in the stems tips of short-day plants, and very shortly thereafter the primordia of the flower parts can be detected. Because the leaves and stem tips are some distance from each other, it is inferred that under the appropriate day length the leaves synthesize a flowering hormone that travels to the stem tips and there induces RNA synthesis. Although the evidence for the existence of this hormone is strong, to date it has eluded the attempts of biochemists to isolate it.

Control over chemical reactions in the organism thus lies at many different levels. Some possible sequences of control are the following:

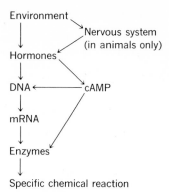

This simplified chain of events should not be construed to mean that this is the only sequence possible. Environmental temperature might act at almost any level, slowing or hastening the rate of any of the chemical reactions. Hormones may exert their effects in ways other than stimulating DNA. There is evidence that some hormones may affect the permeability of cell membranes; ADH, the antidiuretic hormone, is one such. Reactants in a chemical reaction may turn on the gene controlling that reaction, or the products may turn it off. Negative feedback mechanisms involving two or more components of the preceding scheme also play their roles and prevent any one component from exerting too great an effect.

## Questions

1. Define the following terms:
   nucleic acid
   nucleotide
   complementary strand
   template
   triplet
   genetic code
   ribosome

2. What is the chemical difference between DNA and RNA?

3. How many types of RNA are there? What are they?

*391*

**4.** What are the functions of DNA and the various types of RNA?

**5.** Indicate step by step how the genetic code in DNA in the nucleus controls the synthesis of an enzyme in the cytoplasm.

**6.** Given the following sequence in a messenger RNA molecule at a ribosome,

UUGGCACGAUAA

what sequence of amino acids would you expect in the polypeptide forming there?

**7.** What turns a gene "on" and "off"?

**8.** What is cyclic AMP, and what is its function?

**9.** What is the value of not having all genes function continuously?

## Suggested Readings

Baker, J. J. W., and G. E. Allen. *Matter, Energy, and Life.* Reading, Mass.: Addison-Wesley Publishing Co., Inc., 1965.

Beadle, M., and G. Beadle *The Language of Life.* Garden City, N.Y.: Doubleday & Company, Inc., 1966.

Beerman, W., and U. Clever. "Chromosome Puffs," *Scientific American* 210(4): 50–58 (April 1964).

Benzer, S. "The Fine Structure of the Gene," *Scientific American* 206(1): 70–84 (January 1962).

Bonner, D. M., and S. E. Mills. *Heredity,* 2nd ed. Englewood Cliffs, N.J.: Prentice-Hall, Inc., 1946.

Changeaux, J.-P. "The Control of Biochemical Reactions," *Scientific American* 212(4): 36–45 (April 1965).

Davidson, E. "Hormones and Genes," *Scientific American* 212(6): 36–45 (June 1965).

Gamow, G. "Information Transfer in the Living Cell," *Scientific American* 193(4): 70–78 (October 1955).

Gay, H. "Nuclear Control of the Cell," *Scientific American* 202(1): 126–136 (January 1960).

Gurdon, J. B. "Transplanted Nuclei and Cell Differentiation," *Scientific American* 219(6): 24–35 (December 1968).

Hoagland, M. R. "Nucleic Acids and Proteins," *Scientific American* 201(6): 55–61 (December 1959).

Hurwitz, J., and J. J. Furth. "Messenger RNA," *Scientific American* 206(2): 41–49 (February 1962).

Kerr, N. S. *Principles of Development.* Dubuque, Iowa: William C. Brown Company, Publishers, 1967.

Ptashne, M., and W. Gilbert. "Genetic Repressors," *Scientific American* 222(6): 36–44 (June 1970).

Rich, A. "Polyribosomes," *Scientific American* 209(6): 44–53 (December 1963).

Yanofsky, C. "Gene Structure and Protein Structure," *Scientific American* 216(4): 80–94 (April 1967).

# The Biological Clock

Part IX

Many persons boast that their pet dogs and cats can tell time, for the animals eagerly await their masters' daily homecoming, the dinner hour, or the time for an evening run. The same persons might receive with skepticism the information that plants can "tell time," yet the evidence is strong that all eucaryotic organisms possess internal timers called **biological clocks.** Such timers have not been discovered in procaryotic organisms.

## Circadian Rhythms

Daily rhythms in organisms are commonplace. Human beings usually are active by day and sleep by night; with mice and flying squirrels the schedule is reversed. Tulip flowers open by day and close at night. Bean plants raise their leaves by day, and at night they lower them. In a given species, cell divisions usually occur at a time of day characteristic for that species. Human beings experience a daily body temperature cycle, late afternoon and early evening usually bringing the highest temperatures of the day; lowest temperatures usually come in the morning just before rising from sleep. Blood viscosity is greatest in the late morning hours and lowest about midnight. The concentration of plasma proteins is highest in the afternoon and lowest during the hours after midnight. Enzymes exhibit daily cycles in their levels of activity, and biochemical processes do not proceed at the same rate throughout a 24-hour period.

Before biologists seriously investigated daily rhythms, they generally felt that the rising of the sun (or its setting) was a signal initiating a train of activity that ran its course and then stopped until the next sunrise (or sunset) caused the onset of activity again. Thus the organism would act like an hourglass; a particular activity would be analogous to the running of sand from the upper chamber to the lower one. This flow of sand would

# 44

# Circadian Rhythms and the Biological Clock

1 Most living organisms exhibit natural rhythms called circadian rhythms; these are approximately 24 hours in length.

2 Circadian rhythms are controlled by mechanisms called biological clocks.

3 Biological clocks are believed to be internal mechanisms, but there is some evidence that they may be external to the organism.

continue until all of it had run through, and then there would be no activity until something turned the hourglass over. The change from darkness to light (or vice versa) then would be analogous to the turning of the hourglass. Activity would run down in about 12 hours (or 10 or 14 hours or whatever the length of an observed activity or process might be) and start again only when the next rising (or setting) of the sun turned it on. But soon a few experiments showed that the situation is not so simple. Rather there seem to exist innate rhythms internally timed but nonetheless affected by external factors.

**44.1** Luminescence rhythms in *Gonyaulax polyedra*. (a) In cycles of alternating light and dark (gray bars) periods, the algae show maximum luminescence about the middle of the dark period and minimum luminescence during the light period. (b) In constant dim illumination, the algae maintain their luminescence rhythm, but the average period length is 24.4 hours. Notice that the peaks are shifted slightly to the right of those in the upper curve. (J. W. Hastings and B. M. Sweeney, "The *Gonyaulax* Clock." In *Photoperiodism and Related Phenomena in Plants and Animals*, Publication No. 55, p. 569. Copyright 1959 by the American Association for the Advancement of Science.)

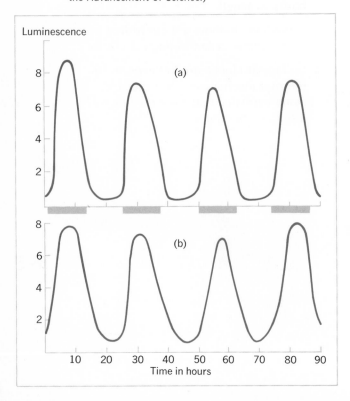

To determine whether daily rhythms are imposed by the environment or are innate, researchers kept organisms in closed chambers in which environmental factors such as light, temperature, and humidity can be controlled accurately. One or more of these factors can be kept constant or changed at desired times. One important fact learned from early experiments is that most organisms maintain daily rhythms even when kept under what appear to be constant environmental conditions. Of the organisms investigated, only bacteria and blue-green algae fail to show these rhythms.

The single-celled alga *Gonyaulax polyedra* has been a useful experimental organism for researchers in the field of biological clocks. This marine alga possesses the property of luminescence and is responsible for the glowing of some ocean waters at night. *Gonyaulax* luminesces when disturbed at any time of day, but the cells shine much more brightly at night. They exhibit a daily rhythm in the intensity of the light they emit (Fig. 44.1a). If *Gonyaulax* is maintained in constant dim light at a constant temperature, it retains its luminescence rhythm (Fig. 44.1b). Plants and animals kept in constant environmental conditions usually adhere to similar rhythms of a variety of processes—photosynthesis, cell division, running activity, and many others—for weeks, months, or even years. Animals generally maintain their rhythms indefinitely in constant darkness. Green plants usually lose their rhythms after a few days in constant darkness (possibly because of a need for their photosynthetic product), but in dim light they, too, maintain their rhythms indefinitely.

**Table 44.1**

Lengths of Some Circadian Periods

| | |
|---|---|
| Luminescence of *Gonyaulax* | 24.4 hours |
| Activity of flying squirrels | 23–24.5 hours |
| Activity of mice | 25–25.5 hours |
| Growth rate in oat coleoptiles | 23.3 hours |
| Spore production in *Oedogonium* (a green alga) | 22.0 hours |

Constant bright light, however, generally causes the loss of rhythmicity in both plants and animals. Because these rhythms can continue even in the absence of clues that might indicate the time of day, it seems unlikely that sunrise and sunset or daily temperature changes could be the timing device that regulates rhythmic activities of organisms in nature (however, these environmental factors can reset the rhythms).

Under constant conditions, the periods of rhythmicity exhibited by organisms rarely are exactly 24 hours. Some are longer, some shorter. They vary not only from species to species, but between individuals of the same species. A few experimentally determined period lengths are given in Table 44.1. There seem to be no physical phenomena in the environment that correspond to these and the many other experimentally determined period lengths. This strongly suggests (but does not prove) that the rhythms are timed by an internal timer.

The varying lengths of period of about one day's length, but not exactly that, have given rise to the term **circadian** rhythm. This term is compounded from the Latin *circa* (about) and *diem* (day).

Other evidence that biological clocks are internal mechanisms comes from the fact that circadian rhythms can be reset to any time the experimenter desires regardless of local time. By keeping organisms in bright light from 6 P.M. to 6 A.M. and in darkness from 6 A.M. to 6 P.M., the experimenter reverses their circadian cycles. *Gonyaulax*, for example, fluoresces more during the experimental night (local day) than during the experimental day (local night). Bean plants lower their leaves and flying squirrels become active during the experimental night; bean plants raise their leaves and flying squirrels become quiet during experimental day. Again, the results suggest that an internal timer has been reset by the changed light regime.

Furthermore, a single change from light to darkness can reset circadian rhythms to any desired time. In one experiment, cultures of *Gonyaulax* were exposed for one year to constant bright light to cause them to lose their circadian luminescence rhythm. Then eight cultures were transferred to constant darkenss; each transfer was made at a different time of day. Immediately on entering darkness, each culture began to show an increase in luminescence that reached a peak about four to six hours after the transfer and then began to decline (Fig. 44.2). Thus each culture had its phase determined by the time at which the transfer was

**44.2** Luminescence rhythms of eight cultures of *Gonyaulax* brought from constant bright light to constant darkness. Each curve represents a different culture. The beginning of each curve indicates the time of transfer. The peak of luminescence comes four to six hours after the transfer. To avoid the confusion of too much overlapping of curves, only the first part of the first cycle is shown for each culture. (B. M. Sweeney and J. W. Hastings, "Characteristics of the Diurnal Rhythm of Luminescence in *Gonyaulax polyedra*," J. Cell. Comp. Physiol. 49:115–128 (1958).)

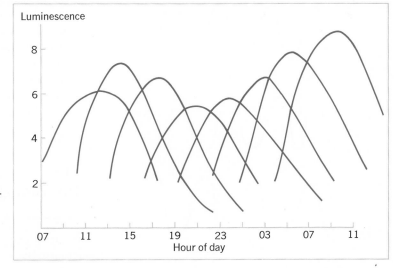

made, not by local time. This indicates that circadian rhythms can be reset to any time of day.

Single flashes of light provided to organisms kept in otherwise constant darkness also reset circadian rhythms. Light coming in subjective day (time corresponding to what would have been the organisms' light period if their earlier regime of alternating light and dark periods had been continued) shifts the phase very little. Light coming during the first half of the subjective night ad-

**44.3** Light flashes interrupting otherwise constant darkness at different times may shift the phase of a circadian rhythm in different ways. (a) Luminescence rhythm of *Gonyaulax* culture kept in constant darkness. Lettered arrows indicate the time of light flashes, the results of which are given in the remaining curves. (b) Light flash in subjective day does not shift rhythm. (c) Light flash early in subjective night advances rhythm. (d) Light flash late in subjective night delays rhythm. Gray bars indicate subjective night. The vertical lines are guide lines.

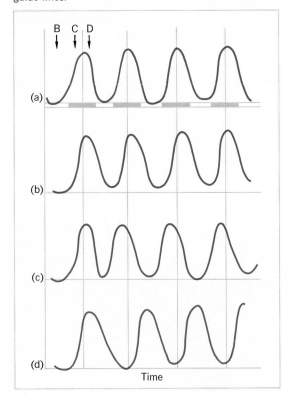

vances the phase several hours, and light coming during the second half of the subjective night delays the phase several hours (Fig. 44.3). Thus the state of the clock must vary during a normal 24-hour period, for it is relatively insensitive to light during daylight hours and sensitive to it in different ways during different parts of the night.

At first one might wonder why organisms have evolved rhythms only approaching 24 hours in length rather than rhythms exactly 24 hours long. Although the natural day is 24 hours long, the sun does not rise the same time every morning nor does it set the same time every evening. The ability of biological clocks to be reset by light and dark regimes has the advantage of permitting organisms to adapt to the changing lengths of light and dark periods.

A good clock should be temperature independent; that is, temperature changes should not affect the rate at which the mechanism works. A watch that runs fast in summer and slow in winter is of little use as a timekeeper. In this sense, biological clocks are remarkably independent of temperature, a very large temperature change usually lengthening or shortening the circadian rhythm only very slightly.

The biological clock is not completely independent of temperature effects, however. In the absence of other environmental clues, a change in temperature can reset the clock. In general, relatively high temperatures have the same effect as light, and relatively low temperatures the same effect as darkenss. Continuous exposure to extremely high or low temperatures causes organisms to lose their circadian rhythms.

# Theories of the Nature of the Biological Clock

The term *biological clock* covers a multitude of ignorance. We are in a position something like that of a man who is trying to understand how his watch works but for some reason is unable to open it. He could subject the watch to various treatments: heating, cooling, gentle tapping, hitting it with a hammer, soaking it in water, and so on. All that he could observe is whether the hands continue to go around or not, and if they continue to move, whether the rate or direction of movement is altered. From these observations he must draw conclusions about the working mechanism of the watch. Furthermore, if a particular treatment has altered the movement of the hands, he must decide somehow if he has stopped the mechanism itself or if the hands have been disengaged from the mechanism which still may be working perfectly.

We really do not know for certain that we have altered or reset a biological clock when we reset a single circadian rhythm, for if the visible rhythm (luminescence rhythm in *Gonyaulax* or activity rhythm in flying squirrels, for instance) is analogous to the hands of the clock, perhaps we merely have disengaged the circadian rhythm from the biological clock, which continues to run in perfect phase with local time. However, the literature on biological clocks is full of evidence that the phases of many diverse circadian processes in many species can be reset similarly by similar light treatments. Not only that, but a particular light treatment usually produces the same shifts in several different circadian processes within an organism or a culture of organisms. *Gonyaulax* has circadian rhythms of cell division, photosynthesis, and luminescence, each with a peak of activity at a different time during an ordinary 24-hour cycle. A single light treatment that shifts the phase of one of these shifts the phases of the others in a corresponding manner. It seems unlikely that light would have the same effect on all these processes if it acted on each of them directly; surely some would respond differently. Many biologists conclude, therefore, that the experimental treatments described above alter the biological clock itself (Fig. 44.4).

Although the few experiments described here

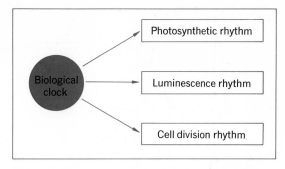

**44.4** Relation of biological clock to visible (or measurable) rhythms. If a given experimental treatment acts on the clock itself and resets it or speeds or slows it, then similar changes would be expected in all the rhythms controlled by the clock. If, on the other hand, the treatment acts directly on the individual rhythms or on the reactions (indicated by arrows) by which the clock controls the rhythms, then we would expect that these rhythms might respond differently to the treatment.

and many more have been interpreted to indicate that the experimenters have been tinkering with an internal biological clock, the nature of the clock and its location within the organism have not yet been determined. If the clock is within the organism, presumably it is a chemical one and should be susceptible to alteration by chemical treatments. Yet substance after substance has been found to have no effect on the clock. Of the very few that have some effect, actinomycin D, an inhibitor of messenger RNA synthesis, strongly inhibits the luminescence rhythm of *Gonyaulax* (Fig. 44.5). This has led to the very tentative theory that the biological clock involves DNA-mediated mRNA synthesis and well might be a negative feedback mechanism involving DNA, mRNA, and some of the enzymes synthesized in the cytoplasm. If this is so, then every eucaryotic cell may contain its own clock.

The difficulty in locating the clock might stem from the fact that the clock is not even within the living organism, but external to it. One theory holds that the small changes in the earth's magnetic field may be the timer. Once thought to be unresponsive to weak magnetic fields, several organisms recently have been found to be sensitive

399

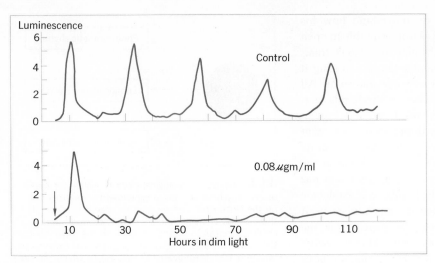

**44.5** The effect of actinomycin D on the circadian luminescence rhythm of *Gonyaulax*. Upper curve: untreated culture. Lower curve: treated culture; actinomycin D added at time indicated by arrow. The treated culture exhibited one cycle of luminescence and then became arrhythmic. (M. W. Karakashian and J. W. Hastings, "The Inhibition of a Biological Clock by Actinomycin D," *Proc. Nat. Acad. Sci.* U.S.A. 48:2130–2137 (1962).)

to them. This and some additional evidence support the external clock theory, which is defended enthusiastically by some biologists and rejected with equal enthusiasm by others. Both theories require a great deal more experimental work before either one can be accepted, or perhaps both will be joined in a common theory.

## Significance of the Clock

Regardless of their points of view on the location and nature of the biological clock, biologists generally agree that it does exist. Its importance in our daily lives is just beginning to become appreciated. We all have certain times of day when we feel better than at others. Should we plan to do our most important work at that time of day? Rats given light from 7 A.M. to 7 P.M. and darkness from 7 P.M to 7 A.M. are much more susceptible to damage from X-rays at 2 A.M. than at other times. Do human beings vary in their susceptibility to radiation, poisons, or even irritating acquaintances throughout the day? Is there a better chance of recovery from surgery performed at one time of day? Some insects have circadian rhythms of susceptibility to insecticides; perhaps the quantities of these substances used by farmers and foresters could be reduced if applications were made at the appropriate time of day.

Travelers whose biological clocks still are in phase with local time in their home towns often experience fatigue and some general discomfort for a few days after a long east or west flight. At the very least, they find they are awaking too early if the trip was toward the west or awaking too late if the trip was toward the east. During the first few days, their biological clocks gradually shift to the new local time as they are reset by the new light and dark schedule. Businessmen and statesmen now frequently are cautioned to make no important decisions the first day or two after a long international trip. Is the efficiency of airline pilots impaired if they accommodate their activities to local time at every stopover? Should a pilot maintain a daily activity rhythm that is always in phase with the city in which he spends most of his time? Noncircadian schedules such as eight hours of activity alternating with four hours of sleep were found unsatisfactory by early astronauts. Later astronauts maintained schedules that approximated those of their earthbound fellows.

In some species, some events occur only at a

400

certain time of year—reproduction, spring and autumn migrations, onset and breaking of dormancy, dropping of leaves—and their occurrence at the wrong time of the year could be disastrous for many species. The next chapter discusses the role of biological clocks in annual rhythms.

# Questions

**1.** Define the following terms:
circadian rhythm
biological clock
**2.** Give some examples of circadian rhythms.
**3.** In what respect are biological clocks independent of environmental factors such as light and temperature?
**4.** In what way do environmental factors influence biological clocks?
**5.** Why is it important that temperature not affect the rate at which the biological clock runs?
**6.** How could you easily advance the phase of an organism's circadian rhythm? How could you delay it?
**7.** Can you explain "jet lag"?
**8.** What is the value of having a biological clock?

# Suggested Readings

Aschoff, J. "Circadian Rhythms in Man," *Science* 148: 1427–1432 (1965).

Brown, F. A., Jr., J. W. Hastings, and J. D. Palmer. *The Biological Clock, Two Views.* New York: Academic Press, Inc., 1970.

Bünning, E. *The Physiological Clock,* rev. 2nd ed. New York: Springer-Verlag, 1967.

Clayton, R. K. *Light and Living Matter,* Vol. 2. New York: McGraw-Hill Book Company, 1971.

Hawking, F. "The Clock of the Malaria Parasite," *Scientific American* 222(6): 123–131 (June 1972).

Palmer, J. D. "How a Bird Tells the Time of Day," *Natural History* 75(3): 48–53 (March 1966).

Sweeney, B. M. *Rhythmic Phenomena in Plants.* New York: Academic Press, Inc., 1969.

Wurtman, R. J., and J. Axelrod. "The Pineal Gland," *Scientific American* 213(1): 50–60 (July 1965).

# 45

# The Biological Clock and Annual Rhythms

1 Day and night lengths control the seasonal timing of many annual processes.

2 The biological clock plays a role in annual rhythms.

The study of biological clocks received its initial impetus from an examination not of daily rhythms, but of an annual one. In 1920 it was learned that the length of daylight controls flowering in the Maryland Mammoth variety of tobacco. This led to the study of the effects of light on other plants, and over the years it was discovered that flowering plants fall into three general categories: short-day plants, long-day plants, and day-neutral plants.

A **short-day plant** flowers if the length of the light period each day is shorter than a critical length, but not if it is longer (Fig. 45.1). The critical length of the light period varies with the species, but for many short-day plants it is about 15 hours.

A **long-day plant** flowers if the length of the light period each day is longer than a critical length, but not if it is shorter (Fig. 45.2). The critical length for many long-day plants is about nine hours.

**Day-neutral plants** flower under either short or long light periods.

In the field of biological clocks, the term *short-day* means a short light period—*not* a calendar day somehow shortened to less than 24 hours. Corresponding to this, a *long day* is a long light period. In experiments short days often are arranged to be six to eight hours long because they permit short-day plants but not long-day plants to flower. Long days often are arranged to be 16 to 18 hours long because they permit long-day plants but not short-day plants to flower.

Responses of plants and animals to short and long days is called **photoperiodism.**

## Reproduction and Day Length

One thing that becomes apparent with only a little thought is that in a 24-hour calendar day, as the light period varies, the dark period varies inversely; that is, if the light period is long, the dark period is short, and vice versa. Researchers soon realized that the length of the dark period and not

**45.1** Chrysanthemum is a short-day plant. The plant at the left was grown under short days (8L:16D), the plant at the right under long days (16L:8D). (Courtesy of Ornamentals Branch, Plant Science Research Division, Agricultural Research Service, U.S.D.A.)

the length of the light period might be the important factor in flowering.

In one experiment both short-day plants and long-day plants were exposed to short days and long nights, but the middle of the night was interrupted by a short light exposure. Both groups of plants reacted as if they had been given long days and short nights: the short-day plants remained vegetative, and the long-day plants flowered (Fig. 45.3). From the results, it was decided that the length of the dark period was more important than the length of the light period in determining

whether a plant will flower. Had the terms *short-day plant* and *long-day plant* not been so firmly entrenched in the biological literature by this time, their names well might have been changed to long-night plants and short-night plants, respectively. However, we will see later that neither set of names is particularly appropriate, but custom has decreed the continued use of the old names.

Animals display similar responses to day length. Birds such as house finches that breed in spring have very small ovaries and testes in the winter. At this time the reproductive organs produce neither eggs nor sperms. As the days grow longer, the gonads enlarge and produce fully mature eggs and sperms. With respect to development of gonads, the birds respond much like long-day plants. If the

**45.2** Henbane is a long-day plant. The plant at the left was grown under short days (8L:16D), the plant at the right under long days (16L:8D). Compare the flowering responses of these plants with those of the chrysanthemums in the previous figure. (Courtesy of Ornamentals Branch, Plant Science Research Division, Agricultural Research Service, U.S.D.A.)

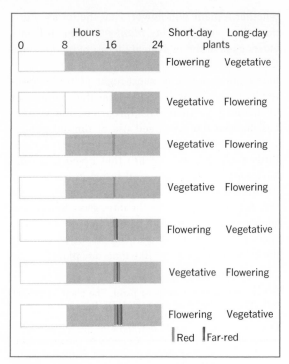

**45.3** Responses of short-day and long-day plants to different lighting schedules. Notice that the two types of plants give opposite responses. White bars: light periods. Gray bars: dark periods.

birds are kept under artificial short days and long nights early in the year, the gonads do not develop or do so extremely slowly; but a brief light interruption of the long nights gives a typical long-day reaction. In migratory species, the lengthening of the light period also brings about a migratory urge. In addition, it determines the direction of the migratory urge (north in many Northern Hemisphere species).

Following the breeding season, the birds enter a refractory period during which time the gonads regress in size, and no amount of exposure to long days can induce enlargement. The refractory period begins in July or August and continues until October or November. Termination of the refractory period is essential for the continuation of the annual cycle, and the termination is a short-day response. Birds kept under artificial long days after summer remain in the refractory period. Short days in autumn bring about migratory urge in migratory species and also determine its direction (south in many Northern Hemisphere species).

We have considered only reproduction in flowering plants and birds, but it is likely that not only reproduction, but other seasonal processes in many species of plants and animals are controlled at least in part by day length. Many insect larvae form cocoons or other dormant structures in autumn in response to the short days, and the breaking of dormancy in spring is a long-day response. In many plants, short days bring on the onset of dormancy of buds in autumn, and long days trigger the renewal of growth in spring. The autumnal dropping of leaves is a short-day response; trees near street lamps often retain their leaves longer than do trees of the same species that receive no nighttime illumination.

This discussion should not imply that day length influences all reproductive processes. Some varieties of corn, tomatoes, onions, and cotton are day neutral; that is, the plants flower regardless of day length. In some of these cases particular temperatures are required before flowering is initiated. Some animals with very short reproductive periods, such as small rodents or insects, produce several generations from spring to autumn as long as the temperature is suitable and food plentiful.

## Phytochrome and Flowering

In the night-interruption experiment described earlier, white light was used. Not all the components of white light reverse the effects of long nights. Later experiments revealed that only red light of wavelengths of about 580 to 660 nanometers is effective. A flash of red light in the middle of a long night has the same effect as a flash of white light.

The suppression of flowering in a short-day plant by a red-light interruption can be reversed if it is followed immediately by a short exposure to light of wavelengths of 720 to 760 nanometers. This light is called far-red; and it is just beyond the end of the spectrum visible to most human eyes. A red, far-red interruption followed immediately by a brief exposure to red inhibits flowering of short-day plants. If the sequence red, far-red, red, far-red is given, there is no inhibition provided that each flash follows the preceding one immediately or with only a short intervening period. Thus red and far-red can be alternated several times, with the last irradiation being the one that determines whether there is to be flowering or not. Figure 45.3 summarizes these results. The illustration also gives the responses of long-day plants to the same treatments, and it is interesting to note that their responses are the opposite of those of short-day plants.

To bring about a chemical reaction, light must be absorbed by a pigment; in this case the receptor pigment is **phytochrome.** It exists in two forms: a blue pigment, which absorbs red light and is designated $P_r$, and a pale yellowish, almost colorless pigment, which absorbs far-red light and is desig-

nated $P_{fr}$. The reversibility of the red and far-red reactions depends on their ability to change $P_r$ to $P_{fr}$ and $P_{fr}$ to $P_r$. The absorption of red light by $P_r$ converts this pigment immediately to $P_{fr}$; similarly, when $P_{fr}$ absorbs far-red light, it is converted to $P_r$. The conversions are virtually instantaneous. White light also causes the immediate conversion of $P_r$ to $P_{fr}$ but not the reverse reaction, even though white light also contains some far-red light. In darkness, $P_{fr}$ spontaneously slowly converts to $P_r$. Figure 45.4 summarizes these reactions.

## Phytochrome and the Biological Clock

Investigators once thought that the dark conversion of phytochrome might be the biological clock that "tells" the plant how long the night is. The theory ran as follows: $P_{fr}$ catalyzes a reaction that suppresses flowering in short-day plants. Throughout all the daylight hours, phytochrome would exist only in the $P_{fr}$ form because of the action of white light. At night the conversion to $P_r$ would occur slowly. During short nights, the conversion to $P_r$ would not be complete, and all night long some $P_{fr}$ would be present and inhibit flowering. if the night is sufficiently long, however, at some time the $P_{fr}$ concentration would fall below a critical level, and for the rest of the night there would be no inhibition and flowering would result. However, the dark conversion is temperature-dependent; the higher the temperature, the faster the conversion. Because biological clocks are virtually temperature-independent, the phytochrome conversion could not be the clock.

A more serious objection to the dark conversion as the clock arose when short-day plants were subjected to a series of light and dark cycles of varying lengths ranging from 18 to 72 hours. In this experiment each cycle had 8 hours of light, but the lengths of the dark periods varied. One group of plants received 8L : 10D cycles, another 8L : 16D cycles, another 8L : 22D, and so on up to 8L : 64D. The results are shown in Fig. 45.5. It is obvious that just a long night is not sufficient to

**45.4** Interconversion of the two forms of phytochrome, $P_r$ and $P_{fr}$. All except the dark conversion are virtually instantaneous.

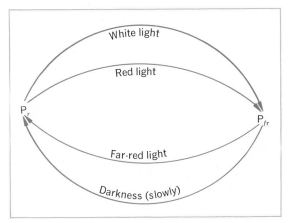

405

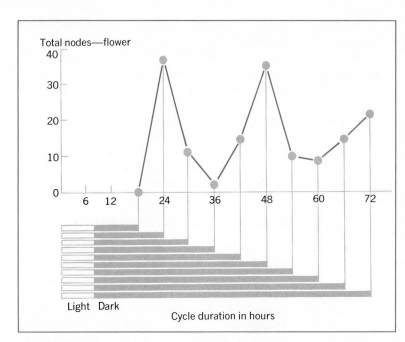

**45.5** The effect of cycle length on flowering in Biloxi soybean, a short-day plant. Ten groups of plants received light–dark cycles of varying lengths. All light periods were eight hours long; dark periods were as indicated in the horizontal bars below the graph. Notice the circadian nature of the response curve. Notice also that a long dark period does not ensure that a short-day plant will flower; an eight-hour "day" coupled with a 28-hour "night" (36-hour cycle) produced almost no flowering. (Karl C. Hamner, "Photoperiodism and Circadian Rhythms," In *Biological Clocks, Cold Spring Harbor Symposia on Quantitative Biology* 25: 269–277 (1960). Copyright © 1961 by the Biological Laboratory, Long Island Biological Association, Inc.)

cause flowering. In this experiment plants flowered with dark periods of 16 and 40 hours, but at an intermediate value (28 hours of darkenss) hardly any flowers were initiated. The graph shows a circadian response by the plants; peaks in the curve occur 24 hours apart (cycle lengths of 24 and 48 hours); troughs also occur 24 hours apart (cycle lengths of 36 and 60 hours).

**45.6** The effect of cycle length on testis enlargement (a long-day response) in the house finch. Six groups of birds received light–dark cycles of varying lengths. All light periods were six hours long; dark periods were as indicated in the horizontal bars below the graph. Notice that this curve, like the one in Fig. 45.5, has a circadian nature but appears to be inverted. The curve for flowering response in long-day plants is very similar to this curve. (Graph by W. Hamner. In K. C. Hamner, and A. Takimoto, "Circadian Rhythms and Plant Photoperiodism," *American Naturalist* 98: 296–322 (1964). Copyright by University of Chicago Press 1964.

406

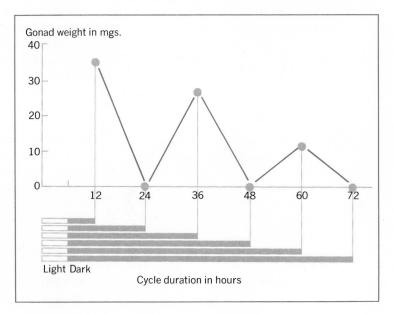

A similar experiment performed with house finches produced similar results, but the curve is reversed (Fig. 45.6). In this case it was a long-day response that was studied—the enlargement of testes. The light periods were 6 hours long in this experiment, and the total cycle lengths ranged from 12 to 72 hours. On cycles of 12, 36, and 60 hours, the testes enlarged, but on cycles of 24, 48, and 72 hours there was no enlargement.

These experiments demonstrate that night length alone is not any more important in flower induction or gonad enlargement than is day length alone. Notice that it is even possible to get a long-day response with a short light period and a long light period. The 36- and 60-hour cycles (6L:30D and 6L:54D, respectively) both produced testis enlargement.

It would appear from these experiments that some cyclic process—probably the biological clock or some process immediately controlled by it—occurs with a circadian rhythm in organisms, and that if light shines on the organism when the rhythm is in a particular phase a particular result will occur; if it shines during another phase, another result will happen. For plants to flower, for example, either $P_r$ or $P_{fr}$ must be present (or perhaps absent) when the biological clock is in a particular phase. Recall the experiment described in the preceding chapter which indicates that the state of the clock varies with the time of day. The optimum time of light exposure varies with short-day and long-day processes.

## Phytochrome and Florigen

The portion of the flowering plant receptive to light and dark cycles is not the stem tip that produces the flower but rather the leaf. Under the appropriate light schedule, the phytochrome reactions presumably initiate the production of a flowering hormone. Named **florigen,** this hormone has not been isolated, nor is its chemical nature known. It remains as elusive as the biological clock itself. Nonetheless there is evidence that it exists, for some messenger must span the distance between the leaves and vegetative buds. Furthermore, a short-day plant will flower under long days, if it is grafted to a short-day plant that receives short days. Evidently florigen even moves through grafts from plant to plant, initiating the floral response.

Upon reaching vegetative stem tips, florigen turns on the genes for flowering. After its arrival the amount of RNA in the stem tip increases greatly, and an increase in the amount of protein soon follows. At least some of this protein must be the enzymes catalyzing the reactions that produce a flower rather than more stem and leaf tissue.

## A Comparable System in Animals

A receptor pigment comparable to phytochrome undoubtedly exists in animals, but it has not been discovered yet. Because blinded animals of many species are unresponsive to day length, the pigment well may be a retinal one, although there is evidence in some species that light penetrating the skull initiates responses. As in plants, red light is most effective in evoking responses.

The usual pathway of gonadal response to day length seems to be partly nervous and partly hormonal. Sympathetic nervous system connections between the eyes and the hypothalamus (via the central nervous system) probably convey to the hypothalamus information regarding day length. When the light period reaches a critical length in spring, the hypothalamus is stimulated to secrete its releasing factors specific for gonadotropins into the hypothalamopituitary portal system (Chapter 36). When the releasing factors reach the anterior pituitary, this latter gland secretes into the general circulation its gonadotropins, which in their turn stimulate enlargement of the gonads.

Although the nature of the biological clock still remains to be discovered, its existence seems to be well established. It is likely that the mechanism evolved relatively early in the history of life on earth, for all but the most primitive organisms have biological clocks. The responses of both plants and animals to light and dark cycles are similar. It is thought, therefore, that the clock is an ancient one—one that for millions of years has taken advantage of the perpetual alternation of day and night and the continuing progress of the seasons in the economical regulation of biological processes.

## Questions

**1.** What is the relationship of biological clocks to annual rhythms?

**2.** Give examples of short-day responses in plants and animals.

**3.** Give examples of long-day responses in plants and animals.

**4.** What are short-day plants and long-day plants? In what respects are their names inadequate?

**5.** Why is it that when some northern species of plants are transported to southern regions, they fail to reproduce?

**6.** How does phytochrome function in flowering? Why is it unlikely that phytochrome is part of the biological clock?

**7.** Explain why the following three light regimes will cause a long-day response (flowering in long-day plants

or testis enlargement in house finches) even though only one of them as a long light period.

$$16L : 8D$$
$$8L : 7\tfrac{1}{2}D : 1L : 7\tfrac{1}{2}D$$
$$8L : 28D$$

**8.** The light-sensitive portion of an organism is not necessarily the part that responds to a particular light regime. What is known about the nervous or hormonal mechanisms that must be involved here?

## Suggested Readings

Butler, W. L., and R. J. Downs, "Light and Plant Development," *Scientific American* 203(6): 56–63 (December 1960).

Farner, D. S. "Time Measurement in Vertebrate Photoperiodism," *American Naturalist* 98: 375–386 (1964).

———. "Control Systems in Bird Reproduction," *Natural History* 75(7): 22–27 (August–September 1966).

Naylor, A. "The Control of Flowering," *Scientific American* 186(5): 49–56 (May 1952).

Pengelley, E. T., and S. J. Asmundson. "Annual Biological Clocks," *Scientific American* 224(4): 72–79 (April 1971).

Salisbury, F. B. "The Flowering Process," *Scientific American* 198(4): 108–117 (April 1958).

———. *The Flowering Process.* New York: Macmillan Publishing Co., Inc., 1963.

Sweeney, B. M. *Rhythmic Phenomena in Plants.* New York: Academic Press, Inc., 1969.

# Evolution

# Part X

In the first half of the nineteenth century, an English naturalist named Charles Darwin conceived and refined his theory of the origin of species from pre-existing species. He was not the first man to propose a theory of evolution, but he was the first to document his views with observations of many species of plants and animals from around the world and with many careful experiments performed over a period of more than 20 years. The thoroughness of his work convinced many biologists that species were not all created at one time, but that new species continue to evolve from pre-existing species.

Darwin's theory can be summarized in a few statements:

1. In each species more offspring are produced than can possibly survive.
2. Within each species the individuals vary, and some of their variations are inherited.
3. Those offspring with favorable variations are the most likely to survive and pass these qualities to their own offspring; this process he called **natural selection.**

One deficiency in Darwin's theory was that he could not demonstrate the physical inheritance of traits, nor could he explain how variations originated. Although Darwin and Mendel were contemporaries, Darwin never learned of the Austrian monk's work, which would have helped to confirm his own theories. Both men died before the rediscovery of Mendel's work by the scientific community and before the discovery of mutations, which provide the variation on which evolution is based.

## Mutation

Few processes, if any, work perfectly all the time. Duplication of DNA molecules, which has occurred perfectly untold trillions of times since

# 46
# Mutation and Evolution

1 Evolution proceeds by natural selection; those individuals with favorable genes are more likely to live longer and produce more offspring than those with unfavorable genes.

2 Mutations are changes in the nucleotide sequences of DNA molecules, and these changes alter the amino acid sequences of the proteins coded for by the DNA.

3 Alteration of the amino acid sequence in an enzyme molecule may cause it to lose or modify its enzymatic properties.

4 New species evolve from pre-existing species by the accumulation of mutant genes.

5 Man has become an effective agent influencing the direction of evolution of other species.

411

life appeared on earth, is nonetheless subject to occasional error. Some of these errors are harmful, some are beneficial.

## Mutation and DNA

Any change in a gene is called a **mutation.** In terms of the amount of material that is altered, a mutation is a small thing; but in terms of its effects, it can be so important as to determine the health or illness of the individual carrying it or even his life or death.

Mutations may occur during the duplication of DNA molecules. As the nucleotides forming the new strand assemble on the old strand serving as a template, a slight error sometimes occurs. A wrong nucleotide may move into position, a nucleotide with adenine taking the place of one with guanine, for instance. An extra nucleotide may slip between two that "belong" there, or a nucleotide may fail to move into its proper place. In all these cases the order of at least some nucleotides changes. Assuming that the change is not so damaging as to be lethal to the organism, the mutant form of DNA then faithfully duplicates thereafter, for the error is not corrected.

We might draw an analogy with a typist making many copies of a letter, each time copying the preceding letter rather than always copying the original. If she is an excellent typist she makes many perfect copies, but eventually she makes an error and does not notice it. Because she uses this imperfect letter as a model for the next, all the rest of the letters will carry the same error. If she makes another error, all subsequent letters will bear two errors and so on. In a similar way, living things have accumulated mutant genes since the origin of life. These mutations account for much of the genetic diversity among living things.

Not only is a mutation faithfully duplicated in all subsequently formed DNA molecules, it also is reflected in the structure of mRNA formed from all the mutant DNA molecules. This may alter the enzyme molecules coded for by the mRNA, and

finally the metabolism of the cell can be affected. Depending on the particular error duplicated in the mRNA, the consequences may be trivial or extremely serious. If a typist types "recieved" rather than "received" in a letter, in all likelihood the reader will understand and act exactly as he would if the word had been spelled correctly. On the other hand, if the typist types "$1,000" in place of "$10.00," the recipient probably will act much differently than he would otherwise. So, too, the cell may be able to read the correct message in an altered mRNA or it may not.

## Effect of Mutation on Enzymatic Activity

Let us assume that the DNA molecule that gives rise to the mRNA molecule in Fig. 43.7 is the one affected by a mutation. If, as a consequence, the first triplet in the mRNA is changed from AAU to AAC (an exchange of cytosine for uracil), the cell will be unaffected, for both of these triplets code for the same amino acid, asparagine. On the other hand, if the first triplet become CAU, the resulting polypeptide chain will begin with histidine rather than asparagine. This might have considerable effect on the enzymatic properties of the final protein molecule, though not necessarily so. Only certain amino acids in an enzyme molecule are responsible for its three-dimensional geometry or form its active portion. If asparagine in this position were one of these, the altered protein with histidine in its place probably would have no enzymatic properties, and the cell would be unable to perform the chemical process that would have been catalyzed by the unaltered protein. The seriousness of this situation depends on the importance of that chemical process. The lack of amylase may be of little consequence to a cell well supplied with glucose, but it would be disastrous to one with starch as its only food supply.

An example of the seriousness of some single-nucleotide substitutions is sickle-cell anemia (Chapter 42). Persons with this disease have abnormal hemoglobin that differs from normal hemoglobin only by the substitution of valine for glutamic acid in the protein portion of the molecule. A glance at the genetic code in Table 43.2 shows that this can come about by the substitution of uracil for adenine in the appropriate triplet, GAA becoming GUA or GAG becoming GUG. A small change indeed, but one serious enough to cause premature deaths in homozygotes and beneficial enough to increase the resistance to malaria in heterozygotes.

The addition or deletion of a nucleotide can alter a few or many amino acids in the polypeptide, depending on its position. Let us imagine that the fifth nucleotide, the one with cytosine, is deleted from the mRNA molecule in Fig. 43.7. Because the cell "reads" the genetic code in triplets beginning from one end of the mRNA molecule, the genetic message has changed from

> AAUACUGUACUGCCUGAAA---

to

> AAUAUGUACUGCCUGAAA---

which when translated into a polypeptide becomes

asparagine-methionine-tyrosine-
cysteine-leucine-lysine---.

This is much different than the original

asparagine-threonine-valine-leucine-
proline-glutamic acid---,

and cannot function as the correct enzyme, for nearly the entire length of the molecule is altered. Had the deletion occurred near the other end of the mRNA molecule, only the last few triplets would have been misread, and so only the last few amino acids would have been changed. The chances for serious consequences then would be somewhat reduced.

The addition of a nucleotide produces results very much like those of a deletion; it causes mis-reading of the genetic code beyond the point of the addition. The exact misreading is different, but drastic change in the polypeptide molecule is just as likely to render it useless as an enzyme.

## Causes of Mutations

A number of chemical compounds and several types of radiation possess mutagenic properties. Chemicals that cause mutations generally are highly reactive substances such as mustard gas and the peroxides. Generally the higher the energy content of the radiation, the more harmful it is; these are the radiations with short wavelengths, especially those with wavelengths shorter than those of visible light, which has only very slight mutagenic capability. These high-energy radiations include gamma rays, X-rays, ultraviolet radiation, and cosmic rays, the sources of which include radioactive materials, X-ray machines, some electronic equipment such as color television sets, and sources in outer space.

Organisms always have been exposed to a certain amount of radioactivity, for the earth naturally contains such radioactive elements as, for example, radium and uranium; and cosmic rays and a little ultraviolet radiation reach the earth from space. The mutations caused by these agents must have provided some of the variation that characterizes living things today.

# Evolution

## Natural Selection

Mutations really are the raw material of **evolution.** If a mutation is harmful, it reduces the reproductive capacity of its possessor and so will be passed on to fewer offspring than would a harmless gene. On the other hand an organism with a beneficial gene is likely to have more offspring, and

413

consequently the gene spreads to more organisms in each succeeding generation. For instance, a deer born with some hereditary defect in his legs that prevents him from running as swiftly as his fellows will have a shorter life than they, for he is the most likely one to become a wolf pack's dinner. His shortened life prevents him from having as many offspring as the other deer—or perhaps he does not have any. As the gene is passed to fewer and fewer individuals, it becomes rarer with each generation and finally may disappear from the population. The more harmful the gene, the faster its rate of disappearance; if it is lethal to the embryo or very young offspring, it cannot be passed on to another generation and so stops there. Deleterious recessive genes disappear from a population much more slowly than dominant genes, for the former can be passed from generation to generation in the heterozygous condition without manifesting themselves.

A deer born with the ability to run more swiftly than his fellows well may live to a ripe old age and produce many offspring. If they inherit his ability, they, too, will live long and pass the beneficial gene to a large number of offspring.

By this method, called **natural selection,** a species improves and becomes better adapted to its environment. With each succeeding generation the proportion of favorable genes tends to increase, and the proportion of unfavorable genes tends to decrease. In some cases unfavorable genes disappear completely, and in others the incidence of the gene may remain more or less constant in the population, for the mutation rate producing more of these genes equals their rate of disappearance from the population. In other cases the harmful gene may also have beneficial effects and so remain in the population; the example of the sickle-cell anemia gene, which increases resistance to malaria, is one such. Its incidence remains constant in malarious regions of Africa, but it is disappearing slowly from American Negroes, who descend from African slaves, for the incidence of malaria in the United States is so low that the gene

414

confers no benefit here. Approximately 9 per cent of American Negroes are heterozygous for the gene, but in western Africa, where the ancestors of many American Negroes lived, about 20 per cent of the native population is heterozygous.

With some environmental changes, genes that once were harmful or neutral in their effect may become beneficial. Herein lie the benefits of diploidy and sexual reproduction. Diploidy permits the retention of at least a small proportion of many recessive genes in a population even though some of them might be less than favorable. Sexual reproduction results in the occasional appearance of homozygotes who, in new environmental conditions, might be the most favored.

When, over many generations, many mutations have accumulated and have changed the characteristics of a population of organisms considerably, we say that a new species has evolved. The amount of change is very little from generation to generation, and it is difficult to say exactly when a new species has evolved. There is no absolute criterion for how much change must have occurred, and biologists themselves often disagree on whether two somewhat different groups of organisms represent two distinct species or just variations within a single species. Usually members of different species will not breed with each other, or if they do they will not produce fertile offspring; but this is not always true.

## Isolation and Evolution

Evolution within a group often is hastened if that group is physically separated from a larger group of the same species with which it otherwise might breed, especially if the separation puts the two groups into different environments. For instance, if a few birds were to be transported from their native homeland on a continent to an oceanic island from which they could not escape, they would have the opportunity to breed only with each other and could not exchange genes with the other members of their species left behind. If the

**46.1** Eastern Pacific Ocean near northern South America. The Galápagos Islands lie 600 miles west of Ecuador, the presumed home of the ancestors of the Galápagos finches. Organisms living on the islands are almost completely isolated from those on the mainland by the intervening expanse of ocean.

environment on the island differs from that on the continent, the displaced birds eventually are likely to evolve new anatomical, physiological, and behavioral characteristics that better adapt them to their new situation—assuming that the environment is not so severe as to kill them all.

Several examples similar to this inspired Charles Darwin to develop his theory of evolution. On a trip around the world he stopped at the Galápagos Islands, which lie in the Pacific Ocean about 600 miles west of the coast of South America (Fig. 46.1). There he found several species of finches which he believed to be related to an ancestral species still living in South America. He assumed that storms occasionally had carried some finches to the Galápagos Islands where they managed to survive, different groups evolving in different ways on the individual islands. The ancestral finches were seed-eating birds that lived on the ground. Some of the Galápagos Islands species retained

these habits, but others gradually developed specializations that enabled them to live in trees and to eat insects. The change from herbivorous to carnivorous habit was accompanied by evolutionary change in size and shape of beaks which made these birds more efficient insect catchers. One species, the woodpecker finch, even evolved a new behavior; after pecking a hole in a tree, the bird holds a cactus spine in its beak and inserts the spine into the hole to search for insects (Fig. 46.2).

Several species of tortoise, each unique to one of the Galápagos Islands, has specializations appropriate to that island. For instance, short-necked tortoises live where there is abundant vegetation close to the ground. Those that live on arid islands where they must rely on tall cactuses for food have long necks (Fig. 46.3).

Many other plant and animal species are unique to only one island in the Galápagos group, though several are present on two or more islands. Many are found no where else in the world but are related to species, presumably their ancestral species, in South America. Without isolation on their islands, the evolution of these species undoubtedly never would have happened.

Isolation need not always be so great as a 600-mile expanse of ocean. A mountain range or even a large hill may be sufficient in the case of small animals that travel only small distances at a time and cannot survive the cold temperature that comes with altitude or can find no suitable food to sustain them on the trip over a mountain range. In Hawaii some species of fresh-water snails are unique to single valleys.

Reproductive isolation occurs as well. If some members of a species are sexually active at a different time of day or year, the two groups no longer can interbreed and in time may evolve into different species.

The greater the variety of genetic material in a population, the more likely it is to adapt to a changing environment. Without a pool of genetic diversity, a species may be unable to adapt and so may become extinct. For this reason, when most

415

**46.2** Six of fourteen species of Galápagos and Cocos Island finches, all of which evolved by natural selection from a single ancestral species in South America. *Camarhynchus pallidus*, the tree-dwelling woodpecker finch, eats insects that it obtains by inserting a cactus spine into the bark of trees. *C. psittacula* is another tree-dwelling, insect-eating finch, but it does not utilize spines in catching food. *Certhidea olivacea*, the warbler finch, is a small tree-dwelling, insect-eating bird that resembles warblers more than it does other finches. *Pinaroloxias inornata* is an insect-eating, tree-dwelling finch found only on Cocos Island, about 600 miles northeast of the Galápagos Islands. *Geospira scandens* is a ground-dwelling, cactus-eating finch. *G. magnirostris* is a ground-dwelling seed-eating finch.

**46.3** The short-necked tortoise lives on Albemarle Island in the Galápagos Islands; this island has a moist climate with abundant vegetation near the ground. The long-necked tortoise lives on Indefatigable Island, a more arid island with sparser vegetation; the tortoise must reach upward to obtain sufficient vegetation.

individuals of a species are killed, there is a loss of genetic variety, and the few remaining individuals may not carry a large enough assortment of genes to adapt to any future changes in the environment. Very small numbers also lead to inbreeding resulting in increased homozygosity, which usually is associated with a decline in vigor. Some endangered species of wild animals still may be in danger of extinction even after they become protected by law, for their low numbers and lack of genetic variety may preclude their survival.

Small numbers do not necessarily lead to extinction, however. The few rabbits and cactuses introduced into Australia (Chapter 4) established themselves and succeeded in populating large portions of the continent. The English sparrows that live throughout the United States descend from a few pairs brought to the Boston area about 1850.

## Artificial Selection

Human beings always have influenced the evolution of organisms with which they came in contact. Sometimes the influence was more or less

**46.4** A few of the more than 100 breeds of dog, all belonging to the species *Canis familiaris*. The breeds are the result of artificial selection by human beings.

417

purposeful, sometimes inadvertent. Artificial selection does not differ much in principle from natural selection, but in the case of artificial selection human beings constitute a significant part of the environment that influences the direction of evolution. For instance, the environment of dairy cattle includes men who wish to have cows capable of producing a great deal of milk and so mate high-yielding cows with bulls that are the sons of high-yielding cows or that have the reputation of siring such daughters. Such crosses tend to produce better milk producers. Very few such crosses would happen among wild cattle, which produce much less milk than their domestic sisters. Beef cattle, on the other hand, have been selected by men for their meat production. The many varieties of several species of domestic plants and animals owe their existence to artificial selection by man. Man has directed the evolution of well over 100 breeds of dogs, ranging from hunting dogs to sled dogs to lap dogs (Fig. 46.4). Cabbage, kale, Brussels sprouts, broccoli, and cauliflower are all varieties of the same species (Fig. 46.5). These and many other artificially selected varieties of economically important plants and animals have no existence in nature. Other human activities inadvertently influence evolution, and sometimes they produce some undesirable results.

The accumulation of genetic changes often is an extraordinarily slow one by human standards. Yet it is not impossible for humans to observe the process of evolution during their lifetimes, and in this century several cases of evolution caused by human activity have been recorded. Generally, those organisms that have short life cycles are most likely to evolve rapidly, for selection can act on several generations within a short period of time. Bacteria, some of which reproduce in less than a half-hour under favorable conditions, and insects, some of which produce several generations in a summer, have demonstrated their ability to evolve new strains well adapted to new environmental conditions.

When first discovered, penicillin was hailed as

a boon that soon would eliminate several serious diseases for which no reliable cures heretofore had been found. A dose of penicillin soon raised from their beds persons ill with strep throat, gonorrhea, syphilis, some types of pneumonia, and other diseases that often had proved fatal. The rejoicing was not to continue unalloyed, however, for penicillin-resistant bacteria began to appear with greater and greater frequency in patients, and the doses of penicillin required to cure an illness began to increase. The reason for the development of penicillin-resistant strains is not hard to imagine. A very small percentage of the natural bacterial populations possess the enzyme penicillinase, which catalyzes the digestion of penicillin to compounds harmless to the bacteria; these bacteria are resistant to penicillin and multiply in its presence. Then as doctors began injecting penicillin into their patients, the susceptible bacteria died, leaving only the resistant few. The survivors soon multiplied, producing offspring all of which inherited the resistance. (Bacteria reproduce primarily by cell division, and so all the offspring are genetically like their parents. Mendelian ratios do not apply here.)

In the early days of penicillin use, patients who seemingly had recovered suddenly relapsed and died. When doctors learned that their treatments had selected for resistant bacteria, they had to raise their initial dosages to kill all but a few of the most resistant bacteria. Then the patient's natural defenses (Chapter 34) usually could take care of the few survivors before their offspring became numerous enough to threaten his life again. However, the technique has not prevented the spread of penicillin-resistant bacteria from person to person. People sometimes receive an initial infection of a species of bacteria resistant to all but very high doses of penicillin. When the required doses become so high that they harm the patient, then penicillin becomes useless. The discovery of other antibiotics has overcome the problem partially, but strains of bacteria resistant to new antibiotics tend to evolve after exposure to them.

418

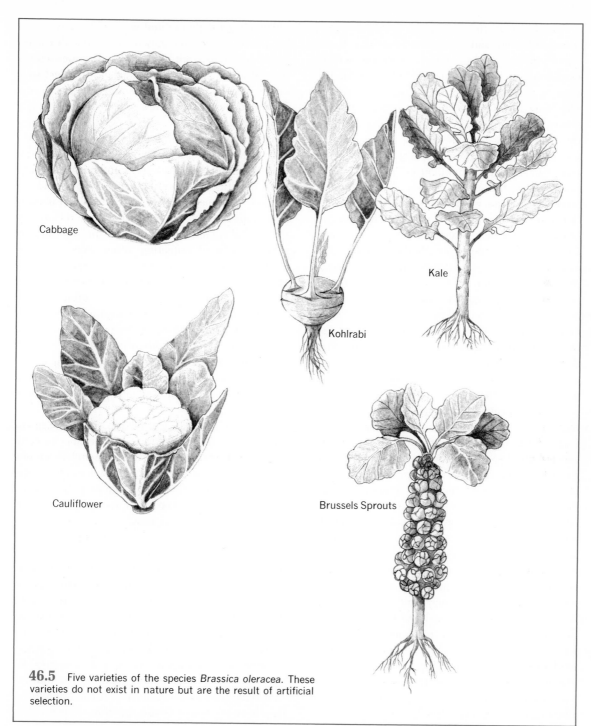

Cabbage

Kohlrabi

Kale

Cauliflower

Brussels Sprouts

**46.5** Five varieties of the species *Brassica oleracea*. These varieties do not exist in nature but are the result of artificial selection.

419

The ability of insects quickly to evolve strains resistant to DDT and other pesticides often results in an outcome directly opposite to the one intended by the users of the pesticide. For the first year or so, applications of pesticides produce a gratifying decrease in the population of insect pests that the farmer wishes to eliminate. During this time, the susceptible insects, which form the major part of the population, die; and the farmer enjoys healthy and abundant crops. But as with the bacteria, the original insect population contains a few resistant members, and these become the base of a population explosion that repopulates the fields with insects resistant to the pesticide. The farmer then uses a higher concentration of the substance, feeling that this will rid him of the new population, but he is merely repeating the process and selecting for insects with yet higher degrees of resistance. In the meantime, the pesticide, which cannot be confined to the target species, moves upward in the food chains and enters those animals—birds, for example—that prey on the target insects. In the predators, the pesticide becomes more highly concentrated (Chapter 14), sometimes causing death of the individuals receiving the pesticide and sometimes lowering their reproductive capacity. In either event the predator species has less chance for evolving a resistant strain than the insects, for the predators produce fewer generations in the same amount of time. Thus while the target species of insects is evolving more and more highly resistant strains, its predators dwindle in numbers and eventually may become extinct in the area. As the natural control by predators disappears, the farmer finds himself in the vicious cycle of applying more and more pesticide with diminishing results and with the unfortunate pollution of soil and streams.

Several new methods of controlling insects without the use of pesticides recently have been used with success against some species. Unlike the use of chemical pesticides, they have the advantage of not contaminating the environment and of affecting only the target species.

The **sterile-male method** generally is successful only in those species in which the females mate only once; it does not work well with promiscuous insects. Large numbers of males of the target species are raised in laboratories and then are irradiated sufficiently to sterilize them but not to kill or maim them. Upon being released into the environment, the sterile males compete with the fertile wild males for females. If the number of sterile males is sufficiently large, most of the females mate with them rather than with the fertile males. These females produce no offspring, and the population size drops rapidly. Several consecutive releases of sterile males can rid the area of the undesirable species without affecting other species. This method had been used to rid southern states of the screw worm fly, the larvae of which burrow into the flesh of livestock and there produce serious sores. Other insect pests currently being attacked by this method or being considered for attack are the pink bollworm and boll weevil, which damage cotton plants; the codling moth, which infects apples, pears, and walnuts; and the cabbage looper, which destroys crops of the cabbage family.

A second method of attacking single species is based on the fact that adults of some species emit a chemical attractant by which members of the opposite sex can find them. A few insects (of appropriate sex) are placed in a trap that other members of the species can enter but not leave. The trapped insects can be destroyed at the convenience of the person who checks the traps regularly. The insects themselves are not even necessary as lures; only a few drops of their attractant often is sufficient to lure prospective mates from a distance of a mile or more. Synthetic sex attractants now are commercially available for use against the pink bollworm, the cabbage looper, and the fall armyworm; the last attacks corn and other grasses.

These two methods probably will be applied to more and more species in the future. In addition to leaving the environment free of contamination, they have the added advantage of not encouraging the evolution of resistant strains. Any members of the species surviving the first attack are just as susceptible to the second as those that succumbed.

Another method of attacking a pest species is to release into the environment large numbers of one of the species preying on or parasitizing it. When used successfully, this method often results in the control of the pest species and not its total irradication, for predator and prey species or parasite and host species often achieve a balance they may maintain indefinitely. If the balance results in a population size of the pest species (prey or host) low enough that it causes only minor damage, the control is considered successful.

California citrus farmers have used several species of ladybird beetles (ladybugs) to prey on scale insects infesting citrus trees. Ladybird beetles also eat aphids and mealybugs, and homeowners sometimes obtain them for release in their gardens. Praying mantises have gained in popularity among gardeners for the same reason.

The introduction of predators and parasites into an area is not without its dangers. The introduced species may prey on or parasitize species other than the target species and so become a pest itself. To reduce the likelihood of this happening, adequate study should precede the introduction of a species into a new area.

## Overspecialization

It is a general rule that overspecialization leads to extinction. Large horns or antlers, for instance, may serve well as weapons, and up to a point the larger they are, the more effective they are. But as they become heavy and unwieldy, their disadvantages offset their advantages. Should they become extremely large, they could be burdens that lead the species to extinction.

The human species boasts relatively few specializations. Man has no fur to keep him warm and no claws or fangs to protect him from predators. His ability to run is exceeded by that of many other animals, some of which find him a toothsome morsel. He possesses, however, two specializations that enabled him to rise above a purely animal existence: a brain clever enough to devise ways of manipulating the environment for his benefit and comfort and a pair of hands capable of executing the brain's ideas. For several thousand years these two specializations have guided the progress of civilization, and in so doing they have brought us the twin problems of overpopulation and pollution. The question arises whether these are overspecializations that will lead the human species to extinction or whether the efforts of brains and hands can be devoted to solving the problems they have created.

## Questions

1. What is a mutation?
2. What consequences could a mutation have at the biochemical level? In what way could a mutation affect an organism? A species?
3. It has been said that mutation is the raw material of evolution. Explain.
4. Summarize Darwin's theory of evolution by natural selection.
5. Differentiate between natural selection and artificial selection. In what way are they alike?
6. Given the following nucleotide sequence in messenger RNA, estimate the relative seriousness of the mutations listed below it.

<u>UCA</u>G<u>UAG</u>GA---<u>AUC</u>

   a. First nucleotide changed to one with guanine.
   b. Third nucleotide changed to one with guanine.
   c. Fourth nucleotide deleted.
   d. Nucleotide with guanine inserted between the last two nucleotides.

7. There are more species of insects than of any other group of animals. Can you give any reason for this?
8. Why is a decrease in the population of a target species by the use of synthetic pesticides likely to be followed by a rise in the population after a few years?
9. In what ways are the sterile-male method and the use of sex attractants superior to the use of synthetic chemical pesticides? What are the disadvantages of these methods?

421

10. Give some evidence that evolution occurs.

11. Are mutations alone responsible for evolution? Explain.

---

# Suggested Readings

Allison, A. C. "Sickle Cells and Evolution," *Scientific American* 195(2): 87–94 (August 1956).

Colbert, E. H. *Evolution of the Vertebrates*. New York: John Wiley & Sons, Inc., 1955.

Conway, G. R. "Pests Follow the Chemicals in the Cocoa of Malaysia," *Natural History* 78(2): 46–51 (February 1969).

Deering, R. A. "Ultraviolet Radiation and Nucleic Acid," *Scientific American* 207(6): 135–144 (December 1962).

Dobzhansky, T. *Evolution, Genetics, and Man*. New York: John Wiley & Sons, Inc., 1955.

Ingram, V. M. "How Do Genes Act?" *Scientific American* 198(1): 68–71 (January 1958).

Jacobson, M., and M. Beroza. "Insect Attractants," *Scientific American* 211(2): 20–27 (August 1964).

Kettlewell, H. B. D. "Darwin's Missing Evidence," *Scientific American* 200(3): 48–53 (March 1959).

———. "Insect Survival and Selection for Pattern," *Science* 148: 1290–1296 (1965).

Knipling, E. F. "Sterile-Male Method of Population Control," *Science* 130: 902–904 (1959).

———. "The Eradication of the Screw-Worm Fly," *Scientific American* 203(4): 54–61 (October 1960).

Lack, D. "Darwin's Finches," *Scientific American* 188(4): 66–72 (April 1953).

Napier, J. "The Evolution of the Hand," *Scientific American* 207(6): 56–62 (December 1962).

Ross, H. H. *Understanding Evolution*. Englewood Cliffs, N.J.: Prentice-Hall, Inc., 1966.

Savage, J. M. *Evolution*, 2nd ed. New York: Holt, Rinehart and Winston, Inc., 1969.

Stebbins, G. L. *Processes of Organic Evolution*. Englewood Cliffs, N.J.: Prentice-Hall, Inc., 1966.

Volpe, E. P. *Understanding Evolution*. Dubuque, Iowa: William C. Brown Company, Publishers, 1967.

Wiesenfeld, S. L. "Sickle Cell Trait in Human Population and Cultural Evolution," *Science* 157: 1134–1140 (1967).

How life originated on earth is a problem that has occupied many minds in the past. Explanations have ranged from the Biblical account of a special six-day creation of the universe and every species of living thing to the ideas of the Greek philosopher Aristotle that animals continue to arise daily from mud, slime, dew, and rotting meat.

Modern scholars look upon the Biblical story of creation as an allegory glorifying God and acknowledging his responsibility for the existence of the universe (something science by its very nature can neither prove nor disprove, for it has no tools with which to probe the supernatural) rather than a step-by-step account of how this was accomplished. Biologists today generally accept that very simple organisms developed from nonliving substances at some unknown time or times in the distant past, but probably more than 3 billion years ago. From these simple living things evolved all organisms alive today, as well as many species that became extinct in the intervening years. Although biologists believe that life once came from nonliving matter, they do not believe that it continues to do so today, for, as we shall see later in this chapter, the environment is much different today than it was at that time. No evidence exists for the spontaneous generation of life from nonliving things today, and several experiments indicate that it does not occur. (See Chapter 46 for experiments by Redi and Pasteur.)

## The Origin of Living Cells

Plants are the only organisms capable of taking carbon dioxide and water and converting them into food, and for this reason animals depend completely on plants for their food. Because of this it would at first seem reasonable to assume that the first living things must have been photosynthetic, for without their products no other organisms would be likely to survive. Yet the photosynthetic

# 47
# The Origin and Early Development of Life on Earth

1 **Life is believed to have arisen from nonliving material in the far distant past.**

2 **The conditions necessary for the origin of life from nonliving matter probably do not exist today, for the metabolism of living things has changed the environment greatly.**

423

apparatus is a complex one, requiring the presence of the light-absorbing pigment chlorophyll, several enzymes, and the assembly of these substances in the correct spacial order on membranes. This would suggest that a simpler, nonphotosynthetic organism must have been the first living thing to appear. But on what could it have fed? And from what were its many parts composed if not from the organic products of a photosynthetic organism? In a very short time we have worked ourselves into a which-came-first-the-chicken-or-the-egg problem. Because each arises from the other, there seems to be no solution.

The apparent insolubility of this problem, which bothered biologists for many years, rested in part on the belief that when life first appeared the physical world contained much the same inorganic substances that it does today. It was assumed that the primitive atmosphere and oceans contained oxygen and carbon dioxide in quantities similar to those present there today. Another misconception long held was that because organic compounds are observed to be formed only by the metabolic activity of living things, this was their only possible origin.

### The Nature of the Primitive Earth

Although there is no consensus about how the earth and the other planets in the solar system formed, it now is believed that the primitive atmosphere of our earth was one that human beings and most of our fellow living things would find extremely poisonous. Besides lacking the free oxygen that we find essential, it contained methane ($CH_4$), ammonia ($NH_3$), and hydrogen ($H_2$); these gases were also dissolved in the waters of the primitive oceans. The element oxygen existed, of course, but then it did not form either molecular oxygen ($O_2$) or ozone ($O_3$). The latter is also a gas, and it exists today in some quantity in the upper layers of the atmosphere where it filters out much of the ultraviolet radiation from the sun. All the oxygen was present in combination with other elements; some of it was in water ($H_2O$) and some of it was in oxides of various elements in the earth's crust (for example, the iron oxide of iron ore and the silicon dioxide of sand).

## Chemical Evolution—Origin of Organic Compounds

It is postulated that simple organic compounds such as amino acids might have arisen by reactions between methane, ammonia, hydrogen, and water. Such syntheses require energy, and several energy sources have been suggested. With no layer of ozone to block it, a great deal of ultraviolet radiation reached the surface of the earth and well might have supplied the necessary energy. Because water also absorbs ultraviolet radiation, if these reactions were energized by this source, the reactions must have occurred in the upper layers of the oceans. Electrical discharges from lightning and heat from lava emitted from erupting volcanoes are other possible energy sources.

During the 1950's and 1960's experimenters working with various combinations of methane, ammonia, hydrogen, and water and some energy source such as ultraviolet radiation, electric sparks, or heat (at a wide variety of temperatures) obtained not only several amino acids, but polypeptides, adenine, guanine, several monosaccharides, several organic acids, formaldehyde, and hydrogen cyanide. The last two are poisonous to most living things, but like ammonia and methane, they can give rise to additional organic compounds. When some of these products were mixed together with phosphates and supplied with more energy, additional substances formed, among them ADP, ATP, ribose, deoxyribose, disaccharides, and polysaccharides. If all these substances can be formed in a test tube in the absence of any living things or their enzymes but under conditions believed to have existed once on the earth, then it seems reasonable to assume that they could and probably did form in the primitive oceans. Among these substances the reader will recognize such energy sources as the sugars, the energy-transferring molecule ATP, the constituents of enzymes and structural proteins, and at least some of the makings of DNA and RNA. In other words, many of the organic compounds we know to be the basic building blocks of life probably formed in the early ocean water, which thus became a sort of dilute soup.

## The First Cells

It is much easier to explain the probable nonbiological synthesis of organic substances from inorganic materials than it is to explain how all these nonliving organic materials might have aggregated together in the complex organization of even the simplest cell we know. Several experimenters working with organic solutions have succeeded in obtaining small spherical objects that superficially resemble microorganisms in appearance. They may also resemble "precells" which probably existed once and gradually evolved into the first cells possessing such characteristics that most biologists would call them alive without question. These experimentally produced spheres all have some of the characteristics of living things—an external membrane that surrounds the structure, for instance—but none of them possesses all the characteristics of living things and so cannot be considered to be alive. Nonetheless some of them show a remarkable degree of internal organization. Perhaps it was on such internal parts that the enzymes of respiration and of other essential processes assembled in an orderly manner allowing for an efficient metabolism that ensured the survival of this object long enough for it to produce new ones like itself. This area of research deserves much more study, for somewhere in this or a similar series of events, the first living cells arose. Figure 47.1 summarizes briefly our present ideas about prebiological evolution.

It is not unlikely that somewhat different living cells arose independently at different times and places. If so, undoubtedly many of them failed to compete successfully with cells of other types and failed to reproduce. Whether all living things descend from a single original cell (or group of nearly identical cells generated under similar circumstances) or from several diverse cells of independent origin has not been answered. However, the basic similarity of the structures and chemical constituents of all living cells—cellular membranes composed of lipoproteins, ribosomes, genetic material composed of nucleic acids, reactions catalyzed by protein enzymes, ATP as an energy-transferring molecule, NAD and NADP as hydrogen-transferring molecules—suggests a common origin for all living things. Even the procaryotes (bacteria and blue-green algae), with their lack of mitochondria, chloroplasts, and true nuclei, share enough features with eucaryotes to be considered as sharing a common origin with them. Only the

**47.1** A simplified scheme of the origin of life. Each horizontal arrow indicates not a single reaction, but a large number of reactions occurring over periods of millions of years.

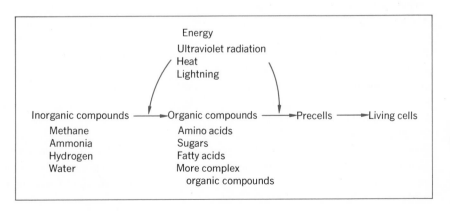

acellular viruses might be interpreted as being of independent origin, and even they have been considered to be the very much reduced descendants of cellular organisms rather than more primitive cellular forms.

## Biological Evolution

Were the first cells plants or animals? Neither, really, but they undoubtedly were nonphotosynthetic and as such required a supply of food. That source of food is not hard to imagine, for the organisms lived in the medium that produced them. Surrounded by energy-rich organic molecules, they had only to absorb what they needed. Conditions were favorable for them to grow and multiply.

With no enemies to devour this first family of living things, the stage was set for a population explosion. Perhaps it was a slow explosion, perhaps it required thousands or millions of years for the habitable waters all to become occupied, but eventually the food supply must have dwindled, and life well might have become extinct if some new form of life had not appeared.

### The Coming of Photosynthesis

The early residents of the earth had wrought changes in the environment. Not only did they use up the food, but they excreted their waste products. Probably at least some of these organisms produced carbon dioxide as a result of their respiration (which perhaps resembled the anaerobic fermentations of some microorganisms today). Among the billions and billions of organisms then inhabiting the world there appeared a mutant form with the ability to manufacture food from the carbon dioxide that had accumulated; the energy for the process came not from other food molecules, but from sunlight. These were the first

photosynthetic organisms. They had an advantage over their nonphotosynthetic relatives for they did not have to depend on the dwindling food supply. Rather, they replenished it, their dead bodies providing nourishment for those that could not make their own. Some less fortunate probably were engulfed by their ravenous relatives who exhibited animal-like qualities.

### The Effects of Molecular Oxygen on Evolution

Not only did photosynthesis bring a potentially endless supply of food into the world, but it also produced a new gas, molecular oxygen ($O_2$). The consequences of this were many-fold. It made possible aerobic respiration; new mutant organisms emerged that now could oxidize their food completely to carbon dioxide and water, and this made their utilization of food more efficient. Larger, more active, and more complex organisms became possible. From the photosynthetic line evolved the algae, some of which remained small and free-living; others of some size clung to rocky shores and assumed the sedentary life we associate with higher plants. From the nonphotosynthetic line evolved the animals, among them the protozoa and many types of marine worms. There were shelled animals, some resembling the squids and snails of today, others resembling lobsters and shrimp. Eventually there were fishes with vertebrae; from some of these there later evolved not only the present-day fishes, but also amphibians, reptiles, birds, and mammals, the last including man himself. The later steps were not to occur until the dry land became more hospitable to life than it originally was.

As oxygen accumulated in the air, some of the oxygen became ozone ($O_3$), which formed a layer in the upper atmosphere and for the first time shielded the bare surface of the land from the destructive effects that high intensities of ultraviolet radiation can have on living things. Until this time no life had ventured onto land. Then algae began to colonize the shores, and over millions of years their descendants evolved into mosses, ferns, coniferous trees, and flowering plants. Once the land was inhabited by plants, animals followed—air-breathing fishes and amphibians. Their descendants today include reptiles, birds, and mammals. The earliest inhabitants of land, both plant and animal, stayed close to the shore, where water was abundant, but as new species evolved they developed adaptations that permitted them to live their lives away from oceans, lakes, and streams. Their independence is not complete, however, for nearly all species require some liquid water, although some, such as the kangaroo rat, can subsist on the water derived from the metabolism of dry foods.

With the appearance of both photosynthesis and aerobic respiration, a balance was apparently struck between the gases produced and consumed by the two processes, the carbon dioxide produced in aerobic respiration being used in photosynthesis and the oxygen produced in photosynthesis being used in aerobic respiration. So nearly equal are the rates of production and use of these two gases that the world's concentration of oxygen does not change perceptibly from one year to the next. However, some slight inequality has existed through the ages, and we are the beneficiaries of it today.

All the organic materials in a plant are synthesized with a certain quantity of oxygen being produced, and eventually the same quantity of oxygen is required to oxidize that organic matter; some of the oxygen is used in the plant's respiration, some in the respiration of the animal that may eat it, and some in the respiration of microorganisms that parasitize the plant or animal or that decompose their bodies when they die. One way or another, all the oxygen a plant produces eventually is used again in respiration. However, the rate of oxidation of dead bodies has lagged slightly behind the rate of photosynthesis for millions of years. Coal deposits are ancient forests that were buried in swamps and never decayed; humus in

topsoil and last year's dead leaves lying on the ground are unoxidized organic material. Because these materials have not decayed, a corresponding amount of oxygen remains in the air. Over millions of years it has accumulated to its present concentration of 21 per cent. So great is the world's oxygen supply that burning of fossil fuels such as coal, gas, and oil will not diminish it appreciably in the foreseeable future. The pollution of air by burning of fuels represents a much more serious threat.

Another consequence of the accumulation of oxygen was the oxidation of any methane and ammonia that might have been left in the air or water. Nearly all the hydrogen gas, being extremely light, escaped from the earth's gravitational pull into space very early, possibly even before the first living cells appeared. With the disappearance of these substances, the environment had changed; no spontaneous formation of organic compounds could occur, and therefore the origin of new life from nonliving organic matter became unlikely. True, living organisms still do continue to excrete organic compounds, and they die and their elements become part of the environment. But bacteria and other microorganisms use these almost as fast as they form, and organic substances also are subject to oxidation by molecular oxygen. The origin of life is considered to have occurred only once or possibly a few times in the far distant past, but because life itself so changed the environment, probably it could not originate on earth again; but the life that did appear had the wonderful capacity of variation.

## The Spreading of Life Throughout the World

It is a general rule that wherever there is an unoccupied habitat some species will move into it. If no species already in existence is suited to occupy it, then a new species eventually evolves and takes over the habitat (the Galápagos Islands finches and tortoises are examples); this, of course, is a slow process. For the past 3 billion years living

EVOLUTION

things have evolved adaptions that enable them to live throughout nearly the entire world.

Some organisms live where life seems relatively "easy." Tropical trees thrive in the warm temperatures and abundant rainfall of equatorial regions, and the lush vegetation provides what by human standards seems to be a veritable paradise for many animals. But such comfortable habitats are not the only ones occupied. Desert plants and animals survive where water is scarce and days are extremely hot. At the other extreme, organisms

**47.2** Gannets nesting on narrow rock ledges. Despite the precarious positions for their nests these birds successfully raise families. Several nearly full-grown young can be distinguished by their fluffy plumage. (Courtesy of Dr. J. B. Nelson.)

like polar bears and penguins survive the rigors of winters in the icy polar regions. Figure 47.2 shows another unlikely habitat successfully occupied. Some animals enjoy more hospitable environments in both summer and winter by braving the dangers of long migrations twice each year.

The spreading of life throughout the seas and over the continents depended on the ability of organisms to adapt to many different environmental conditions by modifying their anatomy, metabolism, and behavior, and these changes arose because of random genetic changes that accumulated slowly over the generations. Mutations produced a wealth of species that occupy almost every conceivable habitat on land, under the sea, and in the air.

## Questions

**1.** Discuss the chemical evolution believed to have occurred before living things originated. How did the environment of that time differ from today's environment?

**2.** Outline the chemical evolution that must have preceded the appearance of the first living things.

**3.** What evidence exists that this chemical evolution could have happened?

**4.** What might have been the energy source that was necessary for the prebiological synthesis of organic molecules?

**5.** Outline the early evolution of life on earth and indicate changes the organisms made in their environment.

**6.** Name several significant consequences of photosynthesis in the evolution of life.

**7.** Why is it unlikely that life could originate today?

## Suggested Readings

Barghoorn, E. S. "The Oldest Fossils," *Scientific American* 224(5): 30–42 (May 1971).

Blum, H. F. "On the Origin and Evolution of Living Machines," *American Scientist* 49: 474–501 (1961).

Calvin, M., and G. J. Calvin. "Atom to Adam," *American Scientist* 52: 163–186 (1964).

Fox, S. W., and R. J. McCauley. "Could Life Originate Now?" *Natural History* 77(7): 26–31 (August–September 1968).

Keosian, J. *The Origin of Life*, 2nd ed. New York: Reinhold Publishing Corporation, 1968.

Wald, G. "The Origin of Life," *Scientific American* 191(2): 44–54 (August 1954).

# Experimental Method

# Part XI

This book touches only briefly on the broad field of biology and merely gives suggestion of its vast scope. How came this large fund of knowledge to be known? By observation—always by careful observation and often by observation that is part of a planned experiment.

Some observations are of the simplest kind. Men living in the appropriate parts of the world must have observed since prehistoric times that the swallows return to Capistrano in spring, that maple leaves change color in autumn, that dogs have four legs, and that birds have but two. It takes a little more effort to catch a dog or pick a plant and to dissect it to see what some of its internal parts might be. With some moderately expensive equipment, today one can make slides from bits of organs and see that these are composed of different kinds of tissues: muscle, nervous tissue, xylem, palisade parenchyma, and so on.

Many years of observations like these have accumulated great masses of information. To be useful, observations must be carefully made, of course. No one knows what animal inspired the myth of the beautiful mermaid, but it is believed to be the sea cow, an aquatic mammal so homely that none but the loneliest sailors could have interpreted it as ravishing and desirable. Scientific observation requires steadier heads than these.

Science and experimental method know no political or racial limitations. At the beginning of the twentieth century, a Dutch botanist, Hugo De Vries, a German botanist, Karl Correns, and an Austrian botanist, Erich von Tschermak, independently and almost simultaneously performed similar genetics experiments and came to the same conclusions about certain patterns of inheritance. When each searched the scientific literature, he found that an Austrian monk named Gregor Mendel had found these same patterns about 30 years earlier.

For the most part, religious or moral considerations do not interfere with the pursuit of scientific knowledge, but occasionally this does happen. The presumed immorality of dissecting a human

# 48
# Experimental Method

**1 Experimental method is a common-sense method of deriving information from nature.**

433

cadaver once retarded the study of anatomy, but obstructions like these usually can be overcome by anyone having sufficient desire to increase our knowledge in a given field.

---

# Experimentation

Simple observations alone, no matter how accurate, can provide us with only a limited amount of information. They can tell us that a dog has one heart and two kidneys, but not what these organs do. **Experiment** is necessary for this. An experiment is merely a way of asking a question of nature. It is also an exercise in logic—common sense, if you will. Despite this, a scientist should be a person of imagination. Before he can perform an experiment, he must recognize that a problem awaiting solution exists, and he must be able to devise an experimental procedure that will answer the question. This often requires imagination and creativity.

## The Steps in Experimentation

The plan of the experiment can be extremely simple or rather complex, but generally the simpler it is, the easier it is to interpret the results. The equipment required may range from unsophisticated materials such as old jelly jars to electron microscopes. The organisms used may be common weeds or specially raised, germ-free animals. Some experiments could have been performed by almost anyone at almost any time in history. Others could be devised only after a great deal of information had accumulated in related fields; the many steps in photosynthesis, respiration, and other biochemical processes, for instance, could not possibly have been worked out until the science of chemistry had reached a certain degree of maturity and provided information on which the biochemists depend.

Experimentation begins with observation that leads to the **recognition of a problem.** From his observations the experimenter formulates a **hypothesis;** this is nothing but a guess, which for the moment he assumes is the answer to the problem. The more familiar he is with the problem, the more likely is his hypothesis to be the correct one, but only experiment can prove this. Once the experimenter states his hypothesis clearly, he can select suitable organisms and other materials to work with and devise an experiment to test the hypothesis.

The biological experimenter usually divides his organisms (or other materials) into two groups: an **experimental group** and a **control group.** The experimental group is given a particular treatment that the control group does not receive. In all other ways, the two groups are treated alike. For instance, both groups receive the same amount of food, unless the experiment is concerned with food; both groups are housed at the same temperature, unless the experiment is concerned with temperature; and so on. The experimental treatment is so devised that the differences in responses of the experimental and control groups hopefully will prove the hypothesis to be right or wrong.

An experiment should be repeated several times to avoid the possibility of some chance occurrences causing abnormal results. If a biologist performs an experiment once with only two cats and one of them does not feel well that day, the results of the experiment are not valid. If he repeats the experiment 50 or 100 times with different cats, one sick animal is not likely to shift the average of the results very much.

A few experiments selected for their simplicity and historical interest will be used to illustrate experimental method.

Aristotle, a Greek philosopher who lived in the fourth century B.C. taught that some animals arose from rotting meat. Maggots, the larval stage in the life cycle of flies, often were seen to grow on several-day-old meat and then to develop into adult flies. The idea of the generation of living things from nonliving matter must have seemed quite reasonable, for it persisted a long time. Then in the seventeenth century A.D., Francesco Redi, an Italian physician and naturalist, observed that adult flies often visited raw meat and that hunters and butchers covered meat with cloth to preserve it. From these simple observations Redi hypothesized that rather than being generated from rotting meat, the maggots actually developed from eggs laid in the meat by adult flies, and that the activities of the maggots contributed to the rotting of the meat.

To test this hypothesis, Redi devised a simple experiment. He placed similar pieces of meat into separate containers. One container was covered to prevent adult flies from reaching the meat; the other remained open, giving adult flies free access to the meat. Numerous maggots appeared in the exposed meat, but none could be found in the covered meat (Fig. 48.1). Repetitions of this experiment at different times of the year, with different kinds of meat, and with different kinds of containers produced the same results. Redi then could consider his hypothesis as a valid one.

The reader may feel that the devising of so simple an experiment required little ingenuity. Yet during the 2,000-year interval between Aristotle and Redi, no one else seems to have thought of it.

For some time it appeared that the origin of living things from nonliving matter had been disproved for all time. But about the time that Redi performed his experiments, a newly discovered group of organisms was being studied with interest. The invention of the microscope had made bacteria and other microorganisms visible to the human eye. The simple, primitive characteristics of these organisms led to the belief that they might arise from nonliving matter even if more complex organisms did not. Microorganisms then were cultivated in broths (as they are today), and some men thought that the broths generated the organisms. Several persons had performed experiments which they interpreted as showing that microorganisms

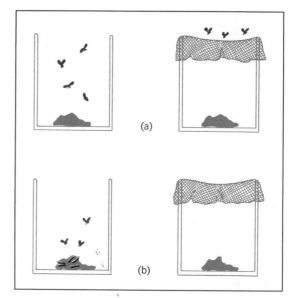

**48.1** Redi's experiment disproving the spontaneous generation of maggots from meat. (a) Two containers, each with a piece of meat; one container is uncovered and flies can approach the meat; the other is covered. (b) Several days later maggots appear in meat exposed to flies but not in the protected meat.

are not spontaneously generated from broths, but other biologists argued that because these broths or the air to which they were exposed had been sterilized by heat or strong acids, the sterilization had destroyed some "vital principle" which they assumed to be present in the broth and/or air and to be essential for the generation of life.

The question remained unanswered until the nineteenth century, when Louis Pasteur, a French chemist, performed another experiment both simple and ingenious in its devising. He hypothesized that broths do not generate live organisms, but that they do support the growth of microorganisms that might reach them from other sources. One such source is the air with its dust particles laden with bacteria and fungus spores.

Pasteur prepared glass flasks containing broths known to support the growth of microorganisms (Fig. 48.2). Then he drew the necks of the flasks into long, slender tubes bent in such a way that

435

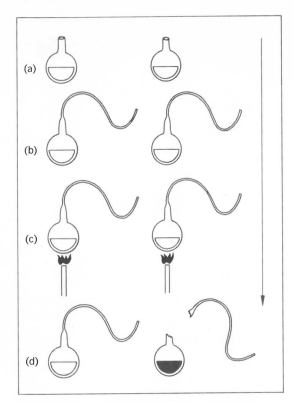

**48.2** Pasteur's experiment disproving spontaneous generation of microorganisms from broth. (a) Broth poured into flasks. (b) Necks of flasks drawn into S-shape so that particles from the air cannot fall into broth. (c) Flasks and their contents sterilized by boiling. (d) Unbroken flasks remained clear and sterile, but broken flasks became turbid from the growth of microorganisms.

dust from the air could not fall into the broth. Then he boiled the broths to sterilize them. After they had cooled they were allowed to stand for days, months, or even years, and no microorganisms appeared in them. Because the flasks were open, all the gases of the air could diffuse through the curved necks to the broth. No one could argue that the lack of growth was due to the destruction of a vital principle in the air.

To prove that boiling had not destroyed another vital principle in the broth, Pasteur divided the flasks into two groups. He allowed one group to remain untouched to serve as controls, and he broke the necks of flasks in the other group in such a way that dust particles from the air could settle onto the broth. Within a day or two, growth of microorganisms was apparent in the second group of flasks. From these results he concluded that the organic materials in a broth possess no vital principle and cannot generate live organisms. The microorganisms growing in a broth arrive there from some other place or are the descendants of introduced organisms. Since the successful conclusion of Pasteur's experiments there has been no serious question about the spontaneous generation of living things from nonliving matter under conditions as they exist in the world today (the origin of life under much different conditions as described in Chapter 47 is another thing).

We might consider another experiment that led to only partially correct conclusions because of a deficiency in the information available to the experimenter. Jean-Baptiste van Helmont, a Flemish physician and chemist who lived during the sixteenth and seventeenth centuries, placed 200 pounds of dried and carefully weighed soil into a pot. In this he planted a willow tree branch weighing exactly 5 pounds. The branch took root, and he watered it with rain water and distilled water. Five years later, van Helmont removed the plant, carefully separating the roots and soil. The plant weighted 169 pounds, 3 ounces, indicating a gain in weight of 164 pounds, 3 ounces. (He did not weigh the leaves that fell each autumn, so the total growth was somewhat more; but this discrepancy is not important to the point under discussion here.) When he dried the soil again and weighed it, he found that it had lost only 2 ounces in the five years.

From these results, van Helmont concluded that all the material in the plant body was derived from water. He may have considered the 2 ounces lost from the soil as due to slight errors in weighing, or he may have assumed that the plants absorbed them; in any event the 2 ounces could not account for the increase in weight of the plant. What he did not consider was the possibility that part of the plant material might have derived from the air surrounding the plant. With the science of chemistry still in its infancy, he could not be expected to conclude that plants derive most of their substance from two compounds: water and carbon dioxide.

For reasons like these, scientists always keep their minds receptive to new ideas, for they know that "facts" that once were "proved" may not be true at all. However, as the collection of more and more data supports the results of earlier experiments, the more nearly certain we are that the conclusions drawn from these experiments are true.

## Publication of Results

Once he successfully completes an experiment, a scientist ordinarily writes a paper describing it. He submits the paper to a scientific journal for publication. Although the forms of published papers vary somewhat, they nearly always contain five major components:

1. Objective of the experiment. Usually this is in the form of a simple statement of the hypothesis that the experiment was devised to prove or disprove.

2. Materials. This is a description of the organisms, chemicals, and/or equipment used in the experiment. If it was important to the experiment, the author gives the age, sex, or other condition of the organisms, the concentration of chemicals, or any other pertinent information.

3. Procedure. The author then describes the steps he followed. He tells the conditions under which the control group was kept and how the treatment of the experimental group(s) differed.

4. Results. The author describes how the control and experimental groups reacted. Often he summarizes the information in the form of tables, charts, or graphs. He may use photographs to document the results.

5. Conclusion. The author explains the conclusions that he draws from the results of the experiment. He may also discuss the significance of these conclusions in relation to facts already known.

The publication of papers exposes the work of scientists to the scrutiny of other scientists. Because the papers describe in detail the materials used and the procedures followed, anyone who wishes to do so can repeat an experiment and so check the accuracy of the author's work. This minimizes the publication of poorly executed or incorrectly interpreted work, and it provides opportunity for the correction of any errors that may have found their way into print.

The publication of scientific papers also disseminates information throughout the scientific community and provides permanent records of experiments performed. Future researchers can use this published information as a foundation on which to base their own experiments. In this way, scientific knowledge accumulates.

## Theories and Laws

When a hypothesis has been tested by experiment many times and found to be generally valid either in its original form or as somewhat amended by later experiments, it is called a **theory.** A theory enjoys a high degree of acceptance among scientists, but there may be some degree of reservation on the part of some persons, or it may be known that the theory has some exceptions. The theory of evolution is an example. Although there are very few biologists who do not accept the theory, it has not been proved beyond all doubt.

A theory that has been tested many times and found to have no exceptions is raised to the level of a **law.** Laws are more likely to be found in the physical sciences than in the biological sciences, for living things are more variable than the individual elements and compounds of which they are

437

composed, and their activities are less accurately predictable. The law of gravity, for example, possesses a high degree of predictability. The chances that a coin tossed into the air will remain there without support are so remote that anyone would expect the coin to fall. We may predict that in a cross of two heterozygous individuals, about three fourths of the offspring will have the dominant trait and about one fourth will have the recessive trait. We cannot predict, however, which individuals among the offspring will have one trait and which individuals will have the other.

## Science and Morality

A scientific fact (or any other fact) possesses no intrinsic morality or immorality. The fact that the chemical bonds between the atoms in a molecule possess a certain amount of energy is neither good nor bad, but the use to which that energy is put may be considered good or bad by society. You may use some wood to build a campfire with which to cook your food and to keep yourself warm on a cold night. On another occasion you may burn down your neighbor's wooden house. Most persons would judge these two uses of chemical bond energy differently.

There are many aspects of nature about which we need more knowledge, and so there is a demand for more research. But we also need an improved morality with regard to our uses of acquired knowledge. Nearly every major religion of the world advocates some variation of the Golden Rule: do unto others as you would have them do unto you. The Golden Rule implies an interdependence among human beings. Admittedly, hate, greed, and other motives have led to the breaking of this rule many times, but its abuse has not been of such proportions that society has dis-

integrated or that the human species has become extinct—at least not so far.

While struggling along with our somewhat tattered Golden Rule, we need a corresponding moral precept for our use of nature on which we also depend. Although we can remove mountains, divert or dam rivers, run our air conditioners with electrical power derived from nuclear energy, and raise large families successfully, a growing question is whether it is morally correct to do this if it destroys our environment. A scientist acting only as a scientist has no interest in the moral problem of what use is made of his discoveries, but a scientist acting as a human being does. So, too, does everyone else.

## Questions

1. Define the following terms:
control group
experimental group
hypothesis
theory
2. Analyze Redi's experiment and determine if he followed the proper experimental procedure.
3. Read an article recently published in a scientific journal and analyze it from the point of view of experimental procedure.

## Suggested Readings

Conant, J. B. *Science and Common Sense.* New Haven: Yale University Press, 1961.

Gabriel, M. L., and S. Fogel, eds. *Great Experiments in Biology.* Englewood Cliffs, N.J.: Prentice-Hall, Inc., 1955.

Taton, R. *Reason and Chance in Scientific Discovery.* New York: John Wiley & Sons, Inc., 1962.

Wightman, W. P. D. *The Growth of Scientific Ideas.* New Haven: Yale University Press, 1951.

# Appendix

## An Outline of Living Things

### Kingdom 1: Viruses

This kingdom includes submicroscopic, noncellular parasites. The smallest viruses consist only of a nucleic acid core surrounded by a protein coat; some of the larger viruses also contain carbohydrates and fats. The nucleic acid of a virus is either DNA or RNA—never both. The RNA-containing viruses are the only organisms known to have their genetic material reside in RNA rather than DNA. Because they have no protein-synthesizing machinery of their own, viruses must "borrow" those of other organisms and so are obligate parasites.

Nearly all cellular organisms are subject to infection by some viruses. In human beings, viruses are responsible for the common cold, flu, poliomyelitis (infantile paralysis), mumps, measles, German measles, chickenpox, smallpox, cowpox, rabies, yellow fever, and warts. Viruses are under suspicion as the possible cause of multiple sclerosis, and they almost certainly cause some human cancers.

One theory has it that viruses in their simplicity resemble the first living things that appeared on earth some 3 billion years ago, "living molecules" synthesized from organic molecules present in the primeval oceans. Another theory is that viruses are descendants of cellular organisms, and that by shedding cytoplasm and nucleus, their remaining genetic material became parasitic on living cells. If the second theory is true, viruses are simple in structure only and highly advanced in their ability to take over the control of the metabolic machinery of intact cells.

### Kingdom 2: Monera

This kingdom includes the procaryotic organisms—the bacteria and blue-green algae. Their cells possess no true nuclei, mitochondria, endoplasmic reticulum, or Golgi bodies; those that are photosynthetic have no chloroplasts. They do have genetic material that occupies a nucleoid of somewhat irregular shape with no membrane to form a distinct boundary between it and the cytoplasm. The cells divide, the nucleoids divide, and the genetic material duplicates itself and becomes distributed to the daughter cells; but no mitotic stages have been observed.

PHYLUM SCHIZOMYCOPHYTA: bacteria. Bacteria are extremely small, most of their cells having diameters of about 0.5 to 1.5 microns. Their small size gives them an extremely large surface area in proportion to their volume (Chapter 20). This, combined with their rapid reproductive rates, which can produce many cells in a very short time, contributes to their effectiveness as agents of illness, spoilers of food, and producers of commercially valuable substances.

Some bacteria-caused diseases of humans are anthrax, tetanus, typhoid fever, cholera, diphtheria, bacillary dysentery, staphylococcal food poisoning, salmonella food poisoning, botulism, scarlet fever, streptococcal sore throat (strep throat), bubonic plague, tuberculosis, leprosy, gonorrhea, syphilis, and some forms of pneumonia.

Other bacteria play invaluable roles as scavengers in nature; their decomposition of dead organisms not only keeps the world from becoming an uninhabitable repository for dead bodies, it also replenishes soils, lakes, streams, and oceans with minerals.

Several bacteria are used in industrial processes where they produce a variety of chemical compounds, including ethyl alcohol, acetic acid, acetone, dextran (a plasma substitute), and several medically important antibiotics (streptomycin, chloramphenicol, aureomycin, terramycin, and others). The chemical activities of bacteria contribute to the production of many cheeses, alcoholic beverages, breads, and pickles.

PHYLUM CYANOPHYTA: blue-green algae. The cells of blue-green algae are somewhat larger than bacterial cells, and many are arranged in filaments or small colonies, some large enough to be seen by the unaided eye. Blue-green algae are photosynthetic and contain chlorophyll, although the cells lack chloroplasts. In addition, they contain two water-soluble pigments, phycocyanin, a blue pigment, and phycoerythrin, a red pigment. The predominance of the green chlorophyll and the blue phycocyanin give many algae of this group a blue-green color, and hence, their name.

Blue-green algae have a world-wide distribution, most of them living in fresh water. Some become pests in reservoirs and give water an unpleasant taste. Some are nitrogen fixers, and those growing in rice paddies in the Orient contribute a significant amount of nitrogen to these fields so important in feeding a large part of the world's population.

## Kingdom 3: Plants

Plants have eucaryotic cells with true nuclei bounded by nuclear membranes, mitochondria, endoplasmic reticulum, and Golgi bodies. Cells of photosynthetic plants have chloroplasts which contain chlorophyll and associated pigments and the photosynthetic enzymes. With some exceptions among the algae and the fungi, plants generally are sessile, and their cells are surrounded by cellulose cell walls.

**Subkingdom Thallophyta:** thallus plants. The algae and fungi possess no true roots, stems, or leaves. Their plant body is called a thallus, and it ranges in complexity from a single cell to a giant seaweed several hundred feet in length. The reproductive organs are single-celled, and no layer of sterile cells surrounds them. The zygote of an alga or fungus is not retained in the parent plant for any length of time, and the embryo develops independently.

PHYLUM CHLOROPHYTA: green algae. Green algae grow primarily in fresh water, but some are marine. They range from single cells to colonial, filamentous, and membranous forms. This is a diverse group with many different kinds of life cycles.

PHYLUM EUGLENOPHYTA: euglenoids. The members of this group are mostly unicellular algae, many of them motile by means of flagella. They reproduce by cell division; sexual reproduction is rare. Though photosynthetic, some of the euglenoids ingest whole food, and for this reason some biologists consider them to be protozoa.

PHYLUM CHRYSOPHYTA: diatoms, yellow-green algae, and golden-brown algae. Perhaps the most important members of this phylum are the diatoms, which form the basis of many aquatic food chains, especially in cold waters; they account for much of the primary productivity of the colder oceans. Unicellular or filamentous, the diatoms have cell walls rich in silicon dioxide ($SiO_2$), which is similar to glass.

PHYLUM PYRROPHYTA: dinoflagellates and cryptomonads. These are single-celled algae motile by flagella. The dinoflagellates provide much of the primary productivity of the warmer oceans. Many of them contain a red pigment, peridinin, and some of them produce a potent toxin responsible for the famous red tides accompanied by tremendous fish kills. Some biologists consider them to be protozoa.

PHYLUM PHAEOPHYTA: brown algae. These are multicellular seaweeds at least a few inches long; the giant kelps reach several hundred feet in length. Most of them are marine and grow in rocky intertidal zones where the plants are covered by water at high tide and are exposed to air at low tide. They are most common in cold waters. They contain a brown pigment, fucoxanthin, which in combination with chlorophyll gives these algae an olive drab to brown color. Many brown algae are edible; they are eaten mostly by Oriental people.

PHYLUM RHODOPHYTA: red algae. Most of the red algae are multicellular seaweeds growing in the intertidal zones; they live mostly in warm tropical and subtropical waters. Like the blue-green algae, they contain the red pigment phycoerythrin and the blue pigment phycocyanin, but the red pigment usually predominates. The combination of phycoerythrin with chlorophyll gives these algae a red to reddish-brown color. Many of them are edible; some produce the substance agar, the standard medium on which many bacteria and fungi are raised in the laboratory. Agar also has excellent laxative properties.

PHYLUM MYXOMYCOPHYTA: slime molds. The slime molds possess a peculiar mixture of animal and plant characteristics. At one stage in their life cycles they exist as separate cells that look like amebas, move like amebas, and ingest food like amebas; at this stage the cells lack cell walls. At another stage in the life cycle, the cells aggregate and form a sessile spore-producing body with cell walls surrounding the cells. At no stage do they possess chlorophyll; they live as saprophytes and parasites. They are common on rotting vegetation, especially in wet weather. Some species parasitize crop plants, but most slime molds have little economic importance.

PHYLUM EUMYCOPHYTA: true fungi. Except for a few unicellular aquatic forms, most true fungi are sessile like most plants. The major constituent of their cell walls is chitin rather than cellulose. A few species are unicellular, but most true fungi consist of filamentous threads called hyphae (singular, hypha). In some species, these are interwoven into relatively large reproductive bodies. All true fungi lack chlorophyll and are saprophytes or parasites.

CLASS PHYCOMYCETES: algalike fungi. This group includes a variety of aquatic and terrestrial species. Perhaps the most familiar to most persons is *Rhizopus*, the black bread mold. Because of the practice of adding chemical preservatives to bread, this mold now is more common on old fruits and vegetables than on bread.

CLASS ASCOMYCETES: sac fungi. This group includes several economically important fungi: yeasts, which produce ethyl alcohol; *Penicillium*, the several species of which ripen some cheeses or produce the antibiotic penicillin; the fungi that cause Dutch elm disease and chestnut blight; and the ergot fungus, which grows on rye plants and causes ergotism in persons eating bread made with infected rye.

CLASS BASIDIOMYCETES: club fungi. The reproductive bodies of some of these fungi are the familiar mushrooms, puffballs, and bracket fungi; some are edible, others deadly poisonous. Wheat rust and corn smut are parasitic on wheat and corn and do considerable damage to these plants.

CLASS DEUTEROMYCETES: imperfect fungi. This is an unnatural group into which are placed all those fungi that cannot be assigned to one of the other classes of fungi because their sexual reproduction has not yet been observed. Among them are the causative organisms of athlete's foot, ringworm, and several other diseases of human beings.

CLASS LICHENES: lichens. Lichens represent a mutualistic relationship between fungi and algae, but they often are classified with the fungi because

441

the fungus is the more obvious member. In most lichens the fungus is an ascomycete, and the alga is a green alga. Lichens form the basis of the food chain in the tundra, where they are eaten by reindeer and caribou, and these animals form a major part of the diet of many Eskimos and Laplanders. Litmus, a pH indicator, is extracted from a lichen.

PHYLUM BRYOPHYTA: mosses and liverworts. Most bryophytes are small terrestrial plants that grow only in damp places. The plants are green and photosynthetic. They have no true roots, stems, and leaves. The reproductive organs are multicellular, the eggs and sperms being surrounded by a layer of sterile cells. The young embryo develops within the female reproductive organ and so is sheltered by the parent plant.

PHYLUM TRACHEOPHYTA: vascular plants. These plants have true roots, stems, and leaves supplied with vascular bundles that transport food, water, and minerals throughout the plant. With a few exceptions, the plants are green and photosynthetic. Some vascular plants are very small, but most of them are large enough to be more or less conspicuous. They comprise most of the plants we see in our everyday lives. The reproductive organs are multicellular, the eggs and sperms being surrounded by a layer of sterile cells. The young embryo develops within the female reproductive organ.

Subphylum Psilopsida: whisk ferns. This is a small group of simple vascular plants of little economic importance. Several fossil members, long extinct, were among the earliest vascular plants.

Subphylum Lycopsida: club mosses. This is another small group of little economic importance; a few plants are sold as Christmas decorations. Present-day club mosses are relatively small, but their fossil relatives were large trees.

Subphylum Sphenopsida: scouring rushes. This is another small group. Because of their high silicon content, the stems of these plants make acceptable substitutes for forgotten scouring pads on camping trips.

Subphylum Pteropsida: pteropsids. This is a large group, the plants ranging in size from the minute duckweed floating on ponds to the giant redwoods.

CLASS FILICINEAE: ferns. This is a large group; their economic value is largely as ornamentals.

CLASS GYMNOSPERMAE: cycads and conifers. These are seed plants that produce their seeds naked on the surface of scales united into cones. Several conifers are important timber trees.

CLASS ANGIOSPERMAE: flowering plants. These are seed plants that produce their seeds within a fruit. This is a large group providing us with a great deal of our food, clothing, and shelter.

Subclass Dicotyledonae: dicots.

Subclass Monocotyledonae: monocots.

## Kingdom 4: Animals

Like the plants, animals have eucaryotic cells with true nuclei bounded by nuclear membranes, mitochondria, endoplasmic reticulum, and Golgi bodies. With the exception of some protozoa frequently classified as algae, members of the animal kingdom have no chlorophyll, and so they require an external source of food. Most animals can move about and search for their food, but a few are sessile.

The animal kingdom is a large one consisting of more than 20 phyla. Some of the smaller, more obscure groups whose lives do not touch directly on ours are omitted here.

PHYLUM PROTOZOA: single-celled animals. This group includes all single-celled animals, including various motile organisms sometimes classified as algae, slime molds, and true fungi.

CLASS MASTIGOPHORA: flagellates. These are single-celled animals moving by means of flagella. Many members of this group are harmless, but it includes the trypanosome that causes African sleeping sickness.

CLASS SARCODINA: amebas. Members of this group move by means of pseudopodia. Many are

442

free-living and harmless, but *Entamoeba histolytica* causes amebic dysentery.

CLASS CILIATA: ciliates. These animals are covered with many fine cilia by which they move.

CLASS SPOROZOA: sporozoans. Members of this class have no means of locomotion. All are parasitic. Several species of *Plasmodium* cause malaria, one of the most serious communicable diseases in the world today.

PHYLUM PORIFERA: sponges. These warm-water, marine plants are sessile like plants. Though multicellular, they show no differentiation into distinct tissues. The skeletons of some species once were popular as bath sponges, but now they have been replaced by plastic products.

PHYLUM COELENTERATA: coelenterates. These are radially symmetrical animals consisting of two distinct layers of tissue surrounding a hollow gastrovascular cavity. This cavity serves as both a digestive and circulatory system.

CLASS HYDROZOA: hydroids. This group includes *Hydra*, some jellyfishes, and the poisonous Portuguese man-of-war.

CLASS SCYPHOZOA: jellyfishes

CLASS ANTHOZOA: sea anemones, corals

PHYLUM PLATYHELMINTHES: flatworms. Animals in this and the following phyla possess distinct tissues organized into distinct organs. The flatworm body is bilaterally symmetrical, the midline of the body separating it into two halves that are mirror images of each other. The body is definitely flattened, and anterior (front) and posterior (rear) ends are obvious. The digestive system has only one opening and doubles as a circulatory system, but there are separate muscular, nervous, and excretory and reproductive systems. These systems are well developed in the free-living flatworms; but in the parasitic flukes and tapeworms, all but the reproductive systems are much reduced.

CLASS TURBELLARIA: free-living flatworms. These nonparasitic flatworms live in oceans and fresh water and on moist ground.

CLASS TREMATODA: flukes. This group includes several parasites of human beings including Chinese liver fluke and the blood flukes (the latter causing schistosomiasis, the most serious communicable disease in the world today).

CLASS CESTODA: tapeworms. Members of this group are all parasites; among them are the pork and beef tapeworms, which can be contracted by people who eat raw or imperfectly cooked pork and beef.

PHYLUM NEMATHELMINTHES: roundworms. These animals possess bilateral symmetry internally, though externally they appear to be almost cylindrical (round in cross section—hence their name). They are widely distributed in many kinds of habitats throughout the world and include both parasitic and free-living forms. They range in size from a fraction of an inch to 4 feet in length. The digestive system has two openings, one serving as a mouth, the other as an anus. Parasitic members of this group include *Ascaris*, hookworms, pinworms, trichina worm (which causes trichinosis in man), and the filaria worm (which causes elephantiasis).

PHYLUM MOLLUSCA: molluscs. These are soft-bodied animals, many of which are protected by hard shells of calcium carbonate. Internally, they possess bilaterally symmetry which is more or less distorted. The digestive system has both mouth and anus. A heart pumps blood through a separate circulatory system. Molluscs inhabit both fresh and salt water as well as land.

CLASS AMPHINEURA: chitons. A shell composed of eight plates protects the soft body.

CLASS GASTROPODA: gastropods. This group includes the common snails and slugs of gardens. The body of the snail is protected by a coiled shell into which it can retreat. Slugs lack shells.

CLASS CEPHALOPODA: cephalopods. These marine animals have eight or ten tentacles near the mouth. The octopus has no shell, and the squid and cuttlefish have only a small shell embedded in the body.

CLASS PELECYPODA: pelecypods. This group includes clams, oysters, scallops, and several other

443

edible aquatic organisms; their bodies are protected by a shell comprised of two halves that the animals can hold tightly clamped together.

PHYLUM ANNELIDA: segmented worms. These bilaterally symmetrical animals are divided into segments that are visible externally as rings around the body. The digestive system has both mouth and anus. Blood is moved through the circulatory system by several "hearts" that are not much more than enlarged muscular vessels. This group includes the common earthworms, sandworms, and leeches.

PHYLUM ARTHROPODA: joint-footed animals. This large phylum is highly diversified and is found in many kinds of habitats throughout the world. Like the annelids, they are segmented and have bilateral symmetry. The digestive system has two openings. The circulatory system is an open one, the blood not being confined only to the heart and blood vessels; it also moves in the body cavity.

An obvious feature of the arthropods is their hard chitinous exoskeleton that not only provides mechanical protection, but also prevents the evaporation of water from the body surface. Though the skeleton is rigid, it is jointed, and therefore mobility is not sacrificed. In some species, as the animal increases in size it sheds its skeleton and grows a new one.

CLASS CRUSTACEA: crayfish, crabs, shrimps, lobsters. These are aquatic organisms that breathe by means of gills. Many are edible.

CLASS CHILOPODA: centipedes. Centipedes have a wormlike appearance, but they have one pair of legs per segment.

CLASS DIPLOPODA: millipedes. These animals resemble the centipedes, but they have two pairs of legs per segment.

CLASS INSECTA: insects. This is the largest group of arthropods, with several thousand species. The segmented body is divided into head, thorax, and abdomen. The head bears one pair of antennae. The thorax has three pairs of legs and may have one or two pairs of wings. Terrestrial insects breathe by a system of tubes called tracheae that communicate with the external air through several openings in the side of the body.

Among the many species of insects are ants, bees, mosquitoes, flies, fleas, lice, beetles, weevils, butterflies, moths, and grasshoppers. Some, such as bees that pollinate flowers and produce honey and beeswax, are beneficial to man. Others, such as the potato beetle and the screw worm, do damage to economically important plants and animals. Still others, such as some mosquitoes and lice, transmit diseases.

CLASS ARACHNOIDEA: arachnids. The arachnids are confused with insects by many persons. This group includes spiders, scorpions, ticks, mites, and king crabs. The arachnid body is divided into only two parts: a cephalothorax and an abdomen. The cephalothorax bears four pairs of legs. Another pair of appendages used as jaws often gives the appearance of a fifth pair of legs. A sixth pair of appendages contains poison glands. Most arachnids are not strong enough to pierce the human skin, or their poison is not potent enough to do much damage; but a few can inflict serious damage. Most species are terrestrial and breathe by specialized lungs.

PHYLUM ECHINODERMATA: spiny-skinned animals. These exclusively marine animals have a five-part radial symmetry in the adult stage. They move by tube feet associated with an internal water vascular system peculiar to these animals. They include the starfishes, sea urchins, sand dollars, sea cucumbers, and sea lilies.

PHYLUM CHORDATA: chordates. Members of this group have three anatomical features in common at some time in their life cycles: a notochord, a hollow dorsal nerve cord, and pharyngeal gill slits. The notochord is a stiff rod that passes longitudinally through the animal and serves as a support. In some animals it lasts through the entire life of the animal; in others, human beings, for example, it is present only in early embryological stages and then is replaced by the vertebrae. The main mass of nervous tissue is dorsally located in chordates, whereas in other animals it is ventral in position (or more or less evenly distributed throughout the body). In chordates it takes the form of a hollow tube (called the spinal cord in many of them) often enlarged at one end into a brain. Young chordate embryos contain gill slits in the pharynx. In aquatic forms the gill slits may develop into gills; in air-breathing species the gill slits develop into eustachian tubes, and lungs develop separately. Chordates are bilaterally symmetrical and possess an endoskeleton of cartilage or of cartilage and bone.

*Subphylum Hemichordata:* acorn worms. These are very primitive chordates; their larvae resemble those of the echinoderms.

*Subphylum Tunicata:* sea squirts. The adult is a sessile animal bearing little resemblance to the typical chordate; the free-swimming larva, however, has all three chordate characteristics.

*Subphylum Cephalochordata:* lancelets. The adult has all three chordate characteristics. The adults resemble the larvae of the lamprey (in the subphylum Vertebrata).

*Subphylum Vertebrata:* vertebrates. These are chordates with a cartilaginous or bony vertebral column that replaces the notochord and protects the spinal cord. The body is divided into a head and trunk, often with a more or less obvious neck between them. Usually there are two pairs of lateral appendages: arms, legs, wings, or fins. Many species have tails.

CLASS AGNATHA: jawless fishes. The lampreys and hagfishes have sucking mouths by which they attach themselves to other aquatic mammals on which they feed. They have no fins.

CLASS CHONDRICHTHYES: cartilaginous fishes. These are marine fishes, the adults having only cartilage and no bone in their skeletons. They include the sharks, skates, and rays. They have fins.

CLASS OSTEICHTHYES: bony fishes. This group includes fresh-water fishes and some salt-water fishes that have bony skeletons. The lateral appendages are fins. This group includes many edible fishes: trout, salmon, perch, bass, cod, and others.

CLASS AMPHIBIA: amphibians. The frogs, toads, and salamanders begin life as aquatic animals breathing through gills. The adults breathe air by lungs, but because their naked skin loses water easily they spend much of their time in or near water.

CLASS REPTILIA: reptiles. Snakes, turtles, lizards, and crocodiles are all air-breathing animals. Some live in or near water; others are adapted to dry habitats such as deserts. Most have skins covered by scales, and their eggs are covered by a shell. The extinct dinosaurs belong to this group.

CLASS AVES: birds. These are warm-blooded animals with the forelimbs modified into wings. With a few exceptions, they have the power of flight. The skin is covered by feathers that help to retain body heat. A hard shell covers the eggs.

CLASS MAMMALIA: mammals. These are warm-blooded animals with all or part of the body covered by hair. With a few exceptions they bear live young and feed them with milk produced in the mammary glands of the mother.

Subclass Prototheria: monotremes. The duck-billed platypus and the spiny anteater are the only mammals that lay eggs rather than bearing live young.

Subclass Metatheria: marsupials. These animals bear live young in a relatively undeveloped state; the young continue to develop in a pouch (marsupium) on the mother's abdomen. Most present-day marsupials are confined to Australia and neighboring islands; they include the kangaroo and koala bear. The opossum is an American marsupial.

Subclass Eutheria: placental mammals. These are animals that bear relatively well-developed live young that were nourished through a placenta in the mother's uterus. Some of the orders of mammals are the following:

445

Insectivora: moles, shrews, hedgehogs
Chiroptera: bats (the only flying mammals)
Rodentia: mice, rats, squirrels, chipmunks
Lagomorpha: rabbits

Carnivora: dogs, cats, foxes, bears, otters, seals

Artiodactyla (even-toed hoofed mammals): pigs, cows, deer, giraffes, camels, hippopotamuses

Perissodactyla (odd-toed hoofed mammals): horses, zebras, tapirs, rhinoceroses

Proboscidea: elephants

Sirenia: sea cows, manatees

Cetacea: whales, dolphins, porpoises

Primates: lemurs, monkeys, baboons, gorillas, chimpanzees, man

## Suggested Readings

Alexopoulos, C. J., and H. C. Bold. *Algae and Fungi.* New York: Macmillan Publishing Co., Inc., 1967.

Barnes, R. D. *Invertebrate Zoology,* 2nd ed. Philadelphia, W. B. Saunders Company, 1968.

Delevoryas, T. *Plant Diversification.* New York: Holt, Rinehart and Winston, Inc., 1966.

Echlin, P. "The Blue-Green Algae," *Scientific American* 214(6): 74–81 (June 1966).

Fraenkel-Conrat, H. "Rebuilding a Virus," *Scientific American* 194(6): 42–47 (June 1956).

Hanson, E. D. *Animal Diversity,* 2nd ed. Englewood Cliffs, N.J.: Prentice-Hall, Inc., 1964.

Lwoff, A. "The Life Cycle of a Virus," *Scientific American* 190(3): 34–37 (March 1954).

Orr, R. T. *The Animal Kingdom.* New York: Macmillan Publishing Co., Inc., 1965.

Wood, W. B., and R. S. Edgar. "Building a Bacterial Virus," *Scientific American* 217(1): 60–74 (July 1967).

## Conversion Table

### Linear Measurements

1 kilometer (km) = 1,000 meters
= 0.62 miles

1 meter (m) = 100 centimeters (cm)
= 1,000 millimeters (mm)
= 1,000,000 microns ($\mu$)
= 1,000,000,000 nanometers (nm) or millimicrons (m$\mu$)
= 10,000,000,000 Angstrom units (A)
= 39.37 inches
= 1.09 yards

1 centimeter = 0.39 inches
1 millimeter = 0.039 inches
1 micron = 0.000039 inches
1 inch = 2.54 centimeters
= 25.4 millimeters
1 mile = 1.61 kilometers
= 5,280 feet

### Volume

1 liter (1) = 1,000 milliliters (ml)
= 1.06 quarts
1 quart = 0.95 liter

### Weight

1 kilogram (kg) = 1,000 grams (g)
= 2.2 pounds
1 gram = 0.035 ounce
1 pound = .454 kilograms
= 454 grams
1 ounce = 28.3 grams

### Temperature

To convert from Celsius (centigrade) to Fahrenheit:

$$F = \left(C \times \frac{9}{5}\right) + 32$$

To convert from Fahrenheit to Celsius:

$$C = (F - 32) \times \frac{5}{9}$$

| | | |
|---|---|---|
| 100°C = 212°F | boiling point of water |
| 37°C = 98.6°F | human body temperature |
| 20°C = 68°F | |
| 0°C = 32°F | freezing point of water |
| −20°C = −4°F | |

## Some Common Functional Groups Found in Biologically Important Molecules

—OH        Hydroxyl

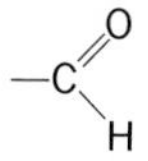

Aldehyde

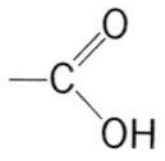

Acid (carboxyl)

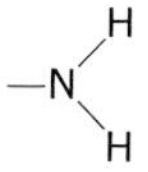

Amino

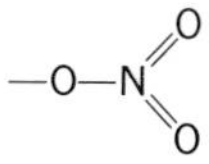

Nitrate

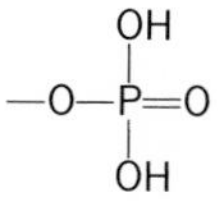

Phosphate

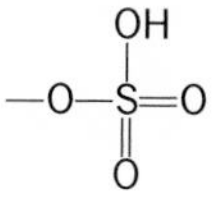

Sulfate

## Answers to Plate I

**A.** Domestic cat in a field of dandelions.

**B.** Earthworm on a clump of moss. The tan objects on the moss are the fruits of elm trees. Several species of weedy plants are somewhat out of focus.

**C.** Fly crawling on the head of a dead trout.

**D.** Mold growing on strawberries.

**E.** Green algae growing on the bark of a maple tree.

**F.** Both the flowers and the leaves belong to the same species—lily-of-the-valley. An undetermined number of species of bacteria cloud the water.

447

# Glossary

**Abscission zone:** a layer of thin-walled cells at the base of a plant organ such as a leaf, fruit, or flower part where it joins the stem. The separation of this layer causes the dropping of the organ.

**Acid:** A substance that releases hydrogen ions (H+) when dissolved in water; having a pH lower than 7.

**Actin:** a muscle protein which, together with the protein myosin, is responsible for the contraction of muscle.

**Active transport:** the movement of a substance across a cell membrane usually against a concentration gradient and therefore requiring the expenditure of metabolic energy.

**Adaptability:** the ability of organisms to adjust to environmental changes.

**Adenosine diphosphate (ADP):** a compound which may react with inorganic phosphate to form ATP.

**Adenosine triphosphate (ATP):** a compound containing a high-energy phosphate bond which is the source of much of the metabolic energy used in living things; the energy is released by the breaking of the high-energy phosphate bond with the resulting production of adenosine diphosphate and inorganic phosphate.

**ADP:** adenosine diphosphate.

**Aerobic respiration:** complete oxidation of a food, often glucose, to carbon dioxide and water in the presence of molecular oxygen in living cells.

**Aleurone:** the outer protein-rich layer of the endosperm of cereal grains.

**Alleles:** alternate forms of a gene occupying the same location on homologous chromosomes.

**Alveolus:** one of the small air sacs of the lung.

**Amino acid:** an organic acid containing an amino group on the carbon atom adjacent to the terminal acid group; proteins are synthesized from amino acids.

**Anabolism:** energy-requiring metabolic processes that convert simple substances into complex substances.

**Anaerobic:** referring to a process or organism that does not require molecular oxygen.

**Anterior:** the front end of an animal.

**Anther:** the pollen-producing portion of a stamen.

**Antibody:** a protein produced in the animal body in response to the presence of an antigen.

**Antigen:** a foreign substance, usually either a protein or a polysaccharide, that elicits the formation of an antibody when injected into an animal.

**Apical dominance:** the inhibition of the growth of lateral buds by the terminal bud of a stem.

**Apical meristem:** the meristem at the tip of a stem or root.

**Arithmetic progression:** a series of numbers in which the difference between any two adjacent numbers is equal.

**Artery:** a blood vessel that conducts blood away from the heart.

**Asexual reproduction:** reproduction not involving the fusion of gametes.

**Atom:** the smallest unit of a chemical element that still retains the characteristics of that element.

**ATP:** adenosine triphosphate.

**Autonomic nervous system:** that part of the nervous system that controls the involuntary action of visceral organs.

**Auxin:** a plant hormone that promotes cells enlargement and inhibits the enlargement of lateral buds.

**Axon:** that part of a neuron that transmits impulses away from the cell body.

**Bark:** the portion of a stem or root external to the vascular cambium; includes cork and phloem.

**Base:** a substance that releases hydroxyl ions (OH−) when dissolved in water; a base has a pH higher than 7.

449

**Biodegradable:** capable of being digested or otherwise decomposed by organisms.

**Biological clock:** a factor or factors responsible for circadian rhythms in eucaryotic organisms.

**Biomass:** the total weight of organisms in an area or at a trophic level.

**Biotic potential:** the inherent capacity of organisms to reproduce and increase in numbers under ideal conditions.

**Blade:** the flat portion of a leaf.

**Bond energy:** the potential energy residing in chemical bonds.

**Bowman's capsule:** the cup-shaped portion of a kidney tubule that surrounds a glomerulus.

**Cambium:** a lateral meristem of stems and roots that contributes to growth in width of these organs. Vascular cambium produces secondary xylem and secondary phloem; cork cambium produces cork and sometimes small amounts of phelloderm.

**cAMP:** cyclic AMP.

**Capillarity:** the movement of water through minute passageways.

**Capillary:** the smallest of the blood vessels across the walls of which substances are exchanged between the blood and body tissues.

**Carbohydrates:** organic compounds consisting of carbon, hydrogen, and oxygen, with the hydrogen and oxygen in a ratio of 2 to 1.

**Carbon cycle:** the cycle of the element carbon moving from the inorganic world to the biological world and back again by the metabolic activities of organisms.

**Carnivore:** an animal that eats other animals.

**Catabolism:** energy-releasing metabolic reactions that convert complex substances into simpler substances.

**Catalyst:** a substance that speeds a chemical reaction but is not consumed in the reaction.

**Cell membrane:** plasma membrane.

**Cell wall:** The outermost layer of most plant cells, consisting primarily of cellulose formed by the protoplasm.

**Cellulose:** an insoluble polysaccharide formed from glucose; the main constituent of many cell walls in plants.

**Central nervous system:** brain and spinal cord.

**Centrioles:** small cytoplasmic bodies usually located near the nucleus of animal cells; centrioles function in cell division of animal cells; most plant cells divide without them.

**Centromere:** portion of a chromosome to which spindle fibers are attached during mitosis and meiosis; also called spindle fiber attachment and kinetochore.

**Chlorophyll:** the green, light-absorbing plant pigment that functions in photosynthesis; higher plants have chlorophylls a and b.

**Chloroplast:** a cytoplasmic organelle containing chlorophyll in association with carotene and xanthophylls; chloroplasts are the site of photosynthesis in green plant tissues.

**Chromoplast:** a cytoplasmic organelle which contains pigments other than chlorophyll, usually carotene and xanthophylls; chromoplasts are responsible for some red and yellow coloration in plants.

**Chromosome:** a threadlike or rod-shaped body in the nucleus; chromosomes consist mostly of nucleoprotein and contain most of the genetic information of an organism. Chromosomes of bacteria and blue-green algae consist only of nucleic acid.

**Cilium:** a small hairlike structure projecting from the surface of some cells; cilia often are numerous and they beat in unison, thus moving the cell or moving materials in the vicinity of the cell.

**Circadian rhythm:** a rhythm of growth or activity that repeats at approximately 24-hour intervals.

**Circular muscle:** muscle that surrounds an organ; the long axis of the muscle cells lies perpendicular to that of the organ. Contraction of circular muscle decreases the circumference of the organ and increases its length.

**Climax community:** the stable, terminal community of a succession; climax communities reproduce themselves indefinitely as long as the climate and the geology of the area do not change significantly.

**Cocarboxylase:** a coenzyme that functions in the removal of the acid (carboxyl) groups from some organic acids; thiamine pyrophosphate.

**CoA:** coenzyme A.

**Coenzyme:** an organic molecule loosely bound to the protein portion of an enzyme; the coenzyme often is the active portion of the complex. Some common coenzymes are NAD, NADP, CoA, and cocarboxylase.

**Coenzyme A:** a coenzyme functioning in aerobic respiration and in the metabolism of fatty acids.

**Coleoptile:** the sheath protecting the stem tip and leaf primordia of the embryo of a cereal grain.

**Coleorrhiza:** the sheath protecting the root tip of the embryo of a cereal grain.

**Commensalism:** a relationship between two species in which one member of the relationship is benefited and the other is neither benefited nor harmed.

**Community:** a group of organisms interacting with each other and living in an area with natural boundaries; a community may be as small as a drop of water or it may extend across a continent.

**Companion cells:** nucleated phloem cells closely associated with sieve tube elements.

**Compound:** a substance usually composed of two or more elements bound together in definite proportions by chemical bonds.

**Cone:** in plants, the seed-bearing structure of coniferous trees such as spruce, pine, and hemlock; in animals, one of the light-sensitive neurons of the retina responsible for color vision.

**Coniferous:** cone-bearing.

**Connective tissues:** in animals, tissues that hold the several parts of the body together; connective tissues consist of isolated cells embedded in a nonliving matrix that they themselves secrete. Examples of connective tissue are bone, cartilage, ligaments, and tendons.

**Consumers:** organisms that cannot manufacture their own food and therefore eat food; consumers are generally the herbivores and carnivores of a food chain.

**Control group:** in experimental procedure, a group of organisms not treated and therefore used as a standard against which to compare the reactions of the experimental or treated group.

**Cork:** the protective outer tissue of stems and roots with secondary growth; the cell walls of cork cells are impermeable to water because of their suberin content.

**Cork cambium:** the lateral meristem producing cork and sometimes phelloderm.

**Cornea:** the transparent tissue covering the pupil of the eye; the cornea together with the lens focuses light rays on the retina.

**Cortex:** the outer portion of some organs that are sharply differentiated into an inner and outer region. In plants, the cortex of roots and stems is storage tissue. Compare *medulla.*

**Cotyledons:** specialized seed leaves of the embryos of seed plants; cotyledons function either as food storage organs or as producers of digestive enzymes.

**Creatine phosphate:** a compound containing a high-energy phosphate bond that may be used in muscle contraction.

**Crista:** one of the invaginations of the inner membrane of the mitochondrion; bears enzymes of the Krebs cycle.

**Cross-pollination:** the transfer of pollen from one plant to the pistil of the flower of another plant.

**Cuticle:** the layer of cutin secreted by the epidermis of leaves and other plant organs.

**Cutin:** a waxy substance impermeable to water and secreted by the leaves and some other plant organs.

**Cyclic AMP:** a cyclic form of adenosine monophosphate that is formed within cells stimulated by certain hormones and that initiates within the cells the specific chemical reactions associated with those hormones.

**Cyclic photophosphorylation:** a cyclic process in photosynthesis that produces ATP, utilizing the energy of light.

**Cytochrome pigments:** pigments that function as electron carriers in photosynthesis and aerobic respiration.

**Cytokinin:** a plant hormone that stimulates cell division.

**Cytoplasm:** the portion of the protoplasm exclusive of the nucleus.

**Dark reactions:** a series of chemical reactions in photosynthesis that reduce carbon dioxide to glucose by utilizing the ATP and NADP produced in the light reactions. The dark reactions are not immediately dependent on light, but do depend on the products of the light reactions.

**Decay:** the decomposition of organic materials.

**Deciduous:** referring to organs that drop at the end of a growing season or at a predetermined time. Deciduous plants are those that lose their leaves once a year.

**Decomposers:** nonphotosynthetic microorganisms, especially bacteria and fungi, that decay dead organic material and convert it to mineral form useful to green plants.

**Dehydration synthesis:** the formation of complex organic compounds from simpler ones through the formation of chemical bonds formed by the removal of water from the reactants.

**Dendrite:** that portion of a neuron that conducts impulses toward the cell body.

**Denitrifying bacteria:** bacteria that reduce nitrates to nitrites and nitrites to molecular nitrogen.

**Deoxyribonucleic acid:** the nucleic acid which is the genetic material of all organisms except some viruses.

**Dermis:** the layer of skin below the epidermis.

**Diaphragm:** a large muscle between the thoracic cavity and the abdominal cavity that functions in breathing movements; also a contraceptive device fitted over the cervix so as to cover the opening to the uterus.

**Diastole:** the relaxation of the atria or the ventricles of the heart, allowing them to fill with blood.

**Dicot:** a plant in which the embryo bears two cotyledons or seed leaves.

**Differentially permeable membrane:** a membrane that permits some substances to pass across it and not others; the differential permeability of a membrane may change from time to time.

**Differentiation:** the maturing of an unspecialized cell or tissue into a specialized cell or tissue.

**Diffusion:** the net movement of a substance from a region of high concentration of that substance to a region of low concentration of that substance.

**Digestion:** the conversion of complex organic substances into the simpler subunits from which they were formed by a dehydration synthesis.

**Dipeptide:** a compound formed from two amino acids by a dehydration synthesis.

**Disaccharide:** a compound formed from two monosaccharides by a dehydration synthesis.

451

**DNA:** deoxyribonucleic acid.

**Dominant:** one of the most abundant species in a community.

**Dominant gene:** a gene that expresses its phenotype regardless of the nature of its allele.

**Dormancy:** a temporary condition of low metabolic rate, reduced physiological activity, and no growth.

**Ecology:** the study of the interrelationships between organisms and their environment.

**Egg:** a female gamete or sex cell; eggs usually are non-motile and are larger and contain more stored food than male gametes.

**Electron:** a negatively charged elementary particle that orbits around the nucleus of an atom.

**Element:** a chemical substance that cannot be broken down into simpler chemical substances.

**Embryo:** an individual produced from a fertilized egg and still in its early stages of development.

**Embryo sac:** that portion of an ovule in which the embryo of a flowering plant develops.

**Endodermis:** a one-celled layer surrounding the vascular bundle and pericycle of roots and some underground stems.

**Endoplasmic reticulum:** a system of double membranes in the cytoplasm usually covered with ribosomes; endoplasmic reticulum is considered to be continuous with the plasma membrane and the nuclear membrane.

**Endosperm:** the polyploid tissue of the seeds of some flowering plants; endosperm arises by the fertilization of two polar nuclei by a sperm nucleus.

**Energy:** the capacity to do work.

**Energy of activation:** energy required for the initiation of a chemical reaction.

**Environmental resistance:** those factors in the environment that prevent organisms from reaching their full biotic potential.

**Enzyme:** an organic catalyst.

**Epidermis:** in plants, the outermost layer of cells of organs with only primary growth; in animals, the several cell layers of the skin external to the dermis.

**Epithelium:** an animal tissue that covers internal organs and lines cavities.

**ER:** endoplasmic reticulum.

**Eucaryotic:** referring to cells or organisms that have true nuclei bounded by a membrane, mitochondria, and Golgi bodies; eucaryotic plants also have plastids. Compare *procaryotic*.

**Eutrophication:** enrichment with dissolved nutrients, said of a lake or other body of water.

**Evaporation:** conversion of a liquid to a vapor or gas.

**Evolution:** the origin of new species from pre-existing species by a process of gradual change.

452

**Excretion:** the removal from a cell or organism of the waste products of its metabolism.

**Experiment:** a method of testing a hypothesis.

**$F_1$:** first generation of offspring.

**$F_2$:** second generation of offspring.

**FAD:** flavine adenine dinucleotide.

**Fat:** an organic compound consisting of carbon and hydrogen and relatively little oxygen; a molecule of fat is formed by a dehydration synthesis involving a glycerol molecule and three fatty acid molecules.

**Fatty acid:** an organic compound consisting mostly of a carbon chain terminated by an acid group.

**Feces:** undigested material passing unabsorbed through the digestive tract.

**Fermentation:** anaerobic oxidation of glucose or other food.

**Fertilization:** the union of two gametes, forming a zygote.

**Fetus:** a late stage of embryological development.

**Flagellum:** a threadlike structure projecting from the protoplasm of a cell and used in locomotion; resembles a cilium but is much longer.

**Flavine adenine dinucleotide:** one of the hydrogen carriers involved in aerobic respiration.

**Florigen:** a hypothetical plant hormone believed to stimulate flowering.

**Flower:** sexual reproductive structure of the flowering plants, usually consisting of four types of organs: sepals, petals, stamens, and pistils.

**Food chain:** a series of organisms in a community, each organism feeding on the one below it in the chain and being fed upon by the one above. The organisms at the base of the chain are photosynthetic plants.

**Food web:** a set of interconnected food chains.

**Fruit:** a ripened ovary; contains seeds.

**Gamete:** a sex cell; in higher plants and animals, the gametes are eggs and sperms; in lower organisms, the gametes produced by both parents may be identical. The fusion of two gametes forms a zygote.

**Gamma globulins:** spherical protein molecules found in the blood; most antibodies are gamma globulins.

**Ganglion:** an enlarged portion of a nerve, containing the cell bodies of its neurons.

**Gene:** a hereditary unit usually located on a chromosome; cytoplasmic genes exist in chloroplasts and mitochondria.

**Genetic code:** the sequence of nucleotides in a DNA molecule that determines the amino acid sequence in protein molecules; a triplet consisting of three nucleotides codes for one amino acid.

**Genetics:** the study of heredity.

**Genotype:** the genetic make-up of an organism. Compare *phenotype*.

**Geotropism:** a growth movement of a plant organ in response to gravity.

**Gibberellin:** a plant hormone that produces elongation of stems.

**Gland:** a cell or organ that produces secretions.

**Glomerulus:** a group of capillaries in Bowman's capsule.

**Glucose:** a common 6-carbon sugar or monosaccharide; the end product of photosynthesis; the chief energy source of many cells.

**Glycogen:** a polysaccharide formed from glucose by dehydration synthesis; it is a common stored food in animals.

**Glycolysis:** the anaerobic conversion of glucose to pyruvic acid; the initial stage of aerobic respiration and fermentations.

**Golgi body:** a membranous cytoplasmic organelle producing secretory material.

**Granum:** one of the disc-shaped bodies in a chloroplast.

**Growth ring:** a ring of secondary xylem formed by variation in the rate of cambium activity throughout the growing season.

**Guard cells:** paired cells surrounding stomata and regulating their opening and closing.

**Heartwood:** nonconducting wood in the center of a tree trunk.

**Hemoglobin:** a red blood pigment consisting of protein and an iron-containing porphyrin ring.

**Herbaceous:** nonwoody.

**Herbivore:** an animal that eats plants.

**Heterozygous:** having different alleles at the same location on paired chromosomes.

**Homeostasis:** the maintaining of a stable condition; frequently applied to the internal condition of organisms.

**Homologous chromosomes:** paired chromosomes identical in size and bearing identical or allelic genes.

**Homozygous:** having identical alleles at the same location on paired chromosomes.

**Hormone:** an organic substance produced in one part of an organism and transported to another part where it exerts its effect.

**Humus:** a mixture of partially decomposed material in soil.

**Hybrid:** the offspring of parents from two species; in genetics, synonymous with *heterozygote*.

**Hydrolysis:** the conversion of complex organic substances into simpler subunits with the formation of water; digestion.

**Hypocotyl:** the portion of a seed plant embryo below the point of attachment of the cotyledon(s).

**Hypodermis:** connective tissue beneath the dermis of the skin.

**Hypothesis:** a tentative explanation of observations to be tested by scientific experiment.

**Immunity:** the ability of an organism to resist invasion by pathogenic organisms.

**Ion:** a charged atom or group of atoms.

**Irritability:** the ability of protoplasm to respond to stimuli.

**Kinetic energy:** energy of motion; energy doing work. Compare *potential energy*.

**Kinin:** cytokinin.

**Krebs cycle:** a cyclic series of chemical reactions by which pyruvic acid is oxidized to carbon dioxide.

**Lacteal:** a lymph vessel in a villus of the small intestine.

**Latent heat of fusion:** the quantity of heat required to change 1 gram of a substance from the solid state to the liquid state.

**Latent heat of vaporization:** the quantity of heat required to change 1 gram of a substance from the liquid state to the vapor state.

**Lateral meristem:** a meristem contributing to growth in width of stems and roots; vascular cambium and cork cambium are lateral meristems.

**Leaf primordium:** a small outgrowth a short distance below the apical meristem of a stem; leaf primordia develop into leaves.

**Leaf scar:** a scar remaining on a stem after the abscission of a leaf.

**Lens:** a portion of the eye that helps to focus light rays on the retina.

**Lenticel:** an opening in the bark of trees by which gases are exchanged with the environment.

**Leucoplast:** a colorless plastid in plants; leucoplasts store food, frequently starch but sometimes oils.

**Ligament:** a connective tissue connecting bones to bones.

**Light reactions:** those chemical reactions of photosynthesis that depend on light and use light energy in the formation of ATP and $NADPH_2$; oxygen is another product of the light reactions.

**Logarithmic progression:** a series of numbers each of which is the same multiple of its preceding number.

**Long-day plant:** a plant that requires a daily light period longer than a critical length to flower.

**Longitudinal muscle:** muscle the cells of which lie parallel to the long axis of an organ; contraction of longitudinal muscle shortens the organ and increases its diameter.

453

**Lymph:** a fluid differing from blood in containing no red blood cells and in being colorless; lymph originates from the passage of fluid from the capillaries into the body tissues.

**Lysosome:** a membrane-bound cytoplasmic organelle in which enzymes are stored.

**Macromolecule:** a large molecule; some macromolecules of living cells are polysaccharides, proteins, and nucleic acids.

**Medulla:** the central portion of some organs. Compare *cortex*.

**Meiosis:** a set of two sequential nuclear divisions that produce four nuclei (and frequently four cells) from an original nucleus (and cell); in the process of meiosis, the chromosome number is reduced from the diploid number to the haploid number.

**Meristem:** an embryonic tissue in plants characterized by frequent cell division.

**Messenger RNA:** a form of RNA that transfers the genetic code from the DNA of the nucleus to the ribosomes of the cytoplasm, where it is translated into the sequence of amino acids in the forming polypeptide molecule.

**Metabolism:** the sum of the chemical reactions occurring in an organism.

**Microvillus:** one of the many minute, fingerlike extensions of the cells lining the villi of the small intestine.

**Middle lamella:** a cementing substance composed of pectic materials between adjacent cells in plants.

**Mitochondrion:** a cytoplasmic organelle, the site of aerobic respiration.

**Mitosis:** the division of a nucleus into two nuclei with the duplication of chromosomes and the exact distribution of one complete set of chromosomes to each daughter cell.

**Molecular motion:** the movement of molecules that results from their kinetic energy.

**Molecule:** the smallest unit of a compound containing two or more atoms in the same proportion.

**Monocot:** a flowering plant the embryo of which bears a single cotyledon or seed leaf.

**Monoculture:** the cultivation of a single crop species.

**Monosaccharide:** a simple carbohydrate, a simple sugar.

**Motor unit:** all the muscle cells served by a single neuron.

**mRNA:** messenger RNA.

**Mutation:** a change in a gene; more specifically, a change in the nucleotide sequence of DNA.

**Mutualism:** a relationship between two species in which both members of the relationship benefit.

**Mycorrhiza:** a mutualistic relationship between a fungus and the root of a higher plant.

**Myelin sheath:** a sheath of fatty material surrounding axons of neurons of the central nervous system.

454

**Myofibril:** one of the longitudinal fibrils in a muscle cell.

**Myosin:** a muscle protein which together with the protein actin is responsible for the contraction of muscle.

**NAD:** nicotinamide adenine dinucleotide.

**NADP:** nicotinamide adenine dinucleotide phosphate.

**Natural selection:** the survival of individuals or species best suited to a particular natural environment.

**Negative feedback mechanism:** a self-regulating mechanism consisting of at least two components, each exercising such control over the other that conditions within the system remain virtually constant or vary only within narrow limits.

**Nephron:** one of the functioning units of a kidney.

**Neurilemma:** a sheath composed of living cells surrounding the axons of neurons in the peripheral nervous system.

**Neuron:** a nerve cell.

**Neutron:** an uncharged elementary particle in the nucleus of an atom.

**Nicotinamide adenine dinucleotide:** a coenzyme that functions as a hydrogen carrier in respiration as well as in some other metabolic processes.

**Nicotinamide adenine dinucleotide phosphate:** a coenzyme that functions as a hydrogen carrier in photosynthesis as well as some other metabolic processes.

**Nitrifying bacteria:** soil bacteria that oxidize ammonia to nitrites and nitrites to nitrates.

**Nitrogen base:** any of several basic, nitrogen-containing organic compounds; specifically, purines and pyrimidines.

**Nitrogen cycle:** the cycle of the element nitrogen moving from the inorganic world to the biological world and back again by the metabolic activities of organisms.

**Nitrogen fixation:** the reduction of molecular nitrogen to an organic form utilizable by higher plants; nitrogen fixation is carried on by a limited number of bacteria and blue-green algae.

**Noncyclic photophosphorylation:** a part of the light reactions in which water molecules split and in which $NADPH_2$ and ATP are formed using the energy of light; oxygen is a by-product.

**Nuclear membrane:** a double membrane surrounding the nucleus of eucaryotic organisms.

**Nuclear sap:** the liquid portion of a nucleus.

**Nucleic acid:** a macromolecule consisting of repeating nucleotide units; DNA and RNA are the nucleic acids of organisms.

**Nucleoid:** that portion of a procaryotic cell that contains the genetic material of the cell; unlike nuclei of eucaryotic cells, the nucleoid is not bound by a membrane.

**Nucleolus:** a spherical, RNA-rich body in the nucleus, believed to be the site of synthesis of ribosomes.

**Nucleotide:** one of the repeating units of a nucleic acid molecule; a nucleotide is formed from a nitrogen base, a 5-carbon sugar, and phosphate.

**Nucleus:** a membrane-bound body present in nearly all eucaryotic cells; the nucleus contains the chromosomes and one or more nucleoli; also the central portion of an atom.

**Omnivore:** an animal that eats both plants and animals.

**Orbit:** the path of an electron as it moves around the nucleus of an atom.

**Organ:** a structure composed of several types of tissue so organized as to perform a particular function or functions; examples are heart and stomach of animals, the leaves and roots of plants.

**Organelle:** a small body within a cell; some cytoplasmic organelles are mitochondria, plastids, and ribosomes.

**Osmosis:** the diffusion of a substance across a differentially permeable membrane.

**Ovary:** in plants, the lower, usually swollen portion of the pistil of flowering plants; in animals, the female, egg-producing reproductive organ.

**Ovule:** an egg-containing structure in an ovary of a flower.

**Oxidative phosphorylation:** the production of ATP in aerobic respiration associated with the formation of water from molecular oxygen and the hydrogen atoms previously removed from the substance being oxidized.

**Palisade parenchyma:** the primary photosynthetic tissue of most leaves, located in the upper part of the mesophyll.

**Parasitism:** a symbiotic relationship between organisms of two different species in which one organism, the parasite, is benefited and the other, the host, is harmed.

**Parenchyma:** a plant tissue composed of thin-walled cells.

**Penis:** an organ of male animals that functions in the conveying of sperm into the female reproductive tract.

**Pepsin:** a protein-digesting enzyme.

**Pericycle:** a single layer of tissue between the vascular bundle and endodermis of a root; pericycle produces branch roots and gives rise to cork cambium.

**Peripheral nervous system:** that portion of the nervous system that consists of cranial and spinal nerves and their ganglia.

**Peristalsis:** successive contractions and relaxations of muscles along the walls of hollow organs such as the esophagus and intestine; peristalsis moves materials through the organ.

**Pesticide:** a substance used to kill pests such as insects or rodents.

**Petal:** a flat, sterile appendage between the sepals and the fertile organs of a flower; many petals are highly colored and serve to attract pollinating insects to the flower.

**Petiole:** a leaf stalk.

**P$_{fr}$:** the form of phytochrome that absorbs far-red light.

**pH:** a symbol indicating the concentration of hydrogen ions in a solution; a pH of 7 is neutral; a pH of less than 7 is acid; a pH of more than 7 is basic.

**Phagocytosis:** the ingestion of solid particles by a cell.

**Pharynx:** a chamber in the neck common to both the digestive and respiratory systems.

**Phelloderm:** secondary cortex produced by the cork cambium in some vascular plants.

**Phenotype:** the appearance or physiological condition of an individual which is the manifestation of its genotype. Compare *genotype*.

**Phloem:** the food-conducting tissue of vascular plants.

**Photoperiodism:** the response of organisms to day length.

**Phototropism:** a growth movement of a plant part in response to unilateral illumination.

**Photosynthesis:** the manufacture of glucose or other food from carbon dioxide and water by green plants utilizing light energy.

**Phytochrome:** a bluish, light-sensitive plant pigment that functions in photoperiodic responses.

**Pinocytotic vesicle:** a cytoplasmic vesicle that takes liquid droplets into the cell without their first passing across the plasma membrane.

**Pioneer community:** the first community in an ecological succession.

**Pistil:** the female organ of a flower; consists of a stigma, style, and ovary.

**Plasma cell:** a type of white blood cell that synthesizes antibodies.

**Plasma membrane:** the external membrane of the cytoplasm; also called cell membrane.

**Plastid:** a cytoplasmic organelle of plant cells; three types of plastids are chloroplasts, chromoplasts, and leucoplasts.

**Platelet:** a type of blood cell functioning in the clotting of blood.

**Pleura:** membrane covering the lungs and lining the thoracic cavity.

**Polar nuclei:** two nuclei at opposite ends of the embryo sac of a flower; the fusion of two polar nuclei with a sperm nucleus produces an endosperm nucleus.

**Pollen grain:** the sperm-bearing structure of flowering plants.

455

**Pollen tube:** a tube produced by a pollen grain and penetrating the stigma, style, and ovary of the pistil on which the pollen grain lands; the pollen tube discharges its sperm nuclei into an embryo sac.

**Pollination:** the transfer of pollen grains from an anther of a stamen to the stigma of a pistil.

**Polypeptide:** a large molecule formed from amino acids by a dehydration synthesis.

**Polyploidy:** possessing three or more complete sets of chromosomes.

**Polysaccharide:** a large molecule formed from monosaccharides or disaccharides by a dehydration synthesis. Some important polysaccharides in living things are starch, glycogen, and cellulose.

**Portal system:** a blood vessel that carries blood between two organs neither of which is the heart.

**Positive feedback mechanism:** a system with at least two components, each component growing and its growth causing further growth of the other component.

**Posterior:** the rear end of an animal.

**Potential energy:** stored energy; energy not doing work. Compare *kinetic energy*.

**P$_r$:** the form of phytochrome that absorbs red light.

**Predation:** the preying of one animal species on another.

**Predator:** an animal that preys on another.

**Prey:** the animal that is the food source of another animal.

**Primary phloem:** the phloem formed by differentiation of cells produced by an apical meristem.

**Primary succession:** an ecological succession beginning on bare rock or in open water.

**Primary xylem:** the xylem formed by differentiation of cells produced by an apical meristem.

**Procaryotic:** having no true nucleus surrounded by a nuclear membrane; procaryotic organisms are the bacteria and blue-green algae.

**Producer:** photosynthetic, green plants which are the basis of food chains.

**Protein:** a macromolecule formed either from several polypeptides or from many amino acids in a dehydration synthesis.

**Proton:** a positively charged elementary particle in the nucleus of an atom.

**Protoplasm:** the living portion of a cell.

**Purine:** a nitrogen base such as adenine or guanine.

**Pyrimidine:** a nitrogen base such as cytosine, thymine, or uracil.

456

**Radioactivity:** the emission of elementary particles from the atoms of certain elements.

**Recessive gene:** a gene that expresses its phenotype only in the homozygous condition and not in the presence of its dominant allele.

**Red blood cell:** blood cells that transport oxygen in the body; red blood cells lack nuclei at maturity, and they are red because they contain the pigment hemoglobin.

**Reflex arc:** a set of at least three neurons, a sensory neuron, an association neuron, and a motor neuron, which permit a quick, involuntary response to stimuli.

**Region of differentiation:** that region of the stem or root in which the undifferentiated cells from the region of elongation undergo maturation; the region of maturation is above the region of elongation in a root and below it in a stem.

**Region of elongation:** that region of the stem or root in which the cells produced by the apical meristem grow in length; the region of elongation is above the apical meristem of a root and below the apical meristem of a stem.

**Reproduction:** the production by an organism of one or more individuals of its own species; reproduction may be asexual, involving only one parent, or it may be sexual, usually involving two parents.

**Respiration:** the oxidation of a food, often glucose, with the release of metabolically useful energy, usually in the form of ATP; aerobic respiration and fermentations are both common types of respiration in living things.

**Respiratory center:** that portion of the brain that functions in controlling involuntary breathing movements.

**Retina:** the tissue of the eye containing light-sensitive neurons; the retina receives light rays focused on it by the cornea and lens.

***Rhizobium:*** a species of mutualistic nitrogen-flixing bacteria that live in the roots of leguminous plants.

**Rhythmic segmentation:** the action of mixing food in the intestine by an alternating series of contractions and relaxations of intestinal muscles.

**Ribonucleic acid:** any of several nucleic acids functioning in the transcription of the genetic code; three types of ribonucleic acid are messenger RNA, transfer RNA, and ribosomal RNA. In some viruses, ribonucleic acid rather than deoxyribonucleic acid is the genetic material.

**Ribosome:** a cytoplasmic organelle that is the site of protein synthesis.

**RNA:** ribonucleic acid.

**Rod:** a light-sensitive neuron of the retina; rods are sensitive in dim light, but unlike cones they are not sensitive to color.

**Root cap:** a tissue protecting the apical meristem of a root.

**Root hair:** a long extension of the epidermal cells in the region of differentiation of a root; root hairs provide most of the absorbing surface of a root.

**Root pressure:** pressure developed within a root when the volume of water absorbed by the root exceeds that lost by transpiration from the aerial parts of a plant.

**Sarcomere:** one of the repeating units of which a myofibril in a muscle cell is composed.

**Sapwood:** the outermost, functioning secondary xylem in a tree.

**Schistosomiasis:** a disease caused by blood flukes; the most prevalent communicable disease in the world today.

**Secondary phloem:** phloem produced by vascular cambium.

**Secondary succession:** an ecological succession originating in an area previously occupied by a community that was disturbed or destroyed but which left some organic material.

**Secondary xylem:** xylem produced by vascular cambium.

**Secretion:** a substance produced within a cell and released from that cell; secretions differ from excretions in being useful to the organisms producing them.

**Seed:** a ripened ovule; a seed contains within it an embryo plant.

**Self-pollination:** the transfer of pollen from the anther of a stamen to the stigma of a pistil in the same flower or in another flower of the same plant.

**Semen:** fluid produced by male reproductive organs and containing sperms.

**Sepal:** one of the flat, leaflike organs that comprise the outermost whorl of a flower; sepals usually are green.

**Seral communities:** the communities following a pioneer community and preceding a climax community of an ecological succession.

**Sere:** an ecological succession.

**Serum:** the liquid portion of the blood remaining after a clot forms.

**Sex chromosomes:** sex-determining chromosomes; in mammals, female animals have two X chromosomes, and males have an X and a Y.

**Sex-linked:** referring to genes located on the X chromosomes or to characteristics determined by those genes.

**Sexual dimorphism:** a condition in which the male and female members of a species are obviously different in appearance.

**Sexual reproduction:** reproduction resulting from the fusion of two gametes; frequently the gametes derive from different parents.

**Short-day plant:** a plant that requires a daily light period shorter than a critical length to flower.

**Sieve tube elements:** the food-conducting cells of the phloem.

**Somatic nervous system:** that part of the nervous system that controls voluntary (skeletal) muscles.

**Species:** a group of organisms capable of interbreeding and producing fertile offspring.

**Specific heat:** the amount of heat required to raise the temperature of 1 gram of a substance 1°C.

**Sperm:** a male gamete or sex cell; sperms usually are motile and are smaller and contain much less stored food than female gametes.

**Sperm nuclei:** the male gametes of flowering plants produced in the pollen grains.

**Sphincter muscle:** a circular muscle regulating the opening and closing of an organ.

**Spindle fibers:** microtubules arranged in the form of a spindle and present during cell division; some spindle fibers connect the centromeres of chromosomes with the poles of the dividing cells; other spindle fibers extend from pole to pole.

**Spongy parenchyma:** the lower part of the mesophyll of a leaf; spongy parenchyma contains large intercellular air spaces through which gases diffuse.

**Spring wood:** the secondary xylem formed early in the growing season.

**Stamen:** the male organ of a flower; consists of a pollen-producing anther and a filament.

**Starch:** a polysaccharide formed from glucose by dehydration synthesis; starch is a common stored food in many plants.

**Sterile-male method:** a method of eradicating insect pests by releasing large numbers of sterilized males into the environment.

**Steroid:** any of several organic compounds composed of four carbon rings; examples are the sex hormones, the hormones of the adrenal cortex, and vitamin D.

**Stigma:** the uppermost portion of a pistil, receptive to pollen.

**Stoma:** an opening in a leaf permitting exchange of gases with the environment.

**Style:** the central portion of a pistil connecting the stigma and ovary.

**Substrate:** the substance upon which an enzyme acts; also, a base, such as rock or soil, to which a stationary organism may be attached.

**Succession:** an orderly, generally predictable sequence of communities that occupies a previously unoccupied or disturbed area.

**Sugar:** a relatively simple carbohydrate, generally a monosaccharide such as glucose or ribose or a disaccharide such as sucrose or maltose.

**Sulfhydryl group:** that portion of a molecule consisting of one sulfur atom and one hydrogen atom ($-SH$); sulfhydryl groups are responsible for some of the cross-linkages between the polypeptide chains of a protein and between the folds of a single polypeptide chain.

457

**Sulfur cycle:** the cycle of the element sulfur moving from the inorganic world to the biological world and back again by the metabolic activities of organisms.

**Summer wood:** secondary xylem formed late in the growing season.

**Symbiosis:** an intimate relationship between members of two species; symbioses include mutualism, commensalism, and parasitism.

**Synapse:** the extremely minute space between the axon of one neuron and the dendrite of another neuron.

**Systole:** contraction of the atria or ventricles of the heart.

**Tendon:** a type of connective tissue connecting muscle to bone.

**Territory:** an area occupied by an individual animal or group of animals and defended by them from intruders.

**Test cross:** a genetic cross made to determine the genotype of an individual showing a dominant phenotype.

**Testis:** the sperm-producing organ of a male animal.

**Theory:** a generalization supported by some experiments but not yet proved to be universally true.

**Tissue:** a group of similar cells performing the same function or functions.

**Tracheid:** a water-conducting xylem cell with tapered ends.

**Transfer RNA:** a form of ribonucleic acid that assembles amino acids at the ribosomes preparatory to polypeptide synthesis.

**Transmitter substance:** a substance that transmits a nervous impulse across a synapse; acetylcholine and noradrenalin are transmitter substances.

**Transpiration:** the loss of water in vapor form from leaves and other aerial plant parts.

**Triplet:** in nucleic acids, three adjacent nucleotides coding for one amino acid.

**tRNA:** transfer RNA.

**Trophic level:** a feeding level in a food chain.

**Tropism:** a growth response of plant parts to external stimuli such as gravity and light.

**Turgor pressure:** pressure developed within a cell because of absorption of water.

**Urea:** an organic waste product of animals formed from two metabolic wastes, carbon dioxide and ammonia.

**Urine:** liquid waste absorbed from the blood by the kidneys and excreted by them; the major waste in urine is urea.

**Uterus:** the organ of the female reproductive system in which the developing embryo and fetus is nourished and protected before birth.

**Vacuole:** a membrane-bound space within the cytoplasm containing water and dissolved materials.

**Vagina:** the organ of the female reproductive system that receives the penis of the male in coitus.

**Valence:** the number of electrons an atom can lose, gain, or share with another atom; the number of chemical bonds an atom can form with other atoms.

**Valve:** a structure that permits the flow of materials in only one direction, as, for example, the valves in veins and the atrioventricular, aortic, and pulmonary valves of the heart.

**Vascular bundle:** a strand of xylem and phloem.

**Vascular cambium:** the lateral meristem producing secondary xylem and secondary phloem.

**Vein:** in plants, a vascular bundle; in animals, a blood vessel that returns blood to the heart from other parts of the body.

**Vessel:** a tube-shaped structure that conducts a fluid; in plants, a vertical series of water-conducting xylem cells; in animals, arteries, veins, and lymph channels.

**Vessel elements:** the individual cells of a vessel in plants; vessel elements have no end walls, but are arranged end to end forming continuous vessels.

**Villus:** one of the finger-shaped protrusions extending from the interior surface of the small intestine into the lumen; the villi greatly increase the absorbing surface of the intestine.

**Vitamin:** any of several organic compounds required in minute quantities by an animal, but which that animal cannot synthesize and therefore must obtain in its food. Many vitamins function as coenzymes.

**White blood cell:** any of several unpigmented cells of the blood, many of which protect the body from invading microorganisms either by ingesting them by phagocytosis or by producing antibodies.

**Wood:** xylem, especially secondary xylem.

**X axis:** the horizontal scale of a graph

**X chromosome:** one of the sex chromosomes; in mammals, individuals with two X chromosomes are female.

**Xylem:** the water-conducting tissue of plants.

**Y axis:** the vertical scale of a graph

**Y chromosome:** one of the sex chromosomes; in mammals, individuals with an X chromosome and a Y chromosome are male.

**Zygote:** a diploid cell resulting from the fusion of two gametes; the first cell in the life of an individual.

458

# Index

Illustrations are indicated by numbers in **boldface.**

459

461

465

water, 47–49
  molecule, 61, 62
  rise of in plants, 227–229
white blood cells, **257, 272, 297,** 299
wind, 53
wood, 218, **219, 220,** 223

X axis, 133, **133**
X chromosome, 377
xanthophyll, 91
xylem, 195–196
  fibers, 220
  parenchyma, 220

primary
  of leaf, **199**
  of root, **199, 208**
  of stem, 211, **212, 213, 217, 218, 219**
ray, 220
secondary, **217,** 218–220, **218, 219**

Y axis, 132, **133**
Y chromosome, 377
yolk sac, **355**

Z line, 278, **281**
zygote, 231, 233, **234**